A new book about numerous aspects of animal welfare merits worldwide attention. – *Psychology Today, 2022.*

Globally, non-human animals – aka animals – need all the help they can get in an increasingly human-dominated world. The *Routledge Handbook of Animal Welfare* comprehensively reviews the welfare of animals in the wide array of settings in which they are used, including emerging issues such as the impacts of intensive farming and the wildlife trade on climate change, biodiversity loss, and antimicrobial resistance and pandemics. A most valuable source of up-to-date information for a global audience.

Marc Bekoff PhD, University of Colorado, USA

Human activity is responsible for serious welfare problems for billions of non-human animals. Identifying the action required to address these issues and securing public support for it demands good evidence, wise reflection, and clear argumentation. This book provides all three and is an essential resource for anyone concerned about animal welfare.

Prof. David L Clough MA (Cantab), MSt (Oxon), PhD (Yale), FHEA, University of Aberdeen, UK

A defining text in the field of animal welfare. Challenging, insightful, and current, this book covers a diversity of subject areas in the rapidly evolving arena of animal welfare. Both informative and balanced, the chapters explore and test our moral perceptions towards animals from a legal, cultural, and ethical perspective. In a time of changing attitudes with so many global issues at the forefront of society, this book provides a welcome and much needed contribution to conceptualisation of animal welfare in the modern day.

Prof. Luke Gamble BVSc, DVM&S, FRCVS, Chief Executive – Worldwide Veterinary Service, UK

The *Routledge Handbook of Animal Welfare* is the most comprehensive and up-to-date coverage of animal welfare issues in all settings in which they are used – or misused. It provides compelling scientific evidence as to why the welfare of animals is important for the sake of millions of sentient beings, and also for our own health. Everyone concerned with the well-being of animals – students and professors, NGO staffers, and legislators – should read and own this book.

Jane Goodall PhD, DBE, Founder – The Jane Goodall Institute & UN Messenger of Peace

This textbook combines both classical and new topics in the rapidly growing discipline of animal welfare. It integrates important neighboring fields, such as animal ethics and animal law. It considers changes in human-animal interaction, and is an invaluable guide for navigating this rapidly changing terrain. The book demonstrates animal welfare science's growing importance for social decision-making and pushes the field forward in a number of important fronts.

Univ.-Prof. Dr Herwig Grimm, Ethics and Human-Animal Studies, Messerli Research Institute, Veterinary University Vienna, Medical University Vienna, University Vienna, Austria

Finally, there is a comprehensive, evidence-based animal welfare text that doesn't shy away from examining the hard questions about how humans use animals, and how the Anthropocene impacts animals. Edited by luminaries, and with a range of expert contributing authors, this text is essential reading for anyone whose study or work relates to animal welfare.

Dr Jennifer Hood BSc Hons, BVMS, PhD, Veterinary Director Animals Australia and Adjunct Senior Lecturer Murdoch Veterinary School, Australia

We are at a critical point in human history. Central to the health, climate, and ecological crises is the way we treat other animals and their habitats. This book brings to the forefront the true impact of animal exploitation in all its forms. I hope the book will persuade many that we can live a kinder, more compassionate life on this planet in harmony with all life forms.

Shireen Kassam MD, PhD, Visiting Professor of Plant-Based Nutrition, University of Winchester, UK

This book on animal welfare is unique in challenging the notion that animals are resources for humans. Kudos to the editors for such an ambitious and ethically oriented book. This book is essential reading for students, and anyone interested in animal welfare and innovative means to improve it. Highly recommended!

Barry Kipperman DVM, DACVIM, MSc, DACAW, Veterinary Specialist in Animal Welfare and Instructor of Veterinary Ethics at the University of California at Davis, USA

Animal welfare law and policy constitute a field of growing importance in society, and in the teaching of law, public ethics, politics and veterinary science. The *Routledge Handbook of Animal Welfare* distils the knowledge and experience of 50 authors, into a comprehensive and rigorous summary of key animal law and welfare issues around the globe. It is a magisterial exposition of the neglectful past; the contemporary awakening; and the future enlightenment concerning our relationship to other animals. This book makes a signal contribution to the fields of animal law and welfare.

The Hon. Michael Kirby AC, CMG, Former Justice of the High Court of Australia and Editor-In-Chief of *The Laws of Australia*

This impressive book folds science with vision on the ethics of animal use and how things need to change. Covering animal welfare and its interdependencies with human health and environmental impact in an accessible way, it is a handbook in the true sense of the word: an authoritative volume that I'll certainly be keeping close to hand.

Prof. Philip Lymbery, Global CEO, Compassion in World Farming International

A uniquely broad collection that describes the varied (and generally harmful) impacts that humans have on other animals, whether by farming them for food or simply encroaching into their wild habitats. Readers will welcome the truly international perspective. A tour de force by a stellar group of animal welfare scientists, experts in animal law and ethics, veterinarians, and animal advocates. Anyone who cares about animals will find this book a comprehensive resource and will be encouraged and inspired by the final section showing how change can be achieved.

Christine Nicol MA, DPhil, Professor of Animal Welfare, Royal Veterinary College, UK

We are witnessing a revolution in how humans view and treat other animals, necessitating a fresh look at the human/non-human animal relationship. The editors of the *Routledge Handbook of Animal Welfare* have assembled experts from various fields, bringing scientific and historical perspectives that are helpful to deepen our understanding of animal protection in the 21st century. This book is a welcome and important contribution to animal welfare literature.

Joyce Tischler JD, Professor of Practice, Animal Law, Center for Animal Law Studies at Lewis & Clark Law School, USA

Provides the scientific case for the multiple (human and non-human) co-benefits of abolishing the economic exploitation of other animals.

Dr Richard Twine, PhD, Co-Director of the Centre for Human-Animal Studies, Edge Hill University, UK

A comprehensive book covering a wide range of important animal welfare issues by fantastic contributors. The inclusion of one welfare topics as well as chapters on animal ethics and law and also social change makes this a brilliant resource. I thoroughly recommend this book to everyone that has an interest in animal welfare and thank all of the wonderful people that made this happen.

Dr Arnja Dale BSc., GDipNFPL, GDipHE, MSc., MSc.(Hons), PhD, Chief Scientific Officer, SPCA NZ

ROUTLEDGE HANDBOOK OF ANIMAL WELFARE

This handbook presents a much-needed and comprehensive exploration of the rapidly growing fields of animal welfare and law.

In recent years there has been increasing attention paid to our complex, multifaceted relationships with other animals, and in particular, the depth and breadth of various societal uses of animals. This has led to a reconsideration of their moral and social status, which has sometimes challenged the interests of those who use animals. In such a contested domain, sound evidence and reasoning become particularly important. Through firm commitment to such principles, this book explores the biological foundations for the moral consideration of animals and for evolving conceptualisations of animal welfare. It reviews in detail the welfare concerns associated with numerous forms of animal use. The inclusion of key recent developments such as climate change, pandemics, and antimicrobial resistance, ensures this text is among the most current in its field. The ethical implications of the various uses of animals by society are considered, and chapters provide important recommendations for reforms of practice, law, or policy. The status of animal law internationally, and in major world regions, is reviewed. Finally, the book considers human behavioural change and strategies for improving stakeholder communication and education.

The handbook is essential reading for students and scholars of animal welfare, animal law and animal ethics everywhere, and for policy-makers and other professionals working in the animal welfare sector.

Andrew Knight (MANZCVS, DipECAWBM (AWSEL), DipACAW, PhD, FRCVS, PFHEA) is a Professor of Animal Welfare and Ethics, and Founding Director of the Centre for Animal Welfare, at the University of Winchester, UK. He is also an Adjunct Professor in the School of Environment and Science at Griffith University, Australia. Prior to working in academia, he practised veterinary medicine for nearly a decade.

Clive JC Phillips (BSc, MA, PhD) was Australia's first Professor of Animal Welfare, at the University of Queensland, Australia, and foundation director of the Centre for Animal Welfare and Ethics. He previously lectured at the University of Cambridge and the University of Wales, UK. His books include *Principles of Cattle Production* (third edition, 2018), *The Animal Trade* (2015), and *The Welfare of Animals: The Silent Majority* (2008). He has authored about four hundred scientific journal articles. He chairs the Queensland and Western Australian Governments' Animal Welfare Boards, is editor-in-chief of the journal *Animals*, edits an Animal Welfare Series, and is Director of the Humane Society International.

Paula Sparks (LLB (Hons), BL) is a Visiting Professor at the University of Winchester, UK, where she teaches animal law. She practised as a barrister at Doughty Street Chambers in London before leaving the bar in 2018 to pursue a full time role with the UK Centre of Animal Law (A-LAW), a charity whose vision is a world where animals are fully protected by law. She frequently lectures and writes about animal-related law and policy.

ROUTLEDGE HANDBOOK OF ANIMAL WELFARE

Edited by
Andrew Knight, Clive Phillips, and Paula Sparks

Routledge
Taylor & Francis Group
LONDON AND NEW YORK

earthscan
from Routledge

COMPASSION in world farming

RSPCA

Albert Schweitzer
Albert Schweitzer Foundation

Cover image: © Getty Images

First published 2023
by Routledge
4 Park Square, Milton Park, Abingdon, Oxon OX14 4RN

and by Routledge
605 Third Avenue, New York, NY 10158

Routledge is an imprint of the Taylor & Francis Group, an informa business

British Library Cataloguing-in-Publication Data
A catalogue record for this book is available from the British Library

Library of Congress Cataloging-in-Publication Data
A catalog record has been requested for this book

ISBN: 978-1-032-02206-2 (hbk)
ISBN: 978-1-032-32575-0 (pbk)
ISBN: 978-1-003-18235-1 (ebk)

DOI: 10.4324/9781003182351

Typeset in Bembo
by Deanta Global Publishing Services, Chennai, India

"Until we extend our circle of compassion to all living things, humanity will not find peace".

– Albert Schweitzer

CONTENTS

CONTRIBUTORS

Cheryl Abbate, PhD, is an Assistant Professor of Philosophy at the University of Nevada, Las Vegas, USA, and specialises in animal ethics. Her recent publications include "A defense of free-roaming cats from a hedonist account of feline well-being" (*Acta Analytica*), "Meat eating and moral responsibility: Exploring the moral distinctions between meat eaters and puppy torturers" (*Utilitas*), and "Valuing animals as they are: Whether they feel it or not" (*European Journal of Philosophy*).

Wladimir Alonso, MSc, DPhil, is a researcher on global health and acted as a consultant to research institutions in several countries, including the US National Institutes of Health, the ICDDR in Bangladesh, Oxford University, and the World Health Organization. He is better known for having pioneered the analyses of latitudinal gradients of seasonality of infectious diseases that revealed that the annual influenza vaccination is administered at the wrong time in the tropics. He has published multiple research papers, co-authored two books and taught workshops on data analysis to audiences around the world. He is currently the executive director of the Welfare Footprint Project.

Pablo Arechavala-Lopez, PhD, is a multidisciplinary ichthyologist, specialising in fish behaviour, ecology, and welfare of farmed and wild marine fishes. A researcher at the Fish Ethology and Welfare Group at CCMAR, Portugal, his studies focus on developing strategies to ensure a more responsible aquaculture activity by investigating and validating behavioural indicators of fish welfare, developing operational tools to monitor fish welfare at farms, and assessing the effects of environmental enrichment on farmed fish species or rising welfare management plans at commercial and global scale. Along with his scientific production, he is also part of the FishEthoGroup, developing fish welfare training, experimentation, and consultancy.

David Arney, PhD, was educated in Britain. He worked in a series of jobs, including as a farm worker, London market stall holder, music club coordinator, administrator in a criminal court, and as a miner, before becoming a Professor of Animal Welfare at the Estonian University of Life Sciences. He has over 20 years' experience of education and research in animal welfare, production, and behaviour. He was a postgraduate student under Clive Phillips. He is primarily interested in the welfare of livestock, but also the ethics of our exploitation of animals in a wider context.

Heather Bacon, OBE, BSc (Hons), BVSc, CertZooMed, SFHEA, MRCVS, is a veterinarian and the academic lead for the veterinary programme at the University of Central Lancashire, UK. She works internationally on veterinary education projects related to animal welfare, behaviour, and veterinary ethics. She collaborates with veterinary and animal welfare organisations such as WSAVA, FVE, and the ACC&D. Previously she worked as the Veterinary Director for the Animals Asia Foundation, an NGO working to improve the lives of animals across Asia. She graduated from the University of Bristol (BVSc), UK, and from the University of Liverpool (BSc Hons), UK. She has been awarded an OBE for services to animal welfare and veterinary education.

David I Bilchitz, BA (Hons), LLB (Wits), MPhil, PhD (Cantab) is a Professor of Fundamental Rights and Constitutional Law at the University of Johannesburg, South Africa, and Professor of Law at the University of Reading, UK. He is also Director of the South African Institute for Advanced Constitutional, Public, Human Rights and International Law (SAIFAC), Vice-President of the International Association of Constitutional Law and a member of the Academy of Science of South Africa. He is a founding director of Animal Law Reform South Africa and has been involved both in his academic work and activism in seeking to improve the status, entitlements, and plight of animals in post-apartheid South Africa.

Harry J Blokhuis, PhD, is a Professor of Ethology at the Swedish University of Agricultural Sciences (SLU). His research focus is on farm animal behaviour and welfare assessment. Harry's studies have involved abnormal behaviours, such as feather pecking and stereotypies, the development of alternative housing systems for poultry, the match between horse and rider, welfare assessment in horses and automatisation of welfare assessment in broilers. He coordinated nine large European projects (LayWel, Welfare Quality, EAWP, PATHWAYS, etc.). He has published more than 120 scientific articles in peer reviewed journals and supervised 10 PhD theses. His H index is 61.

Sabrina Brando, MSc, is a psychologist, changemaker, and director of AnimalConcepts. She has organised over 500 events around the world on a wide variety of topics. She is pursuing a PhD at the University of Stirling, UK, on the topic of human well-being and human–animal interconnectedness, is the Primate Care Training Program Content Coordinator for the Pan African Sanctuary Alliance, and a Research Associate with the Smithsonian Institution. She teaches, conducts research in, and works globally with zoos, aquariums, sanctuaries, and other organisations as well as individuals, in support of optimal animal, human, and planetary well-being.

Donald M Broom, MA, PhD, ScD, Hon DSc, Hon Dr, is Emeritus Professor of Animal Welfare at Cambridge University's Department of Veterinary Medicine, UK. Research interests include scientific assessment of animals' welfare and cognitive abilities, ethics, and sustainable farming. He was Chairman/Vice Chairman of the EU Scientific Committees on Animal Welfare from 1990 to 2012. He has authored 360 refereed papers and 12 books including *Stress and Animal Welfare* (2019), *The Evolution of Morality and Religion* (2003), *Sentience and Animal Welfare* (2014), *Animal Welfare in the European Union* (2017), *Tourism and Animal Welfare* (2018), and *Domestic Animal Behaviour and Welfare* (6th edition, 2022).

Deborah Cao, PhD, LLB (Hons), is a Professor at Griffith University, Australia. She writes about animal law and animal ethics and welfare in China and Chinese legal language and culture. She has also been active in social media in China writing about Chinese culture, society and animals, and was named one of the 200 most influential bloggers in China in 2012. She has written a number of books, including *Animal Law in Australia* (2015), and *Animals in China: Law and Society* (2015). Her recent edited books are *Animal Law and Welfare: International Perspectives* (2016), *Scientific Perspectives to Farm Animal Law and Welfare in China and Beyond* (2022, in Chinese), and *Farm Animal Law and Welfare in China and Beyond* (2022, in Chinese).

Michael S Cockram, BVM, PhD, MRCVS, obtained a veterinary degree and PhD in the UK. After a period of research and teaching in animal welfare at Edinburgh University, UK, he moved to a Chair in Animal Welfare in Canada. He is interested in the interrelationships between animal science, behaviour, physiology, psychology, and health to assess the welfare implications of the management of animals. He has served as the Welfare and Behaviour Section Editor for *Animal: The International Journal of Animal Biosciences*, has participated in the development of several welfare codes of practice, published research papers, and contributed to books on animal welfare, transport, and slaughter.

Sarah Coose, BSc, is a Geography MA student at Brock University, Canada, in the Faculty of Social Sciences. She received a BSc in Biology from Boise State University, USA, with an emphasis in ecology, evolution, and behaviour. Her current research is focused on the ethics and animal welfare practices of animal use in ecotourism contexts and the efficacy of these organisations as vehicles for conservation and multispecies livelihoods.

Ruth De Vere, MRes, QTS, BSc (Hons), previously worked as a high school biology teacher and moved into the charity sector to achieve a greater impact in her work. Working with World Animal Protection (previously the World Society for the Protection of Animals, or WSPA) she managed the global education programmes for eight years. She then led the design, development, and implementation of an innovative Behavioural Insights unit that embedded behaviour science into campaign design. She now works at the animal welfare charity Battersea as the Head of Academy, which provides world-leading professional development and support for rescue centres in the UK and internationally.

Cathy M Dwyer, BSc (Hons), PhD, is a Professor of Animal Behaviour and Welfare at Scotland's Rural College (SRUC), UK, where she leads the Animal Behaviour and Welfare Research Group of about 15 researchers and 20 PhD students. She is also Director of the Jeanne Marchig International Centre for Animal Welfare Education at the Royal (Dick) School of Veterinary Studies at the University of Edinburgh, UK. She has supervised 20 PhD students in behaviour and welfare topics, and teaches on undergraduate and MSc programmes in animal behaviour and welfare at SRUC and the University of Edinburgh. She has written more than 100 academic publications and chairs Scotland's Animal Welfare Commission.

Sandra Edwards, MA, PhD, FRASE, retired from her position as Chair of Agriculture at Newcastle University, UK, in 2017 and took the status of Emerita Professor after working for more than 30 years in applied research on pigs. She has been active in many national and international academic and industry committees, serving as a member of the UK Farm Animal Welfare Council and the Animal Health and Welfare Panel of the European Food Safety Authority. She has also been a Council member of the European Federation of Animal Science and President of the British Society of Animal Science.

David A Fennell, BES, MA, PhD, is a Professor of Geography and Tourism Studies at Brock University, Canada, and researches mainly in the areas of ecotourism, tourism ethics, and moral issues tied to the use of animals in the tourism industry, and sustainability. A major thrust of his research involves the use of theory from other disciplines (e.g., biology, philosophy) to gain traction on many of tourism's most persistent issues and problems. He is the founding Editor-in-Chief of the *Journal of Ecotourism*.

Bob Fischer, PhD, is an Associate Professor of Philosophy at Texas State University, USA. He is the editor of *The Routledge Handbook of Animal Ethics* (2021), the author of *Animal Ethics—A Contemporary Introduction* (2021), and a co-author of *Wildlife Ethics: Animal Ethics in Wildlife Management and Conservation* (forthcoming) and *What We Owe Other Animals* (forthcoming).

Tamzin Furtado, PhD, BA (Hons), is a research associate at the University of Liverpool, UK, where her work focuses on the interconnections between human and animal health and well-being. She completed a PhD at the University of Liverpool studying the management of obesity in horses, particularly horse–human relationships and human behaviour change. She works on projects covering a wide range of aspects of understanding human behaviour in order to improve companion animal welfare, and in using social sciences to find out more about how we can help people to change. She is a member of the team of specialists at Human Behaviour Change for Animals.

Meg Good, BA/LLB (Hons), GDLP, PhD, is the Head of Programs and Legal Counsel at the Australian Alliance for Animals based in Sydney, and an Adjunct Lecturer at the University of Tasmania, Australia, where she coordinates the unit "Animal Law". She was previously the Senior Program Manager and Legal Counsel at Voiceless, the animal protection institute. Meg has 13 years of experience as a tertiary educator and has held senior positions with various animal law organisations, including the Barristers Animal Welfare Panel and the Animal Law Institute. Meg is currently the Secretary of the Australasian Animal Law Teachers' and Researchers' Association.

Jed Goodfellow, BA/LLB (Hons), GDLP, PhD, is the Director, Policy and Government Relations at the Australian Alliance for Animals. He was previously the Senior Policy Officer at RSPCA Australia. Jed holds a BA/LLB (Hons) and completed a PhD in animal welfare regulation in 2015. His research examines governance structures for animal welfare policy. He has taught the Animal Law course annually since 2012 at Macquarie University, and previously practised as a prosecutor for RSPCA South Australia, a lawyer for commercial law firm Clayton Utz, and worked as an inspector for RSPCA Queensland.

Steve Glassey, MEmergMgt, PDipEmergMgt, PGCPM, GCTSS, CertAWI CEM®, is the Patron of Animal Evac New Zealand, a disaster management charity that focuses on advocating for animal-inclusive disaster planning and laws. He has worked as a Disaster Management Officer for the United Nations and been deployed to numerous humanitarian crises around the world. He has a background in animal welfare from being initially an animal protection inspector for the Society for the Prevention of Cruelty to Animals (SPCA) in New Zealand, and later the Chief Executive of Wellington SPCA where he led the redevelopment of the National Rescue Unit and commanded several high-profile animal rescue operations including the 2017 Edgecumbe flood.

Temple Grandin, BA, MS, PhD, is a Professor of Animal Science at Colorado State University, USA. She has served as a consultant to the meat industry on equipment design and was the author of animal welfare guidelines for the North American Meat Institute. These guidelines use objective outcome-based measures to assess animal welfare in slaughter plants. HBO has made a movie about her life, and her books *Animals in Translation* (2005) and *Animals Make Us Human* (2009) were national bestsellers. Facilities she has designed are used in many plants around the world. She has edited and written chapters for three textbooks.

Moira Harris, BSc (Hons), MSc, PhD, PGCertTLHE, SFHEA, is an Independent Research Consultant and Visiting Lecturer at the University of Winchester, UK. She spent 13 years at

Harper Adams University, UK, as a Lecturer/Senior Lecturer and Programme Manager and has a background in research and teaching of applied animal behaviour and welfare, specialising in farmed pigs, poultry, and fish as well as zoo and wild animals (elephants in particular). She is a Series Editor for the Springer *Animal Welfare* book series, a member of the European Union Platform on Animal Welfare fish welfare subgroup, and co-author of the Eurogroup for Animals report *Catching Up: Fish Welfare in Wild Capture Fisheries*.

Sophie Hill, MA, Vet MB, PGCE, AFHEA, MRCVS, is an equine veterinary surgeon based in the UK. Sophie has worked at the Department of Veterinary Medicine, University of Cambridge, and at Anglia Ruskin University, UK. She has also worked in private practice and at the national and international equine welfare charity organisations Redwings Horse Sanctuary and Brooke.

Yashprada Joglekar, BBA, LLB, LLM, is a qualified Advocate from India. She earned her MA in Law from Nirma University, Ahmedabad, Gujarat, India, and BBA and LLB with Criminal Hons from the School of Law, KIIT Deemed to Be University, Bhubaneswar, Odisha, India. As a volunteer with the Wildlife Crime Control Bureau, Government of India (WCCB) she generates awareness towards the conservation of wildlife and organises webinars in collaboration with stakeholders to promote collaborative and interdisciplinary efforts. Her recent work includes the booklet *Wildlife Laws for Students* (2021).

Andrew Knight, BSc (Vet Biol), BVMS, CertAW, PhD, MANZCVS (Animal Welfare), DipECAWBM (AWSEL), DipACAW, FRCVS, PFHEA, previously worked in small animal veterinary practice. He is now a Professor of Animal Welfare and Ethics, and Founding Director of the Centre for Animal Welfare, at the University of Winchester, UK. Additionally, he is a European and British Veterinary Specialist in Animal Welfare Science, Ethics and Law; an American and New Zealand Veterinary Specialist in Animal Welfare; a Fellow of the Royal College of Veterinary Surgeons; and a Principal Fellow of Advance HE. He has a large number of academic and popular publications, several websites, and an extensive series of social media videos on animal welfare issues, all accessible via www.AndrewKnight.info.

Deborah Legge, LLB (Hons), MA, PhD, studied environmental law as part of her LLB at the University of Sheffield, UK. She set up and taught environmental law and animal law courses whilst at Liverpool John Moores University, UK. She has written on both areas of law. She was a member of the consumer regulatory body for water and has written on water regulation. She is now an Associate Lecturer at the Open University. She co-authored one of the earliest legal textbooks in Britain on animal law, the *Law Relating to Animals* (Brooman and Legge, Cavendish 1997).

Matthew Liebman, JD, is the Chair of the Justice for Animals Program and an Associate Professor at the University of San Francisco School of Law in San Francisco, California, USA, where he teaches courses on animal law and professional responsibility. His scholarship examines the animal protection movement's relationship to the legal system, as well as substantive and doctrinal questions in animal law. Before joining USF, he was the Director of Litigation for the Animal Legal Defense Fund, where he litigated animal protection cases.

Terry L Maple, PhD, is Elizabeth S Watts Professor Emeritus at the Georgia Institute of Technology, USA, and Professor in Residence at the Jacksonville Zoo and Gardens. He also led the renaissance of Zoo Atlanta from 1984 to 2003 as the reform President and CEO. The Association of Zoos and Aquariums presented him with its first Career Achievement Award in Animal Welfare in 2019. He is the author, co-author, editor, and co-editor of more than 250 publications. His most recent book is *Beyond Animal Welfare: The Art and Science of Wellness* (2019). He was the Founding Editor of the scientific journal *Zoo Biology*.

Carla Forte Maiolino Molento, DVM, MSc, PhD, is Head of the Animal Welfare Laboratory at the Federal University of Parana, working for decades with animal welfare issues, especially regarding the assessment of priorities and strategies for welfare improvement, with an animal-centred approach. Her main areas of expertise are dog welfare in relation to dog population management, issues regarding the welfare of laboratory animals, the contribution of animal welfare assessment to the recognition of animal abuse and, most of all, the animal welfare issues that emerge from using animals as food sources.

Laetitia Nunny, BA, MSc, has a MSc in International Animal Welfare, Ethics and Law from the University of Edinburgh, UK, and works as a freelance animal welfare consultant. She has published a number of articles on marine mammal welfare including on seal management and hunting, and solitary-sociable dolphin welfare. She also has a particular interest in how marine and terrestrial carnivores are managed to prevent human–wildlife conflict.

Bonnie M Perdue, MS, PhD, earned her PhD at the Georgia Institute of Technology, USA, in Cognition and Brain Science. She then completed a postdoctoral position at Georgia State University, USA, focused on non-human primate cognition. In 2013, she joined the faculty of Agnes Scott College, USA, and is currently an Associate Professor of Psychology and Neuroscience and serves as the Chair of the Psychology Department and co-Director of the Neuroscience program. She has conducted research in collaboration with Zoo Atlanta since 2005. Her research interests focus on human and non-human animal cognition, as well as animal welfare.

Clive Phillips, BSc, MA, PhD, was Australia's first Professor of Animal Welfare, at the University of Queensland, and foundation director of the Centre for Animal Welfare and Ethics. He previously lectured at the University of Cambridge and the University of Wales, UK, and researched the welfare of farm, zoo, and companion animals, and livestock transported by ship. His books include *Principles of Cattle Production* (third edition, 2018), *The Animal Trade* (2015), and *The Welfare of Animals – the Silent Majority* (2008). He has authored about 400 scientific journal articles. Clive chairs Queensland and Western Australian Governments' Animal Welfare Boards, is editor-in-chief of the journal *Animals* and Springer's *Animal Welfare* Series and Director of Humane Society International.

Ana K Rentsch, BSc, MSc, is a PhD candidate in the Department of Animal Bioscience at the University of Guelph, Canada. She completed an MSc at the Center for Proper Housing: Poultry and Rabbits, ZTHZ Zollikofen, Switzerland, on the influence of keel bone fractures on laying hen behaviour in non-cage housing systems. Her research interests are animal behavioural development, strain, and individual variation, and how these traits relate to animal welfare.

Ian A Robertson, LLB, BVSc, MRCVS, is the unusual career combination of veterinarian turned lawyer. He combined those careers working as a Prosecutor in New Zealand's Ministry for Primary Industries, then as the Statewide Specialist in Australia (Animal welfare enforcement), and now works as a Barrister (animal law specialist) in New Zealand and overseas advising, litigating, teaching, and publishing on the subject of animal law. He is the Principal of Guardianz Animal Law (www.guardianz.law) and Co-founder of the Sentient Animal Law Foundation (www.sentientanimallaw.org).

Suzanne Rogers, BSc (Hons), CHBC, has worked in several very different roles that together have paved the way for being the co-Director of Human Behaviour Change for Animals (HBCA). Her first career was in science publishing, she then requalified and founded Learning

About Animals through which she organised educational events. At WSPA (now World Animal Protection) she led the move towards using behaviour change science to improve animal welfare. Since 2011, Suzanne has worked internationally as a consultant for animal welfare and human behaviour change. HBCA provides clients with actionable advice and support to apply the principles of behaviour change to their work.

João L Saraiva, MSc, PhD, is a fish ethologist focused on welfare, with an MSc in Ethology and a PhD in aquatic science. He is the founder and president of the FishEthoGroup Association, leads the Fish Ethology and Welfare Group at CCMAR, Portugal, and lectures on ethology at the University of Algarve, Portugal. He is co-editor of a book on fish welfare and author of many scientific papers and welfare profiles of aquatic species for open access database FishEthoBase. He has participated in many international projects on the welfare of farmed fish and works as a fish welfare consultant for European institutions and other stakeholders.

Cynthia Schuck-Paim, MSc, DPhil, has a PhD in Zoology from Oxford University, UK, and has conducted multiple research projects for research institutions in Europe, the United States and South America on the epidemiology of respiratory diseases and pandemics, animal cognition, and more recently farm animal health and welfare. She has worked as a Research Fellow at Oxford University, the University of Sao Paulo, Brazil, as well as being the founding director of an international consulting company in the field of global health. She is the author of three books, multiple book chapters, and a large number of research publications.

Sonia JM Shad, BA, LLB (Hons), is an animal protection lawyer. She has a BA LLB from the National Law School of India University, Bangalore. She is the founder of the Society for Non-Human Persons, NLSIU first student run animal protection group at a Law School in India. She has worked at Mercy for Animals and the Humane Society International/India. She is currently the Manager of the Animal Law Centre, NALSAR.

Mark P Simmonds, OBE, is a marine biologist and environmental scientist who has spent the better part of his career working on threats to marine wildlife. He has worked in both the non-governmental and university sectors and produced scientific papers spanning a wide range of issues, including assessment of animal welfare, the effects of persistent organic pollution, noise pollution, marine debris, and climate change. He is the author/editor of a number of books about marine mammals and the issues affecting them. He is a Visiting Research Fellow at the University of Bristol, School of Veterinary Sciences, UK.

Eric Slywitch, MD, PhD, is a medical doctor with a PhD in nutrition, having specialised in nutrology, enteral, and parenteral nutrition. He also holds a postgraduate degree in endocrinology, clinical nutrition, and psychoanalysis. He is the author of three books and has published multiple book chapters and technical guidelines for health professionals (medical doctors and nutritionists) and the lay audience. He is a lecturer in three postgraduate courses in medicine, an advisor to multiple organisations and provides training to health professionals in the diagnosis and treatment of nutritional problems.

Lynne U Sneddon, BSc, PhD, is an experimental scientist who addresses questions in fish behaviour and welfare. Whilst at the Roslin Institute, Lynne published the first characterisation of nociceptors in rainbow trout and has investigated the capacity for pain, fear, and stress in fishes using an integrative approach. She is currently a Senior Lecturer at the University of Gothenburg, Sweden, and has published numerous papers, book chapters, and reports in the field of fish welfare. Lynne actively engages with the media and has participated in TV, magazine, and newspaper interviews as well as promoting fish welfare to learned societies, charities, public bodies, and governments.

Paula Sparks, LLB (Hons), BL, is a Visiting Professor at the University of Winchester, UK, where she teaches animal law. She practiced as a barrister at Doughty Street Chambers in London before leaving the bar in 2018 to pursue a full time (voluntary) role with the UK Centre of Animal Law (A-LAW), a charity whose vision is a world where animals are fully protected by law. Paula frequently lectures and writes about animal related law and policy.

Isabelle Veissier, DVM, PhD, is research director from INRAE (a French national institute for research in agriculture, food and environment). Her research focuses on farm animal behaviour and welfare. She has worked on social and maternal behaviour, human-animal relationships, learning and emotional abilities, and the assessment of animal welfare, especially with a multi-criteria approach. Currently she is investigating detection of sickness behaviour by sensor-based technologies. She played a key role in several large European projects (Welfare Quality, EU-PLF, SmartCow). She has published 120 scientific articles in peer reviewed journals and supervised 13 PhD theses. Her H index is 44.

Joy M Verrinder, BA, MBA, MA, PhD, DipT, is Strategic Director of the Animal Welfare League Queensland. She has initiated and led stakeholder coalitions consisting of local and state government, industry, veterinary associations, and animal welfare organisations. She works cooperatively with stakeholders developing legislation, policy and programs to prevent the abandonment and euthanasia of animals. She has been a teacher and education administrator. She has served on boards of state and national animal organisations, government and university Animal Ethics Committees, and state and national government advisory groups. Her PhD focused on developing capacity to address animal ethics issues. Her MA degrees focused on strategic management, professional ethics, and governance.

John Webster, MA, VetMB, PhD, MRCVS, FAfN, is Emeritus Professor of Animal Husbandry at the University of Bristol, UK. After graduating from Cambridge with degrees in physiology and veterinary science, his early research career addressed the energy metabolism of ruminants: nutrition, growth, lactation, and physiological responses to cold stress. In 1978, he moved to Bristol, and with Christine Nicol established the Bristol Animal Welfare and Behaviour Research Unit. He was a founder member of the UK Farm Animal Welfare Council and is a former President of the Nutrition Society and the British Society for Animal Production.

Jo White, BA, MSc, Cert Campaigns, has had a career spanning three decades, focusing on the well-being of animals. From working practically with horses, to leading educational, advocacy and campaigning projects, examples include her work at World Horse Welfare to improve the welfare of horses transported long distances for slaughter. In 2016 she became co-Director of Human Behaviour Change for Animals, supporting organisations to apply human behavioural science to their work. Jo has a MSc in Behaviour Change, a Degree in Equine Studies, among other equine qualifications, and a Certificate in Campaigning. Her current area of study is harnessing human habits to improve animal welfare, including human welfare.

Tina M Widowski, BS, MS, PhD, is a Professor of Applied Animal Behaviour and Welfare in the Department of Animal Biosciences at the University of Guelph, Canada. There she served as Director of the Campbell Centre for the Study of Animal Welfare for 12 years. She currently holds the Egg Farmers of Canada Research Chair in Poultry Welfare, with research mainly focused on how housing and management practices affect the welfare of laying hens and pullets.

Amy P Wilson, BCom, LLB, LLM, is the Executive Director of Animal Law Reform South Africa and a Research Associate with the University of Johannesburg. She is the first South African attorney to graduate with an LLM degree in Animal Law. Amy has over a decade of legal experience and her research focuses on the intersection of the rights of animals, humans and nature in law and policy. She is an independent expert with the United Nations in Harmony with Nature Programme and holds several leadership positions in nonprofits throughout Africa.

Miriam A Zemanova, BSc (Hons), MSc (Hons), PhD, holds degrees in ecology, biological conservation, and natural resources management. Her main research interests are wildlife welfare, humane education, and the development and implementation of non-invasive methods in conservation genetics. She has authored several publications and has provided multiple presentations on animal ethics and the 3Rs principles of responsible animal use. She is the author of the informational website https://3RsWildlife.info, which provides information for wildlife biologists, ecologists, and conservation managers on how to implement the 3Rs within wildlife research.

PREFACE

Recent years have seen a rapid evolution in the social status of animals. Cognitive and behavioural studies have increasingly revealed hitherto unknown characteristics in varying species, such as the existence of cultural behaviour[1] and knowledge,proto-language and other communicative abilities, as well as advanced social relationships, with evidence of manipulative, altruistic, and other complex behaviours, and ever-expanding theories of mind and consciousness. It has become clear that the differences between humans and non-human animals are nearly always differences of degree, rather than kind (Benz-Schwarzburg & Knight, 2011). Such awakening awareness of the richness and diversity of morally relevant characteristics within various animal species has led to a reconsideration of their historical and contemporary exclusion from the anthropocentric circle of moral consideration.

Stimulated by this science-based re-examination of animal characteristics, there has been increasing attention paid to our complex, multifaceted relationships with other animals, both within and outwith our societies; and in particular, the depth and breadth of various societal uses of animals. This has led to a reconsideration of their moral and social status, and to a range of important, associated developments.

In recent years this trend has accelerated, with classical concerns about the welfare of animals in farms, transportation, slaughter, laboratories, zoos, entertainment, as companions, working animals, and free-living animals in the wild, being complemented by concerns about the effects of climate change on animals, animal disaster management, and animal welfare. Most recently the impacts of animals on human health have been a focus – notably the contribution of practices such as intensive farming and the wildlife trade to antimicrobial resistance, pandemics, and other public health concerns.

Many of these concerns are controversial, throwing into sharp relief the differing interests of stakeholders such as industry, government, consumers, researchers, and of course, the animals themselves. In the face of changing attitudes, challenges to some long-accepted practices are inevitable, as both scientific knowledge and social values evolve. In such a contentious domain, sound evidence and reasoning become particularly important.

Through firm commitment to such principles, this book explores the biological foundations for the moral consideration of animals, and for evolving conceptualisations of animal welfare. It reviews in detail the welfare concerns associated with numerous forms of animal use. The inclusion of recent concerns such as climate change, pandemics, antimicrobial resistance, and the others above, ensures this text is among the most current in its field. The ethical implications of the various uses of animals within society are considered, and chapters are frequently accompanied by recommendations for reforms of practice or policy. The status of animal law internationally, and in major world regions, is reviewed. Finally, the book considers human behavioural change and strategies for improving stakeholder communication and education.

This textbook aims to assist and inform policy-makers, researchers, and other professionals in the animal welfare sector, and students in fields such as animal welfare science, ethics and law, veterinary science, animal science, agriculture, anthrozoology, and others. We would never have been able to cover such a wide diversity of topics so concisely within a single volume without the expert contributions of our 50 authors, many of whom are leaders in their fields. We are grateful for their invaluable assistance. And we are particularly grateful to you, the reader, and to the ever-growing body of people around the world, whose concern for animals and their welfare is helping to improve welfare standards in so many domains in which animals are used. With animals increasingly impacted by human activities, the need has never been greater.

Note

1 i.e. transmitted non-genetically

Reference

Benz-Schwarzburg J and Knight A, 2011. Cognitive relatives yet moral strangers? *Journal of Animal Ethics*, 1(1), pp. 9–36.

ACKNOWLEDGEMENTS

When Routledge suggested this book, we were delighted that a major publisher had recognised there was an urgent need for a new textbook in this field. We are grateful to Routledge for realising the importance of producing a book that is based firmly on scientific evidence and sound reasoning. We are also particularly grateful to all the scientists and philosophers who have toiled over many years to understand animal welfare and our attitudes towards the animals with whom we share our planet, and to the many animals (human and non-human) that have endured being part of research to generate this knowledge. As well, this project would never have been possible without the hard work and invaluable contributions of our many chapter authors, who managed to add the writing of excellent chapters to heavy workloads of their own. It would not have been possible to complete this project if our workplaces and managers had not allowed us the flexibility in managing our own workloads. We are grateful to all of these people, and also to our long-suffering partners, families, and friends, who partly lost us to this project for much of 2021–2022.

We are also grateful to Compassion in World Farming International, the Royal Society for the Prevention of Cruelty to Animals and the Albert Schweitzer Foundation, whose financial support allowed this book to be published open access in its entirety. These sponsors had no other role or influence over the conceptualisation, authorship or editing of this book.

PART I

Animal welfare fundamentals

1

THE MORAL STATUS OF ANIMALS

Biological foundations

John Webster

Morality is conventionally defined as the differentiation of intentions, decisions, and actions into those that are proper and improper. This is easy to say but it immediately raises questions as to what is and is not proper. Philosophers, religious leaders, politicians, and folk down the pub have been arguing over these questions forever and we must all be aware of moral standards that have changed even within the last 30 years; our attitudes to animal welfare being among them. There are, however, two moral principles that have stood the test of time. These are the Categorical Imperative and the Golden Rule. The Categorical Imperative of Immanuel Kant requires each individual to act according to the maxim "whereby you can, will that (your actions) should become a universal law" (Knowles and Partington 1999). The Golden Rule is most simply described as "do as you would be done by". These two maxims can be further refined, respectively into respect for the principles of *beneficence* (do good and do no harm) and *autonomy*, give equal respect to the rights of others.

For most of history, the moral concepts of right and wrong were applied only to intentions and actions within the human species. This was challenged by Albert Schweitzer, who wrote

> the great fault of all ethics hitherto has been that they believed themselves to have to deal only with the relations of man to man. In reality, the question is what is his attitude to the world and all that comes within his reach.
>
> *(see Brabazon 2000)*

This was encapsulated in his principle of reverence for life.

The moral status of animals, as perceived by us, has evolved from the days of Descartes (1596–1650) who assumed that non-human animals were unable to think (*non cogitant, ergo non sunt*) so could be considered as automata, not within our moral compass. The utilitarian, Jeremy Bentham (1748–1832) displayed greater empathy when he wrote "the question is not can they think ... *but can they suffer?*" (ODQ 1996). The UK Protection of Animals Act (1911) made it an offence to "cause unnecessary suffering by doing or omitting to do any act". The 1997 Treaty of Amsterdam recognised that "since animals are sentient beings, members should pay full regard to the welfare requirements of animals". Recognition of non-human animals as sentient beings is becoming enshrined in law in other nations, e.g. United Kingdom and New Zealand. These pronouncements reflect the evolution of our moral values but they beg several questions: "what

DOI: 10.4324/9781003182351-2

constitutes suffering, especially *necessary* suffering?" "what are the welfare requirement of animals?", and "what is meant by sentience?" We can only give proper respect to the moral status of animals if we have a clear understanding of the complex biological principles that should underpin these broad assumptions.

The biology of animal welfare

All animals are presented with challenges to their physiological and psychological state. Their welfare depends on their success in coping with these challenges. The sentient animal is motivated to actions designed to avoid suffering and promote a sense of wellbeing (Dawkins 1980). The word in common use when describing challenges to animal welfare is "stress" (Moberg and Mench 2000). This can be unhelpful because it is loosely applied both to stimulus and response and also fails to distinguish between coping and suffering. The pioneer of stress physiology Hans Selye used the word "stressor" to define the challenge, and "stress" to define the response (Selye 1950). He defined this response as the General Adaptation Syndrome. The initial phase of this response is the Alarm Reaction, a definition that recognises both a physiological and psychological component. According to the severity and duration of the challenge, the initial Alarm Reaction may or may not proceed to a state of complete or partial adaptation. An animal that achieves complete adaptation is coping satisfactorily with challenge. Partial adaptation means that the animal is coping but at a continuing physiological and psychological cost. Suffering occurs when an animal fails to cope or has extreme difficulty in coping because the challenge is too severe, complex, or prolonged.

Subsequent chapters will examine in detail approaches to the assessment and management of the welfare of animals within our care. It is necessary to outline some of them briefly at this stage because they help to define our understanding of the biological foundations of the elements of welfare that command our respect. One well-established approach is based on the concept of Five Freedoms and Provisions as described by the UK Farm Animal Welfare Council (FAWC 1994). These are:

- *Freedom from thirst, hunger and malnutrition*: by ready access to fresh water and a diet to maintain health and vigour;
- *Freedom from discomfort*: by providing a suitable environment including shelter and a comfortable resting area.;
- *Freedom from pain, injury and disease*: by prevention, rapid diagnosis, and treatment;
- *Freedom from fear and stress*: by ensuring conditions that avoid mental suffering;
- *Freedom to express normal behaviour*: by providing sufficient space, proper facilities, and the company of the animal's own kind.

These rules are close to being comprehensive and the first four *freedoms from* have stood the test of time. The fifth is a *freedom to*, and as with all such freedoms can create moral problems, often exemplified by the phrase "no one should have the right to shout fire in a crowded cinema". On reflection, I believe the fifth freedom would be better described as *freedom of choice* (Webster 2022).

The pan-European Welfare Quality® programme has established assessment protocols for the welfare of farm animals according to 4 welfare principles: nutrition, environment, health, behaviour, defined by 12 criteria (Welfare Quality® 2009). These, like the five freedoms, have the practical merit that they can readily form a template for quality control programmes operated by government departments, non-governmental associations such as RSPCA, or supermarkets seeking custom on the basis of quality assurance as to animal welfare standards. They can also highlight specific, severe welfare problems (physiological or psychological) requiring immediate attention.

An alternative approach to the characterisation of animal welfare is the Five Domains model (Mellor 2016). This recognises four measurable input categories: nutrition, environment, health, and behaviour. The fifth domain is defined as "mental state", which seeks to estimate the overall effect of these variables on the animal's sense of wellbeing, otherwise defined as quality of life. This approach appeals to our moral sense of duty to understand animal welfare as perceived within their minds, not ours, and is a good template upon which to base future research into animal behaviour. In practice, however, the structure of the five freedoms may be more useful in the identification of specific problems and the implementation of welfare assessment protocols (Webster 2016).

If we are to do right by sentient animals, we need to do more than just protect them from things that may do them physical or emotional harm, we need to get into their minds. In this regard it helps to practise a form of reverse anthropomorphism. Far from being unscientific, this is the basis of motivation analysis (Dawkins 1980). The scientist creates a hypothesis as to how (e.g.) a chicken might cope with a potential problem, then presents it with a set of alternative solutions. Their choice is defined by the preference test, their strength of motivation to action – how much these feelings matter – is defined by the cost they are prepared to pay (Mason et al., 1998). By this approach we can begin to understand how they themselves interpret the meaning of 'quality of life'.

Sentience and consciousness

The words sentience and consciousness are freely employed in discussions of animal welfare, usually without further explanation. This creates problems because they can mean different things to different people. In my new book, "Understanding Sentient Minds" (Webster 2022) I pose five questions.

- *What, indeed, is animal sentience?*
- *Is animal sentience an either/or thing or are there degrees of sentience?*
- *If there are degrees of sentience, at what degree does quality of life matter to the animal* (and so to us)?
- *What, if anything, is the difference between sentience and consciousness?*
- *What do we mean by the sentient mind?*

Search for "animal sentience" on Wikipedia and you are directed to Animal Consciousness, or the state of self-awareness in a non-human animal. It proceeds to define consciousness in humans as "sentience, awareness, subjectivity, the ability to experience or feel, wakefulness, having a sense of self and the executive control of the mind". This summary highlights the problems of definition: the word consciousness is used to describe any and all of these properties: i.e. it can mean what you choose it to mean. It fails to address the obvious variation in the nature of sentience within the animal kingdom, the extent to which it may or may not involve consciousness, and how this might affect our moral duty of care in terms of our actions in regard to (e.g.) a worm and an elephant.

The expression of sentience within the animal kingdom will include sensations and emotions ranging from the primitive (hunger, pain) to the complex (hope, despair, love, hate). The sentient mind is able to perceive and interpret these sensations and emotions through more or less complex cognitive processing of incoming information in the light of past experience (which may or may not involve feelings). At some stage, depending on your definition, this may involve consciousness.

The five Buddhist *skandhas* of sentience

If we are to meet our duty of care with respect to all sentient animals, we need a comprehensive understanding of the nature of sentience itself and the operation of the sentient mind. I believe the most satisfactory *scientific* analysis of the biological principles that determine the nature of sentience is given by the five degrees (*skandha*) of Buddhist philosophy. These are matter, sensation, perception, mental formulation, and consciousness, illustrated in Figure 1.1 as five concentric circles of increasing depth, signifying increasing complexity from the outer, superficial circle of matter to the deepest circle of consciousness. Figure 1.1 also presents estimates, based on evidence relating to animal behaviour and motivation, of the degrees of sentience involved in the interpretation of primitive sensations like hunger and pain and expressions of more complex behaviours and emotions such as companionship, altruism, hope, and despair.

Matter describes living organisms as defined by their physical structure, chemical composition, and processes that enable them to operate within a complex environment. This category embraces all plants and animals. It includes the ability to react to environmental stimuli, like the

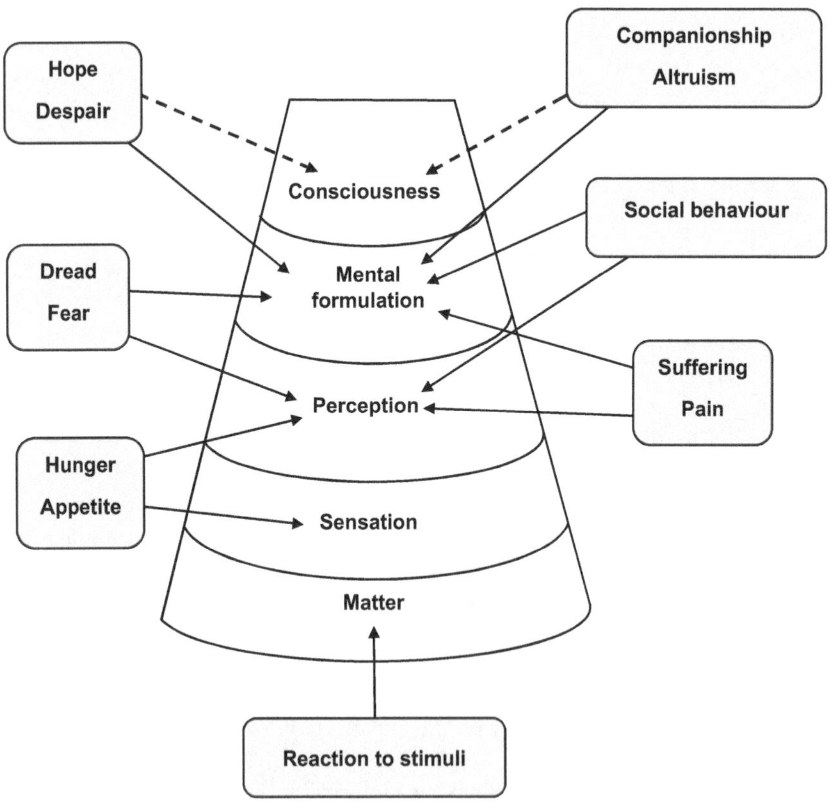

Figure 1.1 The five *skandhas* of sentience. Solid arrows indicate the proven extent of sentience involved in different forms of experience and social behaviour. Dotted lines indicate possible but unproven extension of sentience into the inner circles.

movement of sunflowers towards the sun, or the movement of amoebae away from an acid solution, without necessarily involving sensation as we would define it.

Sensation describes the ability of living creatures to experience feelings, and the intensity of feelings that take them out of their comfort zone. These clearly include physiological sensations such as hunger, thirst, pain, severe heat and cold, and may include hard-wired acute responses to threat that we may interpret as fear. At this depth of sentience, animals interpret these sensations as unpleasant (aversive), pleasant (attractive), or unimportant (indifferent), and these sensations will motivate them to take action to avoid or reduce the threat to their wellbeing.

Perception describes the ability to register, recognise and remember objects, experiences, and emotions. Species with the property of perception do not just live in the present, they can learn from experience. This enhances their capacity to cope by adapting to the challenges of life but increases the potential for suffering if the challenges are too severe, too prolonged, or if they are in an environment that restricts their ability to perform coping behaviour.

Mental formulation describes the ability to create mental images that integrate and interpret complex information, experiences, sensations, and emotions. This enables animals to learn from experience so increases their capacity to cope with challenge, but equally increases the potential for suffering if they find themselves unable to cope. The ability to create mental pictures also creates the capacity to develop the mind through education, given and received.

Consciousness In the Buddhist *skandhas* the word consciousness is restricted to the deepest circle of sentience and equates to the most precise definition of human consciousness, best described as "being aware that we are aware". This carries the potential for advanced forms of social behaviour, both good and bad, such as empathy, compassion, and cheating.

The Schweitzer principle of reverence for life requires us to respect all degrees of sentience and this is entirely consistent with our new moral and practical imperative to practise planet husbandry: to sustain and conserve the balance of nature for the welfare of all life. It does not, however, compel us to apply the same set of rules to a dandelion as to a horse. Animals whose degree of sentience extends only to the property of sensation will respond to primitive sensations such as pain, malaise, hunger, and sex in a way that may be intense and probably adaptive but, by this definition, hard-wired and not necessarily involving what we might understand as emotion. However, the UK Animals (Scientific Procedures) Act (1986) recognises that the property of sensation is sufficient to give animals protected status in regard to actions likely to cause pain, suffering, distress or lasting harm, and requires these actions to be set against possible benefits to society. Species given protected status by the Act currently include all vertebrates and the invertebrate cephalopods. In the light of new research, this may have to be extended to other invertebrates (Smith 2020). In a broader moral context it accepts that a primitive sensation such as pain may feel the same to a fish as to a dog.

In almost all the animal species whose lives are affected by human contact, the expression of sentience is not limited to primitive sensation. The rules that govern our moral duty to respect their welfare must take account of the biological evidence as to the extent to which they demonstrate the three inner circles of sentience, namely perception, mental formulation, and consciousness. These are summarised in Table 1.1.

Species that have the power of perception can learn from experience. This increases their ability to mount an effective immediate response and improves the chances of doing things better next time. It also carries the potential to increase distress and anxiety if they learn that they cannot cope. Species who demonstrate the property of mental formulation, the ability to create mental pictures (or diagrams) that integrate and interpret complex experiences, sensations, and emotions are even better equipped to deal with challenges because they do more than recognise the associations between cause and effect, they can understand them. This gives them the poten-

Table 1.1 Emotional and cognitive expressions of sentience with welfare implications

	Emotion	*Cognition*
Perception	Pain and fear	Avoidance
	Hunger and thirst	Food selection
	Comfort	Nest building
	Curiosity and security	Interpret simple social signals
Mental formulation	Anxiety and depression	
	Pleasure, joy, hope, grief	Recognition of social signals
Consciousness	Affiliative behaviour	Awareness of self and non-self
	Altruism and compassion	Deceit

tial to communicate their understanding with others. All species with the power of perception, whether or not they can demonstrate the capacity for mental formulation, have the ability to make decisions, based on experience, as to how best to cope with challenge. For all these species, our moral responsibility must extend to provision of the fifth freedom; best expressed as freedom of action to engage in appropriate coping behaviour.

The inner *skandha* of higher consciousness is applied only to mental formulations described in scientific terms as metarepresentation, or "theory of mind" (Frith and Frith 2005). These derive from having a sense of self and non-self and can give rise to affiliative behaviours such as altruism and compassion, but equally to anti-social behaviours such as deceit. The number of species for which we have good evidence for theory of mind is limited and largely restricted to social mammals, e.g. great apes, dolphins, and other cetaceae (Krupenye and Call 2019) but the list is growing. It should probably include social corvids (e.g. rooks, Clayton and Emery 2007) and possibly some invertebrates (e.g. cephalopods, Smith 2020). Our duty to social species with these powers should respect and understand their need to communicate and respond appropriately to their social signals.

Our duty of care: The ethical matrix

The *raison d'etre* for this chapter and for this book is to convey an understanding of the biological principles that determine the sentience and the welfare of animals as a sound basis for ethical judgements as to their moral status as seen through our eyes and, more importantly, our actions as seen through theirs. I have used the biologically valid structure of the Buddhist *skandhas* to describe the range of expression of sentience from simple sensation to the full expression of consciousness as revealed by evidence of theory of mind, or metarepresentation. With increasing depth of sentience, sensation is augmented by perception, emotion, cognition, understanding, and awareness. The description of our fellow mortals as "Cognitive relations yet moral strangers" (Benz-Schwarzburg and Knight 2011) is an eloquent expression of the problem we face. However, I would add that cognition is only one facet of sentience. The emotional response to the challenges of life, ranging from perception to full awareness, is the critical determinant of wellbeing. Nevertheless, the extent to which an animal displays the cognitive abilities necessary for higher emotions such as hope and despair or social graces such as companionship and affiliative behaviour must determine the practical expression of our duty of care. I repeat, we need not apply the same rules to the worm as to the elephant.

What then is our duty of care? Starting from the general principle of reverence for life, our actions in regard to the animals with whom we share the planet should be defined by our understanding of their need to promote their own wellbeing, so far as possible through their own actions. In this regard we must reject the characterisation of species as domestic and wild, food animals and pet animals, game and vermin. So far as the species we define as wild are concerned, the most moral course of action is to preserve their habitat, then leave them alone.

For domesticated animals for whom we have a direct duty of care (the great majority of which are the farmed animals), the ethics get more complicated. There are two approaches to questions of ethics, top-down and bottom-up. The top-down approach asks the question: "which moral norms for the evaluation and guidance of conduct should we accept and why?" The bottom-up approach first identifies a specific practical problem then constructs an analysis of relevant ethical issues by a process of induction. This latter approach may conflict with some of the precepts of high morality, particularly when it addresses such topics as killing and necessary suffering. Beauchamp and Childress (1994) outlined a practical approach to problems of medical bioethics in the form of an "ethical matrix", and this has been adapted by Mepham (1996) to address our attitudes and actions with respect to the food animals. These should be based on the two principles of beneficence and autonomy, defined, respectively, in my introduction as "do good and do no harm" and "respect the rights of others". The aim of beneficence is to promote wellbeing, which loosely equates to the first four freedoms *from*. Autonomy can be achieved through the fifth: freedom of choice.

The moral basis of our approach to animal welfare is not, however, something that we can consider in isolation. In the case of farmed animals, we must incorporate it within the broader context of respect for the needs of farmers and consumers, the farmed animals, and the living environment; the aim being to achieve a fair compromise that equates to justice for all. Table 1.2 employs the ethical matrix to examine the moral issues associated with farming animals for food (Webster 2013). The four groups with rights to justice are the producers, consumers, farmed animals, and the living environment. Farmers and consumers (i.e. all humans) who set the standards are the moral agents; farmed animals and the living environment are the moral patients. Because they can have no input to the debate, the responsibility to ensure justice for all is entirely in our hands.

Table 1.2 Food and farming: the ethical matrix

	Wellbeing	*Autonomy*	*Justice*
Moral agents			
Producers and land owners	Financial reward Pride in work	Free competition	Fair trade Good husbandry Support for environmental schemes
Human society	Wholesome, safe, affordable food Access to the countryside	Freedom of choice	Added value for good husbandry
Moral patients			
Farmed animals	Competent and humane husbandry	Environmental enrichment Freedom of choice	"A life worth living"
The living environment	Conservation Sustainability	Biodiversity "Live and let live"	Respect for environment and stewards

The practical expressions of the principles of wellbeing, autonomy, and justice listed in Table 1.2 are largely self-explanatory. Farmers as moral agents have the responsibility to promote the wellbeing of their animals and their land. In return, they have the right to adequate financial reward, fair competition, and pride in their work. Moreover, farmers are not only food producers but major stewards of the living environment. Justice for them and for the environment requires that they receive fair reward for services to the environment: conservation of habitat, soil and water management, carbon sequestration. This equates to public money for public goods. Consumers, i.e. everybody, whatever our incomes and our eating habits, have the right to wholesome, safe, affordable food. In return, we have the responsibility to recognise the added value of products that can guarantee high standards of animal husbandry and environmental protection and reward the producers accordingly. Nevertheless, the principle of freedom of choice implies that those of us who can afford to pay more should not seek to impose standards that put good, honest food out of the reach of the poor.

Human society cannot, of course, fully apply the principle of beneficence, "do good and do no harm" to our treatment of other sentient animals. We kill farmed animals for food, we harm laboratory animals in the pursuit of science and safety testing, we cull individual wild animals for reasons of population management, disease control and protection of habitat. This practice is strictly utilitarian and, as such, morally imperfect. It is, however, a fact of life so we can do no better than operate according to the principles of the great and compassionate utilitarian Jeremy Bentham: always pose the question "can they suffer?" This principle forms the basis of the UK Animals (Scientific Procedures) Act (1986), which "regulates the use of protected animals in any experimental or other scientific procedure that may cause pain, suffering, distress or lasting harm to the animal" (UK Govt 1986). This regulation requires that any harm to the animal must be justified in terms of its likely benefit to the welfare of human society (or other animals). This is, again, a utilitarian principle but it has undoubtedly been a force for good, especially as a driver for the application of the three Rs, replacement, reduction, and refinement, in the design and conduct of experiments with animals (Russel 1995). It is also widely recognised that these principles do not relate only to the procedures themselves but to the day-to-day management of the animals to promote wellbeing, including, wherever possible, freedom of choice through provision of an enriched environment.

Conclusions

Wherever we exert control over the lives of other sentient animals we have a moral responsibility to promote their wellbeing, based on a sound understanding of the biological principles that underpin their physiological and psychological needs. We can do much to meet these needs, as we understand them, by the practice of beneficence. To meet these needs as understood by the animals themselves, we also need to respect the principle of autonomy and respect their need for freedom of choice. We must, however, never let respect for animal welfare become the whole story. It must be put within the context of justice for all: the sustained wellbeing of humans, other sentient animals, and the entire living environment.

References

Beauchamp TL and Childress JF, 1994. *Principles of Biomedical Ethics*. Oxford: Oxford University Press.

Benz-Schwarzburg J and Knight A, 2011. Cognitive relations yet moral strangers? *Journal of Animal Ethics* 1, pp. 9–36.

Brabazon J. 2000. *Albert Schweitzer: A Biography*, 2nd edn. New York: Syracuse University Press.

Clayton NS and Emery NJ, 2007. The social life of corvids. *Current Biology*, 17(16), pp. R652–656.
Dawkins M, 1980. *Animal Suffering: The Science of Animal Welfare*. London: Chapman and Hall.
Farm Animal Welfare Council, 1994. *Second Report on Priorities for Research and Development in Farm Animal Welfare*. Tolworth: MAFF.
Frith C and Frith U, 2005. Theory of mind. *Current Biology*, 15, pp. 644–645.
Knowles EM and Partington A (eds), 1999. *The Oxford dictionary of quotations*. Oxford [England]: Oxford University Press.
Krupenye C and Call J, 2019. Theory of mind in animals: Current and future directions. *WIREs Cognition Science* 10, pp. 76–87.
Mason GJ, McFarland D and Garner J, 1998. A demanding task: Using economic tests to assess animal priorities. *Animal Behaviour*, 55, p. 1070.
Mellor DJ, 2016. Updating animal welfare thinking: Moving beyond the five freedoms to 'A Life worth living'. *Animals*, 6, p. 21.
Mepham B, 1996. Ethical analysis of food biotechnologies: An evaluative framework. In Mepham B, eds., *Food Ethics*. London: Routledge, 101–109.
Moberg G and Mench J (eds), 2000. *The Biology of Animal Stress: Basic Principles and Implications for Animal Welfare*. Egham, UK: CABI Publishing.
Protection of Animals Act 1911. www.legislation.gov.uk/ukpga>Geo5
Russell WMS, 1995. The development of the Three R's Concept. *Alternatives to Laboratory Animals*, 23(3), pp. 298–304. https://doi.org/10.1177/026119299502300306
Selye H, 1950. *Stress: the Physiology and Pathology of Exposure to Stress*. Montreal: Acta Publications.
Smith Godfrey P, 2020. *Metazoa: Animal Minds and the Birth of Consciousness*. Glasgow: William Collins.
Treaty of Amsterdam, 1997. The Lisbon treaty: Recognising animal sentience. www.ciwf.og.uk>news 2009/12
UK Govt. (1986). Animals (Scientific Procedures) Act (1986). London: HM Stationery Office.
Webster J, 2013. *Animal Husbandry Regained: The Place of Farm Animals in Sustainable Agriculture*. London and New York: Earthscan from Routledge.
Webster J, 2016. Animal welfare: Freedoms, dominions and 'a life worth living'. *Animals*, 6, 35.
Webster J, 2022. *Animal Welfare: Understanding Sentient Minds - and Why It Matters*. Oxford: Wiley.
Welfare Quality®, 2009. Welfare quality reports No.12: An overview of the development of welfare quality project assessment systems. www.welfarequality.net>wgr12

2
ANIMAL WELFARE CONCEPTS

Donald M Broom

Welfare as a component of sustainability

Members of the public in most countries now have increasing concerns about the sustainability of systems for producing food and other products (Aland and Madec 2009). In addition to profitability of animal production systems and an acceptable price for the consumer, continuation of production now depends on the ethics of production methods (Broom 2010). Sustainability now has a wider meaning than it had in early writings on the subject (Herrero and Thornton 2012, Broom 2017). A system can be unsustainable, and a product can be considered to be of poor quality, because of negative impacts on human welfare, on animal welfare, or on the environment. A system or procedure is sustainable if it is acceptable now and if its expected future effects are acceptable, in particular in relation to resource availability, consequences of functioning, and morality of action (Broom 2014).

Poor welfare of wild or farmed animals is one of the major reasons why consumers may refuse to buy a product or may lobby governments, retail companies, or producers to insist on changes in production methodology. Scientific studies evaluating animal welfare provide the evidence required for such decisions. Sustainability has many components and consumers need a scoring system based on scientific evidence so that they can consider all components and decide what is sustainable (Broom 2021a).

Humans and other animals

Animals are living beings with a nervous system and mechanisms for obtaining energy, using energy, and reproducing. They derive energy by consuming other organisms and most have an effective means of locomotion and a range of sense organs. It is incorrect to use the word "animal" to mean solely farmed animals, owned animals, mammals, or warm-blooded animals, and also to say "humans and animals" since humans are animals. The idea that non-human animals are more aggressive, less controlled, or more subject to lust than the average human is also wrong (Broom 1998, Hofman 2014).

There is only one biology for all of the animals in the world and almost all mechanisms in humans are identical with those in many other species. It is difficult to find any human quality that is not shared with some other species, examples including: language, emotions, the notion

 DOI: 10.4324/9781003182351-3

of culture or society, cooperation, altruism, tool use, and a concept of the future (Clayton and Emery 2015, McBride and Morton 2018). Human abilities are also possessed by other animal species, at least to some degree, so research on other species is used to better understand humans. Every species has differences from all the others and humans have some more complex brain processing capacities and better mathematical logic, perception of time, complex reasoning, analytical capacity, and prediction of events than most other species. These are differences in degree rather than in absolute capability (Falk and Hofman 2012) and DNA sequencing shows that genetic differences among species, including humans, are small and the similarities large (Boffelli et al., 2004). Although humans favour their own species, the idea that humans are special, or more important than other animals, is not scientifically logical.

Cognition, awareness, emotions, and feelings are important adaptive mechanisms in animals. They occur in the brain and result from, or lead to, sensory mechanisms, muscular responses, glandular responses, and other bodily changes. The organs of the body, such as the heart, influence brain function, but thoughts and feelings are in the brain and not in the heart or any other part of the body. All of the analysis, thought, and emotional aspects of brain functioning are closely interlinked, so it is not useful for the concept of mind to be considered separately from the brain (Panksepp 2005, Broom 2003, 2014, LeDoux 2012). Cognition is having a representation in the brain of an object, event, or process, in relation to its context, where the representation can exist whether or not the object, event, or process is directly detectable or actually occurring at the time (Broom 2014). Awareness is a state during which concepts of environment, of self, and of self in relation to environment result from complex brain analysis of sensory stimuli or constructs based on memory (Broom 2014). There are several levels of awareness: unaware, perceptual awareness, cognitive awareness, assessment awareness, and executive awareness. A conscious individual is one that has the capability to perceive and respond to sensory stimuli (Broom 2014). A feeling is a brain construct involving at least perceptual awareness which is associated with a life regulating system, is recognisable by the individual when it recurs and may change behaviour or act as a reinforcer in learning (Broom 1998). An emotion is a physiologically describable component of a feeling characterised by electrical and neurochemical activity in particular regions of the brain, autonomic nervous system activity, hormone release, and peripheral consequences that may include behaviour (see discussions by Paul and Mendl 2018, Broom and Johnson 2019). Feelings are adaptive mechanisms that have evolved and include pain, fear, anxiety, sexual pleasure, eating pleasure, exhilaration, achievement pleasure, other sensory pleasure, social affection, guilt, anger, rage, malaise, tiredness, hunger, thirst, thermal discomfort, grief, frustration, depression, boredom, loneliness, jealousy, and lust (Broom 1998, 2014). Suffering is one or more bad feelings continuing for more than a few seconds or minutes. The concept of affect concerns emotions, feelings, moods, and affective dispositions (Sander 2013).

As humans and other complex animals develop, there is a stage when they become aware of themselves and of their interactions with their environment. Some of this occurs when an individual learns to avoid a painful action, like chewing their own foot. During development, the ability appears for a human and many other animals to experience pleasurable states such as happiness and aversive states such as pain, fear, and grief. This is when the individual becomes sentient. Sentience means having the capacity to have feelings, including having the levels of awareness and cognitive ability necessary to have feelings. A sentient being is one that, in order to have feelings, has abilities such as the ability to evaluate the actions of others in relation to itself and third parties, to remember some of its own actions and their consequences, to assess risks and benefits, and to have some degree of awareness (Broom 2014). Thus individuals, including humans, are not sentient at early life stages or when they have some brain pathologies or injuries. The current scientific view is that sentient animals include mammals, birds, reptiles, amphibians, fish, cephalopod molluscs, and decapod crustaceans. The decisions about sentience are important because sentient animals are better protected by law.

Welfare

With the exception of studies of animal disease, which can greatly improve welfare, animal welfare science hardly existed 40 years ago but it has developed rapidly since then (Broom 2011). During this time, concepts have been refined and a range of methods of assessment have been developed. Challenges to animal functioning and responses to these are a major biological study area. Examples of challenges and associated inability to control interactions with the environment result from (i) pathogens; (ii) tissue damage; (iii) attack or threat of attack by a conspecific or predator; (iv) other social competition; (v) complexity of information processing in a situation where an individual receives excessive stimulation; (vi) lack of key stimuli such as a teat for a young mammal or social contact cues; and (vii) lack of overall stimulation (Broom and Johnson 2019, Broom 2021b). Systems that respond to or prepare for challenges are coping systems and coping means having control of mental and bodily stability (Broom and Johnson 2019). Coping requires the functioning of the nervous system, including the brain, so it is limited to animals. Adaptation can occur without nervous system involvement. Potentially damaging challenges may come from the environment outside the body or from the environment of systems within it.

Coping attempts may be unsuccessful in that control is not achieved but, as soon as there is control, the individual is coping. Coping systems may respond to short-term or long-term problems, or to both. The responses may involve brain activity, endocrine, immunological, or other physiological responses or behaviour but these various types of responses are interdependent. While brain changes regulate bodily coping responses, adrenal changes can have consequences for brain function, lymphocytes have opioid receptors and can alter brain activity, and heart rate changes can be used to regulate mental state and further responses (Broom 2019).

Most coping systems require feelings as a part of their functioning, for example, pain, fear and aspects of pleasure, all of which are adaptive (Broom 1998, Fraser 2008). Coping systems in humans and other species have simple aspects and also complex brain functioning. Investigations of welfare evaluate how easy or difficult it is for the individual to cope with the environment and how great the impact of positive or negative aspects of the environment is on the individual. An individual with no problems to deal with is likely to be in a good state, including good feelings, as indicated by body physiology, brain state, and behaviour. Another individual may be unable to cope with problems in life. Prolonged failure to cope results in cessation of growth, reproduction, and eventually life. A further individual facing problems may use its coping mechanisms and cope but only with difficulty. The signs of success in coping, of failure to cope, or difficulty in coping and associated feelings, can be measured. The welfare of an individual is its state as regards its attempts to cope with its environment (Broom 1986), and this includes feelings and health. Welfare is thus a characteristic of an individual animal during a certain time interval and the state of the individual can be assessed. Hence, welfare will vary on a range from very good to very poor. Welfare concerns how well the individual fares, or goes through life, and since welfare is defined as a state, it is not grammatically correct to refer to the "welfare state" of an individual. Whilst the mean welfare in a population can be described, we cannot refer to the welfare of a population or an environment.

This meaning of welfare is widely used by animal welfare scientists and is close to the usage dating back to Shakespeare's time of how well an individual fares, or goes through life. The scientific definition is quite different from referring to welfare as a service or other resource given to an individual such as handouts to the poor. Welfare scientists all agree that animal welfare is measurable and hence is a scientific concept (Fraser 2008, Broom and Johnson 2019). Welfare involves mental aspects so research on welfare involves measurements of brain function and of its consequences for behaviour and physiology. Animal welfare indicators give information about positive and negative feelings and other coping mechanisms such as those that affect health. The OIE follows this definition when writing about what is meant by animal welfare although some of their explanatory wording is not precise.

Some other attempts to define welfare have placed sole emphasis on feelings (Duncan and Petherick 1991). However, feelings only comprise part of the mechanisms used by individuals to cope with their environment. Those mechanisms that do not involve feelings, for example, those that are used to cope with pathology and other negative or positive impacts of the environment, are an important part of welfare. Fraser (1999) pointed out that when members of the public talk about animal welfare, their ideas often include the functioning of the animals, the feelings of the animals and the naturalness of the environment. Rollin (1995) advocated that "animals should be able to lead reasonably natural lives" and both Rollin and Fraser (Fraser 2008) refer to the importance of understanding animal needs. These authors did not say that naturalness contributes to a definition of welfare or should be part of welfare assessment. Appleby and Hughes (1997) explain what they mean by welfare using a diagram in which naturalness is a circle partly overlapping with two other circles labelled function and feelings. I consider this diagram to be misleading. The state of an individual trying to cope with its environment depends on its biological functioning, i.e. on its nature. Natural conditions have affected the evolution of coping mechanisms and the needs of each species. Gygax and Hillman (2018) state "Natural behaviour in this sense involves reaching adequate goal states for all persistent or recurring wants that arise in a given environment". Any environment provided should meet the needs of the animal but does not have to be the environment in the wild. Conditions in the wild can lead to starvation, disease, predation, and hence very poor welfare (Yeates 2018). The overlapping circles diagram is incorrect because the concept and definition of welfare does not include naturalness and because feelings are a part of function. Hence, if there were three circles, they would have to be superimposed on one another.

The term "well-being" is sometimes used interchangeably with "welfare", but well-being is sometimes less precise in usage and can be taken to refer more to the positive while it is important that the concept of welfare ranges from negative to positive. Welfare is the word used in English versions of European legislation. Despite colloquial usage, most American scientists and the American Veterinary Medical Association now use welfare rather than well-being. A further term, "quality of life" is often used to refer to people, or companion animals, who are ill or recovering from illness. Both quality of life and welfare can be good or poor but, while welfare can refer to short-term situations, quality of life is not normally used for a short time-scale, such as one or two hours or days. Quality of life means welfare during a period of more than a few days (Broom 2007, 2014) and can be assessed using the wide range of welfare indicators. Subjective measures of quality of life should be rigorously verified, both for humans and for non-humans (Green and Mellor 2011).

The concept of "a life worth living" could be based on scientific information but is ethical, or policy-related, rather than scientific. Who decides when it is worth living or not worth living (Broom 2014)? If the individual is not human, it is a human evaluation rather than an evaluation by the subject, so conclusions may be erroneous (Broom 2014, 2021b). The idea of "a life worth living" is not scientifically usable (Green and Mellor 2011).

Health

Health, like welfare, varies over a range from good to poor. Health refers to brain and body systems that combat pathogens, tissue damage, or physiological disorder so health can be defined as the state of an individual as regards its attempts to cope with pathology (Broom 2006). Welfare is a broader term so health is a part of welfare. At the time of the World Health Organization statement "health is a state of complete physical, mental and social well-being and not merely the absence of disease or infirmity" the word welfare was not being used in a scientific way. However, since welfare is essentially the same concept as well-being, WHO was defining health as an aspect of welfare. Most people limit health to conditions related to pathology, physical or mental. The environment has a major impact on human and non-human health, but it is not scientifically correct to refer to the health of the environment. The environment is not self-regulating and no coping is involved.

As pointed out above in relation to there being only one biology, the definition and discussion of the terms health and welfare make it clear that they apply to many kinds of animals. A central principle of the one health concept is that health means exactly the same for non-human animals as it does for humans. The one health strategy encourages interdisciplinary collaboration and communication in relation to all aspects of interactions with the environment and health care for humans and non-human animals (Karesh 2014). It has long been recognised that human psychiatry and medicine could learn from research on farm and other animals and vice versa (Broom 2001a, 2001b) but progress has been slowed by the attitude that human research was quite different from that on other animals. The one welfare approach emphasises that the concept of welfare is identical when applied to humans or to non-human animals (Garcia Pinillos et al., 2015, Broom 2017) and should be incorporated into teaching about both (McGreevy et al., 2020, Broom 2021b). When the welfare of any individual is poor, there is increased susceptibility to disease so improving welfare generally reduces disease. For example, there is similarity between post-partum problems in pigs and humans (Daigle 2018). In order to utilise this approach effectively, all humans and non-humans should be considered as individuals so herd treatment is not always sufficient.

A consequence when concepts such as biology, health, welfare, stress, pain, etc. have the same meaning for all animals, including humans, is that other words should also have the same meaning for all species. For example, euthanasia means killing an individual for the benefit of that individual and in a humane way (Broom 2007b, 2017). Hence euthanasia does not just mean humane killing and should not be used when pet or laboratory animals are humanely killed for the convenience of their owner.

Stress

When an individual is said to be stressed the normal meaning is that the individual is subjected to a potentially or actually damaging effect of its environment. However, there has been a confusing range of uses of the term stress. As a consequence, some scientists have limited the use to a single physiological response mechanism: hypothalamic–pituitary–adrenal cortex (HPA) activity. Equating stress with HPA axis activity renders the word redundant, because we could just say HPA axis activity, and this is not in accord with usage as such activity is temporarily increased during courtship, mating, active prey catching, and active social interaction. Also, many other responses to challenges can occur. Another meaning of stress equates it with stimulation but if most impacts of the environment on an organism are called stress, then again the term has no value. Stimuli that benefit individuals would never be called stressors by most people. Stress is an environmental effect on an individual which overtaxes its control systems and results in adverse consequences and eventually reduced fitness (Broom and Johnson 1993). Responses to stress and short-term consequences of stress may be called strain. When coping is not possible and failure to cope leads to reduced fitness, the individual is stressed. Fitness reduction occurs if there are indications that fewer offspring bearing the genes of that individual would reach future generations. Brief or long-term effects that would not have such consequences may cause poor welfare but are not called stresses. It is my view that it is better to define stress as above than to subdivide stress into eustress, that does not harm, and distress, that does harm. If there is no harm it is not stress. Stress indicators include substantial immunosuppression, injury, behaviour abnormality, and physiological overload that increases the chances that food acquisition or the ability to avoid dangerous aggression will be reduced. Hence the above definition of stress distinguishes minor disturbances to an animal's equilibrium, perhaps with some necessity for energy usage but with no consequence for fitness, from disturbances that reduce fitness or are likely to do so (Broom and Johnson 2019). The definition of stress could be used for plants or any other living organism, whereas welfare applies only to animals. Distress describes the state of individuals that are stressed but also those subject to other effects.

Needs

A need is a requirement, which is part of the basic biology of an animal, to obtain a particular resource or respond to a particular environmental or bodily stimulus. The need is in the brain and mediates effective functioning of the animal. Needs may be met by physiology or behaviour but the need itself is not physiological or behavioural (Toates and Jensen 1991). Studies of motivation are important for investigating needs. How hard will the individual work for a resource or an opportunity for action? Motivational mechanisms depend on biological functioning in that species. Another approach to finding out what are the needs of individuals is to assess the welfare of individuals whose needs are not satisfied (Hughes and Duncan 1988, Dawkins 1990, Broom and Johnson 2019). In human psychology research, the idea of a hierarchy of needs general to all individuals is not helpful. The evaluation of how well human needs are fulfilled (Taormina and Gao 2013) often involves asking human subjects, a methodology open to systematic bias. Most work on non-human species is more objective.

Control systems in animals, including humans, have evolved in such a way that the means of obtaining an objective have become important to the animal (Toates and Jensen 1991, Broom 2017). Resources such as food and appropriate physical conditions may not be sufficient to fulfil needs as the animal may need to perform a certain behaviour and be seriously affected if unable to carry out the activity, even when the objective of the activity is present. Many species have been shown, by the use of operant and other techniques, to work for food even in the presence of food (Inglis et al., 1997). A pig needs to root and manipulate materials while a chicken needs to search for and find food items. Pigs have a strong preference to root in soil or to chew deformable material such as straw and small branches and will work for the opportunity to do so (Hutson 1989). Chou et al., (2020) found that pigs preferred beech twigs to spruce twigs but used all twigs, and also a rubber toy. More substantial solid materials like metal bars do not meet the needs. Broiler chickens given soldier fly larvae were more active, walked more, and had fewer leg disorders (Ipema et al., 2020). Hens need to dust-bathe (Vestergaard 1980) and hens and sows need to build a nest before giving birth or laying eggs. The terminology used in motivational strength estimation is that developed for micro-economics, for example, demand, price, elasticity, consumer surplus (Matthews and Ladewig 1994, Kirkden et al., 2003).

For welfare evaluations in species whose needs have been investigated, the rather general ideas of freedoms or domains are not now required since the more scientific approach using needs can occur (Broom and Johnson 2019, Broom 2021b). Indeed freedoms, like rights, can be rather questionable concepts leading to harmful consequences. Hence it is better to consider the obligations of each person to animals that they use (Broom 2003, 2014, Mellor 2016). A list of the needs of a species has been the first step in Council of Europe recommendations and EU scientific reports on that species' welfare for over 30 years. For some wild and zoo species, where knowledge of needs is somewhat lacking, consideration of a range of domains is useful as a guideline for the needs of members of a species.

Pain and other feelings

Pain is a significant aspect of poor welfare. The concept that pain is limited to humans or mammals has long been thought improbable. Melzack and Dennis (1980) stated: "The nervous systems of all vertebrates are organized in fundamentally the same way"; and "the experience of pain is often inferred from the behaviours of mammals, and it is not unreasonable to attribute pain experience to birds, amphibia and fish" (and presumably, reptiles). Pain detection and processing mechanisms in fish and other animals are reviewed by Sneddon (2019), who explains precise assessment methods. It is often said that pain is difficult to study in non-human animals because they cannot report pain or its severity. However, human self-reporting of pain, for example on a scale from no pain to very severe pain, may be unreliable. People can lie about their own pain or deceive themselves, so more recent guides to pain assessment in humans include chapters on direct measurement. If this is not done, non-human pain studies are often better than those used for humans.

It is useful to call pain receptors nociceptors but any other distinction between nociception and pain is a consequence of attempts to emphasise differences between humans, or mammals in general, and other animals (Wall 1992, Broom 2014b, Broom and Johnson 2019). In sentient animals, most reactions to output from nociceptors involve high-level brain activity because of the great importance of learning from tissue damage and other harms. Other sensory systems do not have different names for simpler and more complex aspects. It is misleading to make a distinction between nociception and pain in sentient animals.

Other feelings which, like pain, are adaptive mechanisms that have evolved include fear, anxiety, sexual pleasure, eating pleasure, exhilaration, achievement pleasure, other sensory pleasure, social affection, guilt, anger, rage, malaise, tiredness, hunger, thirst, thermal discomfort, grief, frustration, depression, boredom, loneliness, jealousy, and lust (Broom 1998, 2014). Some of these, like fear and depression, can be worse than pain and evidence for them is available from both direct studies, such as oxytocin concentrations indicating forms of pleasure, or indirect studies such as judgement bias that indicate the "glass half-full/glass half-empty" distinction (Kis et al., 2015, Mendl and Paul 2020).

Magnitude of good or poor welfare

When considering the impacts of treatments or conditions on welfare, the duration of the state is important. The magnitude of good or poor welfare is a function of intensity and duration. Since long-term problems of a certain severity are worse than short-term problems, poor living conditions that lead to poor welfare are generally the most important to the animals involved and hence to concerned consumers. A range of welfare assessment studies can be used to consider mean welfare over a long time-scale and to identify individuals whose welfare is poor over much of their life, usually because their living conditions do not meet their needs.

References

Aland A and Madec F (eds), 2009. *Sustainable Animal Production*. Wageningen Academic Publishers, Wageningen, Netherlands.

Appleby MC and Hughes BO (eds), 1997. *Animal Welfare*. CABI, Wallingford, UK.

Boffelli D, Nobrega MA and Rubin EM, 2004. Comparative genomics at the vertebrate extremes. *Nature Reviews Genetics*, 5, pp. 456–465.

Broom DM, 1986. Indicators of poor welfare. *British Veterinary Journal*, 142, pp. 524–526.

Broom DM, 1998. Welfare, stress and the evolution of feelings. *Advances in the Study of Behavior*, 27, pp. 371–403.

Broom DM (ed), 2001a. *Coping with Challenge: Welfare in Animals including Humans*. Dahlem University Press, Berlin.

Broom DM, 2001b. Evolution of pain. In: Soulsby EJL Lord and Morton D (eds) *Pain: Its Nature and Management in Man and Animals*. Royal Society of Medicine International Congress Symposium Series 246, London, UK, pp. 17–25.

Broom DM, 2003. *The Evolution of Morality and Religion*. Cambridge University Press, Cambridge, UK, p. 259.

Broom DM, 2006. Behaviour and welfare in relation to pathology. *Applied Animal Behaviour Science*, 97, pp. 71–83.

Broom DM, 2007. Quality of life means welfare: how is it related to other concepts and assessed? *Animal Welfare*, 16(suppl), pp. 45–53.

Broom DM, 2010. Animal welfare: an aspect of care, sustainability, and food quality required by the public. *Journal of Veterinary Medical Education*, 37,pp. 83–88.

Broom DM, 2011. A history of animal welfare science. *Acta Biotheoretica*, 59, pp. 121–137.

Broom DM, 2014. *Sentience and Animal Welfare*. CABI, Wallingford, UK, p. 200.

Broom DM, 2017. Components of sustainable animal production and the use of silvopastoral systems. *Revista Brasileira Zootecnia*, 46, pp. 683–688. https://doi.org/10.1590/S1806-92902017000800009

Broom DM, 2019. Abnormal behavior and the self-regulation of motivational state. *Journal of Veterinary Behavior*, 29, pp. 1–3. https://doi.org/10.1016/j.jveb.2018.09.001

Broom DM, 2021a. A method for assessing sustainability, with beef production as an example. *Biological Reviews*, 96, 1836–1853. https://doi.org/10.1111/brv.12726

Broom DM, 2021b. *Domestic Animal Behaviour and Welfare*, 6th edn. (in press). Wallingford: CABI.

Broom DM and Johnson KG, 1993. *Stress and Animal Welfare*. Chapman and Hall, London, p. 222.

Broom DM and Johnson KG, 2019. *Stress and Animal Welfare: Key Issues in the Biology of Humans and Other Animals*, 2nd edn. Springer Nature, Cham, Switzerland, p. 230.

Chou JY, D'Eath RB, Sandercock DA and O'Driscoll K, 2020. Enrichment use in finishing pigs and its relationship with damaging behaviours: Comparing three wood species and a rubber floor toy. *Applied Animal Behaviour Science*, 224, 104944. https://doi.org/10.1016/j.applanim.2020.104944

Clayton NS and Emery NJ, 2015. Avian models for human cognitive neuroscience: a proposal. *Neuron*, 86, pp. 1330–1342. https://doi.org/10.1016/j.neuron.2015.04.024

Daigle C, 2018. Parallels between postpartum disorders in humans and preweaning piglet mortality in sows. *Animals*, 8, p. 22. https://doi.org/10.3390/ani8020022

Dawkins MS, 1990. From an animal's point of view: motivation, fitness and animal welfare. *Behavioral and Brain Sciences*, 13, pp. 1–31.

Duncan IJH and Petherick JC, 1991. The implications of cognitive processes for animal welfare. *Journal of Animal Science*, 69, pp. 5017–5022.

Falk D and Hofman MA, 2012. *Evolution of the Primate Brain from Neuron to Behavior*. Elsevier, Amsterdam, The Netherlands.

Fraser D, 1999. Animal ethics and animal welfare science: bridging the two cultures. *Applied Animal Behaviour Science*, 65, pp. 171–189.

Fraser D, 2008. *Understanding Animal Welfare: the Science in its Cultural Context*. Wiley Blackwell, Chichester, UK.

García Pinillos R, Appleby MC, Scott-Park F and Smith CW, 2015. One welfare. *Veterinary Record*, 179, pp. 629–630.
Green TC and Mellor DJ, 2011. Extending ideas about animal welfare assessment to include 'quality of life' and related concepts. *New Zealand Veterinary Journal*, 59, pp. 263–271.
Gygax L and Hillman E, 2018. "Naturalness" and its relation to animal welfare from an ethological perspective. *Agriculture*, 8, p. 136. https://doi.org/10.3390/agriculture8090136
Herrero M and Thornton PK, 2012. Livestock and global change: Emerging issues for sustainable food systems. PNAS, 110, 20878–20881.
Hofman MA, 2014. Evolution of the human brain: when bigger is better. *Frontiers in Neuroanatomy*, 8, p. 15. https://doi.org/10.3389/fnana.2014.00015
Hughes BO and Duncan IJH, 1988. Behavioural needs: can they be explained in terms of motivational models? *Applied Animal Behaviour Science*, 20, pp. 352–355.
Hutson GD, 1989. Operant tests of access to earth as a reinforcement for weaner piglets. *Animal Production*, 48, pp. 561–569.
Inglis IR, Forkman B and Lazarus J, 1997. Free food or earned food? A review and fuzzy model of contrafreeloading. *Animal Behaviour*, 53, pp. 1171–1191.
Ipema AF, Gerrits WJ, Bokkers EA, Kemp B and Bolhuis JE, 2020. Provisioning of live black soldier fly larvae (Hermetia illucens) benefits broiler activity and leg health in a frequency-and dose-dependent manner. *Applied Animal Behaviour Science*, 230, p. 105082. https://doi.org/10.1016/j.applanim.2020.105082
Karesh WB (Ed.) 2014. *One Health. O.I.E. Scientific and Technical Review*, 38. Paris: O.I.E.
Kirkden RD, Edwards JSS and Broom DM, 2003. A theoretical comparison of the consumer surplus and the elasticities of demand as measures of motivational strength. *Animal Behaviour*, 65, pp. 157–178.
Kis A, Hernádi A, KanizsárO, Gácsi M and Topál J, 2015. Oxytocin induces positive expectations about ambivalent stimuli (cognitive bias) in dogs. *Hormones and Behavior*, 69, 1–7. https://doi.org/10.1016/j.yhbeh.2014.12.004
LeDoux J, 2012. Rethinking the emotional brain. *Neuron*, 73, 653–676. https://doi.org/10.1016/j.neuron.2012.02.004
Matthews LR and Ladewig J, 1994. Environmental requirements of pigs measured by behavioural demand functions. *Animal Behaviour*, 47, pp. 713–719.
McBride SD and Morton AJ, 2018. Indices of comparative cognition: Assessing animal models of human brain function. *Experimental Brain Research*, 236, pp. 3379–3390. https://doi.org/10.1007/s00221-018-5370-8
McGreevy PD, Fawcett A, Johnson J, Freire R, Collins T, Degeling C, Fisher AD, Hazel SJ, Hood J, Lloyd JK, Phillips CJ, Stafford K, Hyde ML, Wilson B and Tsioumis V, 2020. Review of the online one welfare portal: shared curriculum resources for veterinary undergraduate learning and teaching in animal welfare and ethics. *Animals*, 10, p. 1341. https://doi.org/10.3390/ani10081341
Mellor DJ, 2016. Updating animal welfare thinking: moving beyond the "five freedoms" towards "a life worth living". *Animals*, 6, p. 21. https://doi.org/10.3390/ani6030021
Melzack R and Dennis SG, 1980. Phylogenetic evolution of pain expression in animals. In: Kosterlitz HW and Terenius LY (eds) *Pain and Society. Report of Dahlem Workshop.* Verlag Chemie, Weinheim, Germany, pp. 13–26.
Mendl M and Paul ES, 2020. Animal affect and decision-making. *Neuroscience and Biobehavioral Reviews*, 112, pp. 144–163. https://doi.org/10.1016/j.neubiorev.2020.01.025
Panksepp, J, 2005. Affective consciousness: core emotional feelings in animals and humans. *Consciousness and Cognition*, 14, pp. 30–80.
Paul ES and Mendl MT, 2018. Animal emotion: descriptive and prescriptive definitions and their implications for a comparative perspective. *Applied Animal Behaviour Science*, 205, pp. 202–209. https://doi.org/10.1016/j.applanim.2018.01.008
Rollin BE, 1995. *Farm Animal Welfare: Social, Bioethical and Research Issues.* Iowa State University Press, Ames, Iowa.
Sander D, 2013. Models of emotion: the affective neuroscience approach. In J Armony and P Vuilleumier (eds) *The Cambridge Handbook of Affective Neuroscience.* Cambridge University Press, Cambridge, UK, pp. 5–53.
Sneddon LU, 2019. Evolution of nociception and pain: evidence from fish models. *Philosophical Transactions of the Royal Society B* 374(1785), p.20190290. https://doi.org/10.1098/rstb.2019.0290
Taormina RJ and Gao JH, 2013. Maslow and the motivation hierarchy: measuring satisfaction of the needs. *The American Journal of Psychology*, 126, 155–177.

Toates F and Jensen P 1991. In: Meyer JA and Wilson S (eds) *Farm Animals to Animats*. MIT Press, Cambridge, Massachusetts, pp. 194–205.

Vestergaard K, 1980. The regulation of dustbathing and other behaviour patterns in the laying hen: a Lorenzian approach. In: Moss R (ed.) *The Laying Hen and its Environment. Current Topics in Veterinary Medicine and Animal Science*, vol. 8. Martinus Nijhoff, The Hague, Netherlands, pp. 101–113.

Wall PD, 1992. Defining 'pain in animals'. In: Short CE and van Poznak A (eds) *Animal Pain*. Churchill Livingstone, New York, pp. 63–79.

Yeates J, 2018. Naturalness and animal welfare. *Animals*, 8, pp. 53–70. https://doi.org/10.3390/ani8040053

3

ANIMAL WELFARE ASSESSMENT

Harry J Blokhuis and Isabelle Veissier

Introduction

Humans keep animals for different purposes: to produce food, to provide labour, for research purposes, for leisure, as companions, etc. Depending on the specific purpose, the external conditions and the animal species and type involved, humans provide care by means of different forms of housing and shelter, opportunities for species-specific behaviour, food, water, and veterinary control and treatment. The welfare of kept animals, i.e. how they perceive the conditions in which they live, is to a large extent determined by the quality of this care.

We focus here on farmed animals kept for food. However, the general principles of welfare assessment described in this chapter are applicable to a wide variety of settings. The welfare of farmed animals is in the first place the responsibility of farmers and those that directly care for the animals. High welfare standards generally result in reduced mortality, lower levels of injury and disease, and often higher productivity. Farmers strive for good levels of animal welfare, either because they want to provide good care for their animals or because of the favourable effects on production and thus farm economy, or both (Bock and van Huik, 2007). Moreover, since consumers nowadays expect good welfare for food producing animals, high welfare standards are more and more a prerequisite to access (international) markets. Thus, international financial institutions like the International Finance Corporation (IFC), whose policies are applied worldwide (Broom, 2017), recognise that businesses that enhance animal welfare are likely to have a competitive advantage in the global marketplace. The IFC has published a Good Practice Note entitled *Animal Welfare in Livestock Operations* (Mousseau et al., 2014) to increase awareness about the relevance of animal welfare and to guide investment practices in the field of livestock.

The interconnections between animal health and welfare and human health and welfare and their relation with environmental factors (climate change, biodiversity), are increasingly recognised by society at large (Pinillos et al., 2016, Olmos Antillón et al., 2021). This also contributes to a large public concern for animal welfare, as illustrated by the results of the special Eurobarometer on the attitudes of Europeans towards animal welfare (European Commission, 2016) that showed that 94% of citizens from the European Union (EU) believe it is important to protect the welfare of farmed animals. The importance of the welfare of farm animals has been gradually affirmed over the last 50 years, and citizens' interest in the way these animals live and die seems to continue to increase (Peyraud and MacLeod, 2020).

 DOI: 10.4324/9781003182351-4

In many countries, and especially in the EU, societal concern about farm animal welfare has been translated into a legislative corpus defining housing conditions and management practices for specific farm animal species (Blokhuis et al., 2008, Buller et al., 2018) as well as in private welfare assurance schemes and related labelling systems (Main et al., 2014).

It is for all of the above reasons that information on the welfare status of farm animals and how to improve related practices is important for all companies in the food chains, from farms to retail, and ultimately for consumers and the general public. Five broad groups of information demand may be distinguished (Blokhuis, 2018):

1. To provide farmers and other chain actors (e.g. transporters, slaughterhouses) with data to manage and improve animal welfare;
2. To give food retailers/restaurants the opportunity to brand products or their corporate identity;
3. To allow consumers to purchase products from animals with assured welfare;
4. To inform society about the welfare status of farm animals;
5. To regulate animal protection and to check compliance with such legislation.

Similar purposes can be found for other types of animals. For instance, zoos may use data on animal welfare to manage their animals (Purpose 1), to demonstrate their corporate responsibility (Purpose 2), to allow visitors to choose zoos according to the welfare of animal (equivalent to Purpose 3), to inform society about the level of animal welfare in zoos (Purpose 4), and such data can serve to develop appropriate legislation (Purpose 5). Because of the different purposes and use, the information in these different categories may have different forms and different levels of integration and detail. However, the information is always derived from the same set of measures by which the welfare of the animals involved is assessed. In this chapter, we address the various welfare assessment frameworks as well as the measures to check animal welfare. We then discuss how the results from measures can be assembled into information to meet the various purposes and use of animal welfare assessment.

Welfare assessment frameworks

Domains of animal welfare

Assessing the welfare of an animal is challenging because it is a multi-dimensional concept. For a holistic assessment of animal welfare one has to address all relevant domains. Domains of welfare can be described under three headings (Webster et al., 2015):

- Good biological functioning (including good health and vigour), here welfare is viewed as the satisfaction of biological needs, which are essential to life;
- Affective state (absence of stress, presence of positive experiences), here the emphasis is on what an animal experiences as being pleasant vs. unpleasant;
- Natural living, here one assumes that the impact of the farm environment is related to the deviation from the natural environment of the species, and that the extent to which natural behaviour can be expressed is an indication of welfare.

Overall measures of animal welfare

Some authors proposed single indicators as overall measures of welfare. Hurnik (1990) suggested using animal longevity as a measure of welfare, assuming that longevity indicates that the animal's health and functioning are not compromised to such an extent that the life span is affected. However, such an indicator remains questionable because the duration of a farm animal's life depends on various reasons not related to health or functioning (e.g. animals reared for meat are slaughtered at an age corresponding to market needs). A second example of a proposed single indicator of welfare comes from Geers et al. (2003), who concluded that the blood concentration of haptoglobin – an Acute Phase Protein (APP) – at slaughter is an integrative measure of a pig's welfare during its lifetime. However, APP seems to be activated only in case of tissue damage; e.g. mixing animals may be stressful but results in increased concentrations of APP only when lesions are due to aggressions between animals (Piñeiro et al., 2005). Indeed, in Geers et al.'s (2003) study, the welfare of pigs on farms was essentially assessed through health and space allowance. The latter is known to affect interactions between animals, so that a high density can result in aggression and tissue damage. APP is thus likely to reflect tissue damage due to diseases or injuries – related to the domain "good biological functioning" – and not the stress experienced by the animals in the absence of lesions.

A third example comes from studies by Barnett and Hemsworth (1990) and Wagner et al. (2021) that address more directly the affective states, with emphasis on stress and behaviour. Barnett and Hemsworth (1990) propose to use free corticosteroid concentrations in blood to detect chronic states of poor welfare. Based on concomitant signs of increase in metabolic cost, immunosuppression, altered reproduction or growth, they set a threshold of 40% increase in free corticosteroid – compared to control environments – to conclude that animals are in a state of chronic stress. However, this is not considering that some diseases might be detrimental for welfare without inducing a corticoid release (i.e. not stressful). Wagner et al. (2021) identified abrupt changes in the circadian rhythm of activity in cows in case of disease or modifications in the environment (handling, mixing, etc.). The changes in rhythm are likely to reflect the malaise perceived by animals when sick or stressed. It is however unsure that a disease developing gradually or environmental conditions deteriorating slowly would lead to a detectable change in the circadian rhythm. Before using changes in corticosteroids or in rhythm as an overall measure of an animal's welfare, it should be shown that these measures are sensitive to a wide range of adverse conditions.

Multi-criteria measurement of animal welfare

As explained above, to date there is no satisfactory single measure that covers the three domains of welfare. Such a measure should be sensitive to the effects of all the various factors in the animal's (internal and external) environment that can affect its welfare. An alternative approach to get an overall picture of an animal's welfare is to define criteria that cover all aspects of good welfare in terms of biological functioning, affective states or lack of naturalness, and then design a measuring framework through which the extent to which the criteria are fulfilled can be evaluated.

Several lists of welfare criteria have been proposed. The most influential is the list of "five freedoms" established by the Farm Animal Welfare Council (1992):

- Freedom from hunger and thirst;
- Freedom from discomfort;
- Freedom from pain, injury, or disease;
- Freedom to express normal behaviour;
- Freedom from fear and distress.

The five freedoms are widely used and form the basis of EU policy to develop legislation (e.g. European Commission, 2007) and by business operators to develop care protocols or certification schemes (e.g. the Freedom Food Scheme, Main et al., 2001). However, the five freedoms are not a fully adequate list of welfare criteria on which a holistic measurement framework can be based. First, some freedoms are partly redundant. For instance, "Freedom from discomfort" and "Freedom from pain, injuries, or disease" overlap because an uncomfortable lying area is often associated with injuries. Second, several freedoms contain independent items that require specific consideration. For example, by grouping hunger and thirst, there is a risk that the assessment framework proposes only one measure for the two items. These considerations led scientists from the Welfare Quality project to propose an adaptation of the five freedoms, resulting in a list of 12 independent criteria (Botreau et al., 2007b):

1. Absence of prolonged hunger: animals should not suffer from prolonged hunger, that is, they should have a suitable and appropriate diet;
2. Absence of prolonged thirst: animals should not suffer from prolonged thirst, that is, they should have a sufficient and accessible water supply;
3. Comfort around resting: animals should have comfort when they are resting;
4. Thermal comfort: animals should be neither too hot nor too cold;
5. Ease of movement: animals should have enough space to be able to move around freely;
6. Absence of injuries: animals should be free of injuries, for example, skin damage and locomotion disorders;
7. Absence of disease: animals should be free from disease;
8. Absence of pain induced by management procedures: animals should not suffer pain induced by inappropriate management, handling, slaughter, or surgical procedures (e.g. castration, dehorning);
9. Expression of social behaviours: animals should be able to express normal, non-harmful, social behaviours (e.g. grooming);
10. Expression of other behaviours: animals should be able to express other normal behaviours, that is, it should be possible to express species-specific natural behaviours such as foraging;
11. Good human–animal relationship: animals should not be afraid of humans and be handled well in all situations, that is, handlers should promote good human–animal relationships;
12. Positive emotional state: negative emotions such as fear, distress, frustration, or apathy should be avoided, whereas positive emotions such as security or contentment should be promoted.

Welfare Quality project partners then designed measuring protocols for different species and types of animal, with precise measures for each criterion, to cover all welfare aspects while eliminating redundancies (Welfare Quality, 2009c, Welfare Quality, 2009b, Welfare Quality, 2009a). Other scientists and stakeholders used the same Welfare Quality criteria to design measurement frameworks for species not covered by the initial project such as sheep, goats, horses, turkeys, rabbits, mice, and dolphins (AWIN, 2015a, AWIN, 2015c, AWIN, 2015b, Clegg et al., 2015, Dalmau et al., 2020, Spangenberg and Keeling, 2016).

In the Welfare Quality protocols, the 12 criteria are grouped into principles. The grouping is done in such a way that, within a given principle, the criteria may compensate one another to some extent, whereas this is not the case between principles. Four principles are distinguished:

- Feeding (Criteria 1 and 2), addressing whether animals are properly fed and supplied with water;
- Housing (Criteria 3 to 5), addressing whether animals are properly housed;

- Health (Criteria 6 to 8), addressing whether animals are healthy;
- Behaviour (Criteria 9 to 12), addressing whether the behaviour of animals reflects optimised emotional states.

Inclusion of positive welfare

Frameworks to measure animal welfare may vary with time because new scientific knowledge allows the use of better or easier-to-apply measuring methods, or scientific studies substantiate changing societal values. The five freedoms focus on the prevention of animal suffering. Only "freedom to express normal behaviour" refers to positive affective states or "positive welfare", i.e. requires that animals are provided with more than what is essential for them not to suffer. Boissy et al. (2007) discussed the existence of positive affective states in animals and provided avenues to detect them from observation of behaviour (play, grooming, and exploration).

Mellor and Beausoleil (2015) proposed a framework for welfare assessment based on five domains. The first four domains – nutrition, environment, health, and behaviour – are close to the four Welfare Quality principles. The fifth domain covers mental states that derive from the resources provided regarding the first four domains (e.g. hunger and thirst when the provision of food and water – domain "nutrition" – is insufficient). Mellor and Beausoleil (2015) insisted on the necessity to minimise negative affective states and at the same time to promote the positive ones (e.g. pleasures due to the taste of food or to satiety) even if not all negative experiences can be removed. It is now recognised that the environment should provide opportunities for animals to experience positive emotional states (Mellor, 2016) and that animals should live "a life worth living" (Farm Animal Welfare Council, 2009). The 12 Welfare Quality criteria can still guide the development of welfare measures. These measures should nevertheless allow more space for addressing positive states rather than only eliminating the poor ones.

Measures of animal welfare

Categories of measures

Once criteria for good welfare are defined, measures are required that can be applied in practice to assess to what extent the criteria are fulfilled. There are many measures available, some of them require specialised equipment and can only be used under experimental or laboratory conditions (like detailed behaviour observations or invasive physiological measures of stress), others can be applied in practice on farms or at slaughter (e.g. skin lesions). It is beyond the scope of this chapter to describe measures in detail. We will only address the main categories of measures and focus on measures to assess animal welfare in practice.

Measures are often divided into two categories. The first category comprises measures related to resources in the animal's environment and to the management of the animals that are crucially important for the quality of the animals' lives. These include, for example, the availability and quality of litter and space, feeding routines, or animal handling.

In contrast to these resource and management measures, the second category focuses on aspects of the animal's health, physiology, and behaviour that are measured in/on the animal. These measures more directly relate to the animal's welfare. Examples are measures of rectal temperature, blood level of cortisol, lameness, fearfulness, or wounds.

The first category is also referred to as "input-based measures" and the second category as "output-based measures". Clearly, these two categories of measures are closely related since the actual status of the animals depends on the quality of available resources and how these are applied and managed. The advantage of input measures is that they are relatively easy to define and assess (especially resources). That is why most requirements in welfare regulations refer to this type of measure. However, there are also difficulties with the use of input measures. Animals differ in their genetic structure, early experience, and temperament, and may therefore experience the same environment in different ways. Even very similar environments may be managed differently by the stockperson, further affecting the animal's experience of a particular situation.

Thus, resource- or management-based measures provide information about the risks for welfare but do not always reliably predict the effect on animal welfare in a particular situation. Therefore, research has been focusing on the further development of animal-based measures, which are considered to show the "outcome" of the interaction between the animal and its environment (housing design and management). To be suitable to assess animal welfare in practice, measures must be specific, in that they measure what they are supposed to measure, and they must be sensitive, repeatable, and feasible in practical conditions (Knierim et al., 2021). For some of the welfare criteria listed above, the available measures do not fulfil these requirements. For example, thirst is difficult to measure with simple methods unless the animal is very dehydrated, in which case a pinch skin test can be used (when the skin of a dehydrated animal is pinched it does not immediately resume its initial shape). Similarly, the pain induced by dehorning cannot be measured at the time of a farm visit because dehorning may have occurred a long time ago. In such cases where animal-based measures are not available, we need to rely on resource- or management-based indicators (Blokhuis et al., 2019).

At present, the protocols to assess animal welfare are based mainly on observations of animals (e.g. clinical signs, body condition) or the environment (e.g. number and cleanliness of drinkers), and to a lesser extent on interviews with farmers (or slaughter plant managers, e.g. about procedures for pain management) and on the collection of data from farm records (e.g. mortality). Gathering all information with a protocol, like that provided by Welfare Quality, may require several hours or even a full day on a single farm. Such welfare assessments are therefore often applied infrequently, e.g. farms may be visited once a year for a certification scheme, or only a sample of farms are visited, e.g. under the cross-compliance scheme the European Union requests that 1% of farms are inspected in each Member State per year. The frequency with which measures are taken depends on the intended use. For example, if a food retailer wants to brand a product based on a design characteristic of the housing (e.g. "cage free" eggs or "pasture based" beef), a yearly or biennial assessment may be appropriate. But, for management purposes, measures have to be taken much more frequently. Indeed some measures are only useful when outcomes are available on a daily basis (e.g. detection of a disease).

Automatic monitoring

Modern technology in the area of Precision Livestock Farming (PLF) can help to make measurements more continuous and thereby more applicable in daily welfare management. PLF entails the automated monitoring of livestock to enable farmers to optimise production, and the health and welfare status of their animals. Sensors are now available to record a number of parameters on animals or their environment. Animal activities can now be detected thanks to accelerometers, locating systems, or image analysis; coughing can be detected from sound analysis and fever from infrared cameras; the animals can be automatically weighed; etc. This offers the possibility to obtain data continuously and at an individual level (at least in large species like cattle and pigs) and allows progression from a periodic assessment to continuous monitoring, with a view to detecting problems as soon as they occur and to be able to remedy them quickly (Faverdin et al., 2021). Similar sensor-based systems are now used on zoo animals and in lab animals to monitor temperature, heart rate, or activity.

To date, sensors provide only partial information about animal welfare. The focus of PLF is primarily on production and health indicators, which have straightforward impacts on farm profitability. There is therefore a risk of redefining welfare solely in terms of production and health, ignoring aspects such as expression of positive behaviours (Buller et al., 2020). To avoid this pitfall one needs to further develop digital tools to more specifically address animal welfare.

Qualitative assessment

In humans, the quality of life is often assessed by questioning people about how they feel regarding a number of items (how is their social life, or their feeling of happiness, their amount of pain, the extent to which they are limited in their everyday life due to disease?) (e.g. Kahneman and Deaton, 2010, de Jong et al., 2012). People answer questions on a visual analogue scale, e.g. from 0 to 100. A similar approach can be used in animals, but with an external observer rating the behaviour of an animal according to predefined terms. For instance, Wiseman-Orr et al. (2006) developed a questionnaire to measure the Health-Related Quality of Life of dogs affected by chronic joint disease. This includes 109 descriptors of the dog's attitude (apathetic, complaining, etc.) to be rated on a 0-6 point scale, where 0 corresponds to a descriptor not appropriate to the dog and 6 corresponds to a descriptor that very well fits to the dog's attitude. In a similar vein, Wemelsfelder and her collaborators developed a Qualitative Behaviour Assessment (QBA) approach whereby the observer describes an animal or a group of animals, using descriptors decided by the observer or pre-defined (Wemelsfelder and Lawrence, 2001). QBA was introduced in the Welfare Quality protocol, with both positive and negative aspects addressed with descriptors such as playful/active and fearful/depressed, respectively. QBA addresses not only what the animals do but also how they do it (e.g. they move in a way that suggests play vs. fear). Qualitative rating and machine learning can be combined to automatically classify an animal behaviour from pictures or videos (Neethirajan et al., 2021). Therefore, in the future PLF techniques might be used to detect animals' activities as well as their internal state.

Implementation: From welfare measurement to assessment

Assessing animal welfare involves much more than providing results obtained from various welfare measures. Assessing the welfare of an animal or a group of animals implies that outcomes from these measures are interpreted – e.g. compared to reference values. Information about the welfare status of animals can be used for different purposes and these may require different levels of integration, from no integration at all (e.g. in the form of a dashboard for farmers and other animal caretakers to have an overview of what is going well or wrong on the farm or other form of animal facility in case of, e.g. zoo or lab animals) to an overall assessment (e.g. for labelling purposes).

Reference values

Farmers and farm advisors may use welfare assessment to highlight positive and negative aspects on a farm, during transport, or at slaughter. Results from welfare measures can identify welfare problems and help farmers and other chain actors identify corrective action and improve animal welfare (the first purpose of a welfare assessment listed in the Introduction). There is no need to aggregate the data. Each welfare aspect is addressed separately and often compared to a reference value. One way to produce reference values is to use the distribution of results of a population of farms. With such a benchmarking approach the results of a farm can be compared to the average or to first, second, or third quartiles of the population. With this approach, farms from the lowest category (e.g. first quartile) may improve over years but remain in the lowest category (Mullan et al., 2021), which may discourage people from trying to improve. A reference value can also be defined *a priori*, as an absolute target to be achieved. For instance, experts may consider that on a dairy farm there should be no more than 5% lame cows. Reference values can also be minimum standards defined by legislation, often based on expert opinion. In that case, comparing the results of a farm with the reference corresponds to checking compliance with legislation (the Purpose 5 listed in the Introduction).

If the frequency or the severity of a problem is above the reference value, corrective action at farm level – and not only at individual level – should be put in place. For instance, if too many cows in a herd are lame, a lameness control plan should be decided (Leach and Whay, 2008). Another approach is to define a target value based on the initial situation. For example, the defined goal may be to reduce the incidence of lameness by 5%.

Producing results in the form of a gradient of welfare

When comparing to reference values, the results are expressed as "the farm is above the reference value for a certain aspect" – meaning that the welfare is good, vs. "the farm is below the reference value" – meaning that the welfare is not good. However, one does not know how good a good farm is or how bad a bad farm is. For this it has to be decided if the parameter that is measured indicates a mild or large welfare impact (positive or negative) on the animal. For example: "how large is the suffering experienced by a lame animal?", or "how pleasant is the presence of a social partner?". There are scientific methods available to study such questions. For instance, to evaluate how pleasant the presence of a social partner is, the animal is placed in an experimental setting where it is measured how much it will work to get access to the social partner (Holm et al., 2002). However, this would give no information about the amount of social reward the animal experiences in its real life, where it will be in contact with various partners. Similarly, one could imagine assessing how painful lameness is to an animal by offering access to a painkiller and measure the uptake (Danbury et al., 2000). Again, this would not tell us how much the lame animal actually suffered during specific lameness episodes in its real life. Therefore, such experimental results are generally complemented with expert opinion: based on the literature or their own experience, experts estimate the consequences in terms of pain, fear, discomfort, pleasure, etc. that are likely to be experienced by animals given the signs observed in them (e.g. lesions, abnormal behaviour) or the environment in which the animals lives (e.g. individual housing, tethering). This approach was used in the Welfare Quality project. Experts in animal behaviour or health were shown sets of results taken from animals or their environment and asked to attribute a value from 0 (very low welfare) to 100 (excellent welfare). Discussions between experts allowed consideration of various points of view and helped in reaching a consensus. Then functions were designed according to these experts' opinion to compute the results of a farm for the various measures into a value-score (between 0 and 100). This was done for the 12 criteria of the Welfare Quality scheme.

Overall welfare assessment

It may be required to not only check a series of welfare items but also to produce an overall assessment. To assess the welfare of animals in various conditions to be able to recommend, or not, those conditions, measures of health, behaviour, stress, vigour, etc. of the animals are generally applied. The information is then often integrated in an informal way, taking into account the severity of problems (e.g. the consequences of a given housing condition on the animal) and the frequency of that problem in a population, before a judgement is made. Such an approach is used by the European Food Safety Authority (EFSA) to prioritise welfare issues (e.g. EFSA AHAW Panel (EFSA Panel on Animal Health and Welfare), 2014). A similar informal aggregation of information is used for official farm inspections. The EU asks its Member States to check that farms comply with the EU legislation to protect animals. The inspectors are provided with checklists to verify each point of the legislation. Then, after a farm visit, the inspector formulates a conclusion in the form of, e.g. minor, moderate, or major non-compliances. This implies some sort of implicit aggregation, which is left open to the inspectors (Lomellini-Dereclenne et al., 2017).

An overall assessment may also be necessary in certification schemes with a view to brand products and to allow consumers to purchase products with assured welfare (Purposes 2 and 3 mentioned in the Introduction). A scheme may focus on some aspects of animal welfare or cover all aspects. Farms may be asked to achieve a certain degree of compliance in all aspects or a percentage of them. Providing many welfare scores (e.g. one for each criterion of welfare) does not ease the communication. Rather a summary information may be delivered in the form of a single score. Aggregation into an overall assessment is often done with a view to categorising farms according to the quality of animal welfare they provide, as in the Welfare Quality protocols (Botreau et al., 2009). Several issues need to be carefully addressed when the information from several criteria (e.g. absence of thirst, absence of hunger) is merged: one should determine if some criteria are considered more important than others, and more importantly, if compensation between criteria is allowed; e.g. can lack of expression of natural behaviour be compensated by good health? (for a review of constraints upon criteria aggregation see Botreau et al., 2007a). In an ideal world, the animal's point of view regarding balancing of criteria should be taken into account. But the methodology for that does not exist yet. Again, we have to rely on expert opinion. In the Welfare Quality project, experts – who could be animal scientists, social scientists, or stakeholders – were asked questions such as "When a farm scores 40 for health and 60 for behaviour (on a 0 = 'very poor welfare' to 100 = 'Excellent welfare' scale), what score do you give for the combination of the two?". Some experts attributed more importance to some criteria than to others. In the example above, veterinarians proposed a summary score lower than ethologists did, whereas the opposite was observed when a farm scored 60 for health and 40 for behaviour. This reflected the higher importance that veterinarians attributed to health and the higher importance attributed by ethologists to behaviour. More importantly, a few experts allowed compensation between criteria, e.g. good health compensating for poor behaviour or vice versa. However, most experts did not allow full compensation between criteria: for them a score of 40/100 added to a score of 60/100 always produced a summary score less than 50/100, so that the summary scores are highly influenced by the lowest criterion-scores. There is no objective truth about this, only diverging points of views, and the mathematical methods to aggregate the data need to reflect the reasoning of experts consulted (Botreau et al., 2008). The process by which data are interpreted and aggregated must then be transparent and explicit, so that users can check if the reasoning matches their own reasoning (Veissier et al., 2011). This is the reason why in the Welfare Quality project all the formulas are given and illustrated by examples.

To inform citizens (Purpose 4 listed in the Introduction), either specific information can be given or a summary information. For instance, the prevalence of lameness in broilers can be of interest for citizens. In general however, more integrated information will be provided, e.g. in the form of a statement like "only a certain percentage of farms present severe non-compliance with the legislation to protect farm animals". In that case a comparison with minimum standards

Table 3.1 Correspondence between the purpose of an assessment and how it can be implemented

Purpose of the assessment	*Scope of the assessment*		*Interpretation of results*		*Aggregation of results for overall assessment*
	Exhaustive	*Focused*	*Comparison to reference values*	*Gradient of welfare (per measure)*	
To help farmers, transporters, or slaughterhouses to manage animal welfare	Yes		Yes	Possible	Not necessary
To brand products					
• on specific welfare aspects		Yes	Possible	Possible	In general, no
• on welfare in general	Yes		Possible	Possible	Yes
To allow consumers to purchase products from animals with assured welfare	Yes		Possible	Possible	Yes
To inform society					
• about specific welfare aspects		Yes	Possible	Possible	In general, no
• about welfare in general	Yes		Possible	Possible	Yes
To check compliance with legislation	Yes		Yes		Yes

is necessary and some integration is made (by farm inspectors, see above). The correspondence between the purpose of the assessment and the way it can be implemented is summarised in Table 3.1.

Conclusions

Animal welfare is a multi-dimensional concept, embracing good health, physical comfort, possibilities to perform natural behaviour, positive experiences, etc. The assessment of welfare can serve different purposes. For example, it can be used to guide production processes and management (on farms, during transport, and at slaughter) or provide consumer information in assurance schemes. The choice of measures depends on the purpose, but a holistic assessment of animal welfare needs to address all of its dimensions.

Assessing the welfare state of an animal requires outcome-based measures, e.g. clinical signs or behaviour, which result from an interaction between the animal and its environment (i.e. the way it is housed, fed, and managed). Resource- or management-based measures are also essential *a posteriori* to identify causes of a poor welfare state or *a priori* to estimate risks for poor welfare.

One should nevertheless distinguish between the *measurement* of poor (or good) welfare (e.g. evidence of illness or disturbed behaviour, or evidence of positive experiences), and *assessment* of welfare. The latter depends not only on the evidence from measures but also on the value attributed to them. Science can provide guidance for such assessment but societal/ethical concerns also play a determining role. Animal welfare scientists (who belong to the field of natural sciences) should engage with people having interest in animal welfare and the scientists studying those people (from social sciences and humanities) to build together animal welfare assessment tools.

References

Awin, 2015a. AWIN welfare assessment protocole for donkeys, *AWIN Consortium*. Available at https://air.unimi.it/retrieve/handle/2434/269100/384805/AWINProtocolDonkeys.pdf.

Awin, 2015b. AWIN welfare assessment protocole for goats, *AWIN Consortium*. Available at https://air.unimi.it/retrieve/handle/2434/269102/384790/AWINProtocolGoats.pdf

Awin, 2015c. AWIN welfare assessment protocole for horses, *AWIN Consortium*. Available at https://air.unimi.it/retrieve/handle/2434/269097/384836/AWINProtocolHorses.pdf

Barnett JL and Hemsworth PH, 1990. The validity of physiological and behavioural measures of animal welfare. *Applied Animal Behaviour Science*, 25, pp. 177–187.

Blokhuis HJ, 2018. Animal welfare information in a changing world. In Butterworth A, ed., *Animal Welfare Challenges: Dilemmas in a Changing World*. Wallingford, UK: CABI.

Blokhuis HJ, Keeling LJ, Gavinelli A and Serratosa J, 2008. Animal welfare's impact on the food chain. *Trends in Food Science & Technology*, 19, pp. S79–87.

Blokhuis HJ, Veissier I, Miele M and Jones RB, 2019. Safeguarding farm animal welfare. In Vogt M., ed., *Sustainability Certification Schemes in the Agricultural and Natural Resource Sectors. Outcomes for Society and the Environment*. London, UK: Routledge.

Bock BB and Van Huik MM, 2007. Animal welfare: the attitudes and behaviour of European pig farmers. *British Food Journal*, 109, pp. 931–944.

Boissy A, Manteuffel G, Jensen MB, Moe RO, Spruijt B, Keeling L, Winckler C, Forkman B, Dimitrov I, Langbein J, Bakken M, Veissier I and Aubert A, 2007. Assessment of positive emotions in animals to improve their welfare. *Physiology and Behavior*, 92, pp. 375–397.

Botreau R, Bracke MBM, Perny P, Butterworth A, Capdeville J, Van Reenen CG and Veissier I, 2007a. Aggregation of measures to produce an overall assessment of animal welfare: part 2: analysis of constraints. *Animal*, 1, pp. 1188–1197.

Botreau R, Veissier I, Butterworth A, Bracke MBM and Keeling LJ, 2007b. Definition of criteria for overall assessment of animal welfare. *Animal Welfare*, 16, pp. 225–228.

Botreau R, Capdeville J, Perny P and Veissier I, 2008. Multicriteria evaluation of animal welfare at farm level: an application to MCDA methodologies. *Foundations of Computing and Decision Sciences*, 33, pp. 287–316.

Botreau R, Veissier I and Perny P, 2009. Overall assessment of cow welfare: strategy adopted in Welfare Quality®. *Animal Welfare*, 18, pp. 363–370.

Broom DM, 2017. *Animal Welfare in the European Union*. Brussels, Belgium: European Parliament.

Buller HJ, Blokhuis H, Jensen P and Keeling L, 2018. Towards farm animal welfare and sustainability. *Animals*, 8, pp. 1–13.

Buller H, Blokhuis H, Lokhorst K, Silberberg M and Veissier I, 2020. Animal welfare management in a digital world. *Animals* 10, p. 1779.

Clegg I, Borger-Turner J and Eskelinen H, 2015. C-Well: the development of a welfare assessment index for captive bottlenose dolphins (Tursiops truncatus). *Animal Welfare*, 24, pp. 267–282.

Dalmau A, Moles X and Pallisera J, 2020. Animal welfare assessment protocol for does, bucks, and kit rabbits reared for production. *Frontiers in Veterinary Science*, 7, pp. 1–18.

Danbury TC, Weeks CA, Chambers JP, Waterman-Pearson AE and Kestin SC, 2000. Self-selection of the analgesic drug carprofen by lame broiler chickens. *Veterinary Record*, 146, pp. 307–311.

De Jong T, Maliepaard M, Raat H and Mathijssen I, 2012. Health-related problems and quality of life in patients with syndromic and complex craniosynostosis. *Child's Nervous System*, 28, pp. 879–882.

Efsa Ahaw Panel (Efsa Panel on Animal Health and Welfare), 2014. Scientific opinion on the welfare risks related to the farming of sheep for wool, meat and milk production. *EFSA Journal*, 12, p. 128.

European Commission, 2007. Animal welfare: factsheet. In *DG Health and Consumer Protection*. Brussels, Belgium: European Commission. https://www.dopharma.com/wp-content/uploads/2016/09/Animal-Welfare-farmed-animals-factsheet-2007_en.pdf

European Commission, 2016. *Special Eurobarometer 442: Attitudes of Europeans towards Animal Welfare*. Brussels, Belgium: Directorate-General for Communication.

Farm Animal Welfare Council, 1992. FAWC updates the five freedoms. *Veterinary Record*, 17, p. 357.

Farm Animal Welfare Council, 2009. *Farm Animal Welfare in Great Britain: Past, Present and Future*. London, UK: Farm Animal Welfare Council.

Faverdin P, Allain C, Guatteo R, Hostiou N and Veissier I, 2021. Élevage de précision : De nouvelles informations utiles pour la décision ? *INRAE Productions Animales*, 33, pp. 223–234.

Geers R, Petersen B, Huysmans K, Knura-Deszczka S, De Becker M, Gymnich S, Henot D, Hiss S and Sauerwein H, 2003. On-farm monitoring of pig welfare by assessment of housing, management, health records and plasma haptoglobin. *Animal Welfare*, 12, pp. 643–647.

Holm L, Jensen MB and Jeppesen LL, 2002. Calves' motivation for access to two different types of social contact measured by operant conditioning. *Applied Animal Behaviour Science*, 79, pp. 175–194.

Hurnik JF, 1990. World's poultry science association invited lecture: animal welfare: ethical aspects and practical considerations. *Poultry Science*, 69, pp. 1827–1834.

Kahneman D and Deaton A, 2010. High income improves evaluation of life but not emotional well-being. *Proceedings of the National Academy of Sciences of the United States of America*, 107, pp. 16489–16493.

Knierim U, Winckler C, Mounier L and Veissier I, 2021. Developing effective welfare measures for cattle. In: *Understanding the Behaviour and Improving the Welfare of Dairy Cattle*. Cambridge, Burleigh Dodds. 23 pp.

Leach KA and Whay R, 2008. The Welfare quality lameness control programme for dairy cattle: resources to help farmers and advisors tackle lameness problems in dairy herds. *Wefare Quality Consortium*. Available at http://www.welfarequalitynetwork.net/media/1122/wqr14.pdf

Lomellini-Dereclenne AC, Miele M, Mounier L and Veissier I, 2017. Implementation of the European legislation to protect farm animals: a case study on French inspections to find solutions to improve compliance. *Animal Welfare*, 26, 311–321.

Main DC, Webster AJF and Green LE, 2001. Animal welfare assessment in farm assurance schemes. *Acta Agriculturae Scandinavica, Section A, Animal Science*, 30(Suppl.), pp. 108–113.

Main DCJ, Mullan S, Atkinson C, Cooper M, Wrathall JHM and Blokhuis HJ, 2014. Best practice framework for animal welfare certification schemes. *Trends in Food Science & Technology*, 37, pp. 127–136.

Mellor DJ and Beausoleil NJ, 2015. Extending the 'Five Domains' model for animal welfare assessment to incorporate positive welfare states. *Animal Welfare*, 24, pp. 241–253.

Mellor JD, 2016. Updating animal welfare thinking: moving beyond the "five freedoms" towards "A Life Worth Living". *Animals*, 6(3), p. 21. https://doi.org/10.3390/ani6030021

Mousseau L-P, Hatton RJ, Constantine M, Botha S, Hui Xin Goh A and Plante C, 2014. *Good Practice Note: Improving Animal Welfare in Livestock Operations*, Washington, DC: International Finance Corporation, World Bank Group.

Mullan S, Stuijfzand B and Butterworth A, 2021. Longitudinal national-level monitoring of on-farm broiler welfare identifies consistently poorly performing farms. *Scientific Reports*, 11, pp. 11928.

Neethirajan S, Reimert I and Kemp B, 2021. Measuring farm animal emotions: sensor-based approaches. *Sensors*, 21, p. 553.

Olmos Antillón G, Tunón H, De Oliveira D, Jones M, Wallenbeck A, Swanson J, Blokhuis H and Keeling L, 2021. Education to meet the challenges of our time: reinforcing links between Animal Welfare and the United Nations' Sustainable Development Goals. *Sustainability Science*, 13, pp. 3328.

Peyraud J-L and Macleod M, 2020. *Future of EU Livestock: How to Contribute to a Sustainable Agricultural Sector?* Brussels, Belgium: European Commission.

Piñeiro C, Morales J, Piñeiro M, Ruiz De La Torre JL, Mateos GG and Manteca X, 2005. *Psychological stress caused by changes in feeding management did not affect the concentration of Acute Phase Proteins in pigs*. In 5th International Colloquium on Animal Acute Phase Proteins, 14–15 March 2005, Dublin, Ireland.

Pinillos RG, Appleby M, Manteca X, Scott-Park F, Smith C and Velarde A, 2016. One Welfare - a platform for improving human and animal welfare. *Veterinary Record*, 179, pp. 412–413.

Spangenberg EM and Keeling LJ, 2016. Assessing the welfare of laboratory mice in their home environment using animal-based measures: a benchmarking tool. *Lab Anim*, 50, pp. 30–8.

Veissier I, Jensen KK, Botreau R and Sandøe P, 2011. Highlighting ethical choices underlying the scoring of animal welfare in the Welfare Quality® scheme. *Animal Welfare, Special Issue Knowing Animals*, 20, pp. 89–101.

Wagner N, Mialon M, Sloth K, Lardy R, Ledoux DMS, De Boyer Des Roches A and Veissier I, 2021. Detection of changes in the circadian rhythm of cattle in relation to disease, stress, and reproductive events. *Methods*, 186, pp. 14–21.

Webster JR, Schütz KE, Sutherland MA, Stewart M and Mellor DJ, 2015. Different animal welfare orientations towards some key research areas of current relevance to pastoral dairy farming in New Zealand. *New Zealand Veterinary Journal*, 63, pp. 31–36.

Welfare Quality, 2009a. *Welfare Quality® Assessment Protocol for Cattle (Fattening Cattle, Dairy Cows, Veal Calves)*. Lelystad, The Netherlands: Welfare Quality® Consortium.

Welfare Quality, 2009b. *Welfare Quality® Assessment Protocol for Pigs (Sows and Piglets, Growing and Finishing Pigs)*, Lelystad, The Netherlands: Welfare Quality® Consortium.

Welfare Quality, 2009c. *Welfare Quality® Assessment Protocol for Poultry (Broilers, Laying Hens)*. Lelystad, The Netherlands: Welfare Quality® Consortium.

Wemelsfelder F and Lawrence AB, 2001. Qualitative assessment of animal behaviour as an on-farm welfare-monitoring tool. *Acta Agriculturae Scandinavica, Section A, Animal Science*, 30(Supplementum), pp. 21–25.

Wiseman-Orr ML, Scott EM, Reid J and Nolan AM, 2006. Validation of a structured questionnaire as an instrument to measure chronic pain in dogs on the basis of effects on health-related quality of life. *American Journal of Veterinary Research*, 67, pp. 1826–1836.

PART II

Animal farming, transportation, and killing

4

CONTEMPORARY ANIMAL FARMING

Carla Forte Maiolino Molento and Clive Julian Christie Phillips

Introduction

In the first agricultural systems, beginning about 10,000 years ago, animals were raised so that their products could be consumed in individual homesteads. Greater yields of animal products were possible compared with the traditional system of hunting and gathering. Because surpluses at certain times of year and in favourable locations were an inevitable part of settled agriculture, trading in animal products developed, facilitated by humans' strong social and cognitive skills (Phillips, 2015). This was the start of the intensification of animal use, which humans have perpetuated since that time. Products such as milk and wool that could be continually produced were preferred so that the production unit (the animal) was maintained. Killing animals deprived their keepers of the production unit, unless it reproduced before it was killed. Animals, such as cattle, requiring large amounts of fodder every day, had to be killed at the end of the growing season. Then technology was developed to store forage, as hay and then more efficiently as silage, for use when there was a shortage, usually due to cold weather or dry conditions. Hence animals with a short reproductive cycle were used for meat, such as pigs and chickens. Those with longer reproduction cycles, such as cattle and sheep, were used to generate replenishable products – milk and traction (cattle) and wool (sheep).

Later food production enterprises developed in response to demand from concentrations of the human population. In the evolution of our food production systems from animals, natural resources were often limiting – feed or water supply, nutrient deficiencies, water quality issues. The different commodities were concentrated into regions that most suited the production system. Over time, techniques were developed to overcome most of these problems to allow animals to be economically reared. For example, in some regions mineral deficiencies, particularly cobalt, iodine, copper, and selenium, prevented satisfactory growth of animals on pastures, but slow release supplements were developed to overcome this problem (Grace and Knowles, 2012). Fodder is also now transported for farm animal consumption around the world, to overcome feed shortages in countries where demand for animal products is high, such as China (Phillips, 2015).

Agricultural revolutions

In the agricultural revolution that has lasted from the mid-18th century until today, animal use for traction and transport has gradually been replaced by machines (Mazoyer and Roudart,

DOI: 10.4324/9781003182351-6

2012). Cattle and horses seemed originally well suited to these functions, from the perspective of human interests, because of their size. However, for work they had the disadvantage that their digestive system was adapted to allow coarse grasses and forbs to be consumed slowly. Because these grasses and forbs are of low nutritional quality, large quantities have to be consumed over 8–12 hours daily, which gives the animals little time to rest. Even though using animals for draught work has considerably decreased in many countries, an argument for using farm animals in a broad spectrum of farm activities has emerged, associated with the concept of "agricultural exceptionalism" that excludes agriculture from legislation addressing cruelty to animals (Eisen, 2020).

At the same time as mechanisation replaced animals on the land, synthetic fertilisers and selected high-yielding genotypes of plants and animals allowed a dramatic increase in the output per animal. However, some farms failed to modernise in this way, because of constraints of lack of capital, or unwillingness of farmers to adopt the new technologies. This facilitated the amalgamation of smaller farms into larger units. As the successful farms grew to sell large volumes of animal products for public consumption, their revenue increased to the point that the children were well educated and often lacked the motivation to spend their lives in the hard physical work that animal farming entails. Thus, the knowledge of how to successfully manage animals failed to be passed down through the generations, creating a shortage of skilled stockpeople. This accelerated the transition from small farms to large, labour-efficient units, sparking the industrialisation of animal production. Such trends have continued throughout the 20th century, but particularly in the second half of the century, following the shortage of male labourers available for agricultural work after the world wars of the first half of the century.

The expansion of cereal production with new, high-yielding genotypes, the use of synthetic fertilisers and the advent of herbicides and pesticides, allowed surpluses to be fed to farm animals. This trend expanded until nowadays, when animal production uses the majority of agricultural land on the planet (van Zanten et al., 2016). Deforestation, originally in Europe for the development of navies, continues to this day, particularly in South America, but with the new purpose of providing grazing lands for cattle as well as land for soy production, more than three-quarters of which is used as animal feed (Ritchie and Roser, 2021). Other clearances have included the removal of small farmers from the land to allow large landowners or the government to take control and establish big herds and flocks of animals. This occurred most famously in Scotland in the 18th and 19th centuries.

The agricultural revolution developed at different rates in different parts of the world. Some countries, such as France and Germany, chose to protect their many small farmers from the pace of development, as a social policy. Support was provided in the form of subsidies, which were originally provided to increase productivity, stabilise markets, and ensure fair prices following food shortages in the world wars of the first half of the 20th century. The fastest expansion of farm size was during 20th century collectivisation of farms in the Soviet Union and communist countries of Eastern Europe. Here the motivation was to extract farming land from private ownership by small-scale producers and place it into the hands of the local community or the state, creating collectives and state farms, respectively. It continues to this day in China, where concern about food security and quality has led the government to forcibly shut down small farms and replace them with large industrial farms with state support. The collapse of communist rule in Eastern Europe and the Soviet states brought about a reversal of communal ownership of farms, with land being returned to their former owners in the 1990s. However, such small units were often unviable economically and amalgamation of farms followed this transition.

In Britain, the industrial revolution and migration into the cities of the 18th and 19th centuries sparked an intensification of animal production to provide nutrient-dense food preferred by those working in heavy industry. As well, importation of animal products from former colonies, such as Canada, Australia, and New Zealand, was encouraged, which increased the competition for British producers. At the same time, the creation of an extensive rail, and later road, network allowed transport of the animals or their products around the country. In the case of live animals, long-distance transport creates considerable welfare challenges, as the animals have little space to engage in their normal behaviour. Originally animals were moved on foot, which limited the distance they could travel and required food to be provided en route, but today there is a growing movement of animals in ships, planes and by truck, with all the concomitant welfare problems (Chapter 11).

Welfare problems of intensive animal farming

These changes facilitated a concentration of animals on farms at stocking densities much greater even than when animals were used for traction in place of machines. Poultry have been increasingly favoured by the consumer for reasons of health and low cost of production, but because the animals are small, many more animals have to be grown compared with large livestock, like cattle. Thus more animals are facing a life of poor welfare. In the 1960s poultry and pigs began to be kept in units of several thousand animals, which coincided with changes in housing so that animals were kept in small cages. Housing for laying chickens involved keeping them in groups of five or six in a battery cage, which because of their small size failed to provide for some essentials for good welfare: opportunities for nesting, resting on a perch, flapping their wings, and scratching. Similarly, sows came to be kept in individual crates during pregnancy and for farrowing and suckling their young piglets. This severely limited their freedom of movement and gave them sores and abrasions on their limbs. Intensive housing for cattle and sheep was less common because of their reliance on pasture for feed. However, in the 1950s and 1960s intensive methods of rearing cattle in feedlots were developed in the United States of America to allow large numbers of cattle to be fed concentrated feed in small outdoor pens. Outdoor feeding pens create welfare problems of dustiness of the substrate, causing respiratory disease, especially during dry weather, and an inability of the animals to shelter from the sun unless shade is provided, failure to supply adequate fibre in the ration (highly concentrated diets are usually fed, leading to digestive and hoof disorders), cattle aggressively riding each other, and bulls mounting each other and damaging their penises (Phillips, 2018, pp. 25–26).

In wetter climates, outdoor pens become waterlogged, and cattle become dirty, hence they are often housed during the winter or wet season. This allows them to be provided with grass that has been preserved as hay or silage from the growing season, as well as concentrates to maximise production.

Originally accommodated in individual stalls, tended by many labourers, since the 1960s intensification has led to dairy cattle being kept in groups in large open pens. Lactating dairy cows need to avoid lying on faeces, otherwise they readily contract mastitis in their udders, so individual cubicles, known as free stalls in the Americas, were developed, which cows could walk into forwards and back out of. The excreta are then deposited at the end of the stall or in a passageway beside the stalls, which can then be kept relatively clean by scraping any faeces away each day. This also allows the cows to be stocked at a higher rate and avoids their teats being trodden on by other cows. In China, concern about the quality of milk from small dairy farmers has led to expansion of year-round housing systems with tens of thousands of cows. Year-round housing has also expanded in the West, due to the greater efficiency and milk production potential of preserved forages and concentrated feed, compared to grazed pasture. Pasture management is also more dependent on the weather than conserved feed. Nevertheless, the capital requirements of intensive dairying have often been beyond the reach of many developing countries. In the emerging economies, the need to site dairy production close to the cities has become even more important with the growing urbanisation that is occurring. Thus, peri-urban dairies, with cattle, goats, and buffalo, are common in sub-Saharan Africa, to meet the needs of the city dwellers. However, fodder supply and suitable housing are often restricted, giving rise to welfare problems of unsuitable diets and lack of exercise for the animals. By-products are often used to good effect but are of variable quality and have a short-term life.

Beef cattle are also kept inside in wet climates, usually loose housed in barns bedded with a suitable material to absorb their urine and provide for some comfort whilst lying. If bedding is in short supply, some farmers keep cattle on slatted floors, in which the excreta fall through slits in the concrete floor into a pit below. This provides an uncomfortable surface for cows to walk and lie on. Depending on the relative costs of stored forages and concentrated feed, cattle may be offered only the latter or a mixture of the two.

In integrated production systems, cattle may be fed on arable by-products or water weeds, not cereals that could be fed to humans, and used for multiple purposes, pulling farm implements, transport, milk and beef production, not just producing unnaturally large amounts of milk. With growing demand for animal products because of expanding world population and greater affordability in emerging economies, especially in Asia and Latin America, many countries have been transitioning away from ruminant animals and towards chickens.

In some parts of the world, animals are farmed under similar geographical and climatic conditions to those used for intensive systems in China, the USA, and Europe, but without any financial support from government. Such systems are common in sub-Saharan Africa and South America. Farms are smaller and animal productivity less. Without the subsidies, farmers are more focused on sustainable production at low cost, rather than forcing animals to grow as fast as possible or to give as much milk from as little land as possible. Farmland is often integrated with forestry into production systems that are more sustainable, with less risk of land degradation. The way in which we force the animal production in intensive systems reflects the model of the capitalist striving to produce more from less, regardless of who or what it exploits. Profit is the key consideration.

In the agroforestry systems used in South America, there are benefits to both animals and trees. Trees provide shade and sometimes feed for animals, their deep roots can harvest water in dry times so that they can survive when all the grass around seems dead, the low stocking density of the animals reduces disease transmission, and there are more resources for wildlife. In return, trees are fertilised by the grazing animals. Nothing is forced, it mimics a more natural ecosystem.

The need for change

Such trends towards intensification are a response to the steady increase in world protein demand. However, the situation regarding food production using animals is more complex than a straightforward relationship between supply and demand. An important complicating factor, albeit not the only one, is that without changes to the animal production systems currently in use, the projected global animal-sourced protein demand cannot be sustainably satisfied (Henchion et al., 2021). In other words, if we are concerned with human food security and life on earth, things must change. The worsening global burden of non-communicable diseases and the effects of food production from animals on greenhouse gas emissions, nitrogen and phosphorus pollution, biodiversity loss, and water and land use tend to reduce the stability of the

Earth system (Willett et al., 2019). Animal-sourced foods are the most resource-intensive, widely criticised for their contribution to antimicrobial resistance, their potential for zoonotic diseases, as well as their negative impact on animal health and welfare (Willet et al., 2019). Further intensifying the same systems seems hardly a proper solution to the problems.

Thus, widespread changes are needed, including a substantial global shift towards healthy dietary patterns, large reductions in food loss and waste and major improvements in food production practices, so that a "Great Food Transformation" is achieved (Willett et al., 2019). Echoing the call for major changes, Herrero et al. (2020) proposed that future technologies and systemic innovation are critical for the profound transformation the food system needs. They have identified these technologies, assessed their readiness, and proposed action points that may accelerate the transition towards a more sustainable food system, emphasising the importance of constructive stakeholder dialogue and clear transition pathways (Herrero et al., 2020).

Regarding animal welfare, specialised scientists and field staff have been working for decades to improve the lives of animals involved in food production systems. However, it has not been an easy road. Therefore, the call for changes in the current paradigm driving animal protein production sounds to be good news for billions of animals produced in farming systems, in which their lives seem hardly worth living (Mellor, 2016). A major convergence between anthropocentric goals and animal protection finally illuminates the difficult road: the recognition of the need to move away from intensive factory farming. The suffering and killing of animals for food receives insufficient attention in contrast to food security, overshadowed by humanity's fear of compromising human public health or causing a decline in resource availability from planet Earth. For self-centred reasons, it seems we will increasingly see changes in food production systems and major shifts of the paradigm soon. A more mature animal ethics debate will likely have to wait longer. However, the changes are most welcome from an animal-centred point of view.

The changes required to decrease the negative environmental impact and other human-centred goals will probably be context-specific (Henchion et al., 2021). If animal interests are not amongst the major drivers for such modifications, there may be a variety of unplanned consequences for the animals. Questions related to farm animal welfare traverse all regions, and best mitigation strategies for animal suffering are also likely not the same for all contexts. Thus, different approaches are of interest, both for their independent values as well as for the possible combinations that may better achieve the goals in terms of food system transitions. We have selected two potential changes in the food production systems for discussion, on the basis that they may offer some degree of improvement for animal welfare. They are, first, the avoidance of suffering through incremental improvements in farm animal welfare and, second, the radical innovations which may completely release animals from the position of being valued as food-producing devices.

Welfare impacts and animal ethics of changes in food production

Until recently, incremental gains were the only path for increasing the welfare of animals used for food production, as long as humanity maintained its demand for animal proteins. In fact, improvements in productivity through incremental innovations have been the rule in agriculture for more than 10,000 years. However, using our knowledge for the benefit of farm animals themselves is a more recent event, essentially after the seminal book by Ruth Harrison, *Animal Machines* (Harrison, 1964). Decades of research in animal welfare science have offered major advancements in our knowledge in a range of relevant topics, such as sentience distribution across farm animal species, animal welfare assessment, and strategies to improve welfare status within food production systems.

We know that the resolution of some welfare problems brings improvements in production indices and, accordingly, are beneficial to the economy of animal food chains as well. These may be called convergent issues, as animal welfare improvements become aligned with increased profits. We also know that this is not always the case, otherwise the push for ever more profitable animal production practices would have produced the best living conditions for animals involved in all major production chains. Thus, some issues are divergent, as increases in animal welfare require increases in production costs. Both convergent and divergent types of issues may, of course, be improved by advancements in knowledge, followed by their adoption in the field. In addition, these types of problems cannot always be neatly identified in real life and require study as an applied area of farm animal welfare science. However, when stakeholders need to decide between decreasing animal suffering or either improving or maintaining profit margins, history shows that animals tend to lose.

Finally, there is controversy in relation to whether the killing of animals is a welfare issue or not. The more traditional schools of animal welfare science tend to consider the killing of animals as a separate issue of animal ethics, unrelated to animal welfare science which, in turn, would concern itself only with abolishing the suffering throughout any killing process, to the point that the World Organisation for Animal Health includes humane slaughter in the definition of good animal welfare (OIE, 2021). A different perspective on this issue recognises the evident losses to welfare brought by early death of a healthy sentient being (Yeates, 2010). As animals have the potential for both good and bad experiences, and the balance describes their welfare, depriving them of the opportunity for the former is contrary to their opportunity for good welfare (Phillips, 2009, pp 8–9). In fact, life is of the utmost instrumental value to animals, as it enables them to realise their autonomy and is thus in their highest interests; consequently, killing may be understood as the greatest harm that can be done to them, as stated by Balluch (2006). Thus, besides avoiding the suffering of animals, it is also important to reflect on their systematic killing for human purposes.

The avoidance of suffering within animal farming

The majority of animal welfare science, when considering farm animals, has been directed to the study of the animals' suffering and ways to mitigate it. Decades of efforts have produced knowledge to support many regulations around the world, which have undoubtedly reduced the suffering of animals engaged in the food production systems. Considering for instance the European case, science is the basis for many European Food Safety Authority Scientific Reports on issues regarding animal welfare, and scientific knowledge has informed the many European Directives and Regulations relevant to animal welfare (Broom, 2017).

There are numerous examples of scientific advancements that have improved animal welfare. As the type of animal raising system can have major welfare consequences, there are many efforts in animal welfare science to understand the impacts on the animals. These studies have been supporting transitions to farming systems that are more animal-friendly, such as those which allow animals to go outside and enjoy fresh air, direct sunlight, a variety of choices, more possibilities for social activities and more room to move around. These are called free-range farming systems, and their use for the benefit of animals requires detailed knowledge of each specific context. For instance, free-range systems for laying hens provide positive gains for their welfare but this may be limited by how much use the birds make of the opportunity to go outside, which in turn depends on many different aspects – the environment inside and outside, the birds' genetics, etc. (Stadig et al., 2017). Thus, it seems essential to understand and manage aspects which are specific to both animal and environmental characteristics of free-ranging systems to effectively create all the potential welfare benefits. Another interesting example of changing farming systems for the benefit of animals is the transition from open pasture to silvopastoral systems. When compared with the most used animal production systems, silvopastoral systems can provide efficient feed conversion, increased biodiversity, enhanced connectivity between habitat patches, and better animal welfare (Broom et al., 2013).

A different approach from that concerning system choice is the so-called precision animal science, which seeks to include high technology in the monitoring and managing of farm animals. This approach has created some controversy as to whether it will improve the lives of animals or leave them worse off, as there is a historical basis to fear that efficiency gains offered by new technologies will be used to increase production still further. According to Dawkins (2021), the answer to the question of whether smart farming improves or damages animal welfare depends on three main factors: how welfare is defined, whether high welfare standards will be a priority within smart farming systems, and whether smart farming can actually deliver its promised improvements when applied in the real world. When looking at on-farm possibilities, precision animal farming has potential for the improvement of animal welfare. For example, the application of precision animal science in the dairy cow sector allows for better assessment of lameness, mastitis, and body condition scores, which are important animal-based indicators of cow welfare (Silva et al., 2021). As assessment is the first requisite for the management of animal welfare, facilitating proper assessment is a key factor. More precise evaluation of welfare indicators, which can also become more frequent as it becomes less time-consuming for farmers or veterinarians, will tend to benefit animal welfare.

More recently, animal welfare researchers have been investigating positive emotions for farm animals. This approach has inspired studies on how to think of ways to improve the lives of animals that are made to be born within food production chains, beyond the focus on the avoidance of suffering. How can the practices assure the consumer that all these animals have at least a life worth living, if a good life may fall out of reach? According to Mellor (2016), a "life worth living" is characterised as a favourable balance of salient positive and negative experiences, but less so than in a "good life", where such balance is strongly positive. It is likely that most farm animals, confined within factory farms, lead a "life worth avoiding", as the balance of salient positive and negative experiences is unfavourable. This way of thinking about animal welfare requires a deeper understanding of positive emotions and how to promote them. This field of study seems extremely relevant, as the interests of animals may not be fully supported if our decisions are focused exclusively on avoiding suffering.

We have cited examples of approaches to improve farm animal welfare and some of their shortcomings. Support for impartial research on, and implementation of, farm animal welfare advances remains imperative. The incremental gains in farm animal welfare are necessary and will remain so for as long as animals are involved in the food industry. Animal welfare gains provide relief for the animals brought into farming scenarios, regardless of their preferences and choices. However, incremental gains cannot liberate farm animals from being seen as a commodity, with a planned, precocious death imposed by all production chains. Their inherent individual value as unique sentient beings with interests, particularly the interest of remaining alive, is invariably violated.

Radical innovation: the avoidance of killing

Is it possible to eat well without killing animals? Yes it is, according to the Academy of Nutrition and Dietetics of the United States (Melina et al., 2016) and the National Health Service in

England (NHS, 2018), amongst others, and as proven by the increasing numbers of healthy vegan people in the world. It is also true that many humans, probably most of us, like to eat animal products. Thus, the research to produce meat and other animal products, either real animal products obtained by cell growth in bioreactors, or simulations in innovative and alternative ways seem a good strategy to give people what they want, while at the same time releasing animals from farming practices. In this context there emerges a new generation of vegan substitutes for animal products, the plant-based meats, milk, and dairy, as well as eggs. The consumption of tofu and pea proteins as meat substitutes is not innovative, but the new products such as the plant-based burgers are. Their molecular-based recipes have exponentially improved product taste, texture, and other sensorial characteristics, increasing the resemblance to the conventional products and resulting in a rapidly increasing market share over the last ten years. This is excellent news for the animals, as well as for environmental concerns, as the impact of even the lowest-impact animal products typically exceed those of vegetable substitutes, providing evidence for the importance of such a dietary change (Poore and Nemecek, 2018). Yet, it is common to hear people say that these are not real animal products, a thought many people seem to consider important.

Then, more recently, came cellular animal science, with the promise of producing animal protein without causing suffering to animals and without killing them. This uncoupling of meat from animal slaughter has, of course, a radical positive benefit for animals, removing them from the instrumental value of their bodies as food for humans. There are also at least two indirect benefits for animals: (1) the liberation from the meat paradox (Loughnan et al., 2014), which will likely transform human–animal interactions very profoundly; and (2) the higher efficiency of cell-based meat production as compared to conventional meat, which means that more land and water will be available (Tuomisto and Teixeira de Mattos, 2011). Part of the resources released may be used for the recovery of natural habitats, whose destruction has been a major challenge for wild animals (for a more detailed discussion, see Chapter 23 and Heidemann et al., 2020). This is important because, if denying farm animals their life, and potentially periods of good welfare, is compensated by providing a good life for wildlife, cellular animal science is more justifiable.

The greater efficiency of cell-based animal production systems as compared to the conventional ones also brings relevant benefits to the disease environment. Cell-based animal production is expected to reduce the significant contribution of animal agriculture to the development of antibiotic resistance, as well as the risk of foodborne and new zoonotic diseases, allowing for a safer escape from the two major human-mediated risk factors for the next pandemic: (1) the increasing human demand for animal protein, and (2) unsustainable agricultural intensification (UNEP, 2020).

Conclusions

Intensification of animal production systems has been a feature of agricultural systems since people first transitioned from hunting and gathering. It is a response to human expansion, in terms of both population and the complexity of our management of the land and its resources. The biomass of cattle is now twice that of people on the planet, and the largest of any animal on earth (Phillips, 2018, p. 6). This continued intensification of animal production has forced farmers to adopt systems which consistently fail to provide animals with opportunities for normal behaviour, appropriate nutrition and reproduction, and, in many cases, a life worth living.

If we asked a cow, a chicken, a sow, and a jaguar which food production system they would choose, they would likely choose the one in which there is the least impact on animals. Rapid changes in food production technologies have meant that this is increasingly seen as systems in which animals are not used on an ongoing basis. Living cells may be cultured or plants may be processed to make products that have the attractiveness of those from animals, but none of the welfare or other disadvantages. Animals are key stakeholders in the production system (Chapter 34) and their view must be recognised as important. This is not the only question, but it does seem a major one for those interested in the protection and welfare of animals.

References

Balluch M, 2006. Animals have a right to life. *ALTEX: Alternatives to Animal Experimentation*, 23(4), pp. 281–293. Available at: https://www.altex.org/index.php/altex/article/view/804

Broom DM, 2017. *Animal Welfare in the European Union: Study for the Peti Committee.* Directorate General For Internal Policies EU, 78. http://www.europarl.europa.eu/RegData/etudes/STUD/2017/583114/IPOL_STU(2017)583114_EN.pdf

Broom DM, Galindo FA, Murgueitio E, 2013. Sustainable, efficient livestock production with high biodiversity and good welfare for animals. *Proceedings of the Royal Society B*, 280, 20132025. https://doi.org/10.1098/rspb.2013.2025.

Dawkins MS, 2021. Does smart farming improve or damage animal welfare? Technology and what animals want. *Frontiers in Animal Science*, 2, 736536. https://doi.org/10.3389/fanim.2021.736536.

Eisen J, 2020. Down on the farm: status, exploitation and agricultural exceptionalism. In Blattner C, Coulter K and Kymlicka W, eds., *Animal Labour: A New Frontier of Interspecies Justice?* Oxford University Press, Oxford, UK.

Grace ND and Knowles SO, 2012. Trace element supplementation of livestock in New Zealand: meeting the challenges of free-range grazing systems. *Veterinary Medicine International*, 2012, 639472. https://doi.org/10.1155/2012/639472.

Harrison R, 1964. *Animal Machines.* Methuen and Company, London, UK.

Heidemann MS, Molento CFM, Reis GG and Phillips CJC, 2020. Uncoupling meat from animal slaughter and its impacts on human-animal relationships. *Frontiers in Psychology* 11, p. 1824. https://doi.org/10.3389/fpsyg.2020.01824.

Henchion M, Moloney AP, Hyland J et al., 2021. Review: trends for meat, milk and egg consumption for the next decades and the role played by livestock systems in the global production of proteins. *Animal.* https://doi.org/10.1016/j.animal.2021.100287.

Herrero M, Thornton PK, Mason-D'Croz D, Palmer J, Bento, TG, Bodirsky BL et al., 2020. Innovation can accelerate the transition towards a sustainable food system. *Nature Food* 1, pp. 266–272. https://doi.org/10.1038/s43016-020-0074-1.

Loughnan S, Bastian B and Haslam N, 2014. The psychology of eating animals. *Current Directions in Psychological Science* 23, pp. 104–108. https://doi.org/10.1177/0963721414525781.

Mazoyer M and Roudart L, 2012. *A History of World Agriculture: From the Neolithic Age to the Current Crisis.* Monthly Review Press, New York.

Melina V, Craig W, Levin S, 2016. Position of the Academy of Nutrition and Dietetics: vegetarian diets. *Journal of the Academy of Nutrition and Dietetics*. 116(12) ,pp. 1970–1980. https://doi.org/10.1016/j.jand.2016.09.025.

Mellor DJ, 2016. Updating animal welfare thinking: moving beyond the "Five Freedoms" towards "A Life Worth Living". *Animals* 6, p. 21. https://doi.org/10.3390/ani6030021.

National Health Service, 2018. The vegetarian diet. Eat well. https://www.nhs.uk/live-well/eat-well/the-vegetarian-diet/. Accessed 11 October, 2021.

OIE, 2021. World Organisation for animal health: Terrestrial animal health code, section 7. *Animal Welfare*, Chapter 7.1. Introduction to the Recommendations for Animal Welfare, Article 7.1.1. Available at https://www.oie.int/en/what-we-do/standards/codes-and-manuals/terrestrial-code-online-access/?id=169&L=1&htmfile=chapitre_aw_introduction.htm.

Phillips CJC, 2009. *The Welfare of Animals: The Silent Majority.* Springer, Dordrecht.

Phillips CJC, 2015. *The Animal Trade.* CABI, Wallingford, UK.

Phillips CJC, 2018. *Principles of Cattle Production*, 3rd edn. CABI, Wallingford, UK.

Poore J and Nemecek T, 2018. Reducing food's environmental impacts through producers and consumers. *Science*, 360(6392), pp. 987–992. https://doi.org/10.1126/science.aaq0216.

Ritchie, H. and Roser, M, 2021. Forests and deforestation. Published online at *OurWorldInData.org*. Retrieved from: https://ourworldindata.org/forests-and-deforestation (Accessed 11 October, 2021).

Silva SR, Araujo JP, Guedes C, Silva F, Almeida M, Cerqueira JL, 2021. Precision technologies to address dairy cattle welfare: focus on lameness, mastitis and body condition. *Animals*, 11, p. 2253. https://doi.org/10.3390/ani11082253.

Stadig LM, Rodenburg TB, Ampe B, Reubens B, Tuyttens FAM, 2017. Effect of free-range access, shelter type and weather conditions on free-range use and welfare of slow-growing broiler chickens. *Applied Animal Behaviour Science*, 192, pp. 15–23, https://doi.org/10.1016/j.applanim.2016.11.008.

Tuomisto HL and Teixeira de Mattos MJ, 2011. Environmental impacts of cultured meat production. *Environmental Science and Technology* 45, pp. 6117–6123. https://doi.org/10.1021/es200130u.

UNEP: United Nations Environment Programme and International Livestock Research Institute, 2020. *Preventing the Next Pandemic: Zoonotic Diseases and How to Break the Chain of Transmission*. UNEP/ILRI, Nairobi, Kenya.

van Zanten HHE, Mollenhorst H, Klootwijk CW et al., 2016. Global food supply: land use efficiency of livestock systems. *International Journal of Life Cycle Assessment* 21, pp. 747–758. https://doi.org/10.1007/s11367-015-0944-1.

Willett W, Rockström J, Loken B, Springmann M, Lang T, Vermeulen S, Garnett T, Tilman D, DeClerck F, Wood A et al., 2019. Food in the anthropocene: the EAT–Lancet Commission on healthy diets from sustainable food systems. *The Lancet*, 393, pp. 447–492.

Yeates JW, 2010. Death is a welfare issue. *Journal of Agricultural and Environmental Ethics*, 23 pp. 229–241. https://doi.org/10.1007/s10806-009-9199-9.

5
FARMING POULTRY

Tina M Widowski and Ana K Rentsch

Introduction

Domestic fowl (poultry) are either land-fowl, belonging to the order *Galliformes* (chicken and turkeys), or waterfowl that belong to the order *Anseriformes* (ducks and geese). Several species of poultry are farmed intensively all around the world with a substantial increase in the last five decades. The number of chickens alone doubled between 1970 and 1990 and then more than doubled again to almost 26 billion chickens worldwide by 2019 (FAO, 2021). Most poultry production systems are characterised by large group sizes and high stocking density. They are managed to maximise production and feed efficiency with minimal labour at low cost. The absence of any cultural or religious restrictions on eating poultry products as well as affordability of poultry produce are drivers for intensive production, though in recent years consumer awareness of welfare concerns in intensive poultry farming systems has increased. The various segments of the poultry industry are highly specialised, and each has its own welfare concerns which are listed in Table 5.1. In this chapter we discuss key welfare concerns in farmed chickens (meat and egg production), turkeys, and ducks. Welfare concerns arise from genetic selection for high production in birds selected for fast growth or high output of eggs. We discuss how housing systems meant for manageability affect the animals living in them, and how management practices meant to alleviate problematic behaviour cause new issues.

Effects of selection for productivity on poultry health and welfare

Techniques for quantitative genetic selection of economically important production traits were developed in the 1940s (Anderson et al., 2013), and by the 1950s, these selection methods were being used in earnest by poultry breeding companies. This resulted in the development of separate genetic lines and industries for chicken meat and table egg production. The reproductive output and short generation times of poultry species, coupled with intensive selection for production traits, have resulted in tremendous changes in the birds and increases in their levels of productivity (Anderson et al., 2013; Zuidhof et al., 2014). The emphasis on production traits, however, has resulted in unintended but serious consequences for health and welfare.

DOI: 10.4324/9781003182351-7

Table 5.1 Key welfare concerns in farmed poultry

Chickens – Laying hens (Layers) Rearing: 16–19 weeks, Laying: 55–75 weeks		
Eggs	Behavioural restriction	Housing does not allow for the performance of behavioural needs, particularly in conventional cages
	Osteoporosis	Fragile bones prone to deformation or fracturing due to high calcium demands
	Keel bone damage (KBD)	Deformities and/or fractures of the keel bone
	Injurious pecking	Severe feather pecking, resulting in plumage or tissue damage, vent pecking, cannibalism
	Beak treatment	Shortening and removal of the sharp tips of beaks
	Piling	Pile up of birds that can lead to heat stress and suffocation (smothering)
	Induced moulting	Controlled loss and replacement of feathers to prolong laying cycle through forced weight loss
	Comb dubbing	Trimming of comb in layer pullets to preserve feed conversion efficiency
Chickens – Layer breeders Rearing: 16–19 weeks, Laying: 55–75 weeks		
Layer parent stock	Osteoporosis	Fragile bones prone to deformation or fracturing due to high calcium demands
	Keel bone damage (KBD)	Deformities and/or fractures of the keel bone
	Behavioural restriction	Housing does not allow for the performance of behavioural needs
	Injurious pecking	Severe feather pecking, resulting in plumage or tissue damage, vent pecking, cannibalism
	Beak treatment	Shortening and removal of the sharp tips of beaks
	Toe removal	Removal of the spur in roosters
Chickens – Broilers 5–7 weeks		
Meat	Lameness	Conformational changes or pain from musculoskeletal disorders reduce mobility
	Muscle disorders	Breast myopathies (wooden breast, white striping and spaghetti meat)
	Contact dermatitis	Foot lesions, breast blisters, hock burns
	Behavioural restriction	Barren environment or poor litter quality do not allow for performance of behavioural needs
Chickens – Broiler breeders Rearing: 16–19 weeks, Laying: 55–75 weeks		
Broiler parent stock	Feed restriction	Reduced feed (quality or quantity) to slow growth and preserve reproductive function
	Toe removal	Removal of the spur of roosters
	Injurious pecking	Severe feather pecking, resulting in plumage or tissue damage, including cannibalism
	Beak treatment	Shortening and removal of the sharp tips of beaks
	Aggression	Rooster aggression towards females and forced mating
	Contact dermatitis	Foot lesions, breast blisters, hock burns
	Lameness	Reduced mobility as the flock ages
Turkeys 12 weeks for hens – up to 20 weeks for toms		
Meat	Lameness	Conformational changes or pain from musculoskeletal disorders reduce mobility

(*Continued*)

Table 5.1 (Continued)

	Injurious pecking	Head pecking in turkey toms and severe feather pecking in hens and toms
	Beak trimming	Shortening and removal of the sharp tips of beak
	Contact dermatitis	Foot lesions, breast blisters, hock burns
	Claw removal	Removal of the claws of forward-facing toes
Turkey – Breeders Rearing:~30 weeks, Laying:~28 weeks		
Turkey parent stock	Feed restriction	Reduced feed (quality or quantity) to slow growth and preserve reproductive function
	Claw removal	Removal of the claws of forward-facing toes
	Lameness	Leg weakness in turkey toms
	Injurious pecking	Head pecking in turkey toms and severe feather pecking in hens and toms
	Beak trimming	Shortening and removal of the sharp tips of beak
	Artificial mating	Semen collection and artificial insemination, potential welfare concerns due to restraint
Ducks – Pekin and Muscovy ducks Pekin: 5–7 weeks, Muscovy: 10–12 weeks		
Meat	Behavioural restriction	Housing does not allow for the performance of behavioural needs, especially water bathing
	Injurious pecking	Severe feather pecking resulting in plumage damage
	Bill trimming	Shortening of the bill
Ducks – Mule ducks~14 weeks		
Foie gras and meat	Behavioural restriction	Housing does not allow for the performance of behavioural needs, especially water bathing
	Force feeding	Intubation and forced feeding to increase liver size
Ducks – Breeders Rearing: 18–22 weeks, Laying:~40 weeks, 47 weeks for Pekin		
Duck parent stock	Feed restriction	Feed restriction to slow growth and preserve reproductive function.
	Injurious pecking	Severe feather pecking resulting in plumage damage
	Beak treatment	Shortening of the bill
All farmed poultry		
High stocking density	High number of birds in restricted space	
Transport	Transportation of hatchlings and catching and transport at end of production and of pullets in layers and breeders	

Meat production

Broiler chickens have been mainly selected for fast growth rates, better feed efficiency (g body weight per g feed consumed) and high breast yield, currently reaching a market weight of 2.1 kg at 35 days of age (Torrey et al., 2021). A comparison of two commercial broiler strains from 1950 and 2005 showed a 400% increase in growth rate, a 50% improvement in conversion of feed to meat, and an 80% increase in breast yield (Zuidhof et al., 2014). These changes in growth rate and body structure have been associated with skeletal, cardiovascular, and metabolic disorders, as well as reduced immune function. Skeletal and leg disorders are of particular concern for welfare since they often cause pain and reduce mobility, thereby affecting the birds' ability to walk, perch, and perform other motivated behaviour. Selection for better leg and cardiovascular health has been successful in reducing some of the early identified disorders, such as long bone deformities, tibial dyschondroplasia and ascites, but recent studies still report moderate to severe lameness in 15–25% of broilers (see Torrey et al., 2021). Reduced mobility may be due to pain, leg strength that cannot support a heavy body weight, or conformational changes (e.g., breast to leg ratio) that make walking difficult or energetically expensive; or may be due to combinations of these factors. The relevance of low levels of activity to welfare is sometimes debated, since selection for better feed efficiency may have led to more sedentary temperaments. However, reduced mobility is associated with other potentially painful foot and leg health problems such as foot pad dermatitis and hock burns, as the birds spend more time sitting on litter (see litter quality below). A more recent development in the broiler industry is the increasing prevalence of breast myopathies (wooden breast, white striping, and spaghetti meat), which are muscle disorders that affect appearance and eating quality of the meat (Santos et al., 2021). While these disorders have received attention because of the implications for meat quality, whether and how they affect muscle function and bird welfare is currently unknown.

Selection for fast growth also affects the welfare of parent stock, the broiler breeders. Fast growth is coupled with large appetites, and allowing breeders to eat to satiety results in poor health, and impaired reproductive performance (de Jong and Guémené, 2011). Therefore, the growth rates of broiler breeders are carefully controlled by feed restriction to achieve optimal health and reproductive performance. Broiler breeder hens are allotted less than 50% of *ad libitum* ("at will") intake during rearing (Arrazola et al., 2019a) and 50–90% of *ad libitum* intake during lay (de Jong and Guémené, 2011). These levels of feed restriction result in signs of chronic hunger and frustration, including stereotypic behaviour (repeated pecking at objects or an empty feeder), pacing, over-drinking, aggression and feather pecking, as well as physiological indicators of stress (Arrazola et al., 2019a). Research focused on alternative feeding methods to mitigate the negative effects of feed restriction are covered in the section on feeding practices.

There is increasing attention to the use of "slow" growing genetic strains of broilers. Some higher welfare food assurance schemes only permit slow growing breeds of chicken, and an increasing number of food retailer and food service companies are making commitments to sourcing birds with better welfare, which include the use of slow-growing strains (Fernyhough et al., 2020). Slow-growing strains generally have less lameness (better gait scores), better foot health, higher levels of activity and greater use of environmental enrichment. Additionally, prevalence of breast muscle myopathies is less in the slow-growing strains. However, the definition of "slow growing" varies considerably, with time to market weight taking anywhere between 7 and 53 days longer than conventional fast-growing birds (Torrey et al., 2021). A comparison of 12 strains of slow-growing chickens to two conventional fast strains, indicated differences in activity and enrichment use (Dawson et al., 2021), mobility, and prevalence of muscle myopathies (Santos et al., 2021). These myopathies were related to rate of growth and breast yield, such that the slow-growing strains with the fastest growth rates were more similar to conventional birds than those that grew more slowly and had smaller breast yields.

Slow-growing genetics also benefits the parents. Slow-growing pullets are feed-restricted at a later age, show fewer signs of feeding frustration (reduced feeding motivation, less over-drinking, less feather pecking) and fewer signs of chronic stress (Arrazola and Torrey, 2021).

Genetic selection for meat production in turkeys has also resulted in a heavy, fast-growing bird with exceptionally large breast muscles. Turkeys are unable to mate naturally which requires the use of artificial insemination for breeding. Changes in growth rate and body conformation have resulted in many of the same welfare issues that occur in broiler chickens, including abnormalities of the musculoskeletal system, lameness and the need to feed restrict turkey breeders in order to maintain health and fertility (Erasmus, 2018). Genetic correlations of various traits indicate that walking ability, soundness of hip and leg structures and contact dermatitis (of the foot pad and breast) are all negatively correlated with body weight, and walking ability is positively correlated with survival and longevity (Quinton et al., 2011). Because male tom turkeys grow much faster and are usually marketed at a much heavier weight than hens, welfare outcomes tend to be worse for toms than for hens (Marchewka et al., 2019).

Meat-type ducks have also been selected for rapid growth and experience many of the same welfare problems as broiler chickens, including impaired leg health and foot lesions, which can be exacerbated by housing them on slatted floors. Breeding ducks also need to be feed-restricted to prevent excessive weight gain (Chen et al., 2021).

Egg production

Selection for production traits in laying hens has resulted in earlier sexual maturity (average age at 50% production has decreased by ~30 days since the 1970s), better feed efficiency (g egg produced per g feed consumed), reduced mortality, lower body weight, larger eggs with better shell quality, and higher and more persistent levels of production (Anderson et al., 2013). The major consequences for health and welfare are metabolic diseases associated with high egg output, and these impact the hen differently, depending on how she is housed. Egg production requires large amounts of lipid to be deposited into egg yolks, with the liver being the site of most lipid synthesis and storage. Fatty liver hemorrhagic syndrome (fatty liver disease) is a metabolic disorder of high producing laying hens that can result in mortality. The condition is associated with high body weight (obesity) and can be reduced when hens are provided space for movement and exercise; therefore, it is mainly a problem in caged layers (Shini et al., 2019).

Laying hens have medullary bone in addition to structural bone tissues (cortical and trabecular bone) which acts as calcium storage for eggshell formation (Kim et al., 2012). During the laying period, calcium is primarily deposited in medullary bone, but it is drawn from all types of bone, leading to net loss of structural bone. Consequently, high calcium demand for eggs leads to osteoporosis. While cage layer fatigue (collapsing of the skeleton) has been mostly eliminated through improving calcium:phosphorus ratios and vitamin D content in feed, today's laying hens are still highly prone to bone fractures and deformities (Widowski et al., 2017).

Keel bone fractures (common in laying hens) are associated with pain, impaired mobility, and chronic stress in afflicted hens (Armstrong et al., 2020). Housing can be designed to reduce such injuries by allowing for load-bearing exercise (especially during skeletal development), by adding ramps or optimising perch position for easier navigation, and by providing softer perch material which can decrease impact severity (Rufener and Makagon, 2020). Even with concentrated efforts from researchers, feed and genetic companies and farmers, osteoporosis in laying hens remains a serious welfare concern.

The biological limit of laying one egg per day has practically been achieved at peak production (laying >98%) in most commercial lines of laying hens. Current goals of genetics companies are to extend the persistency in lay with stability in egg quality so that commercial laying hens will remain in production for 90 to 100 weeks, producing 500 eggs in their lifetime (Bain et al., 2016). Extending the laying cycle has several advantages in terms of environmental impact, economics and potentially animal ethics, since the overall number of laying hens needed to support the world's consumption of eggs will be significantly lower (Fernyhough et al., 2020). However, it is acknowledged that selection for extended production must be accompanied by selection for increased skeletal integrity (Preisinger, 2018). For decades, selection in pedigree and breeding flocks has been done in individual cages for ease of quantifying egg traits. However, with much of the world moving away from housing hens in cages, layer genetics companies are increasingly adding traits related to behaviour into their selection programmes. Several approaches are being used to address the welfare problems of injurious behaviour (i.e. feather pecking), including identifying phenotypes with predisposition to perform injurious behaviour, group selection for survivability in birds with intact beaks, and selection for beak shapes that cause less damage.

Behavioural needs of farmed poultry

Behavioural needs are instinctual behaviours that are highly motivated and performed under circumstances even when the intended function is no longer necessary (Cooper and Albentosa, 2003; Weeks and Nicol, 2006). Their performance can either induce a positive affective state or an inability to perform such behaviour may induce a negative affective state. Consideration of such behaviours has increased in recent years when it comes to housing design (Widowski et al., 2017). Concerns about behavioural deprivation of laying hens housed in conventional cages sparked research into behavioural needs. Hence, the literature is extensive for laying hens, while it is lacking for other poultry species. This section describes the main four behavioural needs known for laying hens: nesting, perching, dustbathing, and foraging, and includes other species where information is available.

Nesting

Nesting behaviour in laying hens is known to be a behaviour of high priority (Weeks and Nicol, 2006). Pre-laying behaviour in laying hens starts hours before oviposition (egg laying) and is triggered by the hormones of ovulation a day prior. It consists of nest searching and inspection, nest building, and finally, sitting. Hens are highly motivated to lay in a discrete enclosed nest as is shown by their willingness to work for access to a nest, by showing signs of frustration when nest boxes are absent, and by delaying oviposition when the behaviour is interrupted at crucial timepoints.

For cage-free egg production as well as breeding flocks of layers, turkeys and ducks, producers rely on nesting motivation of the hens to lay their eggs in the nest. Pekin ducks perform the same pre-laying behaviours as laying hens, and nest boxes are a commodity that is competed for (Barrett et al., 2019). There is also evidence for duck hens' willingness to work for access to nest boxes and stress when access is hindered. Therefore, provision of sufficient nest space is important both for welfare and for reducing eggs laid outside of the nest.

Perching and roosting

For predator avoidance, fowls seek safe roosting spots at night when they are resting and vulnerable (Mench, 2009). High motivation for elevated roosting in hens has been shown through their willingness to work for elevated perches. In the absence of perches, hens will roost on the highest housing fixtures (Weeks and Nicol, 2006). Given the opportunity, laying hens spend about a quarter of their time on perches during the day and all of their time during the night.

Broiler breeders, too, show motivation for perching in both fast and slow-growing strains (Gebhardt-Henrich et al., 2018). Like laying hens, broiler breeders perch on any elevated structures, not just intended perches. Initially, perching increases with age, though as age and body weight increase, perching propensity declines, most likely due to mobility issues. Hence, housing design for broiler breeders should take their heavy body weights into account.

Broiler chickens are motivated to perch, though vertical navigation and balancing on perches are problematic with their heavy body weights, especially in fast-growing broilers (Riber et al.,

2018). Use of perches depends on broiler genotype, age, flock size, temperature, and perch design. In high temperatures, cooling perches and a greater distance from the litter allow for better thermoregulation. Elevated platforms can be used as alternative roosting structures that are more accessible and show promising results in improving foot and leg health. It has also been suggested that structures in the barn, such as panels, barriers, and bales of straw can satisfy motivation for predator avoidance in broilers, with additional positive effects on leg health (Pedersen et al., 2020) and bird distribution (Riber et al., 2018).

Dustbathing and water bathing

To maintain feather condition, ducks perform water bathing and chickens and turkeys perform dustbathing (Mench, 2009). By dustbathing, hens work small substrate particles through their feathers removing excess oil and parasites, and the behaviour may be considered pleasurable. In caged systems, hens perform "sham" dustbathing in which the "bathing" motions are performed on wire flooring. Whether or not sham dustbathing is satisfying for the hens is still unknown (Widowski et al., 2017).

While ducks do not dustbathe, they need a water source to keep their eyes, nostrils, and feathers clean and also use water for dabbling and swimming (Rodenburg et al., 2005). Water bathing includes dipping their head under water and splashing it over their wings when given water sources in form of a pond, trough, or overhead showers. Motivation for water bathing in ducks has been measured by their readiness to work for access to water baths and troughs (Jones et al., 2009). Water from nipple drinkers is not enough satisfy motivation as shown through rebound behaviour when subsequently given access to an open water source. However, open water sources present a high risk for bacterial contamination and poor hygiene, and research findings in regard to preening, body condition and cleanliness are contradictory (Chen et al., 2021). Currently, the welfare implications for ducks housed without access to open water are controversial.

Foraging

Foraging, consisting of ground-pecking, scratching, and grazing, fills a large portion of the day for wild-living fowl species (Mench, 2009). Although most farmed poultry are provided with adequate feed *ad libitum*, and genetic selection and food availability have reduced the time budget allocated to foraging, the behaviour remains a priority. In conventional cages with no foraging material accessible, hens perform foraging behaviour by raking their beak through feed while scratching the wire floor (Widowski et al., 2017). Hens are willing to work for access to foraging material and there is convincing evidence that hens perform contra-free-loading, i.e., when given a choice, they prefer to work for food rather than "free-loading" and eating at the feeder (Weeks and Nicol, 2006). Absence of foraging opportunities may lead redirected pecking behaviour and is a risk factor for feather pecking (van Staaveren and Harlander, 2020).

Injurious pecking in farmed poultry

Injurious pecking refers to bird-to-bird pecking that results in damage to plumage or tissues and is a serious welfare problem in nearly all types of poultry, regardless of housing system. Injurious pecking can be divided into 1) **severe feather pecking** which results in damage or removal of feathers, usually from the back, tail, and wings; 2) **tissue pecking** directed at featherless areas of the body including vent and toes; and 3) **aggressive pecking** which is usually directed at the head or neck (Dalton et al., 2013; van Staaveren and Harlander, 2020). Injurious pecking is both a welfare and economic problem as it can result in pain, fear, morbidity and mortality, and loss of feather cover results in increased requirements for feed or supplemental heat because of reduced thermoregulatory ability. Although all forms of injurious pecking are multi-factorial in nature, severe feather pecking in laying hens has received the most study, with nutrient composition (e.g. protein and fiber), diet form (e.g. mash versus pellets), availability of foraging substrate, environmental stress, social stress, gut microflora, and genetics all known to play a role. Although severe feather pecking increases with age, early experience can predispose flocks to outbreaks, and it is generally considered to be more effective to prevent severe feather pecking by using appropriate management strategies and providing environmental enrichment throughout life, rather than attempt to mitigate the problem after it has developed (van Staaveren et al., 2021). Much less is known about the risk factors for vent and toe pecking, which can lead to outbreaks of cannibalism in laying hens. Unlike feather and tissue pecking, aggressive pecking can be considered a normal, although unwanted, behaviour associated with social competition. Aggressive head pecking is a major cause of injury for maturing male turkey flocks. Although there are many recommendations for prevention and control of injurious pecking, the unpredictability of the behaviours and the severe consequences of their occurrence make beak trimming the most common form of prevention in most parts of world.

Housing

Laying hen housing

Detailed descriptions of commercial laying hen housing systems can be found in Karcher and Mench (2018). These are summarised in the following.

Conventional cages

Conventional cages (CC) were developed in the 1950s to optimise egg production. A CC consists of a small wire cube with an internal line of nipple drinkers and a feed trough outside the cage, preventing fecal contamination. Gravity and a tilted wire floor help eggs roll out and onto a conveyer belt for automated collection. Hens are usually housed in small groups of 5-10 birds at a high stocking density. Strict biosecurity coupled with absence of litter and separation from manure results in generally good health, few injuries and low mortality for hens in CC. However, spatial restriction and cage simplicity do not allow for natural species-specific behaviour such as nesting, perching, foraging, and dustbathing. Bones are weakest in CCs compared to other systems and prone to fracture during depopulation (Lay et al., 2011). Due to welfare concerns, CCs are increasingly being replaced by alternative housing systems in many regions of the world.

Furnished cages

Furnished cages (FC; also enriched, modified, or colony cages) were developed to address welfare issues of CCs. FCs include perches, scratch mats, an enclosed nesting area, and sometimes nail shorteners. Group size differs between countries and cage designs but generally range between 10 and 100 birds per cage, with higher space allowance than CCs. These cages represent a compromise between CC and non-cage systems since they allow for more behavioural freedom, while still maintaining cleanliness and manageability of CC (Lay et al., 2011). These cages have been criticised as providing inadequate enrichment for so many birds. A single nesting and scratching area is common. While nests are well used, perches are lower than preferred by hens, and foraging and dustbathing are not fully supported (see Foraging and Dustbathing). Injurious pecking can be a problem in FC with increased group sizes allowing for more victims. Bone-loading behaviour from perching and more space for locomotion increases bone strength, but structural elements can lead to deformed or fractured keel bones during production.

Non-cage systems

Non-cage (NC) or cage-free housing systems can either be barn systems (single-tier) or multi-tiered aviaries. All offer nest boxes, and in many regions, perches are also required. All multi-tier aviaries offer a litter floor for foraging and dustbathing below or adjacent to the main structure, while single-tier systems have wire or slatted floors, litter floors or some combination (Lay et al., 2011). Group sizes in NC layer houses are usually in the thousands, while stocking density depends on regional or customer requirements. Free-range hens are additionally given outdoor access.

The increased access to space and availability of elevated roosting places, secluded nesting areas and litter on the floor offer more opportunities for motivated behaviour compared to cage systems. However, the presence of litter reduces hygiene and air quality, with high levels of dust and greater risks of parasites, foot pad dermatitis and bumblefoot. Hens in NC housing are at risk of piling on top of one another; if not interrupted, piling can lead to smothering and death. More feather pecking is observed in NC systems if they have high stocking densities, and compared to cage systems, the large group sizes in NC allow for more victims and more severe consequences (van Staaveren and Harlander, 2020). Outdoor access increases the risk of predation and avian diseases from wild birds. Often, outdoor areas are underutilised because of lack of early experience, individual differences in ranging behaviour or range design, especially provision of structures for predator avoidance (Campbell et al., 2021). Mortality is generally considered to be higher in NC housing compared to cage housing (Weeks et al., 2016). However, mortality can be reduced with improved management practices and is reported to match that of CC within the first 20 years of NC housing in Europe (Schuck-Paim et al., 2021).

While the added spatial complexity of aviaries leads to hens with stronger bones due to increased bone-loading exercise, the increased complexity also offers more opportunities for bone fractures and deformities. Keel bone damage is a prevalent issue in NC housing and can affect the majority of a flock by the end of the laying period (Rufener and Makagon, 2020).

Galliformes are ground-dwelling species with limited flight ability, making multi-tiered structures challenging for hens. For this reason, it is imperative that the rearing environment matches the structural complexity of the adult housing. A hen's spatial navigation skills need to be acquired within the first few weeks of life to facilitate successful adaptation to complex vertical structures as an adult (Widowski and Torrey, 2018).

Stocking density

Poultry farming systems typically comprise groups of thousands of birds housed in the same barn, and public perception is that most farmed poultry are extremely crowded. There is a direct conflict between providing birds more space and economic efficiency, since stocking birds at higher rates generally reduces the cost of production. Floor space requirements, often expressed as area per bird for adult birds (e.g. cm^2/bird for laying hens and breeders) or area per unit of body weight for growing birds (e.g. kg/m^2 for meat birds), are found in most animal care regulations and guidelines. Often, there are additional space requirements for resources in the barn, such as feeders, perches and nests.

Determining the amount of space birds need to ensure welfare is difficult because thresholds vary for different biological and behavioural outcomes. At a certain point, high density results in increased mortality and reduced production performance due to stress, and this is often the "cut off" used to set requirements. However, even when birds are provided enough space to maintain health and production levels, behaviour is often impaired. This is further complicated by the fact that stocking density affects the barn environment, as birds produce heat and moisture. When temperature and humidity are well-controlled by optimal ventilation, poultry can tolerate higher stocking densities, at least in terms of health and performance.

Although most poultry species evolved living in small groups with stable social hierarchies, or pecking orders, they adapt to living in large groups. As group size increases and individual recognition becomes difficult, birds seem to adopt a strategy of "social tolerance" rather than attempting to maintain a social hierarchy (Estevez et al., 2007). Evidence suggests that aggression decreases as group sizes increase, as long as resources are readily available to prevent competition.

When birds are housed in cages, overall cage size relates to group size such that birds housed in larger groups have more total space available for locomotion. Since birds tend to cluster around resources such as feeders, housing birds in larger groups also results in more "free space" to perform behaviours such wing flapping, (Mench and Blatchford, 2014), but it may also result in crowding in some areas.

Litter quality

Most poultry meat birds are housed on some form of litter or deep bedding. Breeder flocks, layers in non-cage systems and some ducks are housed on litter or some combination of slats or wire floors and litter. Litter quality, assessed in terms of moisture, pH and ammonia, has significant effects on bird health and welfare. This is particularly important for meat birds who spend extended periods of time sitting or lying on litter which commonly does not get replaced during the production cycle. Poor litter conditions, with high moisture and ammonia levels, increase the risk of contact dermatitis, which includes footpad dermatitis, hock lesions and breast blisters, known to be painful conditions. Wet or caked litter also reduces opportunities for dust bathing. In contrast, dry litter can lead to extremely dusty conditions which impairs respiratory health. For these reasons, litter quality is often evaluated in animal welfare assessments.

Husbandry procedures

A number of procedures are routinely performed on poultry species in order to modify various anatomical structures. These include beak shortening (beak trimming or beak treatment) of laying hens, broiler breeders and turkeys, bill-trimming of ducks, toe and snood removal of turkeys, and toe or spur removal of breeder roosters. Removal of the comb (dubbing), while no longer common, may be performed in laying hens and layer breeding stock (Fiks and de Jong, 2007). Most of these procedures are done to prevent injuries that occur during production, but most have welfare implications in and of themselves. Although the literature on beak trimming is extensive, few studies have addressed welfare implications of other procedures.

Beak treatment

There is considerable evidence that shortening and removal of the sharp tips of beaks/bills is highly effective at reducing morbidity and mortality caused by injurious pecking in most poultry species (reviewed by McKeegan, 2017). However, the bills and beaks of birds are highly innervated with nociceptors, chemoreceptors, thermoreceptors and mechanoreceptors, making them functionally important as sensory organs. Additionally, beaks and bills serve a vital functional role in daily activities including feeding, drinking, grasping objects, mating, preening, and parasite removal. The risks of beak shortening procedures to welfare include acute pain during and shortly following the procedure, chronic pain that can last long after the beak tissues have healed, and impaired feeding behaviour and parasite removal due to alteration of size, shape and conformation of the beak, as well as uneven mandible length or crossed mandibles following regrowth.[1]

The potential for acute and chronic pain depends on the specific procedure used, the severity of the trim, and the age at which it is performed. The most common procedures for beak shortening are hot blade (HB) trimming and infrared beak treatment (IRBT). Most research on these procedures has been done on laying hens, although it is generally assumed that the effects on welfare are similar for other birds. Hot blade (HB) trimming involves cutting and cauterising the beak using a device with a guillotine-style heated blade. The bird's beak is manually placed on a guard bar or through an orifice that determines the proportion of upper and lower mandibles that are removed. The severity and amount of tissue damage incurred are affected by the design and adjustment of the device, blade temperature and duration of cautery. Hot blade trimming may be performed at the hatchery or on the farm at various ages and results in an open wound for several days. HB trimming results in reduced growth rate and behavioural changes (reduced activity, reduced feeding, reduced pecking, guarding behaviour) that are suggestive of acute pain. Evidence for chronic pain suggests that when HB trimming is performed on older laying hens it causes neuromas, with spontaneous discharge of pain signals. The likelihood of chronic pain appears to be highly dependent on age and extent of the tissue removed, whereas acute pain appears to result from HB trimming at any age.

Infrared beak treatment (IRBT) involves exposing the beak tip to an infrared beam using a specialised automated device. The machine can be adjusted for plate (orifice) size and power settings so that specific protocols can be used for size and strain of the chick (beak pigmentation absorbs energy at different rates), amount of beak tissue exposed (determining severity of the trim) and amount of energy applied (Schwean-Lardner, 2018). The infrared energy penetrates the outer surface of the beak and damages the underlying dermis resulting in necrosis and sloughing of the tissue within a few weeks. There is evidence suggesting that reduced feeding efficiency and acute pain also result from this procedure. However, in most studies that compared IRBT to HB, the negative effects of IRBT are usually intermediate between HB and untreated birds (reviewed by McKeegan, 2017). Most studies have reported better long-term outcomes following IRBT compared to HB in terms of beak length, beak abnormalities and uneven regrowth. There is little evidence for neuroma formation or chronic pain resulting from IRBT, although few studies have addressed this, in part, because the procedure is only done at hatching. There are some inconsistencies in results reported in the literature that suggest that IRBT machine settings are critical (Schwean-Lardner, 2018; Struthers et al., 2019). Some clear advantages of IRBT over HB include elimination of an open wound, and more gradual erosion of the beak during sloughing, which presumably helps chicks adapt to changes in beak shape. The procedure is also done at the same time as sexing and vaccination, so its use results in less handling stress at the hatchery than HB. These advantages coupled with the differences in evidence for degree of acute and chronic pain, suggest IRBT has reduced impact on welfare compared to HB. This is important since it has largely replaced HB in many regions of the world (McKeegan, 2017).

Procedures used for bill-trimming of ducks are usually done at the hatchery or up to three weeks of age and include IRBT, trimming by cold-cutting with scissors, hot blade trimming with cautery and tip searing, which involves holding the bill against a hot cautery blade for a few seconds (Gustafson et al., 2007a). One study comparing HB and searing methods on Pekin ducks indicated that both methods reduced feeding, drinking and preening and increased resting compared to untrimmed controls during the first two weeks, but no evidence of neuromas and substantially improved feather condition in the long term (Gustafson et al., 2007a). Searing resulted in less weight loss and less scar tissue. Cold-cutting the bills of Muscovy ducks at 20 days post-hatch decreased bill-related behaviour, increased resting, and reduced body weight for one week following the procedure (Gustafson et al., 2007b). It seems likely that bill-trimming causes acute but not chronic pain and is effective at preventing feather loss and cannibalism.

Claw and toe removal

In the turkey industry, the claws of the forward-facing toes are routinely reduced or removed to prevent scratching injury and improve carcass condition. Toes are exposed to microwave energy using an automated system at the hatchery which damages tissues, causing them to slough within 1-3 weeks (Fournier et al., 2015). Both tom and hen poults show short-term behavioural changes indicative of acute pain. In one study, trimmed toms walked less than untrimmed toms, but stance and gait scores did not differ at 133 days. Trimmed turkeys show substantially reduced prevalence of scratches measured at processing.

In breeder roosters, removal of the spur and/or outermost joint of the backward facing claw is performed to reduce injuries to hens during mating (Fiks and de Jong, 2007). Although a significant reduction in injuries to hens is attributed to this procedure (Riber, 2017), any short- or long-term effects on pain and mobility in the males are currently unknown.

In many countries the various procedures described above are standard practice, although there are considerable regional differences in their application (Fiks and de Jong, 2007). There is substantial evidence that most of the procedures are effective at reducing injuries, and thus improve overall flock welfare. However, most cause acute pain and some can affect long-term function; all are routinely conducted without any form of pain mitigation. Therefore, there is increasing pressure to find alternative management practices to prevent injuries and legislation banning these procedures is continuously evolving.

Feeding practices that compromise welfare

Moulting of laying hens

As laying hens age, their rates of production and egg quality typically decline. Induced (forced) moulting is a practice used to extend the period of egg production with improvements in egg quality. Moults were traditionally induced by a 10 - 14 day period of total feed withdrawal, a reduced photoperiod, and sometimes, a 2 - 3 day period of water deprivation (Glatz and Tilbrook, 2021). During this time, hens lose bodyweight and feathers, egg laying ceases and mortality rates are high. Over the next couple of weeks, the reproductive tract regenerates, plumage is replaced, and hens come back into lay with relatively high rates of production and improved egg quality. Because of welfare concerns, moulting has been banned in some countries and in others only non-feed withdrawal methods that involve nutrient restriction by limiting feed intake or by reducing nutrient density are allowed. Non-feed withdrawal methods are also controversial, in that a significant loss of weight is necessary for a moult to be effective, hens still appear to be highly feed motivated, and because mortality still occurs (Nicol et al., 2017). Both fasting and non-fasting moults have deleterious effects on bone mineralisation.

Feed restriction of broiler breeders

The parents of meat birds are feed-restricted to slow growth and control weight, with the level of feed restriction being most severe during rearing. Broiler breeders are typically fed a limited amount of feed daily, competition for feed is intense, and flocks can have poor body weight uniformity, with smaller birds at a disadvantage. Alternative feeding strategies, in which the birds are fed larger allocations of feed on alternate days (skip a day) or by reducing feeding days per week with larger allotments on the on-feed days, are used to improve flock uniformity. These feeding schedules are prohibited in some European countries because prolonged feed withdrawal is detrimental for welfare. However, there is experimental evidence that these feeding strategies may be better for welfare, possibly by allowing birds to achieve satiety, at least on some days. Broiler breeder pullets on skip a day (Morrissey et al., 2014) and other non-daily feeding schedules (Arrazola et al., 2019b) have better feather condition, reduced feeding motivation and lower indicators of stress compared to daily fed birds.

Qualitative restriction strategies are also being investigated to mitigate the negative effects of feed restriction by feeding low density, high fibre diets or by including an appetite suppressant in the feed to increase satiety or reduce hunger, respectively. Results from studies are mixed, with some indicating less stereotypic behaviour, better feather condition, reduced feeding motivation and lower indicators of stress (Nicol et al., 2017). However, there is also evidence that breeder pullets find appetite suppressants aversive, and the reduction in hunger that they cause may be due to a feeling of malaise (Arrazola and Torrey, 2019).

Although feed restriction of breeders is also practiced in the turkey and duck industries, no research has evaluated the effects on welfare. It is likely that the birds' experience is similar to that of broiler breeders, with severe feed restriction resulting in chronic hunger, but with improvements in health. Overall, research suggests that some feeding strategies are better than others, but none alleviate the problem. Genetic selection for slower growth leading to reduced appetite will likely be the most effective solution.

Force feeding ducks for foie gras

Foie gras (French: fatty liver) is a gourmet food product traditionally derived from the livers of geese, but now more commonly produced from ducks (Rochlitz and Broom, 2017).. It is induced by force feeding male mule ducks (hybrids of Pekin and Muscovy ducks) a high carbohydrate diet which dramatically increases the size and fat content of the liver. The procedure involves restraining ducks and inserting a tube into the esophagus to dispense the feed, 2 - 3 times per day. The ducks are feed-restricted for a part of their 14-week growing period and force fed during the last 2–3 weeks. Welfare concerns include distress and sometimes injuries caused by restraint and intubation, compromised health due to liver dysfunction and mortality (Rochlitz and Broom, 2017). This practice is highly contentious and prohibited in many countries, although the product is still often imported.

Routine killing of male chicks

Male chicks of egg-producing lines have no use in egg production and are not economically viable for use in meat production. Therefore, billions of newly hatched chicks are routinely killed at hatcheries each year, which is a highly controversial practice. The two main methods for killing these chicks are maceration with a specially designed mechanical device and immersion in carbon dioxide. When conducted properly, maceration is reported to cause instantaneous death and is considered to be a humane killing method by many authorities, although it is aesthetically displeasing (Jongman and Fisher, 2021). Immersion in carbon dioxide causes rapid (<10s), but not immediate, loss of consciousness, and there are concerns that exposure to the high concentration of gas required for the newly hatched chick may cause a brief period of pain and distress. Regardless of method, the routine killing of viable, healthy chicks is of major ethical concern. A considerable amount of research and development into practical methods for sexing of chick embryos in the egg, ideally prior to incubation, are underway. This technology will eliminate the need to incubate and hatch male chicks altogether and may even allow for using those eggs for consumption.

Conclusions

Genetic selection for highly productive traits coupled with barren, high-density housing has resulted in numerous welfare issues for poultry. Increasingly, public concern for poultry welfare is leading to legislation or market demands that are changing poultry farming practices. For example, housing systems for laying hens have been in the public eye for decades, leading to the increasing adoption of housing systems that provide appropriate furnishings that better support the behavioural needs of the birds and increase their quality of life. Public attention on welfare problems caused by genetic selection for productivity is much more recent, resulting in commitments by food service and retailers to only purchase broiler chickens with better welfare, which includes slower growth as well as enriched housing. How these concessions towards animal welfare evolve in the future is yet to be seen. Some husbandry procedures have been eliminated or refined to less painful methods (e.g., infrared beak trimming). Although beak trimming has been banned in some countries, the consequences of not using beak trimming on hen welfare requires increased research in ways to mitigate feather pecking, including breeding and management. As management practices change, we can expect new welfare issues to emerge. It must also be acknowledged that most practices to improve welfare come with increased cost of production and therefore an increase in cost of product.

Note

1 In the avian/poultry literature, the mandible and maxilla are referred to as upper/lower mandibles. This terminology is used here.

References

Anderson KE, Havenstein GB, Jenkins PK and Osborne J, 2013. Changes in commercial laying stock performance, 1958–2011: Thirty-seven flocks of the North Carolina random sample and subsequent layer performance and management tests. *World's Poultry Science Journal*, 69(3), pp. 489–514, https://doi.org/10.1017/S0043933913000536.

Armstrong EA, Rufener C, Toscano MJ, Eastham JE, Guy JH, Sandilands V, Boswell T and Smulders TV, 2020. Keel bone fractures induce a depressive-like state in laying hens. *Scientific Reports*, pp. 1–14, https://doi.org/10.1038/s41598-020-59940-1.

Arrazola A and Torrey S, 2019. Conditioned place avoidance using encapsulated calcium propionate as an appetite suppressant for broiler breeders. *PLoS ONE*, 14(7), pp. 1–15.

Arrazola A and Torrey S, 2021. Welfare and Performance of Slower growing Broiler Breeders during Rearing. *Poultry Science*, p. 101434, https://doi.org/10.1016/j.psj.2021.101434.

Arrazola A, Mosco E, Widowski TM, Guerin MT, Kiarie EG and Torrey S, 2019a. The effect of alternative feeding strategies for broiler breeder pullets: 1. Welfare and performance during rearing. *Poultry Science*, 98(9), pp. 3377–3390, https://doi.org/10.3382/ps/pez170.

Arrazola A, Mosco E, Widowski TMM, Guerin MTT, Kiarie EGG and Torrey S, 2019b. The effect of alternative feeding strategies for broiler breeder pullets: 2. Welfare and performance during lay. *Poultry Science*, 98(9), pp. 6205–6216, https://doi.org/10.3382/ps/pez447.

Bain MM, Nys Y and Dunn IC, 2016. Increasing persistency in lay and stabilising egg quality in longer laying cycles. What are the challenges? *British Poultry Science*, 57(3), pp. 330–338, https://doi.org/10.1080/00071668.2016.1161727.

Barrett L, Malecki I and Blache D, 2019. Differences in pre-laying behavior between floor-laying and nest-laying pekin ducks. *Animals*, 9(2), pp. 1–14, https://doi.org/10.3390/ani9020040.

Campbell DLM, Bari MS and Rault JL, 2021. Free-range egg production: its implications for hen welfare, *Animal Production Science*, 61(10), pp. 848–855, https://doi.org/10.1071/AN19576.

Chen X, Shafer D, Sifri M, Lilburn M, Karcher D, Cherry P, Wakenell P, Fraley S, Turk M and Fraley GS, 2021. Centennial review: History and husbandry recommendations for raising Pekin ducks in research or commercial production. *Poultry Science*, 100(8), p. 101241, https://doi.org/10.1016/j.psj.2021.101241.

Cooper JJ and Albentosa MJ, 2003. Behavioural priorities of laying hens. *Avian and Poultry Biology Reviews*, 14(3), pp. 127–149, http://eprints.lincoln.ac.uk/1007/%5Cnhttp://openurl.ingenta.com/content/xref?genre=article&issn=1470-2061&volume=14&issue=3&spage=127.

Dalton HA, Wood BJ and Torrey S, 2013. Injurious pecking in domestic turkeys: Development, causes, and potential solutions. *World's Poultry Science Journal*, 69(4), pp. 865–876, https://doi.org/10.1017/S004393391300086X.

Dawson LC, Widowski TM, Liu Z, Edwards. AM and Torrey S, 2021. In pursuit of a better broiler: A comparison of the inactivity, behavior, and enrichment use of fast- and slower-growing broiler chickens. *Poultry Science*, https://doi.org/10.1101/2021.05.18.444545.

Erasmus MA, 2018. Welfare issues in turkey production. In Mench JA, ed., *Advances in Poultry Welfare*. Duxford: Woodhead Publishing, pp. 263–291, https://doi.org/10.1016/B978-0-08-100915-4.00013-0.

Estevez I, Andersen IL and Nævdal E, 2007. Group size, density and social dynamics in farm animals. *Applied Animal Behaviour Science*, 103(3–4), pp. 185–204, https://doi.org/10.1016/j.applanim.2006.05.025.

Fernyhough M, Nicol CJ, van de Braak T, Toscano MJ and Tønnessen M, 2020. The Ethics of Laying Hen Genetics. *Journal of Agricultural and Environmental Ethics*, 33(1), pp. 15–36, https://doi.org/10.1007/s10806-019-09810-2.

Fiks T and de Jong I, 2007. Mutilations in poultry in European poultry production systems. *Lohmann Information*, 42(1), pp. 35–46, https://www.lohmann-information.com/content/l_i_42_2007-04_artikel5.pdf.

Food and Agriculture Organisation of the United Nations (FAO), 2021. Crops and livestock products. *FAOSTAT Survey Data*, September 28, 2021, http://www.fao.org/faostat/en/#data/QCL (Accessed 2021-09-29).

Fournier J, Schwean-Lardner K, Knezacek TD, Gomis S and Classen HL, 2015. The effect of toe trimming on behavior, mobility, toe length and other indicators of welfare in tom turkeys. *Poultry Science*, 94(7), pp. 1446–1453, https://doi.org/10.3382/ps/pev112.

Gebhardt-Henrich SG, Toscano MJ and Würbel H, 2018. Use of aerial perches and perches on aviary tiers by broiler breeders. *Applied Animal Behaviour Science*, 203(November 2017), pp. 24–33, https://doi.org/10.1016/j.applanim.2018.02.013.

Glatz PC and Tilbrook AJ, 2021. Welfare issues associated with moulting of laying hens. *Animal Production Science*, 61(10), pp. 1006–1012, https://doi.org/10.1071/AN19700.

Gustafson LA, Cheng HW, Garner JP, Pajor EA and Mench JA, 2007a. Effects of bill-trimming Muscovy ducks on behavior, body weight gain, and bill morphopathology. *Applied Animal Behaviour Science*, 103(1–2), pp. 59–74.

Gustafson LA, Cheng HW, Garner JP, Pajor EA and Mench JA, 2007b. The effects of different bill-trimming methods on the well-being of Pekin ducks. *Poultry Science*, 86(9), pp. 1831–1839, https://doi.org/10.1093/ps/86.9.1831.

Jones TA, Waitt CD and Dawkins MS, 2009. Water off a duck's back: Showers and troughs match ponds for improving duck welfare. *Applied Animal Behaviour Science*, 116(1), pp. 52–57, https://doi.org/10.1016/j.applanim.2008.07.008.

de Jong IC and Guémené D, 2011. Major welfare issues in broiler breeders. *World's Poultry Science Journal*, 67(1), pp. 73–82, https://doi.org/10.1017/S0043933911000067.

Jongman EC and Fisher AD, 2021. Euthanasia of laying hens: an overview. *Animal Production Science*, 61(10), pp. 1042–1047, https://doi.org/10.1071/AN20224.

Karcher DM and Mench JA, 2018. Overview of commercial poultry production systems and their main welfare challenges. In Mench JA, ed., *Advances in Poultry Welfare*. Woodhead Publishing, Duxford, UK, pp. 3–25, https://doi.org/10.1016/C2015-0-04880-8.

Kim WK, Bloomfield SA, Sugiyama T and Ricke SC, 2012. Concepts and methods for understanding bone metabolism in laying hens. *World's Poultry Science Journal*, 68(1), pp. 71–82, https://doi.org/10.1017/S0043933912000086.

Lay Jr. DC, Fulton RM, Hester PY, Karcher DM, Kjaer JB, Mench JA, Mullens BA, Newberry RC, Nicol CJ, O'Sullivan NP and Porter RE, 2011. Hen welfare in different housing systems. *Poultry Science*, 90, pp. 278–294, https://doi.org/10.3382/ps.2010-00962.

Marchewka J, Vasdal G and Moe RO, 2019. Identifying welfare issues in turkey hen and tom flocks applying the transect walk method. *Poultry Science*, 98(9), pp. 3391–3399, https://doi.org/10.3382/ps/pez211.

McKeegan D, 2017. Beak trimming of laying hens: welfare costs and benefits. In *Achieving Sustainable Production of Eggs Volume 2*. Cambridge: Burleigh Dodds Science Publishing Limited, pp. 125–144, https://doi.org/10.19103/AS.2016.0012.26.

Mench JA, 2009. Behaviour of fowl and other domesticated birds. In Jensen P, ed., *The Ethology of Domestic Animals: An Introductory Text*. Wallingford: CABI, pp. 121–136.

Mench JA and Blatchford RA, 2014. Determination of space use by laying hens using kinematic analysis. *Poultry Science*, 93(4), pp. 794–798, https://doi.org/10.3382/ps.2013-03549.

Morrissey KLH, Widowski T, Leeson S, Sandilands V, Arnone A and Torrey S, 2014. The effect of dietary alterations during rearing on feather condition in broiler breeder females. *Poultry Science*, 93(7), pp. 1636–1643, https://doi.org/ 10.3382/ps.2013-03822.

Nicol CJ, Bouwsema J, Caplen G, Davies AC, Hockenhull J, Lambton SL, Lines JA, Mullan S and Weeks CA, 2017. *Farmed Bird Welfare Science Review*. Melbourne, Victoria, https://agriculture.vic.gov.au/__data/assets/pdf_file/0008/529829/Farmed-Bird-Welfare-Science-Review.pdf.

Pedersen IJ, Tahamtani FM, Forkman B, Young JF, Poulsen HD and Riber AB, 2020. Effects of environmental enrichment on health and bone characteristics of fast growing broiler chickens. *Poultry Science*, 99(4), pp. 1946–1955, https://doi.org/10.1016/j.psj.2019.11.061.

Preisinger R, 2018. Innovative layer genetics to handle global challenges in egg production. *British Poultry Science*, 59(1), pp. 1–6, https://doi.org/10.1080/00071668.2018.1401828.

Quinton CD, Wood BJ and Miller SP, 2011. Genetic analysis of survival and fitness in Turkeys with multiple-trait animal models. *Poultry Science*, 90(11), pp. 2479–2486, https://doi.org/10.3382/ps.2011-01604.

Riber AB, 2017. *Alternatives to Mutilation of the Outermost Joint of the Backward-facing Toe in Broiler Breeder Males*. Aarhus, Denmark: DCA: Nationalt Center for Fødevarer og Jordbrug, https://doi.org/10.7120/09627286.26.2.135.

Riber AB, van de Weerd HA, de Jong IC and Steenfeldt S, 2018. Review of environmental enrichment for broiler chickens. *Poultry Science*, 97(2), pp. 378–396, https://doi.org/10.3382/ps/pex344.

Rochlitz I and Broom DM, 2017. The welfare of ducks during foie gras production. *Animal Welfare*, 26(2), pp. 135–149, https://doi.org/10.7120/09627286.26.2.135.

Rodenburg TB, Bracke MBM, Berk J, Cooper J and Faure JM, (2005). The welfare of ducks in European duck husbandry systems. *Worlds Poultry Science*, 61, 633–646.

Rufener C and Makagon MM, 2020. Keel bone fractures in laying hens: a systematic review of prevalence across age, housing systems, and strains. *Journal of Animal Science*, 98(1), pp. S36–S51, https://doi.org/10.1093/jas/skaa145.

Santos MN, Rothschild D, Widowski TM, Barbut S, Kiarie EG, Mandell I, Guerin MT, Edwards AM and Torrey S, 2021. In pursuit of a better broiler: Carcass traits and muscle myopathies in conventional and slower-growing strains of broiler chickens. *Poultry Science*, 100(9), p. 101309, https://doi.org/10.1016/j.psj.2021.101309.

Schuck-Paim C, Negro-Calduch E and Alonso WJ, 2021. Laying hen mortality in different indoor housing systems: a meta-analysis of data from commercial farms in 16 countries. *Scientific Reports*, 11(1), https://doi.org/10.1038/s41598-021-81868-3.

Schwean-Lardner K, 2018. The effects of hatchery practices on the welfare of poultry. In *Advances in Poultry Welfare*. Duxford: Woodhead Publishing, pp. 29–48, https://doi.org/10.1016/B978-0-08-100915-4.00002-6.

Shini A, Shini S and Bryden WL, 2019. Fatty liver haemorrhagic syndrome occurrence in laying hens: impact of production system. *Avian Pathology*, 48(1), pp. 25–34, https://doi.org/10.1080/03079457.2018.1538550.

van Staaveren N and Harlander A, 2020. Cause and prevention of injurious pecking in chickens. In Nicol C, ed., *Understanding the Behaviour and Improving the Welfare of Chickens*. Cambridge: Burleigh Dodds Science Publishing Limited, pp. 509–566.

van Staaveren N, Ellis J, Baes CF and Harlander-Matauschek A, 2021. A meta-analysis on the effect of environmental enrichment on feather pecking and feather damage in laying hens. *Poultry Science*, 100(2), pp. 397–411, https://doi.org/10.1016/j.psj.2020.11.006.

Struthers S, Classen HL, Gomis S and Schwean-Lardner K, 2019. The effect of beak tissue sloughing and post-treatment beak shape on the productivity of infrared beak-treated layer pullets and hens. *Poultry Science*, 98(9), pp. 3637–3646, https://doi.org/10.3382/ps/pez230.

Torrey S, Mohammadigheisar M, Nascimento dos Santos M, Rothschild D, Dawson LC, Liu Z, Kiarie EG, Edwards AM, Mandell I, Karrow N, Tulpan D and Widowski TM, 2021. In pursuit of a better broiler: growth, efficiency, and mortality of 16 strains of broiler chickens. *Poultry Science*, 100(3), p. 100955, https://doi.org/10.1016/j.psj.2020.12.052.

Weeks CA and Nicol CJ, 2006. Behavioural needs, priorities and preferences of laying hens. *World's Poultry Science Journal*, 62(2), pp. 296–307, https://doi.org/10.1079/WPS200598.

Weeks CA, Lambton SL and Williams AG, 2016. Implications for welfare, productivity and sustainability of the variation in reported levels of mortality for laying hen flocks kept in different housing systems: A meta-analysis of ten studies. *PLoS ONE*, 11(1), pp. 1–15, https://doi.org/10.1371/journal.pone.0146394%0A.

Widowski T and Torrey S, 2018. Rearing young birds for adaptability. In Mench, JA, ed., *Advances in Poultry Welfare*. Duxforn, UK: Woodhead Publishing, pp. 49–76, https://linkinghub.elsevier.com/retrieve/pii/B9780081009154000038.

Widowski TM, Casey-Trott TM and Morrissey K, 2017. Welfare of laying hens: an overview. In Roberts, J, ed., *Achieving Sustainable Production of Eggs Volume 2: Animal Welfare and Sustainability*. Cambridge: Burleigh Dodds Science Publishing Limited, pp. 57–84, https://doi.org/10.19103/AS.2016.0012.22.

Zuidhof MJ, Schneider BL, Carney VL, Korver DR and Robinson FE, 2014. Growth, efficiency, and yield of commercial broilers from 1957, 1978, and 20051. *Poultry Science*, 93(12), pp. 2970–2982, https://doi.org/10.3382/ps.2014-04291.

6
FARMING PIGS

Sandra Edwards

Introduction

Pigs provide the greatest weight of meat produced worldwide for human consumption and are farmed on all continents, with the largest populations found in China, Europe, and North America (FAO, 2020). They may be produced in a variety of different systems, depending on the climatic and socioeconomic circumstances of the region. However, they are increasingly farmed in intensive systems which are relatively similar in basic characteristics in all countries (Cameron, 2000). The evolutionary biology of the pig, as an omnivore ranging across many different habitats, has given it great flexibility and adaptability. However, it has also given it genetically engrained behavioural dispositions relevant to its welfare in farmed conditions (Edwards and Grand, 2021). This chapter will describe the reasons for the development of current pig farming methods, the different welfare challenges which can exist in extensive and intensive systems, and the practical ways in which these are being addressed.

Extensive farming systems

A minor part of the world pig population is still produced in more extensive systems which might be considered close to the evolutionary habitat of the species. In subsistence farming, animals may wander freely as scavengers. In Mediterranean countries, traditional silvopastoral systems with indigenous breeds are used for the production of high-value cured product (García-Gudiño et al., 2020). Even in modern production, the reintroduction of agro-forestry systems is being investigated to give benefits of dual land use for energy biomass and meat production (Jørgensen et al., 2018). Extensive systems are also adopted in organic farming, which represents a small but growing sector (Früh et al., 2014). Organic production standards stipulate that animals must have access to the outdoors, although this may sometimes be a concrete outrun rather than pasture. The keeping of pigs at pasture is also seen in some conventional farms, most commonly for breeding animals but also for rearing animals in some smallholder systems or in specialised label production (Edwards, 2005).

Outdoor living

Whilst such extensive systems offer the pig great behavioural freedom, with the ability to exercise choice amongst diverse locations and exhibit the important species-specific behaviours of

DOI: 10.4324/9781003182351-8

exploration and natural foraging, they are not without their welfare issues (Edwards et al., 2014). High amongst these is the exposure to climatic extremes, which in different seasons may subject the animals to extreme cold with freezing of water supply, extreme heat with risk of sunburn, or extreme rainfall with muddy underfoot conditions and lack of dry resting areas (Edwards, 2005). To counter these challenges, the provision of appropriate shelter is essential. For cold or wet weather this needs to allow the animal to access a dry, draught-free microclimate for rest, usually provided by a weatherproof structure with plentiful bedding such as straw. Such shelter is particularly important for farrowing sows, as their newborn piglets are very susceptible to cold stress. To counter high temperatures, the provision of shade from the radiant heat of the sun is very important as the thin hair coat of pigs makes them very susceptible to sunburn. The provision of muddy areas also allows them to express their natural behaviour of wallowing to coat the skin with wet mud, promoting evaporative cooling and leaving a protective layer of sunscreen.

A second area of welfare challenge for outdoor pigs relates to biosecurity and the maintenance of health (Delsart et al., 2020). Disease agents (those for African Swine Fever and Salmonella being topical examples) can be carried by wild animals and birds, whose ingress into open pig areas is difficult to prevent. Many of the endemic diseases seen in intensive indoor farming (for example, enzootic pneumonia) may be less prevalent in outdoor conditions due to decreased animal density. However, if an animal does become sick it is more challenging for stockpeople to detect this at an early stage and to catch and restrain an individual animal for treatment. Suffering due to untreated health problems is therefore a significant welfare risk in extensive systems. A particular issue for outdoor animals is infestation with parasites, which can survive for extended periods in the environment and set up a cycle of continuous reinfection. Regular antiparasitic treatment can control the level of infestation, although such an intervention is restricted in organic production and frequent movement to fresh uncontaminated pasture is consequently very important. Similarly, the build-up of disease agents inside farrowing huts which might infect vulnerable piglets can be minimised by relocation of the huts between each farrowing and removal or burning of old bedding.

A third area of welfare challenge is that of predation (Pietrosemoli and Tang, 2020). Whilst in many parts of the world exposure to large carnivores is rare, small carnivores such as foxes can prey upon piglets and, by disturbing farrowing sows, can also increase the risk of neonatal crushing. Scavenging and predatory birds can inflict pecking damage or even carry off small piglets venturing into exposed open areas.

A final area worthy of mention is that of the limitations which can be inadvertently or deliberately imposed on behavioural freedom. Although it may appear that many behaviours are possible in extensive conditions, their expression may be constrained by unregulated social competition or human intervention. Thus, access to resources may be difficult for low-ranking animals if they can be guarded by a more dominant individual. For example, if the size of the shelter provided is too small or the entrance poorly designed, or if food is not distributed over a sufficiently wide area, the more timid individuals within a group may suffer both fear and deprivation. Behavioural restriction may also be deliberately imposed by the contentious human intervention of nose-ringing to prevent rooting behaviour (Horrell et al., 2001). This is a highly motivated behaviour of pigs, but can cause extensive damage to pasture, giving rise to bare, overturned ground in a very short time when stocking density is high. Such paddocks can rapidly become a sea of mud in wet weather, giving poor living conditions for the animals and poor working conditions for stockpeople. This removal of vegetation also results in detrimental consequences for environmental pollution and for production from the following crop. These include soil compaction and increased loss of nutrients through gaseous emissions from the unprotected surface and leaching in rainwater, with consequent eutrophication of nearby watercourses. Since nose-ringing causes both acute pain and chronic behavioural frustration, the balance of the different welfare and environmental harms needs to be carefully considered in each individual situation.

Intensive farming systems

When farmed pigs are kept indoors, they are subject to much greater control over all aspects of their life. This can have welfare benefits, since they can be provided with a carefully regulated and balanced diet, a controlled thermal environment, biosecurity measures to minimise risk of disease ingress and easy treatment of sick individuals. However, since the animals now have little autonomy in behavioural choices, the correct delivery of all these benefits is dependent on the management and husbandry skills of the humans in control. In subsistence farming situations, the necessary resources for optimal provision may be lacking, whilst in small hobby or part-time farms, human skills may be lacking through poor knowledge or limited experience. Benefits should be better realised as pig production enterprises become more professional but, as farms become ever larger, the ability to recruit, train, and retain good staff can again be a serious limitation on welfare. To overcome labour shortage and reduce cost, modern farms increasingly seek to automate many of the processes necessary to rear pigs. Whilst automation can reduce the need for many heavy and routine jobs, such as climate control, feed and water provision, and manure removal, it cannot yet replace the skilled stockperson for rapid problem detection and solution. A high reliance on automation leads to serious problems in many animals when breakdowns occur. Furthermore, when many animals are looked after by few people, they may have insufficient time to give adequate attention to all individuals, leading to increased suffering associated with undetected problems. The continual market pressure on pig farmers to produce cheap meat has resulted in widespread adoption of genetic, housing, and management innovations that increase financial efficiency. This can be beneficial for welfare, as animals which are in good health and without stress will be most biologically productive (Edwards et al., 2006). However, there are also many situations that bring profitability into conflict with animal welfare.

Space restriction in breeding sows

Perhaps the greatest area of welfare concern in intensive housing is the minimisation of expensive building space and the consequent restriction of movement to which animals are subject. This is most pronounced in the case of the breeding sow, where the desire to make efficient use of space while ensuring each animal received its own feed ration and preventing injurious aggression between animals led to the widespread adoption of individual gestation stalls. Whilst these deliver their specific production and welfare objectives, they impose great physical and behavioural restrictions on the sow (Marchant-Forde, 2009). The lack of movement weakens bones and muscles, and unstimulated sows may develop characteristics of apathy or depression. Most noticeably, the prevention of motivated foraging behaviours can result in behavioural abnormalities which develop into stereotypies such as bar biting and sham chewing. This problem is particularly severe in the case of the gestating sow, since she is normally fed a concentrate diet at a level which is optimal for good health and production but does not induce satiety. In consequence, foraging behaviours are induced but cannot be functionally expressed in restrictive stall housing, resulting in a situation where anomalous oral behaviours can become channelled into a neural pathology (Lawrence and Terlouw, 1993). Whilst some alleviating measures can be introduced in stalled conditions, by providing foraging substrate to permit more appropriate behavioural expression or increasing dietary fibre to promote satiety (Meunier-Salaun et al., 2001), this gives only a partial solution. In consequence, a growing number of countries have now implemented legislation to ban or restrict the use of gestation stalls. Exemptions are often given for the first month of gestation, on the basis that this allows a recovery period for sows in poor body condition after lactation, provides protection during the oestrus period to weak sows which might otherwise be injured by the riding activity of bigger animals, and protects sows from social competition stress during the critical implantation period when developing embryos may be adversely affected. However, as knowledge and experience on the optimal management for group-housed sows increases (Spoolder et al., 2009), such exemptions are now being questioned.

A second period when close confinement is still widespread, but being increasingly questioned, is for the farrowing sow (Baxter and Edwards, 2021). Newborn piglets are very vulnerable, with crushing and hypothermia as major contributory causes to the typical mortality levels of 10–15% in the first days after birth. Confinement of the sow in a farrowing crate limits her movement and slows dangerous posture changes which might crush piglets and cause injury or death. Confinement of the sow also allows the provision of supplementary heating close to the birth site and allows safe access to the newborn piglets for stockperson tasks. These considerations have become increasingly important as genetic selection of modern sows for prolificacy has resulted in greater numbers of piglets born in a litter, with individual piglets being less physiologically mature and therefore more susceptible. Despite these production and welfare benefits for piglets, the farrowing crate does impose significant welfare detriments to the sow. Foremost amongst these is the prevention of proper expression of nest building behaviour. In the days prior to farrowing, a sow in the wild will wander widely while seeking a suitable nest site, and then gather branches and grasses and arrange these into a nest in which to give birth. Effective nest construction has a high survival value in these circumstances, and the associated behaviours have been genetically conserved and are triggered by the hormonal changes preceding parturition. Inability to express these behaviours, due to physical restriction and lack of substrate, induces frustration and stress in the sow which may be detrimental to the farrowing process and subsequent maternal behaviour (Yun and Valros, 2015). Furthermore, as lactation proceeds, normal interaction between the sow and her litter is hindered by the farrowing crate and the sow is unable to escape the increasingly importuning piglets. This can be stressful for the sow and hinders the natural behavioural process of gradual weaning for the piglets. In consequence, the abrupt early weaning which now predominates in commercial practice is much more stressful for piglets (Edwards et al., 2020). Attempts to develop alternatives to the farrowing crate which can increase behavioural freedom for the sow whilst safeguarding piglet survival have proved challenging for economic or other reasons. The imposition of temporary crating for only the farrowing period and a few days thereafter, before allowing greater space for the sow in an individual or group pen, can often give benefits for both welfare and performance, although the problem of frustrated nest building still remains. A few countries have implemented legislation to ban the farrowing crate and have developed appropriate management solutions to still operate successfully. Stimulated by current political pressure, research programmes are now developing and testing new pen designs (Baxter and Edwards, 2021). However, it is evident that these will only succeed when accompanied by appropriate genetic and stockperson inputs, and any concurrent increase in litter size from selection for hyperprolific sows will continue to pose major problems.

Social living in restricted space

Lack of space is not just a welfare issue in individual confinement, but also poses a challenge for group living. In the wild, pigs live in stable family groups which have access to widely spread resources. In contrast, group composition in farms is often of uniform age structure and changed by remixing during the life of the pig, while resources are more localised. Pigs will compete aggressively to establish dominance when encountering new individuals. This usually poses only minor injury risk if there is space to flee, but it can result in more severe injury or even death in older and heavier animals (Marchant-Forde and Marchant-Forde, 2005). Whilst strategies to minimise aggression at mixing exist, they are seldom fully effective and are infrequently adopted in practice (Peden et al., 2018). Maintaining animals in stable groups for as much of their life as possible is the best strategy. This can be difficult when adopting group-housing systems for sows, as they must be regrouped at weaning to ensure contemporary batches and efficient use of space. The alternative housing strategy of large dynamic groups involves regular addition and removal of animals. In this situation, designing partitioned resting areas to accommodate sub-grouping behaviour can assist with gradual integration (Marchant-Forde, 2009). In growing pigs, keeping litters together throughout their lives, whilst behaviourally desirable, is inconsistent with achieving efficient use of pen space and matching feed specifications to pig weight. For this reason, pigs are frequently regrouped into weight matched pens when weaned at 3–5 weeks of age. Pigs in the wild integrate with other litters at about two weeks of age with relatively little aggression, and allowing co-mingling of suckling piglets can reduce their fighting after weaning. However, repeated subsequent mixing is detrimental to both welfare and performance (Peden et al., 2018).

Even in stable groups, aggression can be triggered by competition for resources. Space is an important resource, as pigs need sufficient space to rest undisturbed, to carry out their normal range of behaviours whilst avoiding encounters with aggressive dominants, and to show the signals of submission which avert attack (EFSA, 2005). However, the most important sources of aggressive competition are those associated with resources critical for survival – food and water. Whilst pigs are usually given continuous access to water, inadequacy in the location and number of drinking places, or the flow rate of water delivery, can result in frustration and aggression (Turner et al., 1999). Competition for food is generally the most potent source of aggression. When the amount of food is restricted and provided in discrete daily meals, dominant animals will aggressively displace subordinates and monopolise the feed supply. This is especially problematic in groups of pregnant sows, where the volume of feed provided is small relative to their appetite and a degree of chronic hunger is always present. To allow adequate feed intake by subordinate animals, very wide distribution of feed is the minimum preventative strategy, but protected individual feeding facilities are preferable to ensure equitable feed intake with minimal aggression. These may be provided by lockable individual feeding stalls or computer controlled individual feeding stations (Marchant-Forde, 2009). The latter system allows automated individual recognition and rationing, but requires animals to feed sequentially and must therefore have well-designed gating systems to prevent aggression at entry and exit points. Rearing pigs are also sometimes restricted in feed, but usually at a much less severe level than pregnant sows as the production objective is to achieve fast growth. However, any restriction on access to feed still represents a potential source of aggression and good design of feeding facilities is necessary (Manteca and Edwards, 2009). Where food is rationed, it must be provided in long troughs which allow all animals to eat simultaneously. This requires a trough length of at least 10% more than the combined shoulder width of the pigs, while the provision of head partitions preventing lateral movement gives further benefit. It is most common for growing pigs, especially younger ones, to be fed *ad libitum* ("at will"), with food always present in a dispenser. In this situation they can feed in turn throughout the day and many pigs can theoretically share a single feeding place. However, the existence of diurnal rhythms of activity, and of social facilitation which stimulates animals to feed simultaneously, mean that feeding space is not used with full efficiency. As a result, lower ranking individuals may still have suboptimal access despite theoretically adequate provision, and frustration can cause increased aggression.

Thermal and physical comfort

When animals are maintained in indoor conditions with limited space, another challenge is the maintenance of hygiene. Pigs in the wild will spatially separate their areas for resting, feeding and excretion. However, at high stocking density this becomes impossible and fouling of the resting area will occur. To maintain pen hygiene in these conditions, the keeping of pigs on fully or partially slatted flooring is very common (EFSA, 2005). This offers the welfare benefits of good drainage to maintain dry flooring and of separating pigs from their faeces and urine to reduce infectious challenge. However, other welfare implications of the use of slatted flooring are less positive, since pigs often show a preference for lying on solid floors. In fact, it may be that many other properties of the floor are viewed as more important by the pig (Ducreux et al., 2002). Pigs are very temperature-sensitive and will choose flooring with appropriate thermal characteristics which aid them to conserve or lose heat. Thus, in cold conditions they will seek insulating bedding or thermally resistant flooring, will lie in sternal posture with minimal floor contact and will huddle together to conserve heat. In hot conditions, they will seek heat conducting floors, lie in lateral recumbency to maximise surface area for heat loss and look for wet areas where they can wallow to increase evaporative cooling. In pens where these possibilities are absent, they will wallow in their own excreta as the only cooling option. Such pen fouling will also increase ammonia emission and give impaired air quality, which can cause irritation of the eyes and respiratory tract and exacerbate respiratory diseases in both pigs and stockpeople. Strategies for water misting or cooling the floor surface in hot ambient conditions are therefore beneficial for welfare (Opderbeck et al., 2020).

The nature of flooring is also very important for the physical well-being of the pigs (KilBride et al., 2009). Inadequately designed or maintained flooring is a major cause of injury, particularly lameness caused by physical damage to the legs and claws and exacerbated by associated ingress of infectious agents. Slatted flooring can be injurious when the surface area of the slat is too narrow to support the weight of the pig without causing excessive pressure on the sole of the foot, or when the void area is too large so that claws or dew claws are trapped between slats and twisted or torn. Sharp or abrasive edges to the slats result in even more severe penetrative injuries. Solid floors can also be injurious if rough and abrasive, causing claw and skin lesions when moving or changing posture, or very smooth and slippery, increasing the risk of falling and twisting joints during locomotion or vigorous social interaction (EFSA, 2005). In sows, deep-bedded flooring or lack of exercise may lead to lameness from hoof overgrowth due to lack of wear. Hoof disorders in breeding animals, which result in gait abnormality, discomfort, reduced mobility, and production loss, can be remedied by hoof trimming, although this is seldom a routine practice in sow herds (Tinkle et al., 2017). Hard flooring can also cause pressure injuries during prolonged lying. Adventitious bursitis is characterised by a fluid filled swelling which develops when pressure on lymphatic vessels and capillaries causes trauma. In more severe cases, these swellings may become infected or lesioned and develop into deep ulcers. In growing pigs, the most common locations to observe such pressure injuries are around the hock. They occur with high prevalence when pigs are kept on unbedded and slatted flooring (KilBride et al., 2009) and, whilst they seldom cause clinical problems or apparent pain, the extent to which they are associated with discomfort during resting on hard floors is uncertain. A more severe manifestation of a pressure injury associated with hard flooring is seen in the development of deeply ulcerated lesions on the shoulder in lactating sows (Rioja-Lang et al., 2018). These are more likely to develop in situations where poor body condition reduces the cushioning fat layer over the bony prominence of the scapula, or where ill health or lameness increases the time spent in lateral recumbency. The pain associated with such lesions is likely to depend on their degree of severity, and the extent of any associated local or systemic infection.

Barren environments

Another consideration of very great significance if pigs are kept on slatted flooring for liquid manure handling is the inability to provide substrate on the ground for bedding or enrichment. Whilst bedding can aid thermal and physical comfort, these needs can be alleviated by appropriate environmental temperature control and floor choice. However, bedding, even if sparse, also provides a substrate for exploration and foraging behaviours. These behaviours, involving rooting, chewing, and oral manipulation, have had an important evolutionary function and are genetically engrained in pigs (Studnitz et al., 2007). Pigs in the wild can spend more than 50% of daylight hours in such activities, which continue even when they are fully fed. If the pen provides no appropriate substrate for their expression, the behaviours may be redirected to inappropriate pen components or to other animals in the group. This can lead to serious adverse outcomes as discussed below. In part-slatted pens, small amounts of chopped straw or similar material can be provided daily on the solid area without too much risk of entering the slurry and disrupting the function of automated pumping systems. In fully slatted pens, substrate can be offered from racks or foraging towers, but to reduce cost and minimise labour demand it is more common to offer alternative enrichment in the form of balls or hanging items such as wood, plastic toys, or chains. These often fail to incorporate the desirable characteristics for sustaining pig interest (EC, 2016). Items provided on the floor soon become soiled, while even hanging items rapidly loose attraction as the pigs become habituated to them. Changing items frequently can help to maintain pig interest, but it is items which are easily destructible and thus need more frequent replacement which provide the most sustained occupation.

One of the reasons that enrichment is considered so important for pig welfare is that inadequate provision, which fails to satisfy the need for foraging and exploration, can lead to the damaging behaviour of tail biting (Valros, 2018). This is a behaviour which occurs widely in growing pigs of all ages, in which the tail may initially be gently chewed before progressing to more vigorous biting, drawing blood and developing into cannibalism of tissue extending up into the spinal cord. Once blood is present, other pigs are attracted and also develop the behaviour, until an uncontrollable outbreak spreads across the whole group, giving rise to serious injury and even death. Whilst the motivational basis for such behaviour is still poorly understood, it is now apparent that different forms of onset can be observed (Taylor et al., 2010). The causation is also highly multifactorial and more than 80 different risk factors have been identified, including barren housing conditions, nutritional inadequacy, thermal discomfort, frustration arising from social competition, genetic predisposition, and poor health. Other injurious behaviours of ear biting and flank biting may also occur and, while less well studied, are believed to share some of the same causal factors.

Harmful human interventions

For reasons of production enhancement or avoidance of undesirable behaviours with detrimental outcomes for animal welfare, a number of invasive procedures are commonly carried out on farmed pigs. These may be applied in all farming systems but tend to be more common in intensive production. One group of these are often referred to collectively as mutilations or piglet processing (Prunier et al., 2020). They are usually carried out in the first days of life, when the piglet is small and easy to handle, although the reasons for doing them may relate to much later events. The first procedure is tooth resection. Newborn piglets will compete aggressively for the best teats until a stable teat order is formed in the first day of life and, if the settled teat order is subsequently disrupted due to a shortage of milk or to cross-fostering of piglets between litters, this competition will be rekindled. To aid them in procuring and defending a teat, piglets are born with sharp incisor and canine teeth, which can inflict serious wounds on the faces of littermates, cause lesions and discomfort to the udder of the sow, and disturb nursing. To prevent such consequences, it is common to remove the upper portion of these teeth by clipping or grinding. It is uncertain to what extent this causes the piglet pain if done with care, as behavioural and physiological indicators have proved inconsistent. However, molecular markers do suggest pain from both resection methods, while incorrectly applied procedures can cause splintering of the teeth, gum damage, and provide a route for infection. It is therefore recommended that tooth resection is selectively applied only when high-risk litters are identified and it can be justified by a welfare balance.

A second procedure commonly performed in the first week of life is docking of the piglet's tail, typically removing between one and two-thirds of the tail using clippers or thermal cautery. This is done to reduce the risk of injury from tail biting in later life, as it has been demonstrated that a short tail is less likely to be bitten. However, cutting of the tail is usually done without anaesthetic and has been demonstrated to be acutely painful to the piglet. There is also evidence to suggest the possibility of long-term pain from the amputated tail stump, although this is less certain (Prunier et al., 2020). Many farmers believe that, given the current inability to reliably prevent tail biting outbreaks, the harms of early life docking are much less than those from serious biting lesions in later life, and that the procedure is therefore justified. However, others contend that it merely absolves the farmer from addressing the root causes of tail biting problems, which are in themselves welfare problems for the animal (Valros, 2018).

A third type of procedure which may be done in young piglets, or sometimes later in life, relates to animal identification. Young piglets may be identified by ear notching, ear tattooing or the insertion of ear tags. The insertion of ear tags may also be carried out when pigs are older, since breeding animals need to be individually identified for efficient record keeping, or may need an ear tag transponder for electronic feeding systems. Finally, pigs require identification of farm of origin when transported for slaughter. This is usually done by slap marking to give a shoulder tattoo shortly before leaving the farm. All of these identification procedures are again carried out without pain relief and, whilst less severe than other interventions, do cause some short-term pain and handling stress to the pig (Prunier et al., 2020).

Finally, the most painful of the surgical interventions is the castration of male animals. Whilst a few countries produce pigmeat from entire males, this carries the risk of boar taint – an unpleasant odour and taste associated with compounds resulting from production of male sex hormones once the animal reaches puberty. The risk of taint increases as the animal ages and, since slaughtering at a young age is economically (and environmentally) undesirable, it is commercially preferable to castrate male piglets. There are also welfare considerations which promote this practice. Since many traditional breeds show puberty at a young age, unplanned breeding among finishing pigs can be a problem. Furthermore, entire male pigs at puberty will show increased aggression and riding behaviour which can cause disruption within the group and injury to recipient animals. Historically, the castration of male piglets has been done in the first week of life by surgical removal of the testes without anaesthesia or analgesia (Prunier et al., 2020). More recently, concerns about the very significant short- and medium-term pain associated with this process have led to the requirement in a growing number of countries to provide some pain relief and to seek alternatives causing less welfare detriment. One such alternative, in situations where rearing entire males is deemed impractical, is the use of immunocastration (Borell et al., 2020). This involves immunisation against the endogenous gonadotrophin releasing hormone (GnRH) of the pig using a "vaccine", typically given as two injections with an interval of at least four weeks and with the second dose given four to six weeks prior to slaughter. With this procedure, the animal assumes the characteristics of a castrate only after the second injection, giving the production benefits of an efficient entire male in early life, but the subsequent behavioural and meat quality benefits of a castrate. The stress from the handling and injection procedures in older pigs would seem preferable in comparison to the pain of surgical castration, but consumer concerns about pharmacological interventions in food producing animals have limited acceptance of this technology in some markets.

In the drive to alleviate welfare problems without need to employ invasive procedures such as those described above, genetic selection tools are being increasingly explored (Rydhmer, 2021). The pig industry has very effectively harnessed the power of genetic selection over many decades to improve production traits. The development of specialist international breeding companies has facilitated selection within large pig populations using sophisticated statistical tools to make simultaneous genetic progress in desirable production traits such as growth rate, feed efficiency, carcass leanness, and sow prolificacy. Such intense focus on high production has sometimes resulted in undesirable side effects, such as increased risk of skeletal, cardiovascular, and immunological impairment in very fast-growing young animals, or increased risk of neonatal compromise and mortality in very prolific sows (Knap and Rauw, 2009). However, health and welfare traits are now receiving much more attention, with exploration of possibilities to select for disease resistance, robustness, low boar taint and good maternal behaviour, and select against aggressiveness and tail-biting predisposition. Whilst of lower heritability than many of the previous production traits, this possibility is now aided by the increasing use of genomic selection tools.

In the longer term, genetic selection may negate the need for other pharmaceutical interventions to improve growth. Whilst exogenous hormones are not used commercially for growth promotion in pigs, some countries outside the EU routinely use beta-agonists which direct the nutrients from food to muscle growth rather than fat deposition, and thus improve economic efficiency and carcass leanness. The physiological effects of these agents have led some to question the welfare implications of their routine use, although few detrimental effects at appropriate dose levels under good management have been proven (Ritter et al., 2017). The prophylactic use of in-feed antibiotics to reduce disease risk and enhance growth, especially in newly weaned pigs, is still widespread in many counties. However, increasing consumer concern about the risk of development of antibiotic resistance within the human food chain, in combination with concern that prophylactic antibiotics may be used to mask other animal welfare problems, has seen the banning or restriction of their use in a growing number of countries or market outlets (de Briyne et al., 2020), and this trend seems likely to continue.

Monitoring and assuring good pig welfare

Historically, it was the empathy that carers have for their animals and their professional pride that motivated attention to the welfare of farm animals. In many situations, this still pertains and is of paramount importance. However, in some cases, cruelty, ignorance, or the profit motive may still result in welfare impairment. To guard against such situations, legislation has been enacted in many countries and, as a result of societal concern, is increasing in detail and scope. A more recent development is the emergence of Farm Assurance schemes, driven by the market, which require farms to operate according to specified standards or codes of practice exceeding legal requirements. Both legislation and farm assurance have generally focussed on production systems and the specification of resource requirements. Whilst these provide important safeguards against known environmental risks, they often fail to adequately consider the important influences of management and husbandry across diverse farming systems. Ultimately, it is the welfare outcome for each individual animal which is of importance, and ways in which this might be objectively assessed using animal-based measures are receiving increasing emphasis. Initiatives

Table 6.1 Animal-based measures which can be used to indicate the welfare state of farmed pigs (modified from Welfare Quality®, 2009, with further additions). Welfare Quality® measures have standardised assessment protocols, but other measures can provide useful indications during stockperson inspections

Welfare principles	*Welfare criteria*	*Animal-based assessment measures*
Good feeding	Absence of prolonged hunger	Body condition score, stereotyped oral behaviours, competition at feeding points
	Absence of prolonged thirst	Competition at drinking points
Good housing	Comfort around resting	Bursitis, shoulder sores, skin cleanliness
	Thermal comfort	Lying posture, panting, skin soiling, shivering, huddling
	Ease of movement	Gait score, slipping
Good health	Absence of injuries	Lameness, body wounds, tail lesions, ear lesions, vulva lesions
	Absence of disease	Coughing, sneezing, laboured breathing, faecal consistency, hernia, prolapse, signs of inflammation (swellings), and infection (discharges)
	Absence of pain induced by management procedures	Docked tails, resected teeth, removed testicles, nose ring, injection abscesses
Appropriate behaviour	Expression of social behaviours	Skin lesions from aggression, wounds from repetitive massage
	Expression of other behaviours	Stereotyped behaviour, interaction with enrichment
	Good human–animal relationship	Approach/withdrawal response to humans
	Positive emotional state	Play behaviour, Qualitative Behaviour Assessment

such as the EU Welfare Quality® project have developed animal-based measures for pigs to inform observers about each of the domains of welfare (Table 6.1). These approaches are now being translated into practical farm assessment protocols for pigs (Pandolfi et al., 2017). Such initiatives are likely to be facilitated in the future by the increasingly sophisticated surveillance tools being developed for farms and abattoirs in Precision Livestock Farming.

No consideration of animal welfare in any production system is complete without consideration of the day-to-day practicalities of human–animal interactions, the benefits of positive interactions and the management of compromised individuals (Hemsworth and Coleman, 2010). Inconsiderate handling makes pigs fearful and induces chronic stress. The provision of appropriate handling facilities for moving and restraining pigs is therefore essential for low-stress operation. The optimal design for raceways, crushes, and loading ramps is well researched (Grandin, 2007), and the monitoring of handling practice by such simple outcome measures as the frequency of distress vocalisations can be very revealing. Further benefits, particularly in breeding animals, can be given by engaging in regular positive interaction, such as brief stroking and gentle speech, which produces quieter and more productive animals. Finally, it must be acknowledged that in any biological system illness and injury will occur in some individuals. The rapid identification of such compromised individuals and appropriate treatment, including pain relief when required, is essential. This necessitates provision of appropriate hospital accommodation where their particular needs for physical and thermal comfort, access to feed and water and protection from social challenge can be ensured. A rapid and objective decision on their response to treatment and their current and future quality of life is necessary, and humane euthanasia should be promptly applied if required.

Conclusions

Pigs can be farmed in many different systems, each of which brings its own set of risks to animal welfare. The level of risk is greater in intensive production systems, where pigs have less autonomy to exercise their innate behavioural predispositions and deal with welfare challenges. Good management and husbandry are therefore essential to identify and control these risks or to alleviate their welfare consequences. To aid in this task and demonstrate successful achievement of good welfare, the assessment of animal-based measures of welfare outcome for all individuals in the population provides an important tool.

References

Baxter EM and Edwards SA, 2021. Optimising sow and piglet welfare during farrowing and lactation. In Edwards SA, *Understanding the Behaviour and Improving the Welfare of Pigs*. Burleigh Dodds, Cambridge, UK, pp. 121–176.

Borell EV, Bonneau M, Holinger M, Prunier A, Stefanski V, Zöls S and Weiler U, 2020. Welfare aspects of raising entire male pigs and immunocastrates. *Animals*, 10, p. 2140.

Cameron RDA, 2000. *A Review of the Industrialisation of Pig Production Worldwide with Particular Reference to the Asian Region*. Food and Agriculture Organization of the United Nations, Rome.

De Briyne N, Iatridou D, Vanderhaeghen W and Ignate K, 2020. Use of antimicrobials in practice (targeted on cattle, pigs, poultry, horses). In Pokludová L, ed., *Antimicrobials in Livestock 1: Regulation, Science, Practice*. Springer, Cham, pp. 43–79.

Delsart M, Pol F, Dufour B, Rose N and Fablet C, 2020. Pig farming in alternative systems: strengths and challenges in terms of animal welfare, biosecurity, animal health and pork safety. *Agriculture*, 10, p. 261.

Ducreux E, Aloui B, Robin P, Dourmad JY, Courboulay V and Meunier-Salaün MC, 2002. Ambient temperature influences the choice made by pigs for certain types of floor. *Journées de la Recherche Porcine en France*, 34, pp. 211–216.

EC, 2016. *Commission Staff Working Document on Best Practices with a View to the Prevention of Routine Tail-docking and the Provision of Enrichment Materials to Pigs*. Brussels, Belgium: European Commission.

Edwards SA, 2005. Product quality attributes associated with outdoor pig production. *Livestock Production Science*, 94, pp. 5–14.

Edwards SA and Grand N, 2021. Behavioral biology of pigs and minipigs. In Coleman K and Schapiro SJ, ed., *Behavioural Biology of Laboratory Animals*. CRC Press, Boca Raton, FL, pp. 243–259.

Edwards SA, English PR and Fraser D, 2006. Animal welfare. In Straw BE, Zimmerman JJ, D'Allaire S, Taylor DJ, eds., *Diseases of Swine*, 9th edn. Blackwell Publishing, Ames, IA, pp. 1065–1073.

Edwards SA, Prunier A, Bonde M and Stockdale EA, 2014. Organic pig production in Europe: animal health, welfare and production challenges. *Organic Agriculture*, 4, pp. 79–81.

Edwards SA, Turpin DL and Pluske JR, 2020. Weaning age and its long-term influence on health and performance. In Farmer C, ed., *The Suckling and Weaned Piglet*. Wageningen Academic Publishers, Wageningen, The Netherlands, pp 225–250.

European Food Safety Authority (EFSA), 2005. The welfare of weaners and rearing pigs: effects of different space allowances and floor types. *The EFSA Journal*, 268, pp. 1–19.

FAO, 2020. *World Food and Agriculture: Statistical Yearbook 2020*. Food and Agriculture Organization of the United Nations, Rome.

Früh B, Bochicchio D, Edwards S, Hegelund L, Leeb C, Sundrum A, Werne S, Wiberg S and Prunier A, 2014. Description of organic pig production in Europe. *Organic Agriculture*, 4, pp. 83–92.

García-Gudiño J, Monteiro ANTR, Espagnol S, Blanco-Penedo I and Garcia-Launay F, 2020. Life cycle assessment of iberian traditional pig production system in Spain. *Sustainability*, 12, p. 627.

Grandin T (ed.), 2007. *Livestock Handling and Transport*. CABI, Wallingford, UK.

Hemsworth PH and Coleman GJ, 2010. *Human-Livestock Interactions: The Stockperson and the Productivity of Intensively Farmed Animals*. CABI, Wallingford, UK.

Horrell RI, A'Ness PJ, Edwards SA and Eddison JC, 2001. The use of nose-rings in pigs: consequences for rooting, other functional activities, and welfare. *Animal Welfare*, 10, pp. 3–22.
Jørgensen U, Thuesen J, Eriksen J, Horsted K, Hermansen JE, Kristensen K and Kongsted AG, 2018. Nitrogen distribution as affected by stocking density in a combined production system of energy crops and free-range pigs. *Agroforestry Systems*, 92, pp. 987–999.
KilBride A, Gillman C, Ossent P and Green L, 2009. Impact of flooring on the health and welfare of pigs. *In Practice*, 31, pp. 390–395.
Knap PW and Rauw WM, 2009. Selection for high production in pigs. In Rauw WM, ed., *Resource Allocation Theory Applied to Farm Animal Production*. CABI, Wallingford, UK, pp. 210–229.
Lawrence AB and Terlouw EMC, 1993. A review of behavioral factors involved in the development and continued performance of stereotypic behaviors in pigs. *Journal of Animal Science*, 71, pp. 2815–2825.
Manteca X and Edwards S, 2009. Feeding behaviour and social influences on feed intake. In Torrallardona D and Roura E, eds., *Voluntary Feed Intake in Pigs*. Wageningen Academic Publishers, Wageningen, pp. 293–306.
Marchant-Forde JN, 2009. Welfare of dry sows. In Marchant-Forde JN, ed., *The Welfare of Pigs*. Springer, Dordrecht, pp. 95–139.
Marchant-Forde JN and Marchant-Forde RN, 2005. Minimizing inter-pig aggression during mixing. *Pig News and Information*, 26, pp. 63N–71N.
Meunier-Salaun MC, Edwards SA and Robert S, 2001. Effect of dietary fibre on the behaviour and health of the restricted-fed sow. *Animal Feed Science and Technology*, 90, pp. 53–69.
Opderbeck S, Keßler B, Gordillio W, Schrade H, Piepho HP and Gallmann E, 2020. Influence of a cooled, solid lying area on the pen fouling and lying behavior of fattening pigs. *Agriculture*, 10, pp. 307.
Pandolfi F, Stoddart K, Wainwright N, Kyriazakis I and Edwards SA, 2017. The "Real Welfare" Scheme: benchmarking welfare outcomes for commercially farmed pigs. *Animal*, 11, pp. 1816–1824.
Peden RSE, Turner SP, Boyle LA and Camerlink I, 2018. The translation of animal welfare research into practice: the case of mixing aggression between pigs. *Applied Animal Behaviour Science*, 204, pp. 1–9.
Pietrosemoli S and Tang C, 2020. Animal welfare and production challenges associated with pasture pig systems: a review. *Agriculture*, 10, p. 223.
Prunier A, Devillers N, Herskin MS, Sandercock DA, Sinclair ARL, Tallet C and von Borell E, 2020. Husbandry interventions in suckling piglets, painful consequences and mitigation. In Farmer C, eds., *The Suckling and Weaned Piglet*. Wageningen Academic Publishers, Wageningen, The Netherlands, pp. 107–138.
Rioja-Lang FC, Seddon YM and Brown JA, 2018. Shoulder lesions in sows: a review of their causes, prevention, and treatment. *Journal of Swine Health and Production*, 26, pp. 101–107.
Ritter MJ, Johnson AK, Benjamin ME, Carr SN, Ellis M, Faucitano L, Grandin T, Salak-Johnson JL, Thomson DU, Goldhawk C and Calvo-Lorenzo MS, 2017. Effects of ractopamine hydrochloride (Paylean) on welfare indicators for market weight pigs. *Translational Animal Science*, 1, pp. 533–558.
Rydhmer L, 2021. Advances in understanding the genetics of pig behaviour. In Edwards SA, ed., *Understanding the Behaviour and Improving the Welfare of Pigs*. Burleigh Dodds Science Publishing, Cambridge, UK, pp. 3–25.
Spoolder HAM, Geudeke MJ, Van der Peet-Schwering CMC and Soede NM, 2009. Group housing of sows in early pregnancy: a review of success and risk factors. *Livestock Science*, 125, pp. 1–14.
Studnitz M, Jensen MB and Pedersen LJ, 2007. Why do pigs root and in what will they root?: a review on the exploratory behaviour of pigs in relation to environmental enrichment. *Applied Animal Behaviour Science*, 107, pp. 183–97.
Taylor NR, Main DC, Mendl M and Edwards SA, 2010. Tail-biting: a new perspective. *The Veterinary Journal*, 186, pp. 137–47.
Tinkle AK, Duberstein KJ, Wilson ME, Parsley MA, Beckman MK, Torrison J, Azain MJ and Dove CR, 2017. Functional claw trimming improves the gait and locomotion of sows. *Livestock Science*, 195, pp. 53–57.
Turner SP, Edwards SA and Bland VC, 1999. The influence of drinker allocation and group size on the drinking behaviour, welfare and production of growing pigs. *Animal Science*, 68, pp. 617–624.
Valros A, 2018. Tail biting. In Spinka M, ed., *Advances in Pig Welfare*. Woodhead Publishing, Sawston, UK, pp. 137–166.
Welfare Quality®, 2009. *Welfare Quality® Assessment Protocol for Pigs (Sows and Piglets, Growing and Finishing Pigs)*. Welfare Quality® Consortium, Lelystad, Netherlands.
Yun J and Valros A, 2015. Benefits of prepartum nest-building behaviour on parturition and lactation in sows: a review. *Asian-Australasian Journal of Animal Sciences*, 28, p. 1519.

7
FARMING CATTLE

Clive JC Phillips

Introduction

Cattle evolved as grazing and browsing animals and were domesticated approximately 8–10 thousand years ago to serve humans by providing meat, milk, draught power, and many minor benefits, from buttons to blue cheese (Phillips, 2018, p. 1). In addition to the approximately 1 billion cattle worldwide, there are approximately 200 million buffaloes, mainly in Asia (Shahbandeh, 2021).

Cattle production systems are differentiated in most countries into separate herds for beef and dairy production, but in some, such as the British Isles, the two are integrated. Separate breeds have been developed for high milk production and rapid growth, and it is the high-producing, single-purpose herds that have the most welfare risks because the physiology of the cattle is often strained to their limit. This is especially true for dairy cows, whose longevity in intensive milk-producing herds has declined from about 10 years a few decades ago to just 2–3 years now (Hu et al., 2021). Intensive dairying has practices that cause pain to the cows, and they also suffer from fever, fatigue, anxiety, and stress. Unhealthy conditions, such as during heat stress, impair cows' immune systems, increasing their risks of disease, particularly around the time of parturition. Such welfare problems will make it difficult for dairy farmers to maintain social license in the face of public concerns, not just about welfare, but also emissions, food security, food quality, and water usage. Alternatives are available that reduce welfare concerns, for example restricted suckling systems and sheltering cows after their productive life has ended. These get good public support but require greater resources, with the potential to increase polluting emissions and use of scarce resources, such as water.

The growing demand for cattle products, especially milk and meat, has led to intensification of both dairy and beef production systems. In many countries, farms now operate as large businesses with thousands of cows, whereas only 50 years ago the majority of farms had fewer than 100 cows, run by a family. This has had positive and negative effects on cattle welfare. On the positive side, a large farm can purchase up-to-date equipment, which may enable better rationing to supply the needs of high-yielding cows, sick cows to be detected earlier and cows to be milked when they want, rather than at times determined by the farmer. Considerable potential exists to expand the use of new technologies to improve the welfare of cattle, but the motivation is not always present unless it means that the cattle produce more milk or grow faster. On the

 DOI: 10.4324/9781003182351-9

negative side, intensification of cattle production into large farms has diminished the contact of cattle workers with their animals, and the intimate bond between the stockperson and their animals has largely vanished. Humans are an inherently social species; we thrive on bonds that we create with both humans and animals. This bond helped to create a caring attitude towards the individual animal in small farms, whereas nowadays the bonds are absent and cattle are known by numbers not names. Managers of large farms usually have profit as their overriding objective. Diseases, such as acidosis, may be tolerated, unless profit is adversely affected. Recurrent lameness and mastitis have become commonplace, despite the many scientific advances to reduce their prevalence in the herd. Another sign that cows are being overstressed is reproductive failure, causing high culling rates because cows fail to become pregnant.

Welfare problems arise in all aspects of cattle production systems. In this chapter I describe the role of diseases, feeding, housing, routine procedures, reproductive management and genetics, transport, handling and slaughter, mostly with reference to cow welfare, but also with special reference to bulls and calves.

Common cattle diseases

The most prevalent diseases in cattle herds are usually lameness and mastitis. There are situations in which lameness is rare, but in intensive dairy farming it is common, particularly if cows are kept indoors on concrete floors, because of the pressure walking and standing on concrete for long periods imposes on the cows' feet.

The welfare impact of cattle diseases is determined by the duration and the severity of the disease. Common symptoms of pain in cows include withholding the part of the body generating the pain, such as a limb, bruxism (clenching of the teeth) and cessation of rumination. However, as prey animals cattle do not readily exhibit signs of pain.

High milk-producing cows need diets which have unnaturally high energy and protein concentrations if they are to avoid excessive weight loss. Grass alone is usually insufficient for this purpose; hence cows are fed supplements indoors. Standing on concrete puts more pressure on the hooves than is experienced at pasture, with the result that laminitis, or inflammation of the laminae of the hoof, is common. Lame cows experience pain in the affected limb, making it difficult to walk, stand to feed, and interact with other cows. Extremely lame cows have elevated heart and respiration rates (Tadich et al., 2013). The duration of lameness is usually one to two months (Phillips, 1990; Eriksson et al., 2020), longer for sole ulcers, white line disease, and heel erosion than interdigital phlegmon or digital dermatitis (Whay et al., 1998).

Mastitis, or inflammation of the udder caused by bacterial infection, also usually lasts about one month (Ruegg, 2017), but is often recurring within and between lactations. The udder of infected cows is swollen and sensitive to touch, making them appear lethargic, dejected, and in pain (Frössling et al., 2017). Lying down may put particular pressure on the infected gland, causing more pain.

Acidosis, or accumulation of acid in the rumen following overconsumption of carbohydrates, is common in early- to mid-lactation cows. The resulting inflammatory cytokines may damage the rumen wall (Zhao et al., 2018), and even in its subacute form inflammatory markers are evident in the blood (Gozho et al., 2005). Ketosis is a similar production-related disease, in which cows produce ketone bodies whilst mobilising body tissue to meet their energy requirements for lactation. This makes cows lethargic and withdrawn (Sahar et al., 2020). Less common but potentially fatal are hypocalcaemia and hypomagnesaemia, deficiencies of calcium and magnesium in the diet. The former is most common around the start of lactation, before cows release enough calcium from their bones to sustain output of calcium in milk, and the latter is most commonly due to excessive potassium fertiliser application to stimulate pasture growth, since potassium in the rumen inhibits magnesium absorption. Dairy cows also suffer from metritis, or infection of the uterus, in which the reproductive tract becomes inflamed, and the passage of urine can be painful.

Bovine respiratory diseases are common in stressed cattle, especially during and after transport. Most are caused by viral pathogens, often followed by bacterial infections. Infected cattle experience hypoxia, hypercapnia and eventually respiratory failure. They attempt to compensate by increasing their heart output but usually become anxious and gasp for air. Other infectious diseases include tuberculosis, a bacterial infection mainly of the lungs, Johne's disease, a bacterial infection of the gastrointestinal tract, and Foot and Mouth disease, a viral infection that causes bruxism and irritation to the feet of cattle. Developed countries have eradicated many of these infectious diseases by a rigorous culling policy of infected herds, but they are common in the emerging economies, such as India and Brazil, which together have 56% of the world's cattle.

Feeding

Cattle are naturally herbivorous, characterised by their ability to digest fibrous feeds, with the aid of a modified forestomach, the rumen, which contains microbes that digest the feed. Boluses of partially digested feed are regularly regurgitated into the mouth to be chewed and mixed with saliva, before returning into the rumen. This process is mostly carried out whilst the cattle are lying quietly, particularly at night.

Cattle are well suited to grazing in herds, with careful selection of the best pasture. They will also take forbs and browse, and many rangeland systems for beef cattle rely on these, as well as spinifex grass. Grasslands are prone to variation in grass growth, as a result of fluctuations in circannual temperature and water availability, part of which is both predictable and normal. However, anthropogenic climate change, with its associated increase in temperature and fluctuations in water availability, will create challenges for cattle farming (Giridhar and Samireddypalle, 2015), particularly in the rangelands, where rainfall is naturally variable and there are limited opportunities to mitigate this with, for example, irrigation. More cattle will starve as a result, a condition characterised by a shortage of nutrients causing a reduction in functionality, such as reproductive failure (Hogan and Phillips, 2015). When starving, cattle initially are more active in seeking an alternative feed supply, their appetitive behaviour expands to include non-traditional potential feeds (pica), but eventually if no source of nutrients is found they become lethargic, as energy expenditure outweighs the chance of finding feed. Strategies to manage drought include selling cattle, but only the breeding animals as a last resort, purchasing fodder or moving cattle to areas with available feed. Action is often too slow, but better weather forecasting is facilitating timely action. With rangelands mostly managed by large corporations, profit is all important, and there is a temptation to overstock land, using carrying capacity in good years as a guide. Overstocking degrades pastureland, further depleting feed resources and allowing weed invasion.

Greater control of cattle feed supply is possible on intensively managed grasslands, particularly if irrigation is used. Fertilisers are often used to accelerate pasture growth, although this may create a nutrient imbalance, for example when potassium fertilisers deplete the sodium in plants which is required by grazing animals. Supplements are often necessary to correct nutrient deficiencies in grazing cattle, even on rangelands where phosphorus is commonly deficient, causing cattle to have weak bones. For high-yielding dairy cows, supplements are commonly used to increase energy and protein intakes, but this has the risk of the metabolic disorders referred to above. Concentrate supplements are often mixed with roughage in a Total Mixed Ration, which is offered to the cattle behind a feeding barrier. Thorough mechanical mixing is required, otherwise the diet will not be balanced for some cows. The feeding barrier should be of solid construction, otherwise cattle will get sores on their shoulders when they push forward to reach the feed (Phillips, 2018, p. 143). They often toss the mix into the passage, especially if it is old and stale. On smaller farms, cattle may eat wet conserved grass (silage) directly from a mound of grass clamped between two walls. To stop them eating unevenly from the "clamp", an electrified wire may be suspended between them and the silage. Some young cows (heifers) become frightened and eat little. Also, those losing their milk teeth may have difficulty extracting the silage.

Beef cattle are often raised initially on pasture but transferred to feedlots for the final stages of growth. There they are fed a diet rich in cereals, and sometimes exclusively cereals, which risks metabolic disorders, especially if the cattle are introduced to the feed suddenly after being on rangelands. Feedlot cattle engage in stereotyped oral behaviours, usually tongue rolling, when fed inadequate roughage because the cereal feed is eaten rapidly and there is insufficient stimulation of the mouth. Antimicrobial compounds, such as the ionophore sodium monensin, may be included in the ration as they improve the efficiency of digestion in the rumen, however, they are now banned in many countries because of the development of antimicrobial resistance, which reduces the efficacy of the antimicrobials when used in human medicine.

Housing

Most dairy cows are kept indoors for ease of management, in particular of their diet. Some barns have individual stalls for tying the cattle (tie-stalls), which offer them little opportunity for movement or interaction with other cows. Cows are usually milked in the tie-stalls. However, many farms have advanced to loose (free walk) housing of their cattle, with a feeding barrier, passageway for feeding, and a lying area. The lying area may be simply an earth floor with straw bedding, which gives the cows a comfortable surface on which to lie. At high stocking densities there is a risk of cows' teats being trodden on, which may lead to mastitis. More commonly cows are given raised concrete beds on which to lie, either side of a passageway, with separation dividers creating a cubicle (free stall) for each cow. They have more difficulty lying down and rising in cubicles compared with a straw-bedded yard, and they have often develop lesions on their legs (Blanco-Penedo et al., 2020). The size and shape of the bed, and construction of the divider, are critical if cows are to be comfortable. If the bed is too short, the cows' rear end hangs over into the passageway; too long, and they excrete on the bed, not in the passageway, which can cause mastitis. Too wide and they can turn around; too narrow and they knock their hocks on lying and rising. If cows find the cubicles uncomfortable, they lie in the passageways, get dirty, and contract mastitis. Passageways need to be scraped regularly to remove excreta, either mechanically by blades passing down the passage, or by hand. Mechanical scraping can trap cows' tails unless a suitable trip switch is incorporated. In future, consumer demand for cows to be able to display natural behaviour will mitigate against buildings with cubicles or tie-stalls (Galama et al., 2020).

Dairy cows are milked between one and three times daily. Traditionally this occurs in a parlour, with milkers in a pit so that the cows are at the milkers' waist height to avoid having to bend down to place teat cups on the cows' udders. Each teat cup consists of a solid tube with a flexible liner, and the space between these two is evacuated in a pulsatile manner, about once every second. Thus milk is removed from the udder by creating a vacuum between the liner and the tube, many times a minute, so that milk is sucked from the teat in a manner simulating a calf suckling her mother's teat. The setting of the vacuum level and pulsation frequency is critical in avoiding pain to the cow during milking and development of mastitis (Bramley,

1992). Robotic milking, in which a cow voluntarily enters a stall and a robot attaches a cluster of teat cups to her udder, is increasingly popular as it does not require a milker to get cows into a parlour and attach the teat cups. Although offering cows more freedom, it raises welfare concerns as the milker is no longer present during the milking process to deal with problems, and the cow–herdsperson bond is further diminished in this latest step in the automation of dairy farming (Driessen and Heutinck, 2015). However, the robot represents a consistent approach, whereas humans may treat cows well or badly in the milking parlour.

In the beef industry, feedlots have many welfare disadvantages compared with keeping cattle at pasture. First, cattle are kept at high stocking densities and the ground can become bogged in wet weather. In dry conditions, it is often dusty, leading to respiratory problems. Second, the concentrated diet is conducive to metabolic disorders and stereotyped oral behaviours, described above. Third, because the cattle are stocked at a relatively high rate and spend little time feeding and in other maintenance activities, a small proportion (2–4%, Phillips, 2018, p. 25) engage in riding behaviour, which may exhaust both the ridden and riding steers. Although this mimics sexual behaviour, it is more connected with aggression in steers, but in entire males (bulls) penises may become damaged during riding behaviour. Finally, the absence of natural shade, and often any shade, predisposes cattle to heat stress and mortality often ensues in the increasingly common heat waves.

If floor substrates like straw are unavailable, beef cattle housing may use slatted floors, in which cattle tread their excreta through slots in the floor. Walking on slatted floors is uncomfortable for cattle, and a bedded yard, using sand or other alternatives to straw, gives greater comfort.

Routine procedures

Dehorning – Cattle evolved with horns for defence from predators, but managing cattle with horns in crowded handling or transport situations is dangerous to both other cattle and humans. Hence the horns are often removed, either in a long-term genetic selection programme to produce polled (i.e. hornless) cattle, or more usually physically in each animal by disbudding (removal of the horn bud in a calf before it has attached to the skull) or dehorning (removal of the horn after it has attached to the skull). The horn is a sensitive part of the animal's body that is highly innervated and vascularised. Horn bud removal in calves, usually with a knife, hot iron, or scoop is a painful procedure, and local anaesthesia and analgesia should be provided. Removal of horns in older cattle is a dangerous procedure, with risks of excessive blood loss, infection, and even death, but it is routinely conducted in rangeland cattle because they are only mustered once or twice a year. Often it is combined with castration, branding, and weaning in a single annual operation. The welfare advantages for cattle of being in a rangeland system – freedom to perform natural behaviour, calves remaining with their dams for at least six months, minimal handling by people – must be balanced against the welfare risks – nutrient deficiencies, ectoparasites, predation, and routine procedures being conducted at an older age when they cause more pain and risks to cattle health. Application of anaesthetic is often not practised on rangeland properties because to be most effective it would require cattle to be handled twice, with enough time in between for the anaesthetic to take effect. Increasingly an analgesic/anaesthetic combination is applied at the same time as processing of cattle, when it will have some benefit in pain relief.

It is important to remember that horns are an integral part of the identity of cattle and removing them is ethically questionable. Also, in some parts of the world cattle are attacked by wild dogs and crocodiles, and removing their horns renders them less able to defend themselves and their calves. Sometimes horns have to be removed because they grow into the animal's skulls, but that is rare.

Desexing – Many cattle are desexed at an early age, to reduce the aggressive temperament in male cattle, which may reduce meat quality, and to prevent breeding in both male and some female cattle. In males it also advances the age at which an adequate fat cover for cooking develops, facilitating early marketing. Castration can use rubber rings around the testicles, which then atrophy and die, a Burdizzo clamp that crushes the spermatic cord, or surgery to completely remove the testes (Anon, 2014). However, castrated males grow slower than bulls and the practice is of questionable necessity. In females, desexing is either by a flank incision or per vaginal severance of the ovaries (Anon, 2011b). It is performed to prevent breeding in cows that are destined for slaughter.

Identification – This is necessary for individual recognition and treatment and sometimes to prevent theft. There are several methods – tagging, notching or tattooing of one ear, fire branding, microchipping, or freeze branding. All cause pain, but particularly fire branding, which should be accompanied by pain relief (Anon, 2011a). Analgesics, however, wear off and continued pain is to be expected, making it essential to choose the most pain-free method for identification.

Implants – Cattle may be implanted with slow-release devices. Some contain sex hormones to increase growth rate, in which case they are inserted into the ear, but there is evidence that the hormones make cattle more susceptible to heat stress (Gaughan et al., 2005). These are banned in many jurisdictions. Other implants contain micronutrients, in the form of boluses inserted into the rumen. Implantation is a stressful process for cattle and the necessity must be carefully considered.

Tail docking – In some wet regions of Australia and elsewhere heifers have about two-thirds of their tail removed so that they keep clean in the dirty conditions that they will have to cope with as adult cows. However, a cow's tail is an important signalling device to other cows, as well as being used to remove flies from their hindquarters, and this practice is ethically questionable.

Reproduction management and genetic selection

Cows often have a low reproductive rate because nutrition fails to meet the demands of high-yielding dairy cows or even beef cows. Increasingly, cows are artificially supplied with female hormones in an attempt to overcome this problem. On most dairy farms semen from a bull known to have good genetic characteristics in his offspring is injected into the reproductive tract of the cows, rather than using natural insemination by a bull. The semen is frozen for storage and thawed before use. It can be sexed (Vishwanath and Moreno, 2018), which potentially avoids producing too many unwanted male calves. In intensive dairy production systems, male calves are almost worthless, so they are destroyed at just a few days of age, preventing them from having the opportunity of a life with good welfare. Because they are of minimal value, their removal from the farm and transport to an abattoir is often without consideration for their welfare. They may be transported to slaughter regardless of their condition. They may be deprived of milk for up to 30 hours and, because they do not have the same herding instinct that older calves have (Jongman and Butler, 2013), they can be more difficult to move and are manhandled or thrown with little regard for their welfare. However, even with sexed semen, a use for the increased number of female calves has to be found, given that cows usually survive 2–3 lactations. Some are exported from the intensive dairy industries of New Zealand, Chile, and Australia to Asia, to expand or replenish their dairy herds, but their welfare there must be considered in debating the ethics of this practice. Often countries in Asia are short of fodder and cows of the Holstein-

Friesian breed used in intensive dairies have big feed requirements, as well as much greater susceptibility to heat stress and tropical diseases than cows of local breeds (Moran, 2012).

Cows' reproductive cycles may be controlled artificially by inserting a slow-release hormone treatment implant into their vagina, and insemination carefully timed following removal of the device. Thus natural mating, one of the foundations of normal behaviour and hence good welfare, is obviated in most dairy herds. As well, embryos may be removed from genetically high-producing cows by flushing the reproductive tract post insemination and inserting them into less productive cows to improve the yield potential of the herd. However, high yields predispose cows to diseases such as mastitis, and increasingly the objectives for breeding cattle are focused on disease resistance and extended lactations, rather than a high peak lactation, which is when most of the production diseases occur. A major risk of breeding methods in the past has been the reduction in genetic diversity and focus on just one breed, the Holstein-Friesian, for dairy farming. The selected cows have high milk yields, but they also require large amounts of concentrate feed, based on cereals, and are susceptible to heat stress and disease. As the world's human population grows, cereals will be increasingly required for feeding to humans, not livestock, and breeds resistant to disease and heat stress and able to utilise by-products and other waste products will be required. Beef cattle have been mainly selected for rapid growth, which has favoured large cattle, but selection for cattle with excessive muscling has produced breeds, such as the Belgian Blue, which have difficulty giving birth without using the Caesarean operation because of their unnatural conformation.

Bulls

If bulls are used on dairy farms, their welfare needs careful consideration. Often they are kept in solitary confinement in a pen near to the cows, for example by the collecting yard for milking. They are let out when necessary to serve cows, by giving them access to a service box, or, in less intensively managed farms, they may be given access to cows at pasture. Solitary bulls are potentially dangerous, but aggressive behaviour can also be a problem in groups of bulls. Managing bulls is challenging and their welfare is often sacrificed for the safety of people handling them, by confining them in cramped conditions.

Calves

The welfare of the calf begins during pregnancy, when stress should be kept to a minimum. Calving is a critical time for cow and calf, and specialist facilities are required, including a clean calving box of adequate size. Allowing cows to calve in cubicle houses can lead to a poor welfare outcome for both cow and calf, with little comfort and a high risk of infection. Calves are usually removed from their mothers at a very young age, less than one week, far earlier than would occur under natural conditions. This induces a pessimistic mood in the calves, similar to that experienced after hot iron dehorning (Daros et al., 2014). Some degree of contact is better for the calves, for example contact over a fence, but it often does not fit in with modern management systems. The calves are then reared in some degree of isolation before being put into groups at the age of five or six weeks. This is to reduce the chances of diseases being transmitted, as calves are susceptible to pneumonia and scours, both communicable diseases. However, isolation in a pen, the normal method of containment, restricts calves' opportunities to move and socialise with other calves, that are so desperately needed after the mother young bond is prematurely severed (Phillips, 2002). Feeding them reconstituted milk powder might seem like a circuitous route to offer cows' milk, but it fits well with milking systems for high-yielding cows.

However, in many emerging economy countries, such as in South America, calves are allowed to suckle their mother's milk after some has been taken for human consumption. Care must be taken that calves get enough milk, but their dams do not lose too much weight, since calves extract more milk from the cow's udder than milking machines. In calf houses, milk may be fed twice a day in buckets, but this often leads to stereotyped sucking behaviour, of anything available in their pens – buckets, neighbours' ears, or mouths. It is preferable to offer milk replacer *ad libitum* (i.e. constantly available) via artificial teats attached to the calves' pens, with the teats leading to a vat of acidified milk to stop it spoiling, or at a "milk bar" where milk is constantly available.

After weaning off milk, the best environment for calves is sheltered grassland, provided it is not too cold, as the young calf generates limited heat. However, most weaned calves are kept in pens in barns. As they approach the time of entering the dairy herd, it is important to get heifers used to the new management system, including lying in cubicles and passing through their milking parlour.

Some calves remain on milk until slaughter at 70–150 kg to produce a pale meat called veal. After the public were alerted to serious welfare problems in this system of meat production in the last century, veal calf producers began modifying the systems of production to try to improve calf welfare without changing the characteristics of the meat. As a result, there are now diverse veal production systems (Anon, 2008); for example, some calves are now group housed, but others are still individually housed in crates to stop them from exercising, which is believed to make the meat darker and tougher (Costa et al., 2016). A concentrate supplement is sometimes fed to ensure adequate iron intake, but not always. Bedding may be restricted to prevent consumption and normal rumen development. Iron content of the ration is limited to prevent dark coloured meat, since consumers prefer white coloured meat, although the European Union has placed restrictions on this.

Transport and handling

All journeys cause cattle to become stressed, even if they are transported on foot. In vehicles, cattle are often overstocked and taken long distances without feed or water. Cattle are often required to be transported long distances if they have come from rangelands, which are usually in remote areas. Market price differentials also encourage cattle farmers to transport their animals further to achieve a higher sale price. Religious and cultural preferences for consuming recently killed animals and animals killed according to the standards of a religion, e.g. halal slaughter, are responsible for some cattle having to endure long journeys by sea and road, when logically carcasses would be transported instead. Many cattle travel from southern countries, e.g. Australia and Brazil, to the more populated northern countries to be slaughtered and consumed. Movement of cattle can also spread infectious diseases, such as foot and mouth disease.

Transport brings the risk of heat stress, because there are large numbers of cattle in a small space, and the heat output per animal is high. Overstocking is common in vehicles, exacerbating the heat stress risk and causing animals to knock into each other and the sides of the vehicle, bruising their muscles. Lack of feed and water also stress cattle transported over long distances. Fatigue is possible in cattle after repeated stepping to counteract the movements of the floor, especially on rough or winding roads or in heavy traffic. Some jurisdictions require livestock vehicles to be periodically unloaded to allow cattle to rest. The method of travel varies significantly in different countries; in Australia cattle are transported at high density in open-top trucks with corrugated floors, designed so that cattle do not lie down. The trucks may have several trailers and on dirt roads significant dust can enter the rear trailers. Distances are often long and the risk of heat stress is high. In contrast to this, in Europe cattle are transported shorter distances in closed topped vehicles and unloaded periodically to rest.

Sea transport brings the risks of not only heat stress, which occurs when cattle travel from cool climatic conditions to a hot region in just a few weeks, but also accumulation of ammonia, released from the excreta which accumulates in the pens, and motion sickness (Phillips and Santurtun, 2013). Loading has special risks, particularly if there is a substantial ramp up to the ship. If tides are very variable, cattle are sometimes winched onto the ship, which must frighten them. Much fewer, but a growing number of cattle, travel by air, usually restricted to high value animals. As well as heat stress and high stocking densities, cattle in aeroplanes have to endure significant G force and noise on take-off, neither of which they are accustomed to. Further information about animal transportation is in Chapter 11.

Handling procedures are variable and of paramount concern for the welfare of cattle at all times, but particularly during transport. Moving cattle requires careful handling, respect for their welfare, and an acknowledgement by the handler that the slowest animal should determine the speed for the group. Much has been done in recent years to promote "low stress handling", which allows animals to move under their own speed with better welfare than if people are hurrying them along. The speed at which to move cattle depends on how far they have to travel and the conditions at the time. Cattle mustered on rangelands may initially be encouraged to move towards a collecting yard by an aeroplane or helicopter, which forms them into groups. Then cowboys on horses, motorbikes, or in vehicles will take over and move them at a slower pace to the yards, so that they are not too stressed on entering the yards.

Cattle are taken through a handling race in order to inject them with slow-release micronutrients, vaccinations or to conduct routine procedures. A curved, solid-sided race encourages cattle to move through, and the handler should stand behind the leading cattle, working them around a "point of balance" just behind the shoulder (Grandin, 2014). Electric prods should only be used as a last resort, if the welfare of other cattle in a line is being affected. Cattle should never be shocked or hit on the head, and their tail should never be twisted or raised by the handler to motivate them to move forward. They should not be moved by placing a pincer or fingers in their nose and bulls should not have a ring inserted there to allow them to be safely handled, because it is extremely painful and can lead to inflammation. The basic principle is to move cattle by encouraging natural behaviour, not by causing pain.

Handling facilities in some emerging economy countries are necessarily simple, and sending cattle from rangelands, that have rarely been handled, for slaughter to these countries, has led to some poor welfare outcomes. For many in these countries, human welfare is a priority; animal welfare is rarely considered.

Slaughter

Most cattle are slaughtered as soon as their working life has ended. Only in India is there a system of retiring cattle to sanctuaries for the remainder of their natural life, a practice required by the Hindu faith. Cows' welfare in the shelters may be compromised by overcrowding, poor flooring, and limited feed (Sharma et al., 2019), but conversely they are given the opportunity for extended life, including one with good welfare in the best facilities.

Slaughter can be a painful process for cattle. Of particular concern is whether cattle lose consciousness quickly, and whether the process is terminal (cattle may regain consciousness after stunning if exsanguination is not conducted rapidly enough). Some religions do not allow animals to be stunned, most notably the Jewish faith (shechita slaughter) and, according to some authorities, the Muslim faith (halal slaughter). Halal slaughter is often conducted without stunning in emerging economy countries because of the cost, but there should still be a means of restraining cattle suitably for a rapid knife cut to the throat. Often this is lying them on the ground, rather than being in the standing position (Imlan et al., 2021). Boxes that invert the cattle are not suitable – they cause fear and discomfort in the cattle. If animals are not stunned, rapid and complete severance of the neck arteries is essential to allow the animals to exsanguinate and die as quickly as possible. Stunning cattle is usually achieved by firing a retracting bolt into the forehead of the animal. Most effective are the penetrating bolts, which enter the cranial cavity to inflict injury directly on the brain; less effective are non-penetrating bolts, which often necessitate repeat stuns (Neves et al., 2016). Further information about slaughtering is in Chapter 12.

Environmental stresses

Thermal stress – Cattle are particularly susceptible to thermal stress because the microbial fermentation in their rumen generates large amounts of heat. Many cattle are kept in hot parts of the world, often in small farms with few resources to mitigate the heat stress, and they are important as a source of work and capital for many subsistence farmers. In sophisticated dairy farms, cattle will be cooled by sprinklers and fans, especially at milking time; this helps them to cope with the heat and produce large quantities of milk. The heat stress experienced by cattle is dependent on the ambient temperature, humidity, solar radiation and air velocity. High-yielding dairy cows, rapidly growing beef cattle, and black-coated cattle are most susceptible. They start to take action to reduce their heat load, by sweating and panting, at temperatures as low as 25°C (Berman et al., 1985).

Facilities can be designed to reduce the risk of heat stress, but in the hotter regions of the world, this is not enough to prevent many animals experiencing heat stress and some dying, particularly in feedlots, in which animals may be crowded together without shade. Global climate change is exacerbating the problem and is expected to become progressively worse over the course of this century, perhaps pushing cattle production towards the earth's poles.

Lighting – Cattle see the world differently to humans, with limited binocular vision but a wide field of monocular vision (330°), and an extended visual streak with concentrated light perception cells, in contrast to our own point source of excellent acuity in the form of a fovea. Cattle handlers need to be aware of their wide field of vision and good night vision, but limited ability to judge the distance of things close to them. Artificial lighting in their facilities is important to allow cows to see their way to feed and to be able to engage in social interaction at night.

Noise and vibration – Cattle hear the world differently to humans, with better hearing in the high frequency zone. This means that high pitched parlour noise, for example, may disturb cattle but not us, particularly if it exceeds 75 dB (the level created by an average vacuum cleaner or radio). This can be detected when they defecate after entry, a sign that they are stressed. Vibration from heavy machinery or traffic can also disturb cattle, especially the low frequency vibrations that travel further (Phillips, 2018, p. 156). These stresses are commonly experienced during transport, when there are multiple stresses for cattle to contend with simultaneously.

Electricity – Cattle are easily stressed by electricity, a characteristic exploited when controlling them with electric fences. Stray voltage often exists in poorly earthed barns, in water troughs, for example, or milking parlours, and cattle behaviour should alert good herdspeople to the problem. Lapping water like a dog, rather than full muzzle immersion in water, is one sign; flinching when teat cups are applied is another.

Working cattle

Cattle are used in many emerging economies, especially in South and South-east Asia, for pulling field implements, such as a plough, raising water, crushing seeds, and moving equipment and timber (Ramaswamy, 1994). Just over 25 years ago, Ramaswamy estimated that there were 246 million cattle and 60 million buffalo used for draught purposes worldwide (Ramaswamy, 1994). Pulling heavy loads can place a strain on the backs of cattle, and prolonged periods of work require adequate rest and extra nutrition, especially energy and the salts needed to replace sweating losses. Draught cattle are often beaten to make them work harder. Another welfare issue is an ill-fitting harness, which can cause chafing of the skin and deep sores. Draught cattle are usually restrained and sometimes blindfolded when undertaking this work, preventing them from engaging in normal behaviour.

Conclusions

Cattle production systems have, in the past, helped humans to colonise new lands, generating valuable foods and other resources for humans, such as draught power, from land which was marginal for other purposes. However, cattle themselves require many resources, resources which are increasingly questioned in times of diminishing availability for humans and other animals, when new technologies can allow even marginal land to be used for crops or forestry. The many welfare concerns about the cattle production systems that humans use are encouraging consumers to switch to the ever-growing number of plant-based alternatives to meat and milk. Unless the welfare issues can be addressed, and it can be shown that cattle production systems have a role to play in using by-products of plant production systems, these systems may become superseded by more efficient and less contentious methods of producing food for human consumption. Systems integrating cattle with trees and pasture, for example, offer many welfare benefits, with natural shade, drought and disease resistance, and high overall productivity, but they require careful management. Cattle, the largest animal we have domesticated on a grand scale, are inherently difficult to look after to high welfare standards. Knowledge transfer between generations of farmers is important, but increasingly at risk due to the migration of young people to the cities. High welfare systems of cattle production are attainable, but only with substantially higher costs to consumers, costs which they may not be willing to bear in the face of strong competition.

References

Anon, 2008. Welfare implications of veal calf husbandry. American Veterinary Medical Association, https://www.avma.org/resources-tools/literature-reviews/welfare-implications-veal-calf-husbandry. Accessed 23 September, 2019.

Anon, 2011a. Welfare implications of hot-iron branding and its alternatives. American Veterinary Medical Association, https://www.avma.org/sites/default/files/resources/hot-iron_branding_bgnd.pdf. Accessed 23 September, 2019.

Anon, 2011b. Welfare implications of ovariectomy in cattle. American Veterinary Medical Association. https://www.avma.org/resources-tools/literature-reviews/welfare-implications-ovariectomy-cattle. Accessed 23 September, 2019.

Anon, 2014. Welfare implications of castration of cattle. American Veterinary Medical Association. https://www.avma.org/resources-tools/literature-reviews/welfare-implications-castration-cattle. Accessed 23 September, 2019.

Berman A, Mamen M, Herz Z, Wolfenson D, Arieli, A, 1985. Upper critical temperatures and forced ventilation effects for high-yielding dairy cows in a subtropical climate. *Journal of Dairy Science*, 68, pp. 1488–1495.

Blanco-Penedo I, Ouweltjes W, Ofner-Schroeck E et al. 2020. Symposium review: animal welfare in free-walk systems in Europe. *Journal of Dairy Science*, 103 (6), pp. 5773–5782

Bramley AJ, 1992. Mastitis and machine milking. In *Machine Milking and Lactation*, ed. Bramley AJ, Dodd FH, Mein GA and Bramley JA. Insight Books, Newberry, England, pp. 343–372.

Costa JHC, von Keyserlingk MAG, Weary DM, 2016. Invited review: effects of group housing of dairy calves on behavior, cognition, performance, and health. *Journal of Dairy Science*, 99(4), pp. 2453–2467.

Daros RR, Costa JHC, von Keyserlingk MAG, Hotzel MJ, Weary DM, 2014. Separation from the dam causes negative judgement bias in dairy calves. *PLoS ONE* 9(5), p. e98429. https://doi.org/10.1371/journal.pone.0098429

Driessen C, Heutinck LFM, 2015. Cows desiring to be milked? Milking robots and the co-evolution of ethics and technology on Dutch dairy farms. *Agriculture and Human Values*, 32, pp. 3–20

Eriksson HK, Daros RR, von Keyserlingk MAG, Weary DM, 2020. Effects of case definition and assessment frequency on lameness incidence estimates. *Journal of Dairy Science*, 103, pp. 638–648 https://doi.org/10.3168/jds.2019-16426

Frössling J, Ohlson A, Hallén-Sandgren C, 2017. Incidence and duration of increased somatic cell count in Swedish dairy cows and associations with milking system type. *Journal of Dairy Science*, 100, pp. 7368–7378.

Galama PJ, Ouweltjes W, Endres MI, et al. 2020. Symposium review Future of housing for dairy cattle. *Journal of Dairy Science*, 103, pp. 5759–5772.

Gaughan JB, Kreikemeier WM, Mader TL, 2005. Hormonal growth-promotant effects on grain-fed cattle under different environments. *International Journal of Biometeorology*, 49, pp. 396–402

Giridhar K, Samireddypalle A, 2015. Impact of climate change on forage availability for livestock. In Sejian V, Gaughan J, Baumgard L, Prasad C, eds., *Climate Change Impact on Livestock: Adaptation and Mitigation.* Springer, New Delhi. https://doi.org/10.1007/978-81-322-2265-1_7

Gozho GN, Plaizier JC, Krause DO, Kennedy AD, Wittenberg KM, 2005. Subacute ruminal acidosis induces ruminal lipopolysaccharide endotoxin release and triggers an inflammatory response. *Journal of Dairy Science*, 88, pp. 1399–1403.

Grandin T, 2014. Behavioural principles of handling cattle and other grazing animals under extensive conditions. Chapter 4 in Grandin T, ed., *Livestock Handling and Transport.* CABI, Wallingford, pp. 39–64.

Hogan JP, Phillips CJC, 2015. Starvation of ruminant livestock. Chapter 3 in Phillips CJC, ed., *Nutrition and the Welfare of Farm Animals.* Springer, New York, pp. 19–36.

Hu HH, Mu T, Ma YF, Wang XP, Ma Y, 2021. Analysis of longevity traits in Holstein cattle: a review. *Frontiers in Genetics*, 12, article Number: 695543. https://doi.org/10.3389/fgene.2021.695543

Imlan JC , Kaka U, Goh YM, Idrus Z, Awad EA, Abubakar AA, Ahmad T, Nizamuddin HNQ, Sazili AQ, 2021. Effects of slaughter positions on catecholamine, blood biochemical and electroencephalogram changes in cattle restrained using a modified Mark IV Box. *Animals*, 11, Article Number: 1979. https://doi.org/10.3390/ani11071979

Jongman EC, Butler KI, 2013. Ease of moving young calves at different ages. *Australian Veterinary Journal*, 91(3), pp. 94–8. https://doi.org/10.1111/avj.12014

Moran J, 2012. *Managing High Grade Dairy Cows in the Tropics.* CSIRO Publishing, Canberra.

Neves JEG, Paranhos da Costa MJR, Roca RO et al. 2016. A note comparing the welfare of Zebu cattle following three stunning-slaughter methods. *Meat Science*, 117, pp. 41–43.

Phillips CJC, 1990. Adverse effects on reproductive performance and lameness of feeding grazing dairy cows partially on silage indoors. *Journal of Agricultural Science, Cambridge* 115, pp. 253–258.

Phillips CJC, 2002. The welfare of calves. In *Cattle Behaviour and Welfare*, CABI, Wallingford, UK, pp. 30–37.

Phillips CJC, 2018. *Principles of Cattle Production.* 3rd edn. CAB International, Wallingford, UK

Phillips CJC and Santurtun E, 2013. The welfare of livestock transported by ship. *The Veterinary Journal* 196, pp. 309–314. https://doi.org/10.1016/j.tvjl.2013.01.007

Ramaswamy NS, 1994. Draught animals and welfare. *Revue Scientifique et Technique, Office Internationale Epizootie* 13, pp. 195–216.

Ruegg PL, 2017. A 100-Year Review: Mastitis detection, management, and prevention. *Journal of Dairy Science* 100, pp. 10381–10397. https://doi.org/10.3168/jds.2017-13023

Sahar MW, Beaver A, Weary DM, von Keyserlingk MAG, 2020. Feeding behavior and agonistic interactions at the feed bunk are associated with hyperketonemia and metritis diagnosis in dairy cattle. *Journal of Dairy Science*, 103, pp. 783–790. https://doi.org/10.3168/jds.2019-16278

Shabandeh, M. 2021. *Cattle Population Worldwide, 2012–2021.* Statista. https://www.statista.com/statistics/263979/global-cattle-population-since-1990/ (Accessed 23 September, 2021).

Sharma A, Kennedy U, Schuetze C, CJC, 2019. The welfare of cows in Indian shelters. *Animals*, *9*(4), p. 172. https://doi.org/10.3390//ani9040172

Tadich N, Tejeda C, Bastias S et al. 2013. Nociceptive threshold, blood constituents and physiological values in 213 cows with locomotion scores ranging from normal to severely lame. *The Veterinary Journal* 197, pp. 401–405.

Vishwanath R, JF Moreno, 2018. Review: semen sexing: current state of the art with emphasis on bovine species. In Animal 12 , Supplement s1: Theory to Practice - International Bull Fertility Conference, 27–30 May 2018, Westport, Ireland, June 2018, pp. s85–s96. https://doi.org/10.1017/S1751731118000496

Whay HR, Waterman AE, Webster AJF, O'Brien JK, 1998. The influence of lesion type on the duration of hyperalgesia associated with hindlimb lameness in dairy cattle. *The Veterinary Journal*, 156, pp. 23–29. https://doi.org/10.1016/ S1090 -0233(98)80058-0.

Zhao C, Liu G, Li X, Guan Y, Wang Y, Yuan X, Sun G, Wang Z, Li X, 2018. Inflammatory mechanism of Rumenitis in dairy cows with subacute ruminal acidosis. *BMC Veterinary Research*, 14, p. 135. https://doi.org/10.1186/s12917-018-1463-7

8
FARMING SHEEP AND GOATS

Cathy M Dwyer

Introduction

Sheep and goats are kept for milk, meat, skins, and fibre (wool and cashmere). The world population of these species is nearly 2 billion animals (1.1 billion sheep, 0.87 billion goats, FAOSTAT, 2014). Sheep are widely distributed across Asia, Africa, Europe, and the Americas, with most numerous populations in China, India, and Australia, but tend to be less common in tropical regions. Goat populations are found particularly in sub-Saharan Africa, Asia, Central America, and the Mediterranean. Typically, goats are better adapted to hot and humid environments than sheep, but fare less well in cold and wet climates, where sheep tend to predominate.

Sheep and goats are adaptable, hardy, and robust, able to utilise poor-quality forage through grazing and browsing. These traits continue to make them popular species for farming in some of the harshest environments on the planet, where they sustain subsistence farmers in Low- and Middle-Income Countries as multipurpose species, often managed in mixed-species herds. In Western countries, specialised breeds for meat, milk, or wool/fibre are more commonly used. They are farmed in extensive or very extensive (ranched) conditions in many countries but can also be kept in semi-intensive systems (housed for some parts of the day or for parts of the year) through to intensive (usually dairy) systems of continuous housing or kept on feedlots.

Extensive management systems are often perceived to be good for welfare. However, this does not mean that small ruminants have universally good welfare and there are no contentious welfare problems. With a few exceptions (such as milking), the most important welfare issues confronting small ruminants are associated with systems of management (essentially differences between animals spending all or most of their time outdoors, compared to all or most indoors) rather than production purpose. Extensive management is more common in sheep than goats, and more common in meat or wool production than in dairy but can be seen in all production systems. In these systems, animals may be held in fenced pastures or have access to large, open rangelands without fences. In the UK, unfenced systems make use of the natural habitat and home-ranging behaviour of sheep (termed hefting), in which generations of animals remain on the same area of land, where they are familiar with the location of food, water, or shelter. Although these unconfined systems can allow considerable behavioural freedom, animals are exposed to welfare issues, including environmental extremes (heat, drought, snowfall, wind, and rain), predation, variability in the availability and quality of food and water, and infre-

DOI: 10.4324/9781003182351-10

quent inspection, which can mean that disease or injury may be undetected, undiagnosed, or untreated.

In indoor management, animals can be more readily inspected, individual treatments are possible, and they can be provided with adequate nutrition more easily. Although sheep and goats are rarely subjected to the very close confinement of some other species, stocking density in indoor management is an issue, and aggression and competition can occur at high stocking density or when feeder space is insufficient. The quality of flooring, bedding, and the environment is also a concern, as small ruminants can be susceptible to respiratory disease, and foot and leg problems associated with poor environmental management. The nature of the human–animal relationship is also critical in indoor systems, where fearfulness and rough handling can cause poor welfare.

Several additional issues are common to all systems, including the use of painful management procedures, the need for handling and restraint, and neonatal morbidity and mortality. The following sections will consider more specific welfare aspects of small ruminant management.

Food and water

As ruminants, both sheep and goats are adapted to utilise low-quality roughage as food, and this contributes to their capacity to survive under some of the harshest environments. However, they can experience significant periods of undernutrition when the environment is unable to provide sufficient nutrients. This is more likely in extensive environments, as animals are more dependent on their ability to find food in the environment, than in more enclosed or indoor systems in which animals depend on humans to provide food. Malnutrition is also more commonly experienced by small ruminants in extensive conditions, where the balance of nutrients, including micronutrients, may be inappropriate.

Small ruminants are well adapted to cope with periods of food shortage and naturally reduce their voluntary food intake in the winter (Iason et al., 2000). They show behavioural adaptations, including movements about the home range, to ensure optimal use of the available forage if they are given the opportunity to do so, and an increase in foraging and grazing behaviour to maximise feed intake. However, these adaptations may not be sufficient to prevent sheep and goats experiencing the impacts of undernutrition, such as prolonged hunger or discomfort. In addition, the period of low forage availability often coincides with pregnancy for extensively managed animals, which increases metabolic demand, particularly for highly fecund animals. The impact of available forage on the welfare of sheep and goats is usually assessed by measuring Body Condition Score (BCS). This is an assessment of the amount of fat and muscle covering the lumbar vertebrae at the level of the last rib, often supplemented with an assessment of fat and muscle cover at the sternum in goats, assessed on a 5-point scale, where 1 is emaciated and 5 is obese. These measures are best made by manual palpation, especially in sheep in full fleece as body condition cannot be accurately assessed by visual inspection alone. Ideal body condition is between 3.0 to 3.5, and management should aim to maintain animals at this level year-round, as thin animals may experience prolonged hunger (Verbeek et al., 2011) and are prone to complications, such as pregnancy toxaemia, whereas fat animals are susceptible to obstetric disorders and metabolic disease. In very extensive farms, with unimproved pastures, supplementary feeding in winter can help to maintain body condition, but in lowland and fenced fields the use of improved pastures, regular grass height measurement, multi-species swards and rotational grazing are all management techniques that can be used to help improve nutrient availability. Although most undernutrition is due to inadequate availability of food, small ruminants may also experience hunger, even with adequate provision of feed, due to the loss of dentition (termed "broken-mouthed"), which prevents them from foraging or grazing effectively.

In extensive environments, sheep and goats are often reliant on natural water courses for drinking, which can become contaminated, or show seasonal variation in flow rates. Although well adapted to low water levels and buffered to some extent by water reserves in the rumen, drought conditions can cause very significant welfare problems. In intensive conditions, sheep and goats are much more reliant on humans to provide adequate water supplies and food, as they cannot satisfy their hunger and thirst by food and water seeking. Social factors can inhibit feeding behaviour, as subordinate animals may be prevented from feeding at preferred times, and food competition, for example when animals are supplemented with highly palatable concentrate feeds, can lead to aggression, displacements, and undernutrition in subordinate members of the social group. In these conditions ensuring adequate feeder space for all animals, and appropriate management of the social group size and composition, can reduce competition to access feed. As both small ruminant species can be horned, ensuring that sufficient space is provided to minimise injuries from aggression is important.

Physical environment

For extensively managed animals the welfare impact of the physical environment is generally through exposure to climatic extremes (e.g. extremes of heat, cold, or wet), and whether adequate shade and shelter is available. Northern temperate sheep breeds have dense woolly fleeces, and are well adapted, physically, behaviourally, and physiologically, to a cold, damp climate. Ewes in full fleece can remain within their thermal neutral range even at temperatures below freezing, provided they are dry, and can avoid the impacts of windchill through making use of shelter in the environment (either natural such as rocky outcrops, or manmade). When given a choice, sheep prefer to be outside even at very low temperatures (Piirsalu et al., 2020), although young lambs, and recently shorn sheep, will be less able to cope with low temperatures. Hair sheep (e.g. Blackhead Persian, Santa Inês) and goats are less resistant to cold and damp, and require better protection from wind, rain, and snow (Bøe and Ehrlenbruch, 2013). The ability to find a dry resting area is important for the welfare of both species, as wet and muddy or contaminated coats will significantly reduce their ability to resist cold temperatures.

Heat stress, and exposure to high temperatures, can also be significant issues in outdoor management. Access to shade is an important factor in the ability of animals to resist high temperatures, and competition to remain in the shade can occur if insufficient shade is provided. High temperatures will increase water intake (Silanikove, 2000), and can reduce feed intake and reproductive behaviour in both males and females. Sheep are generally less tolerant of hot and humid environments compared to goats. With climate change, an increase in sudden and extreme weather, such as flooding, snow fall, wildfires, and heat waves, can leave extensively managed animals vulnerable to catastrophic events, leading to very poor welfare and high mortalities, such as drowning, smothering, or burn injuries. These can be difficult for stockpeople to manage, where often human lives may also be at risk, but risk management and emergency planning can help to limit the impact of these events.

For animals maintained indoors, stocking density, quality of flooring, provision of bedding, and air quality are all important factors for physical comfort. Heat stress can also be important indoors, as insufficient ventilation, even at relatively low ambient temperatures, can cause panting and distress, especially in pregnant ewes in full fleece. At stocking densities with less than 1 m^2 per animal, displacements, aggression, and activity increases (Averos et al., 2014), suggesting competition for preferred lying areas. Sheep do not always show overt aggression (although butting and chasing occur at high stocking densities) but may still express dominant behaviour through directed eye gaze, pawing, chin-resting, and displacements. Subordinate animals may, therefore, be regularly moved and have reduced lying and resting times when insufficient space is provided. Small ruminants may be kept on solid or slatted floors, and bedded with straw, wood shavings, or other materials. Goats seem to prefer to lie on solid surfaces and to have access to elevated lying places (Andersen and Bøe, 2007), which may mimic a more mountainous, rocky environment. Access to this environment can also help to wear the hooves and prevent lameness from claw overgrowth in continually housed small ruminants. Sheep have been shown to prefer a bedded surface on which to lie, particularly when shorn (Faerevik et al., 2005), and newborn lambs need a bedded surface to help maintain body temperature. Indoor housed animals may develop calluses on knees and hocks if the bedding is inadequate (Stubsjoen et al., 2011).

Air quality is an important characteristic of small ruminant housing, as sheep and goats are susceptible to respiratory infection and heat stress if the ventilation is insufficient (Navarro et al., 2019). At low ventilation rates the air quality (concentrations of ammonia, carbon dioxide, and particulates) may be poor. This increases physiological stress markers and reduces behavioural activity, including feeding behaviour, immune responses and milk yield in lactating dairy sheep (Sevi et al., 2006).

Health and disease

Sheep and goats share many of the same endemic diseases that affect their welfare: chiefly lameness, internal and external parasitism, mastitis and reproductive disorders, especially dystocia (difficult births). They are also both susceptible to several infectious diseases, such as coccidiosis, Maedi-Visna, paratuberculosis, and Peste des petits ruminants. Some of these diseases can be controlled through vaccination, and concerted efforts have led to regional eradication in some cases. Disease management is challenging in extensive environments where infrequent inspections can reduce the likelihood of animals receiving prompt treatment. An exhaustive account of these different health issues is beyond the scope of this chapter. However, an overview of those issues considered to have the greatest impact on welfare (Rioja-Lang et al., 2020) are presented here.

Lameness

Lameness is a behavioural indicator of foot pain, ranging from mild gait abnormalities to animals ceasing to bear weight on an affected limb or becoming recumbent. The prevalence of lameness in both species can be as high as 9–10%, although this can be reduced by a half in sheep by implementation of best practice foot management (Winter et al., 2015). The main causes of lameness in sheep are infectious micro-organisms, with nearly 90% of lameness relating to footrot or scald caused by infection with *Dichelobacter nodosis*. This bacterium is widespread and can be transmitted between sheep in warm and moist conditions via pasture contamination. Infection causes pain and inflammation and, if untreated, can cause animals to lose condition, reduce lamb survival, growth rates and lactation. Although eradication has been attempted in some places, and vaccines against footrot exist, farmers' main approach is management of cases when they occur. Prompt treatment, with injectable and topical antibiotics, can reduce the incidence and pain associated with infection. Treating each case as it occurs helps reduce the welfare impact and can reduce the spread from one animal to another. However, if animals are seen infrequently, or individual treatment is challenging, animals may be lame for some time before treatment is given. Farmers' acceptance of a certain level of lameness in sheep as "normal" may also contribute to delays in treatment (Dwyer, 2009). Previous approaches to the presence of footrot, particularly excessive paring or foot-trimming, have now been shown to be unhelpful, and may even contribute to the spread of footrot between animals (Wassink et al., 2003). In general, if sheep can walk on hard surfaces, there is adequate natural wear of the hoof. The overgrown hooves seen in footrot result from lameness and pain, preventing the animal from walking properly to wear the hoof, rather than as a cause of lameness.

Although footrot can also affect goats, horn overgrowth and separation are more common causes of lameness in housed dairy goats kept on soft bedding (Can et al., 2016). Up to 90% of dairy goats in Europe may have overgrown claws, and improved walking ability can be seen after trimming (Ajuda et al., 2019). Foot trimming does, however, require capture, handling, and inversion, which are aversive. Foot trimming equipment can also be a means of spreading infection between animals unless these are kept scrupulously clean. Providing for exercise and the opportunity to walk on a hard surface can improve welfare by allowing natural wear of the hoof horn and reducing the need for foot trimming (Gelaskis et al., 2017).

Gastrointestinal parasites

Sheep and goats that are kept outdoors are susceptible to gastrointestinal parasites, through grazing contaminated pastures. These include blood-feeding stomach worms, such as *Haemonchus contortus* particularly in tropical regions, *Teladorsagia circumcincta* and *nematodirus* species, and liver fluke. Many of these parasites cause an anaemic response in the sheep or goat and can be identified through their impact on the colour of mucous membranes. Gastrointestinal parasites also cause discomfort, diarrhoea, dehydration, and loss of condition, and changes in behavioural expression (Grant et al., 2020). In young lambs or kids, when starting to ingest grass, infection can be a significant cause of pre-weaning mortality. The faecal soiling accompanying diarrhoea in infected animals can also cause an increased risk of flystrike or cutaneous myiasis.

High stocking densities contribute to the spread of infection, as well as poor pasture management which increases parasite load. Treatment of gastrointestinal parasites has frequently been by blanket drenching the whole flock or herd. However, a rise in the number of anthelminthic-resistant parasites has led to more targeted, alternative strategies to limit the development of resistance. In particular, the use of alternative forage types or mixed swards, such as chicory or plantain, can provide a more natural approach to reducing worm burdens. There is evidence that sheep and goats infected with parasites will self-medicate by increasing their intake of plants containing condensed tannins, which reduce worm burdens (Villalba et al., 2017).

Ectoparasites

Ectoparasites are organisms that infest the skin, wool, or coat of animals, and can cause lesions (and subsequent secondary infections), and intense discomfort, irritation, and itchiness. The major ectoparasites affecting the small ruminants include mites, lice, ticks, and blowfly larvae. Sheep scab, caused by infestation with mites, is highly contagious and has a major impact on sheep welfare. Scab is an acute or chronic form of allergic dermatitis, where the presence of the mites and its faeces cause the animal to produce a serous exudate at the skin surface on which the mites feed. This is accompanied by intense itchiness, and animals frequently rub against fence posts, pens, or other structures, bite their fleece and break off from feeding, lying, or other behaviours to scratch. Over time, if not treated, these discomfort behaviours occupy more of the animals' time, leading to wool loss and skin lesions, fits, and death (Corke and Broom, 1999). Sheep scab can be treated by plunge-dipping using an organophosphate dip or by injecting with endectocides. Both these procedures carry some risks – organophosphate products are highly toxic to humans, and the recommended endectocides are also used to treat internal worms and can increase anthelminthic resistance. Use of quarantine for all new animals brought onto the farm, and ensuring that sheep flocks do not mix, can reduce the incidence of the disease, minimising the need for treatment and protecting animals from infestation.

Blowflies are one of the most widespread ectoparasites affecting small ruminants, with up to 80% of sheep farms in the UK reporting at least one case each year, causing a condition called cutaneous myiasis or "flystrike". Different species of blowfly (Calliphoridae) are prevalent in different sheep- and goat-producing countries, with varying virulence, however the impact on the welfare of sheep or goats is similar. Female blowflies are attracted to dead animals, or live animals with wounds and soiled wool or hair, and lay their eggs in the warm, moist conditions typically found around the perineal region. Larvae hatch from the eggs and feed on the living tissues. This causes pain, discomfort and itchiness, as well as wool or hair loss at the site of the strike and a route for further infection. Animals with soiled coats around the anus (often called "dags"), through ingestion of rich grazing or gastrointestinal parasites, are more attractive animals for blowflies to attack. Preventative measures, such as clipping away soiled wool and using pour-on products, reduce the likelihood of infestation. Management procedures, such as tail-docking or mulesing (removal of folds of skin from the tail area, only in Australia), have been developed to reduce the risk of flystrike.

Mastitis

Mastitis is a bacterial infection of the udder in lactating animals, caused particularly by *Streptococcus* and *Staphylococcus* species, which results in inflammation, fever, and pain, sometimes severe, for the infected animal. This is more commonly observed in dairy animals, where it might be detected first by animals being restless or trying to avoid attachment of the milking machine, but it can also affect meat ewes or does. In a study in Australia of meat sheep, annually 1% of ewes had clinical mastitis (Munoz et al., 2018), and subclinical infections can also cause welfare concerns. Clinical mastitis involves physical changes in the udder (such as swelling and heat), sickness behaviour (lethargy), and animals may appear lame and reluctant to allow lambs or kids to suckle. In extensively managed animals the only evidence of mastitis might be slower growth rates in the offspring, or increased pre-weaning mortality, and changes in the udders of ewes (such as lumps or hard areas) seen after lactation, although the ewe may have suffered considerable pain earlier in the course of the disease.

Mastitis can be reduced by good management and hygiene. This is particularly important for indoor lambing/kidding pens, to prevent infectious agents passing between animals through contaminated bedding. In dairy animals, infection can also be caused by poor hygiene of milkers and milking machines, with manual milking associated with more mastitis than machine milking (Marogna et al., 2010). Physical injury to the udder or teats can also provide a route for infection. Mastitis is more frequent in ewes/does with high milk yield, and in females raising multiple offspring, where competition for milk may cause stress and physical injury to the udder. Mastitis is readily treatable with antibiotics, and anti-inflammatory drugs to reduce pain.

Dystocia

Dystocia is a prolonged or complicated birth process, that often requires human intervention to deliver the offspring. This can cause pain, haemorrhage, and exhaustion in the mother, and increases the risk of uterine infection and damage through interventions. In the offspring, dystocia causes hypoxia, and birth injuries, including cerebral haemorrhage and central nervous system damage. Birth difficulty is a significant contributor to mortality in both mother and offspring, implicated in the majority of pre-weaning mortalities in lambs and kids (Refshauge et al., 2016; Robertson et al., 2020). Dystocia increases stillbirth in viable offspring, and mortality of liveborn offspring through an increased risk of mothers showing reduced maternal care (Dwyer and Lawrence, 1998), and impacts on neonatal vigour, teat-seeking, and thermoregulation (Dwyer, 2003). Human interventions can help reposition lambs before cervical delivery, although very complicated presentations may require caesarean section. However, the timing of interventions is crucial, as unnecessary obstetric help can cause damage or injury, and may reduce mother-offspring bonding behaviour. Extensively managed animals are less likely to be observed in difficulty at a time when interventions will be able to prevent the deleterious consequences of dystocia. In these situations, often a goal of farm management is to develop a flock or herd where dystocia is less common and animals are more self-reliant.

Dystocia is related to multiple causes, and risk factors for a difficult delivery can be both animal and environmentally based. Dystocia is caused by offspring malpresentation, feto-pelvic disproportion, uterine inertia, delayed or incomplete cervical opening, disease, or congenital malformation in the offspring. Genetic factors (including breed and within breed selection), litter size, maternal nutrition, environmental stress, and exposure to, for example, phytoestrogens can all contribute to the risk of a difficult birth. It is possible to breed for an easier birth process, and this can reduce the risk of dystocia (Matheson et al., 2012) and consequently improve welfare. As prey species, ewes and does are vulnerable when giving birth, and have developed physiological mechanisms to delay giving birth if they feel threatened. If there is constant disturbance, or there is a poor human–animal relationship, parturient females may experience delayed or prolonged births as the effectiveness and frequency of uterine contractions are reduced with stress.

Pregnancy toxaemia

Pregnancy toxaemia occurs in late gestation in ewes and does and is primarily caused by inadequate nutrition in late gestation. This causes mobilisation of fat stores to provide sufficient glucose for the developing foetuses, but at high levels this can overwhelm the capacity of the liver to produce glucose, resulting in the production of ketones. As this is more common in dams carrying larger litters it is often known as twin-lamb disease, or pregnancy ketosis. The presence of ketones in the blood causes lethargy and a reduced appetite, which can exacerbate the condition, neurological symptoms through the poisoning effects of the ketones, and finally recumbency, coma, and death. Pregnancy toxaemia is considered one of the main causes of ewe mortality in some studies (Politis et al., 2021).

Ewes that are very thin (with a BCS of less than 2) or fat ewes (BCS greater than 4) towards the end of pregnancy are most at risk, although sudden loss of feed, stress, or other contributory health conditions, such as lameness or dental disease, also increase the incidence. Ewes in the early stages of pregnancy toxaemia can be treated with oral propylene glycol and encouraged to eat through provision of highly palatable food, such as molasses, and management changes to allow increased feeder space or protection from adverse weather. In the later stages of the disease, treatment is difficult and often euthanasia is required.

Painful management procedures

Small ruminants are subjected to several management procedures that can cause pain, some of which are undertaken for improved health or welfare management of the animal, and some to make management easier for the stockpeople. The most common of these are castration of male lambs and kids, tail docking and mulesing in sheep, and disbudding in kids.

Castration

Castration is usually carried out, within a few days of birth, to reduce unplanned matings, to avoid changes in sensory characteristics of meat in post-pubertal males, and to reduce the risk of injury in managing entire male animals. Several different techniques are routinely used, including use of tight rubber rings (elastration), banding, instruments designed to crush the spermatic cords (known as Burdizzo), and surgical approaches. In many countries the method or timing of the use of some of these methods without appropriate anaesthesia or analgesia may be restricted. For example, in the UK, castration with tight rubber rings without anaesthesia or analgesia is only permitted for lambs or kids under seven days of age and is banned in some European countries. Castration by any method has been shown to be associated with pain behaviours (e.g. rolling, kicking, stamping, abnormal postures: Molony et al., 2002), elevations of plasma cortisol and heart rate (Kells et al., 2020), and altered behavioural expression (Maslowska et al., 2020). These behaviours can persist for several hours after the procedure and can be reduced using local anaesthetics injected into the testes and scrotal neck, but not completely abolished (Kells et al., 2020). Subcutaneous, but not intramuscular, injections of non-steroidal anti-inflammatory drugs (NSAIDs) reduce pain behaviours in the 12 hours after castration (Paull et al., 2012), although not acute pain responses (Kells et al., 2020). More recently, formulations to deliver NSAIDs through a buccal route in small ruminants have been developed and can help to reduce pain expression to a greater degree than is achieved through local anaesthetic alone (Small et al., 2018).

The pain associated with castration in young males can interrupt the bonding between mother and offspring, and the presence of lesions and possible infections can increase the risk of mortality in males. In addition, male lambs and kids may have a growth check associated with pain, and castration can cause slower growth and result in a less commercially valuable carcase. For these reasons, farmers who keep fast-growing breeds of meat sheep or goats, where slaughter weights can be achieved before puberty, are less likely to castrate lambs or kids than previously. However, for hill farmers, with slower growing sheep breeds, their ability to lamb early in the year is restricted by the weather and the need to provide ewes with good grazing during lactation. In addition, a lack of fenced pastures on hill farms may make it impossible to keep entire male lambs away from females. The market for post-pubertal male lambs is a contributory factor in the continuing need for farmers to castrate males, but there is an urgent requirement for a painless method to achieve this.

Tail docking

For sheep, tail docking is routinely carried out to reduce the risk of faecal soiling of the breech area, which can be a risk factor for flystrike (see above). However, the evidence that tail-docking can reduce the incidence is limited (Sutherland and Tucker, 2011), with some studies showing no impact of tail length on the extent of dags or flystrike incidence (Fisher et al., 2004; Soriano et al., 2020). The use of other practices, such as regular shearing of the perineal area, insecticides, and topical applications of deterrents may be as effective and more ethical (Gascoigne et al., 2021).

Tail docking is generally carried out by the same methods as described for castration, or by using hot docking irons. Similar restrictions apply in many countries, and the procedure is associated with behavioural and physiological evidence of pain in the lamb, albeit at a lower level than seen for castration (Molony et al., 2002). The use of subcutaneous local anaesthetic drugs, such as bupivacaine, administered immediately before docking is effective at reducing these responses. However, there is some evidence for long term hyperalgesia and neuroma formation in the tail stump (Larrondo et al., 2019) and that tail docking may have longer lasting impacts on pain sensitivity and behaviour (Clark et al., 2014). Most countries recommend that tails should be docked to retain enough of the tail to cover the vulva and anus. Very short tail docking, where almost no tail is present at all, is associated with an increased risk of bacterial arthritis and rectal prolapse (Thomas et al., 2003; Lloyd et al., 2016), and may increase the risks of flystrike (Fisher et al., 2004).

Mulesing

Mulesing is another practice designed to reduce the incidence of flystrike, and is generally only carried out on Merino sheep, where the very wrinkled skin around the breech area provides an ideal environment for blowflies to lay their eggs. This process requires the removal of skin on either side of the anus, which then heals to a smooth, scar tissue which is less likely to become soiled. The procedure is banned in many countries but is still commonly practised in Australia. Mulesing is carried out by accredited contractors and occurs at the same time as several other procedures, including tail docking, ear marking, and vaccination. Lambs that have experienced this procedure show physiological and behavioural signs of pain that can last for several weeks and altered behavioural responses to humans for up to a month following mulesing (Fell and Shutt, 1989). Although there is no statutory requirement in Australia to provide analgesia or anaesthetic, there is evidence that pain-related behaviour can be partly reduced by use of NSAID and topical anaesthetics (Small et al., 2018). Attempts to develop Merino sheep which have been bred to be less wrinkled in the breech area may provide a permanent solution to the need for mulesing (Scobie et al., 2007).

Disbudding

Disbudding is usually carried out in goat kids soon after birth but is rarely practised with sheep. Disbudding is done to avoid handler or between-animal injury, especially when animals are kept in confined spaces. Disbudding can be carried out using caustic paste, scoops, or thermal cautery (Hempstead et al., 2018b). In general, pastes and scoops are not recommended methods due to the pain associated with these approaches and, with paste, the potential for causing burns to other parts of the animal. In many countries disbudding can only be carried out by a veterinarian, and requires the use of at least local anaesthesia, and often post-operative analgesia. In goat kids, the skull is thin around the site of the horn bud, and disbudding is often carried out under general anaesthesia to reduce the risk of inadvertently causing brain damage. Even with the use of local anaesthesia and analgesia, or general anaesthetic and NSAID, there is evidence of pain in kids post-operatively (e.g. head shaking, reduced growth rate: Hempstead et al., 2018a; Ajuda et al., 2020) for a number of days after the procedure. There are increasing numbers of polled breeds of animals, and cross-breeding or genetic manipulations may make it less likely that these procedures will be required in the future.

Behavioural interactions

Small ruminants are commonly managed in social groups, usually a breeding female flock/herd, with or without their offspring, and a separate male group, except at mating. In general animals are free to express most normal social behaviours and interactions, and, unless kept at high stocking density or with limited resources, conspecific aggression is rare. In animals kept for meat or fibre the offspring remain with their mothers for a relatively long period of time, up to 50% or more of natural lactation. Dairy animals can vary from very intensive systems, which require separation of the offspring from their mothers within a day of birth, to less intensive where the lamb or kid may suck from the mother for up to six weeks before a milking period. Mating in both species is often natural, through exposure of oestrus females to the ram or buck for a period of weeks, which can allow courtship and mating behaviours to be expressed. Stereotypic or abnormal behaviours are rarely seen in animals at pasture, but can occur in housed animals, especially if housed individually. The most frequently reported of these is wool-biting or chewing, where the wool of another animal, sometimes a more subordinate animal, is pulled out. This seems to be related to diet as increased provision of fibre reduces the expression of this behaviour. Other forms of oral stereotypy (licking, biting, or chewing pen fixtures, eating non-food items) and locomotor stereotypy, such as route tracing and repetitive rearing or jumping, also occur, almost always under conditions where animals are confined alone in small pens.

The main causes of welfare concern relating to behavioural interactions come from fear or distress often caused by separation from the social group, interactions with humans or interactions with other animals such as predators.

Fear and distress

As prey species, sheep and goats have specific and highly motivated behavioural adaptations to deal with potential threats from predators. These are maintained, regardless of whether a predator threat is present. This involves highly organised social behaviour, fear, and anxiety when socially isolated or in novel scenarios, and flight from a threat (Dwyer, 2004). All species will also use aggression, particularly head threats and butting, although entire males are more aggressive than females. Sheep are generally more fearful than goats, although less likely to use aggression as a response and more fearful of novel environments or events. Goats can be more curious and less fearful with novelty or potentially threatening situations.

Fear behaviour is expressed by increased vigilance (time spent with the head raised scanning the environment), flight, or panic reactions when flight is prevented. This can increase the chances of injury if panicking animals attempt to climb or jump out of an enclosure to avoid a perceived threat. A potent fear-inducing condition in these species is social isolation. Being part of the social group is an antipredator response, and not being in the social group is extremely stressful. This is usually seen as frequent loud "distress" vocalisations and attempts to re-join social companions, although vocalisations can be suppressed in conditions where there is a perceived predator present (such as a dog or human).

Human–animal relationships

Unless well-handled and familiarised with stockpeople from a young age, small ruminants, especially sheep, regard humans as potential predators. The presence of a stockworker can elicit the same behavioural responses (flight if approached too closely or too quickly) as a predator. For extensively managed animals in some countries this response is utilised for animal movement, often reinforced by using sheepdogs. Small ruminants will tolerate the presence of humans (and dogs) at a distance but maintain a "flight zone" around themselves whereby encroachments into this space will elicit movement away from the threat (Grandin, 2020). The size of the flight zone will vary with species, experience, breed and context, but for both species "low-stress handling" involves working at the outer edge of the flight zone, such that the animal moves away slowly and calmly. Rapid movement into the flight zone will elicit panic and flight, which is counter-productive and can result in injury.

In shepherded management systems, groups of small ruminants can be moved by following the herder, rather than driving the animals from behind. Sheep and goats have a pronounced "following" response, where they tend to follow the animal in front, and this can be used to move animals in a manner that elicits less stress than driving from behind. Dairy animals also encounter humans at close quarters far more frequently than sheep and goats kept for meat or fibre. The quality of the relationship between human and animal is vital for good welfare, and a poor relationship can cause fear, which also affects milk production. Animals milked in a parlour rapidly learn the order of entry and find their position in the parlour, particularly if reinforced by food rewards. Calm and consistent behaviour by stockpeople helps to reinforce this learning, reduces fear and uncooperative behaviour (such as baulking or turning back) and makes the experience more pleasant for both human and animals.

Predation

Small ruminants are often farmed in environments where large predators (wolves, coyotes, bears, etc.) still live. Attacks on small ruminants may also occur from uncontrolled domestic dogs. Sheep and goats have limited defences against predators other than attempting to escape to higher ground (if available) or to run. Small ruminants with horns defend themselves or their young from avian predators to some extent, but in general they have few opportunities to avoid the impact of predation. Predation is therefore still a significant threat to welfare in countries with high predator density, particularly on young lambs.

Predation is obviously a severe welfare issue where sheep or goats are caught and killed or injured by predators. However, the presence of predator also acts as a fear stimulus, increasing vigilance and anxiety, and animals will avoid areas of pasture where predator attacks have occurred for prolonged periods. Prolonged chasing, as can occur particularly with domestic dogs, can lead to exhaustion, injury, and abortion in pregnant animals. Methods of dealing with predators involve fencing, bringing animals in at night, lethal predator control, shepherding, and the use of guardian animals. Although shepherding or bringing animals into housing can be very effective means of dealing with predation, in some systems, it is not practical or possible. Use of guardian animals (usually dogs but also donkeys or llamas) can be effective alternatives (van Bommel and Johnson, 2017). This involves rearing dogs with the flock from a young age, where the dog effectively becomes part of the social group, and the dog will then protect the animals directly through interactions with predators or deter predation and attacks through its presence.

Conclusions

In many systems, small ruminants are kept in extensive outdoor environments, which avoids excessive confinement and replicates to some extent the natural environment in which these species evolved. In these systems animals have considerable behavioural freedom and opportunities to express positive welfare, through social interactions and environmental complexity. The consequences of management in these environments, however, can be a reduction in human–animal contact and an inability to provide individualised health and welfare treatments. This often means that the most severe welfare impacts are through untreated, or sometimes undiagnosed, disease and injury, impacts of predation and through severe environmental changes. This can lead to higher incidence of morbidity and mortality in these systems, compared to indoor management.

In dairy systems animals may be kept more confined, or completely indoors, with daily movements to the milking parlour, although grazing opportunities are still offered in many cases. The quality of the human–animal relationship, space available to each animal and the housing environment are all potential risk factors for poor welfare, as well as early separation of the offspring from the mother. In shepherded, and more pastoral dairy systems, the milking period may be more related to the natural production cycle of the ewe or doe, including a suckling period, and animals may have a positive relationship with the shepherd who leads them to fresh grazing.

Overall, there are considerable opportunities for improving welfare for the management of sheep and goats, since several significant challenges to welfare exist. Many of these can be overcome by good management, veterinary care, and sensitive shepherding. Labour shortages, which may require small ruminants to be more resilient or self-sufficient, and climate change, which can increase the severity of unpredictable environmental events, are increasing risk factors for the welfare of small ruminants.

References

Ajuda IGG, Battini M and Stilwell G, 2019. The role of claw deformation and claw size on goat lameness. *Veterinary and Animal Science*, 8, p. 100080.

Ajuda IGG, Battini M, Mattiello S, Arcuri C and Stillwell G, 2020. Evaluation of pain mitigation strategies in goat kids after cautery disbudding. *Animals*, 10, p. 277.

Andersen, IL and Boe KE, 2007. Resting pattern and social interactions in goats - The impact of size and organisation of lying space. *Applied Animal Behaviour Science*, 108, pp.89–103.

Averos X, Lorea A, de Heredia IB, Ruiz R, Marchewka J, Arranz J and Estevez I, 2014. The behaviour of gestating dairy ewes under different space allowances. *Applied Animal Behaviour Science*, 150, pp.17–26.

Boe KE and Ehrlenbruch R, 2013. Thermoregulatory behavior of dairy goats at low temperatures and the use of outdoor yards. *Canadian Journal of Animal Science*, 93, pp.35–41.

Can E, Vieira A, Battini M, Mattiello S and Stilwell G, 2016. On-farm welfare assessment of dairy goat farms using animal-based indicators: the example of 30 commercial farms in Portugal. *Acta Agriculturae Scandinavica Section A: Animal Science*, 66, pp.43–55.

Clark C, Murrell J, Fernyhough M, O'Rourke T and Mendl M, 2014. Long-term and trans-generational effects of neonatal experience on sheep behaviour. *Biololgy Letters*, 10, p. 20140273.

Corke MJ and Broom DM, 1999. The behaviour of sheep with sheep scab, Psoroptes ovis infestation. *Veterinary Parasitology*, 83, pp.291–300.

Dwyer CM, 2003. Behavioural development in the neonatal lamb: effect of maternal and birth-related factors. *Theriogenology*, 59, pp.1027–1050.

Dwyer CM, 2004. How has the risk of predation shaped the behavioural responses of sheep to fear and predation. *Animal Welfare* 13, pp.269–281.

Dwyer CM, 2009. Welfare of sheep: providing for welfare in an extensive environment. *Small Ruminant Research*, 86, pp.14–21.

Dwyer CM and Lawrence AB, 1998. Variability in the expression of maternal behaviour in primiparous sheep: effects of genotype and litter size. *Applied Animal Behaviour Science*, 58, pp.311–330.

Faerevik G, Andersen IL and Boe KE, 2005. Preferences of sheep for different types of pen flooring. *Applied Animal Behaviour Science*, 90, pp.265–276.

Fell LR and Shutt DA, 1989. Behavioural and hormonal responses to acute surgical stress in sheep. *Applied Animal Behaviour Science*, 22, pp.283–294.

Fisher MW, Gregory NG, Kent JE, Scobie DR, Mellor DJ and Pollard JC, 2004. Justifying the appropriate length for docking lambs' tails: a review of the literature. *Proceedings of the New Zealand Society of Animal Production*, 64, pp.293–296.

Gascoigne E, Mouland C and Lovatt F, 2021. Considering the 3Rs for castration and tail docking in sheep. *In Practice*, 43, pp.152–162.

Gelaskis AI, Valergakis GE and Arsenos G, 2017. Health and welfare of indigenous goat breeds from dairy farms in Greece. In (ed. Simões J., Gutiérrez C.) *Sustainable Goat Production in Adverse Environments: Volume 1*. Springer. Cham, pp.223–246.

Grandin T, 2020. Behavioural principles of handling cattle and other grazing animals under extensive conditions. In: (ed. T. Grandin) *Livestock Handling and Transport*. CAB International, pp.63–86.

Grant EP, Wickham SL, Anderson F, Barnes AL, Fleming PA and Miller DW, 2020. Behavioural assessment of sheep is sensitive to level of gastrointestinal parasite infection. *Applied Animal Behaviour Science*, 223, 104920.

Hempstead MN, Waas JR, Stewart M, Cave VM and Sutherland MA 2018a. Evaluation of alternative to cautery disbudding of dairy goat kids using behavioural measures of post-treatment pain. *Applied Animal Behaviour Science*, 206, pp.32–38.

Hempstead MN, Waas JR, Stewart M, Zobel G, Cave VM, Julian AF and Sutherland MA, 2018b. Pain sensitivity and injury associated with three methods of disbudding goat kids: Cautery, cryosurgical and caustic paste. *The Veterinary Journal*, 239, pp.42–47.

Iason GR, Sim DA and Gordon IJ, 2000. Do endogenous seasonal cycles of food intake influence foraging behaviour and intake by grazing sheep? *Functional Ecology* 14, pp.614–622.

Kells NJ, Beausoleil NJ, Godfrey AJR, Littlewood KE, Ward RN and Johnson CB, 2020. Effect of analgesic strategies on pain behaviour associated with combined ring castration and hot iron tail docking in Merino lambs. *Applied Animal Behaviour Science*, 222, p. 104914.

Larrondo C, Bustamante H, Paredes E and Gallo C, 2019. Long-term hyperalgesia and traumatic neuroma formation in tail-docked lambs. *Animal Welfare*, 28, pp.443–454.

Lloyd J, Kessell A, Barchia I, Schroder J and Rutley D, 2016. Docked tail length is a risk factor for bacterial arthritis in lambs. *Small Ruminant Research*, 144, pp.17–22.

Marogna G, Rolesu S, Lollai S, Tola S and Leori G, 2010. Clinical findings in sheep farms affected by recurrent bacterial mastitis. *Small Ruminant Research*, 88, pp.119–125.

Maslowska K, Mizzoni F, Dwyer CM and Wemelsfelder F, 2020. Qualitative behavioural assessment of pain in castrated lambs. *Applied Animal Behaviour Science*, 233, p. 105143.

Matheson SM, Bunger L and Dwyer CM, 2012. Genetic parameters for fitness and neonatal behavior traits in sheep. *Behavior Genetics*, 42, pp.899–911.

Molony V, Kent JE and McKendrick IJ, 2002. Validation of a method for assessment of an acute pain in lambs. *Applied Animal Behaviour Science*, 76, pp.215–238.

Munoz C, Campbell A, Barber S, Hemsworth P and Doyle R, 2018. Using longitudinal assessment on extensively managed ewes to quantify welfare compromise and risks. *Animals* 8, p. 8.

Navarro T, Ramos JJ, de Arcaute MR and Gonzalez JM, 2019. Predisposing factors inducing ovine respiratory complex in intensive-reared lambs. *Small Ruminant Research*, 180, pp.106–111.

Paull DR, Small AH, Lee C, Palladin P and Colditz IG, 2012. Evaluating a novel analgesic strategy for ring castration of ram lambs. *Veterinary Anesthesia and Analgesia*, 39, pp.539–549.

Piirsalu P, Kaart T, Nutt I, Marcone G and Arney D, 2020. The effect of climate parameters on sheep preferences for outdoors or indoors at low ambient temperatures. *Animals*, 10, p. 1029.

Politis AP, Vasileiou NGC, Cripps PJ, Liagka DV, Boufis PT, Valasi I, Mavrogianni VS and Fthenakis GC, 2021. Mortality of dairy sheep during the peri-parturient period: results of a field investigation in Greece. *Animals*, 11, p. 2172.

Refshauge G, Brien FD, Hinch GN, and van de Ven R, 2016. Neonatal lamb mortality: factors associated with the death of Australian lambs. *Animal Production Science*, 56, pp.726–735.

Rioja-Lang FC, Connor M, Bacon HJ, Lawrence AB and Dwyer CM, 2020. Prioritization of farm animal welfare issues using expert consensus. *Frontiers in Veterinary Science*, 6, p. 495.

Robertson SM, Atkinson T, Friend MA, Allworth MB and Refshauge G, 2020. Reproductive performance in goats and causes of perinatal mortality: a review. *Animal Production Science*, 60, pp.1669–1680.

Scobie DR, O'Connell D, Morris CA and Hickey SM, 2007. A preliminary genetic analysis of breech and tail traits with the aim of improving the welfare of sheep. *Australian Journal of Agricultural Research*, 58, pp.161–167.

Sevi A, Albenzio M, Annicchiarico G, Caroprese M, Marino R and Santillo A, 2006. Effects of dietary protein level on ewe milk yield and nitrogen utilization, and on air quality under different ventilation rates. *Journal of Dairy Research*, 73, pp.197–206.

Silanikove N, 2000. Effects of heat stress on the welfare of extensively managed domestic ruminants. *Livestock Production Science*, 67, pp.1–18.

Small AH, Marini D, Dyall T, Paull D and Lee C, 2018. A randomised field study evaluating the effectiveness of buccal meloxicam and topical local anaesthetic formulations administered singly or in combination at improving welfare of female Merino lambs undergoing surgical mulesing and hot knife tail docking. *Research in Veterinary Science*, 118, pp.305–311.

Soriano VS, Stamm FO, Taconeli CA and Molento CFM, 2020. To dock or not to dock? Faecal soiling measurement in sheep. *Animal Welfare*, 29, pp.81–87.

Stubsjoen SM, Hektoen L, Valle PS, Janczak AM and Zanella AJ, 2011. Assessment of sheep welfare using on-farm registrations and performance data. *Animal Welfare*, 20, pp.239–251.

Sutherland MA and Tucker CB, 2011. The long and short of it: a review of tail docking in farm animals. *Applied Animal Behaviour Science*, 135, pp.179–191.

Thomas DL, Waldron DF, Lowe GD, Morrical DG, Meyer HH, High RA, Berger YM, Clevenger, DD, Fogle GE, Gottfredson RG, Loerch SC, McClure KE, Willingham TD, Zartman DL and Zelinsky RD, 2003. Length of docked tail and the incidence of rectal prolapse in lambs. *Journal of Animal Science*, 81, pp.2725–2732.

Van Bommel L and Johnson C, 2017. Predation Control. In: (eds. D.M. Ferguson, C. Lee, A. Fisher) *Advances in Sheep Welfare*. Woodhead Publishing, Elsevier, pp.177–196.

Verbeek E, Waas JR, McLeay L and Matthews LR, 2011. Measurement of feeding motivation in sheep and the effects of food restriction. *Applied Animal Behaviour Science*, 132, pp.121–130.

Villalba JJ, Costes-Thire M and Giinane C, 2017. Phytochemicals in animal health: diet selection and trade-offs between costs and benefits. *Proceedings of the Nutrition Society*, 76, pp.113–121.

Wassink GJ, Grogono-Thomas R, Moore LJ and Green LE, 2003. Risk factors associated with the prevalence of footrot in sheep from 1999 to 2000. *Veterinary Record*, 152, pp.351–358.

Winter JR, Kaler J, Ferguson E, Kilbride AL and Green LE, 2015. Changes in prevalence of, and risk factors for, lameness in random samples of English sheep flocks: 2004–2013. *Preventive Veterinary Medicine*, 122, pp.121–128.

9

FARMING NON-DOMESTICATED AND SEMI-DOMESTICATED TERRESTRIAL SPECIES

David Arney

Introduction

Domestication is a rather slippery term that needs definition. It is to be regarded not as a state of being, but as a process (Clutton-Brock, 1992). Price (2002) has defined this as "that process by which a population of animals becomes adapted to man and to the captive environment by some combination of genetic changes occurring over generations and environmentally induced developmental events recurring during each generation". According to Price then, and as he identifies, domesticated animals include those invertebrates such as silkworms, oysters, prawns, and honeybees that have been bred and kept captive for our use. If it is agreed that domestication is a process and not a classification of types of species or subspecies, then any animal that is taken from the wild and used for farming over any cycles of generation can be considered to have begun upon the path to domestication. Nevertheless, the terms non-domesticated and partially domesticated are used here to distinguish animals that have been, over many thousands of generations, in close contact with humans (cattle, sheep, goats, and pigs) and those species that are kept for faming purposes but have either been kept for relatively few generations (the silver fox, the musk deer) or have been kept for faming purposes with a light touch of interference or captivity by their human owners (reindeer).

The animals discussed here, and their welfare, are less well studied and reported than other more traditional livestock. Our understanding of their needs and problems is therefore comparatively inadequate, and their requirements may well be different from those of other, better studied species. The rearing systems of these species are described, as well as known problems and how these might be ameliorated. Other problems that are a consequence of their management and that cannot easily be ameliorated are identified, and suggested extant methods to evaluate their welfare are presented.

There are more undomesticated and partially domesticated terrestrial animals that are farmed than are discussed here, such as yak, crocodiles, iguanas, guinea pigs, emus, and ostriches. However, these are of only localised importance and their market influence is limited in global terms, hence they are not considered here. However, it is important to realise that there are some major welfare concerns that are important to individual animals, for example rearing crocodiles in isolation to avoid blemishes on their skin. Invertebrates too are farmed, and can be considered domesticated, such as insects and snails for human food, but although the evidence for sentience

DOI: 10.4324/9781003182351-11

in some invertebrate species is growing, they are excluded here because they are not thought to suffer in the same way as vertebrate species.

Fur-bearing animals

Many production animals produce fibre or pelt products as an important part of their value to the farmer and to the market. Fur animals here are understood to be animals kept for the primary purpose of the production of fur. The production process, at least in Europe, is currently that the animals are bred and reared mainly in northern European countries (principally Denmark, Finland, Poland, Sweden, Lithuania, and Russia), their pelts are sold at auction, mostly at the Copenhagen fur market (Kopenhagen Fur), and the purchased pelts are used to make garments and accessories, commonly in Greece, from where they are exported all over the world. Outside of Europe fur production is important and increasing in China (Sha et al., 2011) and North America (Fur Commission USA, 2019) including Canada.

The legal approval for the farming of fur-bearing animals is changing rapidly, and this has had an impact on fur production numbers, at least in Europe. This has been exacerbated by the COVID-19 pandemic in the early 2020s where, following transmission identified among mink, the Danish farmed mink population was exterminated. Uncertainty as to whether fur farming is likely to be banned has affected producers elsewhere, who are less inclined to fund improvements to their management and housing systems if they are uncertain whether they will be permitted to continue to produce fur animals in the future. Fur farmers in Estonia have voiced frustration at the frequent changes to the fur farming regulations. Meeting demands for new regulations regarding the well-being of animals in their care and also environmental impacts of fur farming, in particular the risks of escapes of animals into the wild environment, causes problems in the future planning of their fur farming operations. An amendment was passed to the Animal Protection Act (Riigiteataja, 2021) that bans fur farming in Estonia, "It is prohibited to keep, breed and propagate animals solely or mainly for the purpose of production of fur" and this will come into force in 2026. Other European countries that have banned fur farming include the United Kingdom, Austria, North Macedonia, Slovenia, Croatia, Luxembourg, Serbia, the Czech Republic and the Netherlands, Belgium, France, Norway, Slovakia, and Bosnia and Herzegovina. While these countries generally had small, if any, numbers of fur producers, it is likely that other countries will follow. The ethics of keeping animals for fur has been reviewed (e.g., Arney and Piirsalu, 2017) and there is strong public support for change. Public understanding and ability to articulate robustly their reasoning for their abhorrence for fur farming may be limited but their feeling that it is wrong is potent and influences policy makers and legislators. And this is not surprising. Wittgenstein tells us, as described and clarified by Rée (2019), that there are some things, such as feelings, love, ideas of rights, for which we do not have the language to adequately define or describe. It is difficult then to argue on the grounds of evidence and philosophical debate against a public that wishes to ban fur farming. For fur farm producers, unlike the farmers of other animals, there is not really the traction in improving welfare for the general public, or indeed policy makers. Improvements to the well-being of pigs by adding straw to pens or adding puzzle toys to laboratory rat cages or describing how distress might be lessened by removing calves from dams early have a point and can help justify their use. But providing access for mink to water to improve their welfare and quality of life does not really help ameliorate the public perception of mink farms. And this is an important distinction. For fur farming, welfare concerns do not really come into it. The general public just think that it is wrong, and that is that. Nevertheless, fur farming continues to be practised and the welfare of these animals remains of concern and should be ameliorated such that the quality of their lives is as good as can be achieved given the constraints of their farming conditions.

Commonly kept animals for fur include the chinchilla (*Chinchilla lanigera*) Figure 9.1, the red fox *(Vulpes vulpes)* Figure 9.2, the silver fox (*Vulpes vulpes*) Figure 9.3, the blue fox (*Vulpes lagopus*), the mink (*Neogale vison*) Figure 9.4, the raccoon dog/Finnraccoon (*Nyctereutes procyonoides*) Figure 9.5, and the rabbit (*Oryctolagus cuniculus*).

The evaluation of the welfare of fur-bearing animals has been examined, tested, and codified into protocols for mink (Møller et al., 2015), foxes (Ahola et al., 2015), and Finnraccoons (Koistinen et al., 2014). These are based on the Welfare Quality protocols for cattle (Welfare Quality® 2009a), pigs (Welfare Quality® 2009b), and poultry (Welfare Quality® 2009c), with four welfare fundamentals: good housing, good feeding, good health, and appropriate behaviour, all of which feed into the fifth fundamental: positive affective (mental) states. These include rel-

Figure 9.1 Chinchilla, photo Peep Piirsalu.

Figure 9.2 Red fox, photo Peep Piirsalu.

Figure 9.3 Silver fox, photo Peep Piirsalu.

Figure 9.4 Mink, photo Peep Piirsalu.

evant content and animal-based measures for each of the species (Mononen et al., 2012). These protocols also include an assessment of stockmanship quality by estimating the quality of the human–animal relationship through a feeding test, temperament test and extent of handling and transportation of the animals. The fur farming industry appears to have adopted these as appropriate and they are recommended for use by fur producers' organisations (Fur Europe, Sagafurs, Furmark). A complication for these assessment protocols in fur production systems is the greater seasonality of the husbandry and management of these species compared to, say, cattle pigs and poultry. The annual production cycle is much less flexible in regards to mating times, whelping times, and slaughter ages. Different welfare outcomes might be expected from the animals at the various stages in the cycle, necessitating repeated visits during the year by assessors.

There are a range of concerns regarding the quality of the life experienced by farmed fur-bearing animals, some of which can be eased by management practices, but some of which appear to be inevitable consequences to these undomesticated animal species of their being

Figure 9.5 Raccoon dog/Finnraccoon, photo Peep Piirsalu.

caged, handled, and restricted in access to resources that are of importance to them. While these animals may thrive as far as health and productivity are concerned, they may nevertheless suffer from an inability in caged systems, and caged conditions are what prevail, to carry out the range of behaviours that are important to them. Few mink farms offer access to water, even though this is important to them (Mason et al., 2001). In Mason et al.'s study, mink were prepared to work (push open weighted gates) harder to get access to water than toys or a raised platform (both of which are known to be desirable resources for them). Deprivation of access to water raised urinary cortisol as much as feed deprivation. While it is not clear whether these deprived mink were water-experienced or not, this does show that the motivation and frustration of denial to perform such behaviour is important to these animals. This does not mean that the provision of other environmental enrichment is unimportant; boredom is a concern and can be relieved by the addition of such enrichments as rubber toys, shelves, and troughs of running water in mink (Meagher and Mason, 2012). Silver foxes too suffer from stress, as evidenced by behavioural and cortisol indicators, but it can, as with other animals (Neely et al., 2018), be reduced by handling in early life (Pedersen and Jeppesen, 1990). Although caging of fur-bearing animals is a problem for foxes, mink, and raccoon dogs, it may be less important for these solitary species, which in the wild spend time resting in confined spaces, than for rabbits, which are social animals. Housing rabbits in small cages restricts their spectrum of behaviours (Dixon et al., 2010), such as burrowing, foraging, and the full range of social interactions with conspecifics.

Stereotypical behaviour, an indicator of poor environmental conditions, has been observed in farmed mink (pacing, somersaults, circular movements of the head (Hansen et al., 2010)), blue foxes (pacing, tail-chasing, cage-biting, tail biting (Korhonen et al., 2001)), chinchillas (fur-chewing bar chewing, cage scratching, and backflipping (Franchi et al., 2016)), raccoon dogs (pacing, scratching at cage, head twirling, and biting or licking the cage (Koistinen et al., 2018)) and rabbits (biting bars and smelling bars (Mugnai et al., 2009) and repetitive hair-chewing, bar-chewing, head-swaying, and pawing (Gunn and Morton, 1995)). In rabbits, the motivation for and expression of these stereotypical behaviours may be reduced by group housing, although this can lead to agonistic behaviour (Mugnai et al., 2009), so their welfare might be jeopardised in group housing too. Genetic selection (Hansen et al., 2010) for lower rates of stereotypical expression may not actually improve the well-being of the mink, as selected animals show signs of increased fear (Svendsen et al., 2007). Fear, particularly fear of the approach of humans, is also a distressing state for any animal that is in regular contact with humans. Mink do demonstrate fear of humans, and less fear of humans is one of the attributes that is thought necessary for the successful domestication of a species. This fear is a predictable consequence of their recent partial domestication, although this can be moderated by genetic selection (Malmkvist and Hansen, 2002). This fear response then seems to be heritable, and therefore may be manageable by producers. Encouragingly, the selection for reduced fear response has no linked negative outcomes on their production values (Thirstrup et al., 2019).

Deer

Species of deer that are farmed are principally Reindeer (*Rangifer turndus*) and Red Deer (*Cervus elaphus*), but also Fallow Deer (*Dama dama*) and Musk Deer (*Moschus moschiferus* and several subspecies). The FAO (de Vos, 1982) also lists the Wapiti (*Cervus canadensis*), Sika (*Cervus nippon*), and Rusa Deer (*Cervus timorensis*) among farmed deer globally. Rearing deer can be an attractive option for farmers as their meat is of high value, regarded as healthy compared to other more traditional meats, and hence the numbers of deer and deer farms is increasing (Proskina and Cerina, 2021). These animals are all scarcely domesticated, if at all, and therefore if they are to be farmed this should be in extensive systems with as little contact with humans as possible.

Red deer and fallow deer

Red Deer and Fallow Deer are farmed extensively and their marketable products are principally venison meat and the soft velvet covering their antlers (for traditional Chinese medicine and other alternative medical offerings). They are mostly kept at pasture but may be housed during the winter (Bartoš and Šiler, 1993), although this is not thought necessary for adults and can lead to aggression and injuries (Pollard and Littlejohn, 1998). This can also be a problem at pasture if stocking densities are too high, which is considered to be over 8 red deer or 16 fallow deer per hectare (see review by Mattiello, 2009). Welfare concerns in deer include predation, which can cause mortality of up to 50% in Italy (Mattiello, 1994), poor fencing entangling individuals, lack of shelter, no access to a wallowing area, handling, restraining, loading, transport, and slaughter. Handling procedures for transport (Waas et al., 1999b; Bornett-Gauci et al., 2006;) and slaughter are particularly distressful to deer, which are only partially domesticated, and are usually infrequently handled. Although a review by Weeks (2000) suggests that deer's experience of transport might not be expressly different from that in other ruminants, she does recognise that deer are more flighty and should be provided with specialist handling and transport facilities. For this reason it is thought that on-farm slaughter, through shooting by a marksmen, is preferable to slaughter at an abattoir or on-farm in a mobile slaughterhouse (Bornett-Gauci et al., 2006).

The feeding of deer is different from cattle and sheep. They have more highly variable seasonal intakes, they are intermediate feeders (they both browse and graze), and will preferentially select browse if it is available, and pastures suitable for cattle and sheep are not necessarily suitable for deer (Mulley, 2003). This difference in feeding behaviour should be recognised by deer farming management systems to maximise welfare and productivity, such as by providing browse and adjusting stocking rates at pasture and possibly offering concentrates in different seasons.

Mattiello (2009) suggests some criteria for the evaluation of deer welfare on farms, and similarly criteria have been described for wild deer (Green, 2016). The evaluation of pelt-biting, as a record of incidences of agonistic behaviour, has been proposed as a stand-alone indicator of poor welfare (Pérez-Barbería et al., 2021). Typical stereotypic behaviours of red deer include wall pacing and vertical/horizontal head movements at the walls of pens (Pollard and Littlejohn, 1996). However, no established protocols have been devised and accepted for the evaluation of the welfare of on-farm deer.

A particular mutilation of deer is the removal of antlers while they are still growing, with severance of nerves and blood vessels, in order to remove their soft velvet covering. This is a product that is valued in the Far East and is peculiar only to those deer farmers, including in New Zealand and North America, that supply these markets (Putman, 1988; Conaglen et al., 2003). Among other claims, deer velvet is said to improve sexual function in human males, a claim which has been tested and for which there is no evidence (Conaglen et al., 2003), and human sport/exercise performance, which has also been tested and for which there is no evidence (Sleivert et al., 2003). To harvest the antler velvet the antlers are surgically removed; best practice includes anaesthesia, for which a range of techniques and drugs are proposed (Johnson et al., 2005), and removal of the antlers by a veterinarian. Where this is not followed and regulated, the suffering experienced by the stags is extremely high. There is concern about the duration of the analgesic effect of administered drugs, the pain experienced by the deer post-operatively, and the distress of the stags when they are isolated, confined, and handled for this procedure (Wilson and Stafford, 2002). Additionally, in regard to human well-being, there is also the concern of drug and drug metabolite residues entering the velvet (Walsh and Wilson, 2002) and subsequently into the humans that consume it.

Reindeer

Reindeer are reared in extensive conditions. They mate without human assistance or selection of mates by herders, except inasmuch as animals planned for slaughter are removed from the breeding herd. They calve on their own, without human intervention; forage for their own feed; are not routinely given supplementary feed except in the winter, when they may be provided with hay and concentrates at pasture; are not housed; and only have close human contact in the summer, when they are collected together for the ear-marking of unmarked animals. The handling and coercion involved in this is known to be particularly stressful for reindeer (Rehbinder et al., 1982). They are then let free again to wander and forage as they please until the autumn when they are again collected together (Figure 9.6) for selection of animals for slaughter, parasitic treatment, and vaccination. Some herders then release them back into the forest for the winter, others keep them in in-by pens. In the latter case they are given hay and sometimes small amounts of concentrates. They are then released into the wild in early spring. Around 100,000 reindeer are slaughtered each year in Finland (Askola SJ, personal communication). Slaughter may be on-farm, but in the EU this must be done, for reasons of hygiene (Rehbinder and Hau, 2006), in a slaughterhouse. Animals for slaughter are inspected by a veterinarian within 24 hours prior to slaughter. The transport of reindeer to slaughterhouses is stressful and leads to impaired welfare and carcase quality, even if the transportation distance is short (Laaksonen et al., 2017).

The welfare of these animals in the wild is not necessarily good, since they can suffer from predation, exposure to the weather, endo- and ecto-parasites, particularly the warble fly (Waller, 2002), and the annoyance of biting and blood-sucking flies in the summer (Kynkäänniemi et al., 2014). Foraging can be difficult, especially in areas where stocking densities are high and availability of feed is poor, which has become more of a problem recently and has led to the necessity for the provision of feed in the winter, which was not part of traditional reindeer herding practice. If, as expected, the climate warms and becomes wetter it might be expected that snow cover will be deeper than previously, making foraging in the winter months even more difficult.

The evaluation of the welfare of reindeer by estimating cortisol metabolites in faeces has been proposed (Özkan et al., 2019), but otherwise no protocols for their welfare are available.

Figure 9.6 Reindeer round-up, photo Aino Põder.

Musk deer

The last of the deer to be considered here, musk deer, are not true deer, being more closely related to bovines than to the cervidae. Their principal product is musk, which is secreted by the males from their scent gland. Their production levels are very small and farmed production is localised in the Far East (Parry-Jones and Wu, 2001). These authors identified problems of high disease and high mortality in these farmed musk deer, but it has been proposed that both of these indicators can be reduced with larger enclosures (Liu et al., 2010). The welfare problems of farmed musk deer include the binding of hind legs to prevent jumping and the clipping of tusks, to prevent injury to conspecifics and stockpeople, see review by He et al. (2014). Another welfare problem is that musk deer are naturally solitary animals in the wild, and this can be incompatible with group living. Other welfare problems identified by He et al. (2014) may be better managed with more understanding, such as inappropriate nutrition (they are often treated as grazers rather than the browsers that they are, with unsuitable feed offered to them, perhaps including concentrates), the lack of shelter from the weather on many farms and little genetic diversity in the farmed population, consequent to genetic drift from the small initial population collected from the wild, possibly leading to high rates of negative traits associated with dystocia (difficulties during birth or its prolongation), infant mortality, and morbidity.

Rabbits

Rabbit farming systems for their meat as the primary product can be attractive for farmers as they have low start-up costs, the animals are precocious and prolific, with a shorter gestation period than other livestock species, and have low husbandry and feed costs. In Europe farming rabbits can often be an adjunct to the production of the main livestock animal of a farm. In which cases it may be that they receive less attention and resources.

The biggest welfare problems for farmed rabbits are due to their partial domestication and incomplete adaptation to confinement in farm conditions (see review by Verga et al., 2009). Breeding animals are usually kept singly, while finishing animals may be kept in small groups. For a social animal this might be assumed to be stressful, and behavioural indicators suggest this is so (Whary et al., 1993), although there can also be high rates of aggression among does in groups (Ruis, 2006). A solution to this, of providing mirrors in single-housing laboratory rabbit cages, has been proposed (Edgar and Seaman, 2010), although the rabbits seem to respond to a mirror not as if to a conspecific it may nevertheless be an environmental enrichment similar to a soft toy (Jones and Phillips, 2005). Other problems discussed by Verga et al. (2009) include high stocking densities, too-early weaning, insufficient environmental enrichments, space allowance, transport, lairage and slaughter, and poor or lack of early handling by their keepers. Some of these, such as stocking rates, will have an economic cost to improve (Verspecht et al., 2011), while others could be readily improved through improved husbandry techniques. Even the provision of a simple environmental enrichment, such as a hanging wooden stick, reduced observed stereotypies in rabbits while also having a positive effect on production performance (Luzi et al., 2003), possibly through the effect gnawing a hard material on reducing incisor overgrowth. Cage sizes are of concern, if they are too small the rabbits will be unable to hop and interact as they would be motivated to do. While on the face of it, it might seem better for the welfare of rabbits to be housed in open-topped pens rather than wire-surround cages, and this is supported by behavioural observations by Podberscek et al. (1991), Rauterberg et al. (2021) found that welfare indicators (lower fertility and more injuries) and health indicators were actually worse in the former. Heat stress may be a problem for rabbits (Liste et al., 2006). The transport of rabbits to slaughter is known to be stressful (Mazzone et al., 2010), and when they arrive at the slaughterhouse, waiting times at lairage longer than six hours raise blood stress indicators (Liste et al., 2009) and so should be kept as brief as possible. Protocols for the evaluation of rabbit welfare have not to date been used in practice on any scale. Verga et al. (2009) suggest some behavioural indicators that could inform such an evaluation, and stereotypies shown by caged rabbits include repetitive hair-chewing, bar-chewing, head-swaying and pawing (Gunn and Morton, 1995), and somersaulting. A tail-biting score has also been proposed (Bill et al., 2019). An evaluation system for rabbit welfare has been suggested and tested by Cerioli et al. (2008) based on three criteria of: management and husbandry, prophylaxis, treatment, and housing, and a welfare evaluation system based on Welfare Quality® (2009a–c) protocols, with the guiding principles of good housing, good feeding, good health, and appropriate behaviour, has also been presented but their use has not to date been widely practised.

Conclusions

A wide range of species and management types have been considered, and there are consequently many different problems that these species face. A common problem is that these animals are undomesticated, or partially domesticated at best, so are likely to be ill-suited to captivity and handling. In addition, there is comparatively little welfare research work that has been undertaken with these species which are less common than the usual domestic species farmed for food. There are few established protocols for the evaluation of the welfare of these animals tested on these animals in farmed conditions compared to more commonly farmed livestock. These should be developed, tested, and provided to farmers and local assessors in a clear, practical way to encourage the best husbandry systems for these animals, and reliably assess their on-farm well-being if we are to continue their use for our purposes.

Acknowledgements

The contributions of Johanna Askola and Aino Põder, erstwhile and current reindeer herders, are acknowledged in providing some of the background on current practice in the section on reindeer farming.

References

Ahola L, Botreau R, Gaborit M, Hovland AL, Koistinen T and Mononen J, 2015. *WelFur: Welfare Assessment Protocol for Foxes*. FurEurope. 978-2-9601617-0-0. hal-02799944

Arney D and Piirsalu P, 2017 The ethics of keeping fur animals, the estonian context. *Proceedings of the Latvian Academy of Sciences Section B Natural Exact and Applied Sciences*. 71 (1–2), pp. 78–80.

Bartoš L and Šiler J, 1993. *Survey of Game Farming in Europe*. FAO Ed., Roma, Italy.

Bill J, Rauterberg SL, Stracke J, Kemper N and Fels M, 2019. Prevalence and severity of tail lesions as a possible welfare indicator for rabbit does. *Animal Welfare*. 28 (4), pp. 511–518.

Bornett-Gauci HLI, Martin JE and Arney DR, 2006. The welfare of low-volume farm animals during transport and at slaughter: a review of current knowledge and recommendations for future research. *Animal Welfare*. 15, pp. 299–308.

Cerioli M, Brivio R, Grilli G, Tittarelli C, Marasciulo V and Lavazza A, 2008. Search for key health and welfare indicators for meat rabbit production and definition of a score method of evaluation. In Proceedings of the 9th World Rabbit Congress, Verona, Italy, 10–13 June 2008, pp. 915–920.

Clutton-Brock J, 1992. The process of domestication. *Mammal Review*. 22 (2), pp. 79–85.

Conaglen HM, Suttie JM and Conaglen JV, 2003. Effect of deer velvet on sexual function in men and their partners: a double-blind, placebo-controlled study. *Archives of Sexual Behavior*. 32 (3), pp. 271–278.

De Vos, 1982. *Deer Farming. FAO Animal Production and Health Paper 27*. Food and Agriculture Organization of the United Nations, Rome.

Dixon LM, Hardiman JR and Cooper JJ, 2010. The effects of spatial restriction on the behavior of rabbits (*Oryctolagus cuniculus*). *Journal of Veterinary Behavior*. 5 (6), pp. 302–308.

Edgar JL and Seaman SC, 2010. The effect of mirrors on the behaviour of singly housed male and female laboratory rabbits. *Animal Welfare*. 19, pp. 461–471.

Franchi V, Aleuy OA and Tadicha TA, 2016. Fur chewing and other abnormal repetitive behaviors in chinchillas (*Chinchilla lanigera*), under commercial fur-farming conditions. *Journal of Veterinary Behavior* 11, pp. 60–64.

Furcommission USA, 2019. *Production Statistics*. Available from https://furcommission.com/production-statistics/ (Accessed 26th October 2021).

Green P, 2016. *Practical Indicators to Assess the Welfare of Wild Deer in Scotland*. Scottish Natural Heritage Commissioned Report No. 944.

Gunn D and Morton DB, 1995. Inventory of the behaviour of New Zealand White rabbits in laboratory cages. *Applied Animal Behaviour Science*. 45 (3–4), pp. 277–292.

Hansen BK, Jeppesen LL and Berg P, 2010. Stereotypic behaviour in farm mink (*Neovison vison*) can be reduced by selection. *Journal of Animal Breeding and Genetics*. 127 (1), pp. 64–73.

He L, Li L-H, Wang WX, Liu G, Liu S-Q, Liu W-H and Hu D-F, 2014. Welfare of farmed musk deer: changes in the biological characteristics of musk deer in farming environments. *Applied Animal Behaviour Science*. 156, pp. 1–5.

Johnson CB, Wilson PR, Woodbury MR and Caulkett NA, 2005. Comparison of analgesic techniques for antler removal in halothane-anaesthetized red deer (Cervus elaphus): electroencephalographic responses. *Veterinary Anaesthesia and Analgesia*. 32 (2), pp. 61–71.

Jones SE and Phillips CJC, 2005. The effects of mirrors on the welfare of caged rabbits. *Animal Welfare*. 14, pp. 195–202.

Koistinen T, Huuki H, Mononen J and Ahola L, 2014. Development of the WelFur On-farm Assessment Protocol for the Finnraccoon. In Proceedings of the 6th International Conference on the Assessment of Animal Welfare at Farm and Group Level, eds. Mounier L and Veissier I. Wageningen Academic Publishers. Wageningen, p. 137.

Koistinen T, Raatikainen S, Sepponen J and Korhonen HT, 2018. Resting preferences and welfare of Finnraccoon (*Nyctereutes procyonoides ussuriensis*) females housed in various housing conditions in winter. *Applied Animal Behaviour Science*. 207, pp. 129–137.

Korhonen HT, Niemelä P and Jauhiainen L, 2001. Effect of space and floor material on the behaviour of farmed blue foxes. *Canadian Journal of Animal Science*. 81 (2), pp. 189–197.

Kynkäänniemi SM, Kettu M, Kortet R, Härkönen L, Kaitala A, Paakkonen T, Mustonen A-M, Nieminen P, Härkönen S, Ylönen H and Laaksonen S, 2014. Acute impacts of the deer ked (Lipoptena cervi) infestation on reindeer (Rangifer tarandus tarandus) behaviour. *Parasitology Research*. 113, pp. 1489–1497.

Laaksonen S, Jokelainen P, Pusenius J and Oksanen A, 2017. Is transport distance correlated with animal welfare and carcass quality of reindeer (Rangifer tarandus tarandus)? *Acta Veterinaria Scandinavica*. 59 (17). https://doi.org/10.1186/s13028-017-0286-z

Liste G, María GA, Buil T, García-Belenguer S, Chacón G, Olleta JL, Sañudo C and Villarroel M, 2006. Journey length and high temperatures: effects on rabbit welfare and meat quality. *Deutsche Tierarztliche Wochenschrift*. 113 (2), pp. 59–64.

Liste G, Villarroel M, Chacón G, Sañudo C, Olleta JL, García-Belenguer S, Alierta S and María GA, 2009. Effect of lairage duration on rabbit welfare and meat quality. *Meat Science*. 82 (1), pp. 71–76

Liu WH, Wang YQ, Li FR, Tang J and Yang Z, 2010. A primary study on breeding the musk deer by enclosure culture in qinling mountains. *Journal of Economic Animal*. 14, pp. 63–66.

Luzi F, Ferrante V, Heinzl E and Verga M, 2003. Effect of environmental enrichment on productive performance and welfare aspects in fattening rabbits. *Italian Journal of Animal Science*. 2 (sup1), pp. 438–440.

Malmkvist J and Hansen SW, 2002. Generalization of fear in farm mink, *Mustela vison*, genetically selected for behaviour towards humans. *Animal Behaviour*. 64 (3), pp. 487–501. https://doi.org/10.1006/anbe.2002.3058

Mason G, Cooper J and Clarebrough C, 2001. Frustrations of fur-farmed mink. *Nature*. 410, pp. 35–36. https://doi.org/10.1038/35065157

Mattiello S, 1994. *L'allevamento di ungulati selvatici: studio di alcuni aspetti produttivi, fisiologici e comportamentali*. PhD Diss., Università di Milano, Italy.

Mattiello S, 2009. Welfare issues of modern deer farming, *Italian Journal of Animal Science*. 8(sup1), pp. 205–217. https://doi.org/10.4081/ijas.2009.s1.205

Mazzone G, Vignola G, Giammarco M, Manetta AC and Lambertini L, 2010. Effects of loading methods on rabbit welfare and meat quality. *Meat Science*. 85 (1), pp. 33–39.

Meagher RK and Mason GJ, 2012. Environmental Enrichment Reduces Signs of Boredom in Caged Mink. *PLoS ONE*. 7 (11), p. e49180.

Møller SH, Hansen SW, Malmkvist J, Vinke CM, Lidfors L, Gaborit M and Botreau R, 2015. *WelFur: Welfare Assessment Protocol for Mink*. Fur Europe.

Mononen J, Møller SH, Hansen SW, Hovland AL, Koistinen T, Lidfors L, Malmkvist J, Vinke CM and Ahola L, 2012. The development of on-farm welfare assessment protocols for foxes and mink: the WelFur project. *Animal Welfare*. 21, pp. 363–371.

Mugnai C, Bosco AD and Castellini C, 2009. Effect of different rearing systems and pre-kindling handling on behaviour and performance of rabbit does. *Applied Animal Behaviour Science*. 118 (1–2), pp. 91–100.

Mulley RC, 2003. The feed requirements of adult red deer. *Journal of New Zealand Grasslands*. 9, pp. 51–55.

Neely C, Lane C, Torres J and Flinn J, 2018. The effect of gentle handling on depressive-like behavior in adult male mice: considerations for human and rodent interactions in the laboratory. *Behavioural Neurology*. 2, pp. 1–7. https://doi.org/10.1155/2018/2976014

Özkan Gülzari Ş, Jørgensen GHM, Eilertsen SM, Hansen I, Hagen SB, Fløystad I and Palme R, 2019. Measuring Faecal Glucocorticoid Metabolites to Assess Adrenocortical Activity in Reindeer. *Animals*. 9 (11), p. 987.

Parry-Jones R and Wu JY, 2001. Musk deer farming as a conservation tool in China. *TRAFFIC East Asia*. pp. 1–3.

Pedersen V and Jeppesen LL, 1990. Effects of early handling on later behaviour and stress responses in the silver fox (*Vulpes vulpes*). *Applied Animal Behaviour Science*. 26, (4), pp. 383–393.

Pérez-Barbería FJ, García AJ, López-Quintanilla M and Landete-Castillejos T, 2021. Pelt Biting as a Practical Indicator of Social and Environment Stress in Farmed Red Deer. Animals. 11 (11), p. 3134.

Podberscek AL, Blackshaw JK, and Beattie AW, 1991. The behaviour of group penned and individually caged laboratory rabbits. *Applied Animal Behaviour Science*. 28, pp. 353–363.

Pollard JC and Littlejohn RP, 1996. The effects of pen size on the behaviour of farmed red deer stags confined in yards. *Applied Animal Behaviour Science*. 47, pp. 247–253.

Pollard JC and Littlejohn RP, 1998. Effects of Winter Housing, Exercise, and Dietary Treatments on the Behaviour and Welfare of Red Deer (Cervus Elaphus) Hinds. *Animal Welfare*. 7 (1), pp. 45–56.

Price EO, 2002. Animal Domestication and Behavior. Pub. CAB international, Wallingford. UK. pp. 10–13.

Proskina L and Cerina S, 2021. Economic assessment of use of pulses in diets for captive red deer *Agronomy Research* 19(S2), pp. 1112–1123. https://doi.org/10.15159/AR.21.049

Putman RJ, 1988. *The Natural History of Deer. Comstock Publishing Associates*. Cornell University Press, New York.

Rauterberg SL, Bill J, Kimm S, Kemper N and Fels M, 2021. Health, performance and soiling of breeding does and their kits kept in two different housing systems on a German rabbit farm. *World Rabbit Science*. 29 (3), pp. 169–182.

Rée J, 2019. The young man one hopes for. *London Review of Books* 41 (22), pp. 7–9.

Rehbinder C and Hau J, 2006. Quantification of cortisol, cortisol immunoreactive metabolites, and immunoglobulin A in serum, saliva, urine, and feces for noninvasive assessment of stress in reindeer. *Canadian Journal of Veterinary Research*. 70 (2), pp. 151–154.

Rehbinder C, Edqvist L-E, Lundström K and Villafane F, 1982. A field study of management stress in reindeer (Rangifer tarandus L). *Rangifer*. 2, pp. 2–21.

Riigiteataja, 2021. Animal protection act. Available from: https://www.riigiteataja.ee/en/eli/ee/Riigikogu/act/530062021002/consolide

Ruis M, 2006. Group housing of breeding does. In: Maertens L. and Coudert P. (eds.) *Recent Advances in Rabbit Sciences*. Plot-it-bvba, Merelbeke, Belgium, pp. 107–111.

Sha L, Xu Y-P and Jin LJ, 2011. A review of mink farming practices in China. *Scientifur*. 35 (3), pp. 27–36

Sleivert G, Burke V, Palmer C, Walmsley A, Gerrard D, Haines S, and Littlejohn R, 2003. The Effects of deer antler velvet extract or powder supplementation on aerobic power, erythropoiesis, and muscular strength and endurance characteristics. *International Journal of Sport Nutrition and Exercise Metabolism*. 13 (3), pp. 251–265

Svendsen PM, Hansen BK, Malmkvist J, Hansen SW, Palme R and Jeppesen LL, 2007. Selection against stereotypic behaviour may have contradictory consequences for the welfare of farm mink (*Mustela vison*). *Applied Animal Behaviour Science*. 107 (1–2), pp. 110–119.

Thirstrup JP, Villumsen TM, Malmkvist J and Lund MS, 2019. Selection for temperament has no negative consequences on important production traits in farmed mink. *Journal of Animal Science*. 97 (5), pp. 1987–1995.

Verga M, Luzi F, Petracci M and Cavani C, 2009. Welfare aspects in rabbit rearing and transport. *Italian Journal of Animal Science*. 8 (sup1), pp. 191–204.

Verspecht A, Maertens L, Vanhonacker F, Tuyttens F, Van Huylenbroeck G and Verbeke W, 2011. Economic impact of decreasing stocking densities in broiler rabbit production based on Belgian farm data. *World Rabbit Science*, 19 (3), pp. 123–132,

Waas JR, Ingram JR and Matthews LR, 1999. Real-time physiological responses of red deer to translocations. *The Journal of Wildlife Management*. 63 (4), pp. 1152–1162.

Waller PJ, 2002. Reindeer (Rangifer tarandus) and yak (Bos (Poephagus) grunniens): disparate animal species—similar environment, management and parasite problems? In: Jianlin H, Richard C, Hanotte O, McVeigh C and Rege JEO, (eds). Yak Production in Central Asian Highlands. Proceedings of the Third International Congress on Yak held in Lhasa, P.R. China, 4–9 September 2000. ILRI (International Livestock Research Institute), Nairobi, Kenya. pp. 429–438.

Walsh VP and Wilson PR, 2002. Chemical analgesia for velvet antler removal in deer. *New Zealand Veterinary Journal*. 50 (6), pp. 237–243.

Weeks CA, 2000. Transport of deer: a review with particular relevance to red deer (Cervus Elaphus) *Animal Welfare*. 9 (1) pp. 63–74.

Welfare Quality®, 2009a. *Welfare Quality® Assessment Protocol for Cattle*. Welfare Quality® Consortium: Lelystad, The Netherlands.

Welfare Quality®, 2009b. *Welfare Quality® Assessment Protocol for Pigs*. Welfare Quality® Consortium, Lelystad, The Netherlands.

Welfare Quality®, 2009c. *Welfare Quality® Assessment Protocol for Poultry*. Welfare Quality® Consortium, Lelystad, The Netherlands.

Whary M, Peper R, Borkowski G, Lawrence W and Ferguson F, 1993. The effects of group housing on the research use of the laboratory rabbit. *Laboratory Animals*. 27, pp. 330–341.

Wilson PR and Stafford KI, 2002. Welfare of farmed deer in New Zealand. 2. Velvet antler removal. *New Zealand Veterinary Journal*. 50 (6), pp. 221–227.

10
FARMING FISH

Joao L Saraiva, Pablo Arechavala-Lopez, and Lynne U Sneddon

Introduction

The demand for food is ever increasing since human populations are growing at a fast rate globally. This has led to a great need for the intensive production of protein and in particular an increasing reliance on farmed animals. Aquaculture, the farming of aquatic animals (fishes and shellfishes), plants, and algae, has flourished since the 1980s (Naylor et al., 2021). The production of fishes and shellfish in aquaculture had risen from 10 million tonnes (Mt) in 1987 to 29 Mt in 1997 (FAO, 2020). More recently, aquaculture supplied more than 80 Mt of fishes and shellfish in 2017. In terms of numbers of fishes, an estimated 51 to 167 billion farmed fishes were slaughtered for food globally in 2017, which represents a 4–6% increase since 2015 (Mood and Brooke, n.d.; Fishcount, 2021). For comparison, total terrestrial farmed animals slaughtered annually is around 75 billion – so aquaculture (i) may exceed this, and (ii) welfare concerns are not as well recognised or addressed for farmed fish including slaughter. Hence, welfare in aquaculture may be a larger concern than for terrestrial farmed animals. Many issues of concern have been highlighted in the growth and practices in aquaculture, including loss of biodiversity, damage to the environment, pollution from waste products and uneaten food, use of antibiotics, poor sustainability, and abuse of human rights (Franks, Ewell and Jacquet, 2021). Additionally, high-profile media campaigns and scientific concern have suggested that fish welfare may be compromised in certain production systems (Brown and Dorey, 2019; Saraiva and Arechavala-Lopez, 2019). Growing scientific evidence has demonstrated the capacity for pain, fear, stress, and distress in fishes, thus there is a need to minimise poor welfare in farmed fish. Further it has been shown that fishes can form relationships within and between species, have complex cognitive abilities including learning and memory skills, assess risks and benefits to inform strategic behavioural decisions, have positive and negative affective states and demonstrate evidence of consciousness (Sneddon and Brown, 2020). Taken together this combined evidence confirms fishes as sentient beings and worthy of improved welfare considerations. With approximately 35,000 species and species specific differences in life history, behaviour, environmental, and nutritional requirements, as well as the developmental stage of the fishes, meeting the welfare needs of fishes in aquaculture is likely to be challenging but achievable with further research. This is even more relevant when we consider that the number of farmed fish species is more than tenfold higher than land animals (Saraiva et al., 2018). Different aquaculture systems will also have welfare

DOI: 10.4324/9781003182351-12

issues that are specific to the type of system. In the following, the main farmed fish species and areas of welfare concern are discussed in relation to the type of aquaculture system they are typically reared in. By using an ethological approach where we consider the behaviour of the wild counterpart, suggestions are made to improve the conditions and practices within these aquaculture systems.

Aquaculture systems and common European farmed fish species

When fish are reared for food production in aquaculture, they can be held in different types of rearing systems, all with different inputs, technology, and outputs (van De Vis et al., 2020). On a global scale, aquaculture is dominated by a relatively small number of rearing systems. There are land-based systems, such as natural rice fields, ponds (natural and artificial, Figure 10.1), various flow-through systems (tanks and raceways) and recirculating aquaculture systems (RAS). Alternatively, fish can be reared in water-based systems, such as freshwater/inshore/offshore floating net pens and cages (Figure 10.1), or semi-closed containment systems (S-CCS). Within every farming system, fishes are subjected to various husbandry routines and operations. Each of these systems or operations can present different welfare challenges or risks to the fish, which in turn are dependent upon both the species and its life stage (van De Vis et al., 2020).

Among the land-based systems, one of the oldest and most natural ways of culturing fish are in rice fields, where fish are raised in the flooded paddy as a supplementary crop in very low (almost natural) densities. This farming system is predominantly developed in seasonally flooded deltas in Asia and Africa (Ottinger, Clauss, and Kuenzer, 2016). Aquaculture ponds can be natural or artificial impoundments that form closed water bodies where fish are mostly reared in extensive conditions, but these can be intensive too. Ponds usually offer little sophisticated technology for water treatment (inlet and outlet) or feeding systems, and are primarily used for freshwater and brackish water aquaculture, although saltwater conditions can be also found in estuarine coastal areas. Tanks and raceways are both flow-through aquaculture systems, artificially constructed (straight-sided or round) and often surrounded with concrete sides and bottom, with water supplied or pumped from rivers, lakes or the sea. Water can be treated before being used, but not reused, and oxygenations and CO_2 degassing systems are commonly used to reduce water demand. Fish can be reared in flow-through systems from semi-intensive to super-intensive conditions, depending on the species and life stage. RASs are typically land-based tanks or raceways with treated and recirculated water. Water is (partially) reused after biological and mechanical treatment and limiting waste compounds are removed. The use of RAS is largely

Figure 10.1 Images of different aquaculture on-growing systems: earthen ponds for seabream and seabass, raceways for rainbow trout, net-pens for Atlantic salmon. Source: authors.

restricted to more high-value species or life stages (especially in hatcheries, where control over environmental conditions is critical and the unit values per individual fish are higher).

Regarding the water-based systems, the most widely used aquaculture systems are floating or suspended cages or net-pens (although they can also be submerged) located in natural aquatic systems such as lakes, rivers, oceans, or artificial water bodies. In floating cages, fish are mostly reared in intensive conditions, enclosed within a net or mesh cage, and subjected to the water current, driving open exchange of water into and out of the rearing system. Offshore net cages are usually exposed to high waves and strong currents. S-CCSs are emergent flow-through systems floating in the sea, where the fish are confined within a watertight or semi-permeable structure. Water must be actively transported into and out of the unit (typically pumped from deeper in the water column) to optimise rearing conditions. These new production methods (S-CCS) have evolved as a response to challenges in open-sea cage production of Atlantic salmon, such as sea-lice infestations, escapees, mortalities, and infections from pathogenic micro-organisms, but are still under development in terms of welfare.

In addition, there is an increased interest in both land-based and nearshore aquaculture systems to combine fed aquaculture species (e.g. finfish), with inorganic extractive aquaculture species (e.g. seaweeds) and organic extractive species (e.g. suspension- and deposit-feeders) cultivated in proximity (OECD and Chopin, 2010). Such systems, described as integrated multi-trophic aquaculture (IMTA), should increase significantly the sustainability of aquaculture, based on a number of potential economic, societal, and environmental benefits, including the recycling of waste nutrients from higher trophic-level species into production of lower trophic-level crops of commercial value.

As mentioned above, aquaculture systems also vary in their degree of production intensity, ranging from extensive to super-intensive systems. In general, extensive systems are characterised by minimal inputs and relatively low yields (close to natural yields), whereas with increasing intensification, additional feed is required to maintain higher stocking rates in semi-intensive conditions. Intensive and super-intensive systems rely to a large extent or even completely on supply of external inputs and technologies. Intensification also implies higher costs for investment and management, be it for the construction of advanced aquaculture technologies (e.g. industrial pond farms, raceways, or offshore cage farms) or for the maintenance of such highly stocked systems (e.g. costs for feed inputs; fuel or electricity for aeration) (Ottinger, Clauss, and Kuenzer, 2016). Aquaculture production is not only diverse in terms of methods of production, sophistication level or intensification, but also in the number of farmed species. Around 100 different species are currently farmed in aquaculture worldwide, but most of the production concentrates on a few species. The most important finfish species farmed in the EU in terms of tonnes of production and value are Atlantic salmon (*Salmo salar*), rainbow trout (*Oncorhynchus mykiss*), gilthead seabream (*Sparus aurata*), European seabass (*Dicentrarchus labrax*), and common carp (*Cyprinus carpio*). The marine finfish sector is the most important economically in the EU, followed by the shellfish sector. In freshwater aquaculture, trout and carp dominate, responsible for 53% and 32% of the total volume produced, whereas in marine finfish production, salmon and trout are responsible for 53% of the total production, and seabass and seabream responsible for a further 38% (FAO, 2020).

Atlantic salmon and rainbow trout are generally farmed intensively and mainly destined for consumption. The most common farming systems used are flow-through systems, RASs, and floating cages (e.g. Jones, 2004; Vandeputte and Labbé, 2012), and the interest in S-CCSs is increasing (van De Vis et al., 2020) (Table 10.1). The intensive system is characterised by high production, at high fish densities, with many parameters under human control. In intensive breeding systems, selected broodstocks are held in large freshwater ponds or tanks (usually flow-

Table 10.1 Farming systems and production phases for the top-five most farmed finfish species in Europe: Atlantic salmon (*Salmo salar*), rainbow trout (*Oncorhynchus mykiss*), gilthead seabream (*Sparus aurata*), European seabass (*Dicentrarchus labrax*), and common carp (*Cyprinus carpio*)

	System	*R. trout*	*A. salmon*	*G. seabream*	*E. seabass*	*C. carp*
Land-based	Polyculture or natural/rice fields	–	–	–	–	on-growing
	Artificial and natural ponds	on-growing	on-growing	on-growing	on-growing	hatchery Nursery on-growing
	Flow-through tanks and raceways	hatchery nursery on-growing	hatchery nursery on-growing	hatchery nursery on-growing	hatchery nursery on-growing	hatchery nursery on-growing
	Recirculating (RAS) tanks and raceways	hatchery nursery on-growing	hatchery nursery on-growing	hatchery nursery on-growing	hatchery nursery on-growing	nursery on-growing
Water-based	Semi-closed containments (S-CCS)	–	on-growing	–	–	–
	Floating cages (marine-freshwater)	on-growing	on-growing	on-growing	on-growing	on-growing
	Off-shore cages	–	on-growing	on-growing	on-growing	–
Others	Integrated multi-trophic farming systems (IMTA)	–	–	on-growing	on-growing	on-growing

through or RAS systems) where they release eggs and milt (i.e. seminal fluid) or these are collected from the fish, which will be mixed to produce fertilised eggs. The fertilised eggs are then placed in purpose-built incubators until hatching. After hatching, the fry absorb nutrients from a yolk sac attached to their bodies, and they remain in the hatching environment until they are able to feed independently. Then, larval fish are directly transferred to the first-feeding tanks. At the nursery stage, Atlantic salmon and rainbow trout have different rearing requirements that will dictate the type of containment, but the source of water available will determine whether flow-through, semi-closed containment systems or recirculation systems are best. Normally, Atlantic salmon are kept on land in freshwater tanks after hatching, before smoltification (i.e. the process of physiological changes that allow salmon to adapt from living in freshwater to living in seawater) starts naturally or is induced artificially. The smolts or post-smolts are then transferred mostly to sea cages, or RAS and S-CCS systems for the final grow-out phase until harvest. In rainbow trout, fry are moved to outdoor grow-out facilities, which can comprise concrete raceways, ponds, RASs, cages in lakes, or sea cages with different sizes and characteristics according to site availabilities, environmental conditions, and specific company targets. Atlantic salmon and rainbow trout are grown to a marketable size, usually within 9 months, in sites dedicated to the production of portion-size trout of 450 g average weight. Some fish, though, are grown to larger sizes over 20 months to be harvested at 3 kg plus. In addition, small-scale rainbow trout farms can use semi-intensive systems for on-growing where young stock are brought in by road and grown-out for either food or re-stocking markets. Extensive salmonids production is quite rare on a commercial scale, and mostly consists of releasing juvenile fish for downstream migration at the smolt stage.

Gilthead seabream and European seabass are each mostly produced in three different aquaculture systems, whose techniques and procedures are very similar for the two species (e.g. Bagni, 2005; Colloca and Cerasi, 2005; Basurco et al., 2011) (Table 10.1). Intensive systems are characterised by a high production at high fish densities, where many parameters are under human control. To secure a reliable and sufficient supply of good quality fish eggs, most hatcheries have established their own broodstock units, where breeders of different age groups are maintained under long-term stocking conditions. Parent animals may come from the wild, but nowadays most of them come from a selective programme at the farm. After hatching, the larvae will absorb their yolk sac and, once they start feeding, weaning usually takes place in a dedicated section of the hatchery (i.e. nursery area) equipped with larger round or rectangular tanks. Juveniles are pre-fattened intensively with a controlled diet and at high densities until they reach the size for the on-growing phase. In intensive production, on-growing units are supplied with juveniles, which may be purchased from separated hatcheries, but large production units normally rear their own. Intensive on-growing phases can be carried out in land-based installations (tanks or raceways) or in coastal floating cages. Semi-intensive farming systems are usually carried out in net enclosures within limited areas of the lagoons or in earthen ponds, where human control of the farming environment is much lower than in intensive systems but greater than in the extensive ones. This technique involves artificial enrichment with fry collected by specialised fishermen or seeding with pre-fattened juveniles in intensive systems to minimise mortality and shorten farming time. Extensive systems are based on the natural migration of euryhaline fish between the open sea and coastal lagoons, brackish ponds, or salt marshes, and they have been widely developed in northern Italy ("vallicoltura") and in southern Spain ("esteros"). This traditional extensive method of lagoon management places special traps or barriers made of reeds, nets, or cement in appropriate lagoon sites to capture fish during their autumn migration to the open sea.

Common carp is a freshwater species that is generally reared in ponds in intensive, semi-intensive, or extensive monoculture or polyculture systems, or in integrated carp culture with other agriculture systems (e.g. Peteri, 2004) (Table 10.1). Spawning can either occur in large ponds, where fry can be harvested or left there until they reach fingerling size and are moved to prepared ponds, or can take place in hatcheries, where ovulation and spermiation are artificially induced by hormonal injections, and eggs are artificially fertilised, then fry are moved from tanks into ponds when they reach the feeding fry stage. The fry are nursed in ponds or alternatively, if predators are present in the ponds (i.e. larger conspecifics, other fish species in polyculture systems, or even potential avian species), in tanks or in industrial raceways or water recirculating systems. Then fingerling production takes place in semi-intensive ponds, and from there they can be moved to on-growing systems, where growing carp to reach market size can take place in 1) extensive monocultural production systems in stagnant water ponds; 2) intensive monocultural production systems in cages, irrigation reservoirs, running water ponds/tanks, or in recirculating systems; 3) polycultural systems with other species; or 4) systems integrated with animal husbandry and/or plant production. From here they can either be transported to be sold live to consumers or restaurants, or to be slaughtered in an abattoir.

Welfare challenges

Welfare constraints exist throughout the production cycle of any farmed fish species. We aim to identify the welfare challenges the fish are exposed to throughout the production cycle from a wider perspective. We depart from the framework proposed by the Fishethobase (Saraiva et al., 2019) and Huntingford (2020): various criteria are used to evaluate the challenges imposed on any farmed fish species, and use the ethology of those species as a standard to compare how well those species may cope with those challenges. We propose a grouping of these major challenges into four main categories (with examples):

1. Ethological
    - Spatial limitations;
    - Reproduction;
    - Density/aggregation/social issues.
2. Physiological
    - Pain;
    - Infectious disease/immunocompetence;*
    - Parasites;*
    - Stress.
3. Environmental
    - Water parameters;
    - Light parameters;
    - Temperature parameters;
    - Environmental complexity.
4. Human-induced/procedural
    - Standard Operational Protocols;
    - Slaughter methods.

The categories marked with an asterisk (*) are mainly health-related issues that are more related to a veterinary approach, and in that sense will generally not be addressed in this chapter. These four main categories are often interlinked (Figure 10.2) and are applicable to all fish farming systems, yet may differ in intensity and severity depending on the combination of species and method. In the following sections we will provide a review of those challenges in a species-

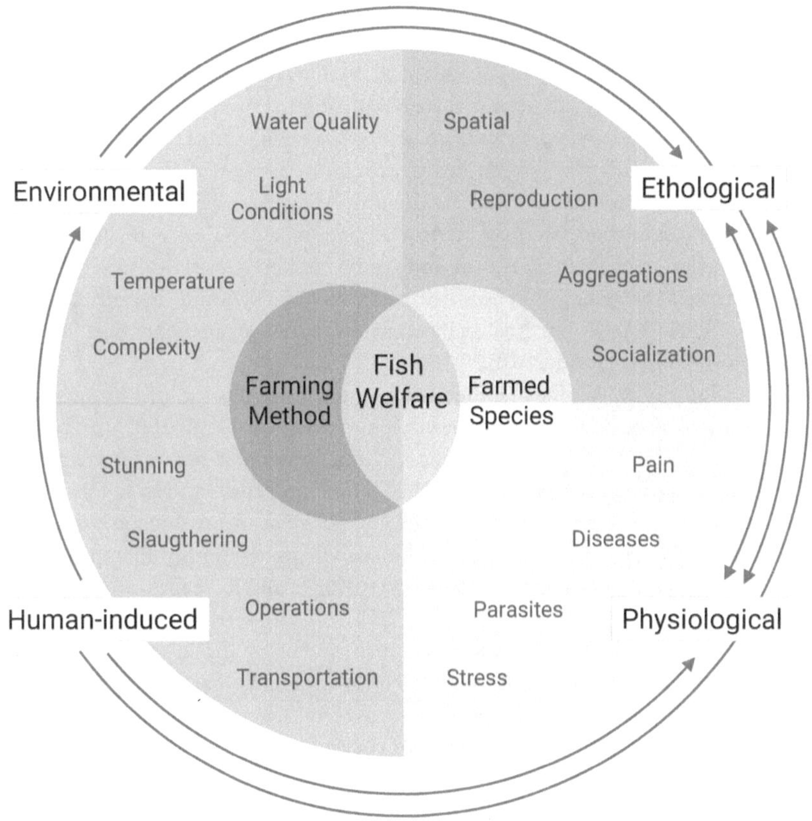

Figure 10.2 Schematic representation of the links between different ethological, physiological, environmental, and human-induced challenges that fishes are exposed to during farming conditions and operations. The arrows simply represent that the four categories are interlinked (as mentioned in the text), and the type of challenges are examples of those explained throughout the text within each category of challenge. Source: authors.

specific approach for the five most commonly farmed fish in Europe, while highlighting possible suggestions for improvement.

Ethological challenges

We define ethological challenges as those that impair behavioural functions directly or indirectly, considering the four classical ethological questions posed by Tinbergen (1963): function, causation, development, and evolution. Captive environments impose constraints that challenge fishes' ability to cope with their environments, for example, restricting free movement of animals. Some species may cope better with such spatial restrictions because they have evolved in restricted areas and are adapted to confined areas. This may be the case with common carp for example, yet this applies only for certain life stages (Flajšhans and Hulata, 2017). Many others, however, are not found in spatially restricted environments and may not be equipped to cope with the spatial challenges of fish farming (Saraiva et al., 2018). This may be especially true for migratory species.

Another immediate aspect of fish farming is the aggregation of animals. Fish under culture conditions are stocked using various methods described in previous sections, sometimes in very high and mostly artificial densities (Saraiva et al., 2018, 2019). This results not only in a technical challenge to maintain water quality (see point three) and monitor disease outbreaks, but also in ethological challenges to cope with proximity to a very large number of conspecifics. Some species encounter near-natural density conditions in specific life stages under some farming methods, such as juveniles of seabream and seabass (Bégout Anras, Lagardére, and Lafaye, 1997; Abecasis and Erzini, 2008), and therefore may be naturally equipped to deal with such a social context. However, in other cases, such as alevins and fry of rainbow trout, this never happens in nature, and such artificial crowding may lead to maladaptive behavioural responses such as aggression (Berejikian et al., 2000), abnormal behaviours, immune impairments, poor feeding, and/or stress (Andersson and Höglund, 2012).

Reproduction is often highly artificial, either because some species are not known to be able to spawn naturally in captivity, for example salmon (Stead and Laird, 2002), or because standard industry procedures dictate artificial spawning inductions as the best means to achieve a regular supply of gametes (Zohar and Mylonas, 2001). These procedures often involve stressful handling, such as prolonged emersion, manipulation, and mechanical damage from stripping (see point four). However, there are species who spawn spontaneously in captivity and there is evidence that egg quality is higher in these cases (for example, seabream) (Forniés et al., 2001).

Cognition in captive environments is challenged mainly due to the absence of stimulation (Korte, Olivier and Koolhaas, 2007). Rearing facilities are usually barren for sanitary and practical reasons, but this may impair cognitive aspects in several species, particularly those that evolved and are adapted to complex environments. For example, sea bream have impaired cognition, brain function, and spatial orientation in barren tanks, which are improved by environmental enrichment (Arechavala-Lopez et al., 2019, 2020). Similarly, positive effects of rearing fish in complex environments have been reported for salmonids, carps, and other species of aquaculture interest (Jones et al., 2021; Näslund and Johnsson, 2016). Finally, life in captivity may induce negative emotional states in fish. It is known that fish may experience emotion-like affective states such as fear (Cerqueira et al., 2017; Tatemoto et al., 2021) or pain (Sneddon, 2015) that may hinder their welfare in ways we are only now starting to unravel.

Physiological challenges

The function-based approach to welfare has been the basis for much of the existing industry standards regarding health (Huntingford and Kadri, 2014). Farmed fish face serious physiological challenges that have significant effects on their health (infections, parasites, etc.). The veterinary and health plans for fish farms are efficient at dealing with these kind of situations and this is why we consider diseases to be mostly beyond the scope of this chapter. However, there are other physiological aspects of fish farming that are indeed within the realm of welfare, for example stress responses and pain. The physiological stress response is adaptive when animals face acutely stressful events, which in the wild are natural, sporadic, short-term, and unforeseen but in farming environments are artificial, prolonged, and repetitive. This may lead to distress and often to chronic stress, with important negative effects on fish welfare. Some of these human-induced stressors (see point four) may in fact induce pain, which not only has an obvious strong immediate impact on welfare but may lead to long lasting negative effects such as avoidance, withdrawal, fasting, immune depression, etc., especially when combined with chronic distress responses to traumatic events (Ashley and Sneddon, 2008). The physiology of farmed fish is subjected to a constant test to maintain homeostasis when we consider that the quality of water, the critical environmental component of fish farming, is influenced by the factors addressed next.

Environmental challenges

The regulation and monitoring of water as a holding medium for captive fish is paramount: water provides the basic life support for farmed fish. Indeed, there are many aspects to address in water quality but we focus on (arguably) the most important ones, salinity, oxygen (O_2), carbon dioxide (CO_2), nitrogen compounds, and pH. Regarding salinity, some of the species addressed in this chapter are euryhaline (i.e. able to adapt to a wide range of salinities) throughout most of their life cycle, such as seabream and seabass, while others have sensitivity windows when they perform migrations from freshwater to saltwater and vice versa (McCormick, 2001). Others are stenohaline (i.e. cannot tolerate a wide fluctuation in the water salinity), such as carp. Providing farmed fish with appropriate salinities in the appropriate life stage will prevent osmotic stress – that can pose a major physiological challenge. Regarding oxygen, while some species can tolerate low oxygen saturations, such as carp (Stecyk and Farrell, 2002), others (e.g. salmon) are very sensitive and experience poor welfare under 50% saturation (Oldham et al., 2019). CO_2 is a by-product of aerobic respiration and accumulates in waters with poor renovation, aeration or flow. It is highly toxic by itself but also because it lowers the pH of the water, it impairs the senses and overall physiology of fish (Ishimatsu et al., 2004). Finally, the accumulation of toxic nitrogen compounds from excretion in poorly filtered water can be deadly for fish (Ip and Chew, 2010).

All these aspects require technical (and, depending on the method and intensification level, often technological) solutions for monitoring and correction. Additionally, they are highly synergistic, so even small changes in any of these variables may have dramatic effects on welfare or even survival of fish. And this becomes even more critical when temperature is entered into the equation. As fish are ectotherms, their physiology is strongly affected by the environmental temperature. In some procedures within usual farming protocols, however, the fish are subjected to temperatures well outside their comfort range, both at the lower end in the common slaughter method of asphyxia on ice, and at the higher end in crowding, harvesting, or other handling events if fish are crowded together in low volumes of water, prone to severe temperature increases. Unfortunately, increasing the water temperature lowers the O_2 saturation, which triggers a cascade of both physical and biological reactions that ultimately lead to severe and rapid degradation of water quality: higher respiratory rates and metabolism cause rapid O_2 depletion and increase in CO_2, as well as increase in release and accumulation of faecal matter and urine, with build-up of ammonia compounds. The accumulation of these bioactive stress signals may also function in a positive feedback mechanism, and this cocktail can be extremely harmful, often deadly for fish in extreme events (Huntingford et al., 2006). Production units must therefore constantly monitor the water parameters and be able to correct them when deviations are found.

Light is also a fundamental aspect to take into account when farming fish. Light intensity for example can have dramatic effects on the physiology and behaviour of farmed species. Photoperiod is also a critical environmental cue that fish use to read their environment. It is therefore not a surprise that manipulation of photoperiod (often combined with temperature) is one the most used techniques to induce spawning, delay maturation, and control the life cycle of farmed species. Some manipulations of photoperiod and light intensity seem to be innocuous or positive, while other highly artificial settings (for example, 24 hour light for extended periods, sometimes in species which never experience such conditions in the wild, or bright lights in species adapted to deeper, darker waters) may hinder the welfare of some of the species in ways yet to be properly evaluated (Huntingford et al., 2006).

It has been mentioned above that barren environments impair ethological aspects of fish welfare. One of the ways to counteract this effect is by environmental enrichment (EE), i.e., the deliberate addition of complexity to the captive environment. For several species, EE improves the overall welfare of farmed fish (Arechavala-Lopez et al., 2019, 2020). However, there are also reports of negative effects that may be due to incorrect interpretations of the ethology of the target species and/or inappropriate deployment of EE measures (e.g. Saraiva et al., 2021). EE measures should not interfere with farming protocols, or the latter should be changed to accommodate EE measures. To summarise, while EE remains a favourable tool to improve the welfare of farmed fish, its implementation must take into account 1) understanding of species-specific requirements and 2) the farming protocols at each facility (Saraiva et al., 2021).

Human-induced challenges

Standard fish farming protocols have been developed and optimised largely from a production perspective. These protocols include human-induced challenges to the welfare of fish, stressors which never occur in nature and therefore fish are not naturally equipped to cope with them. Here we can divide these into two main components: the handling, where we consider all the operations that the fish are subjected to during their lifetimes (transport, grading, vaccinations, moving, crowding, harvesting, etc.) and the slaughter, including stunning (if any), and the procedure that ultimately leads to the killing of the fish.

In the handling component, one important aspect to take into account is the farming method. While culturing fish in ponds may be theoretically less invasive (in the sense that fish experience a more "natural" environment with less handling, as they go through most of their life in the same enclosure at relatively low densities). At the other end of the spectrum are intensive RAS systems where the fish are usually crowded, transported, graded, vaccinated, and treated several times before they reach the end of their production cycle. These procedures may cause mechanical damage to the fish due to contact with other animals' skin and spines, promote excretion and accumulation of urine and faeces due to stress, impair immune functions due to erosion of mucus layers and wounds, which in turn can promote infections and disease outbreaks, and overall severely increase stress due to handling and emersion. There are ways to mitigate the harm inflicted during these procedures: for example, the use of passive methods to grade and move fish, fish pumps instead of brailing, the use of anaesthetics or sedation in transport, manipulation and emersion, and all of these techniques demonstrate good results (van De Vis et al., 2020).

In the slaughter phase, the traditional method of asphyxia on ice without prior stunning has been demonstrated to be the worst, not only in terms of welfare but also in terms of flesh quality. Different species require different technical approaches towards the stunning procedure, however existing evidence shows that, if the fish are effectively stunned prior to slaughter, then the killing occurs painlessly, the flesh quality is better, rigour mortis is delayed, and the shelf life is longer (Poli, 2009). Percussive or electrical stunning solutions exist for all the major species (Saraiva et al., 2019). Regardless of the farming method, both handling and slaughter components depend on staff training and technical capabilities and may sometimes require changes in operational protocols. However, the benefits in terms of welfare and product quality are evident.

Conclusions

Since aquaculture of fishes is growing it is paramount that we seek to reduce the welfare implications of the practices. Welfare issues are system and species-specific and more research is required to fully understand how the industry can overcome these problems and improve quality of life for the fishes. We suggest that knowledge of the life history, environmental conditions, and behavioural needs of the species in nature should inform the assessment and resolution of welfare problems when they arise in captivity. Awareness of the ethology of each species can help inform fish farm design and practices used to ensure the behavioural needs of these animals are being met. Any human-induced disturbance such as handling, size grading, vaccination, transport, and slaughter should be refined to ensure good welfare throughout the fishes' lives. Not only does improved welfare benefit the fish but this also benefits humans by ensuring better growth and higher economic return and also healthy, disease-free food production to ensure public health.

References

Abecasis D and Erzini K 2008. Site fidelity and movements of gilthead sea bream (Sparus aurata) in a coastal lagoon (Ria Formosa, Portugal), *Estuarine, Coastal and Shelf Science*, 79(4), pp. 758–763. doi: 10.1016/j.ecss.2008.06.019.

Andersson MÅ and Höglund E 2012. Linking personality to larval energy reserves in Rainbow Trout (Oncorhynchus mykiss), *PLOS ONE*, 7(11), p. e49247. doi: 10.1371/journal.pone.0049247.

Arechavala-Lopez P et al., 2019. Effects of structural environmental enrichment on welfare of juvenile seabream (*Sparus aurata*), *Aquaculture Reports*, 15, p. 100224. doi: 10.1016/j.aqrep.2019.100224.

Arechavala-Lopez P et al., 2020. Enriched environments enhance cognition, exploratory behaviour and brain physiological functions of Sparus aurata, *Scientific Reports*, 10(1), p. 11252. doi: 10.1038/s41598-020-68306-6.

Ashley PJ and Sneddon LU, 2008. Pain and fear in fish, in *Fish Welfare*. John Wiley & Sons, Ltd, pp. 49–77. doi: 10.1002/9780470697610.ch4.

Bagni M, 2005. *Cultured Aquatic Species Information Programme. Dicentrarchus labrax*. Rome: FAO Fisheries and Aquaculture Department. Available at: http://www.fao.org/fishery/culturedspecies/Dicentrarchus_labrax/en.

Basurco B, Lovatelli A and García B, 2011. Current status of Sparidae Aquaculture, in *Sparidae*. John Wiley & Sons, Ltd, pp. 1–50. doi: 10.1002/9781444392210.ch1.

Bégout Anras M-L, Lagardére J-P and Lafaye J-Y, 1997. Diel activity rhythm of seabass tracked in a natural environment: group effects on swimming patterns and amplitudes, *Canadian Journal of Fisheries and Aquatic Sciences*, 54(1), pp. 162–168. doi: 10.1139/f96-253.

Berejikian BA et al., 2000. Social dominance, growth, and habitat use of age-0 steelhead (Oncorhynchus mykiss) grown in enriched and conventional hatchery rearing environments, *Canadian Journal of Fisheries and Aquatic Sciences*, 57(3), pp. 628–636. doi: 10.1139/f99-288.

Brown C and Dorey C, 2019. Pain and emotion in fishes: fish welfare implications for fisheries and aquaculture, *Animal Studies Journal*, 8(2), pp. 175–201. doi: 10.14453/asj.v8i2.12.

Cerqueira M et al., 2017. Cognitive appraisal of environmental stimuli induces emotion-like states in fish, *Scientific Reports*, 7(1), p. 13181. doi: 10.1038/s41598-017-13173-x.

Colloca F and Cerasi S, 2005. *Cultured Aquatic Species Information Programme. Sparus aurata*. Rome: FAO Fisheries and Aquaculture Department. Available at: http://www.fao.org/fishery/culturedspecies/Sparus_aurata/en.

van De Vis H et al., 2020. Welfare of farmed fish in different production systems and operations, in van de Vis, H et al. (eds) *The Welfare of Fish*. Springer (Animal Welfare), pp. 323–361.

FAO, 2020. *The State of World Fisheries and Aquaculture 2020*. Rome: Food and Agriculture Organization of the United Nations. doi: 10.4060/ca9229en.

Flajšhans M and Hulata G, 2017. *Common Carp - Cyprinus Carpio Biology, Ecology and Genetics*. Genimpact final scientific report. Available at: http://www.sfos.uaf.edu/ksmsc/teaching/courses/fish336/materials/Common%20Carp.pdf (Accessed: 31 March 2017).

Fishcount, 2021. http://fishcount.org.uk/fish-count-estimates-2/numbers-of-farmed-fish-slaughtered-each-year (Accessed: 8 July 2021).

Forniés MA et al., 2001. Spawning induction of individual European sea bass females (*Dicentrarchus labrax*) using different GnRHa-delivery systems, *Aquaculture*, 202(3–4), pp. 221–234. doi: 10.1016/S0044-8486(01)00773-6.

Franks B, Ewell C and Jacquet J, 2021. Animal welfare risks of global aquaculture, *Science Advances*, 7(14), p. eabg0677. doi: 10.1126/sciadv.abg0677.

Huntingford FA et al., 2006. Current issues in fish welfare, *Journal of Fish Biology*, 68(2), pp. 332–372. doi: 10.1111/j.0022-1112.2006.001046.x.

Huntingford FA, 2020. Fish behaviour: determinants and implications for welfare, in Kristiansen, TS et al. (eds) *The Welfare of Fish*. Cham: Springer International Publishing (Animal Welfare), pp. 73–110. doi: 10.1007/978-3-030-41675-1_4.

Huntingford FA and Kadri S, 2014. Defining, assessing and promoting the welfare of farmed fish, *Revue Scientifique et Technique de l'OIE*, 33(1), pp. 233–244. doi: 10.20506/rst.33.1.2286.

Ip YK and Chew SF, 2010. Ammonia production, excretion, toxicity, and defense in fish: a review, *Frontiers in Physiology*, 1, p. 134. doi: 10.3389/fphys.2010.00134.

Ishimatsu A et al., 2004. Effects of co_2 on marine fish: larvae and adults, *Journal of Oceanography*, 60(4), pp. 731–741. doi: 10.1007/s10872-004-5765-y.

Jones M, 2004. *Cultured Aquatic Species Information Programme. Salmo salar*. Rome: FAO Fisheries and Aquaculture Department. Available at: http://www.fao.org/fishery/culturedspecies/Salmo_salar/en.

Jones NAR, Webster M and Vea Salvanes AG, 2021. Physical enrichment research for captive fish: time to focus on the DETAILS, *Journal of Fish Biology*, p. jfb.14773. doi: 10.1111/jfb.14773.

Korte SM, Olivier B and Koolhaas JM, 2007. A new animal welfare concept based on allostasis, *Physiology & Behavior*, 92(3), pp. 422–428. doi: 10.1016/j.physbeh.2006.10.018.

McCormick SD, 2001. Endocrine control of osmoregulation in teleost fish, *American Zoologist*, 41(4), pp. 781–794. doi: 10.1093/icb/41.4.781.

Mood A and Brooke P, n.d. *Fishcount.org.uk*. Available at: www.fishcount.org.uk (Accessed: 8 July 2021).

Näslund J and Johnsson JI, 2016. Environmental enrichment for fish in captive environments: effects of physical structures and substrates, *Fish and Fisheries*, 17(1), pp. 1–30. doi: 10.1111/faf.12088.

Naylor RL et al., 2021. A 20-year retrospective review of global aquaculture, *Nature*, 591(7851), pp. 551–563. doi: 10.1038/s41586-021-03308-6.

OECD and Chopin T, 2010. Integrated multi-trophic aquaculture, in OECD (ed) *Advancing the Aquaculture Agenda*. Paris: OECD, pp. 195–217. doi: 10.1787/9789264088726-15-en.

Oldham T et al., 2019. Metabolic and functional impacts of hypoxia vary with size in Atlantic salmon, *Comparative Biochemistry and Physiology Part A: Molecular & Integrative Physiology*, 231, pp. 30–38. doi: 10.1016/j.cbpa.2019.01.012.

Ottinger M, Clauss K and Kuenzer C, 2016. aquaculture: relevance, distribution, impacts and spatial assessments: a review, *Ocean & Coastal Management*, 119, pp. 244–266. doi: 10.1016/j.ocecoaman.2015.10.015.

Peteri A, 2004. *Cultured Aquatic Species Information Programme. Cyprinus Carpio*. Rome: FAO Fisheries and Aquaculture Department. Available at: http://www.fao.org/fishery/culturedspecies/Cyprinus_carpio/en (Accessed: 31 March 2017).

Poli BM 2009. Farmed fish welfare-suffering assessment and impact on product quality, *Italian Journal of Animal Science*, 8(sup1), pp. 139–160. doi: 10.4081/ijas.2009.s1.139.

Saraiva JL et al., 2018. Domestication and welfare in farmed fish, in Teletchea, F (ed.) *Animal Domestication*. London: IntechOpen, p. 29. doi: 10.5772/intechopen.77251.

Saraiva JL et al., 2019. A global assessment of welfare in farmed fishes: the fishethobase, *Fishes*, 4(2), p. 30. doi: 10.3390/fishes4020030.

Saraiva JL et al., 2021. The effect of tank cover on welfare of farmed Nile tilapia, *Applied Animal Behaviour Science*, 241, p. 105396. doi: 10.1016/j.applanim.2021.105396.

Saraiva JL and Arechavala-Lopez P, 2019. Welfare of fish—no longer the elephant in the room, *Fishes*, 4(3). p. 39. doi: 10.3390/fishes4030039.

Sneddon LU 2015. Pain in aquatic animals. *Journal of Experimental Biology*, 218 (7), pp. 967–976. doi: 10.1242/jeb.088823

Sneddon LU and Brown C, 2020. Mental capacities of fishes, in Johnson LSM, Fenton A, and Shriver A (eds) *Neuroethics and Nonhuman Animals*. Cham: Springer International Publishing (Advances in Neuroethics), pp. 53–71. doi: 10.1007/978-3-030-31011-0_4.

Stead SM and Laird L, 2002. *The Handbook of Salmon Farming*. London: Springer-Verlag (Food Sciences).

Stecyk JAW and Farrell AP, 2002. Cardiorespiratory responses of the common carp (Cyprinus carpio) to severe hypoxia at three acclimation temperatures, *Journal of Experimental Biology*, 205(6), pp. 759–768. doi: 10.1242/jeb.205.6.759.

Tatemoto P et al., 2021. Living with low environmental complexity increases fear indicators in Nile tilapia, *Animal Behaviour*, 174, pp. 169–174. doi: 10.1016/j.anbehav.2021.02.006.

Tinbergen N, 1963. On aims and methods of ethology, *Zeitschrift für Tierpsychologie*, 20(4), pp. 410–433. doi: 10.1111/j.1439-0310.1963.tb01161.x.

Vandeputte M and Labbé L, 2012. *Cultured Aquatic Species Information Programme. Salmo trutta (Berg, 1908).* Rome: FAO Fisheries and Aquaculture Department. Available at: http://www.fao.org/fishery/culturedspecies/Salmo_trutta/en

Zohar Y and Mylonas CC, 2001. Endocrine manipulations of spawning in cultured fish: from hormones to genes, *Aquaculture*, 197(1–4), pp. 99–136. doi: 10.1016/S0044-8486(01)00584-1.

11
TRANSPORTATION

Michael S Cockram

Introduction

The potential welfare issues associated with transporting farmed animals include stress, hyperthermia/hypothermia, injury (pain), fatigue, motion sickness, disease (sickness), hunger, and thirst. Whether welfare issues arise during transportation will depend upon the type of animal (e.g., species, age, and condition), their fitness for transport, the quality of the journey (including vehicle design, stocking density, ventilation, the standard of driving, and quality of the road), journey duration, the environmental conditions, and the associated handling and management of the animals. This chapter reviews the risk factors that can affect the welfare of animals when they are transported and the measures that can be used to mitigate potential welfare issues.

Transportation poses risks to animal welfare, but the prevalence, severity, and types of issues that farm animals experience during transport are contentious. There is a disparity between the judgments of campaigning organisations and those of industry on the perceived welfare consequences of transportation. When animals are transported, they are exposed to multiple factors that can influence their affective state, cause physiological and behavioural changes and sometimes injury and pathological changes. The transport of animals is not a natural process, and in most cases, it will be associated with novelty and stress. However, severe welfare issues are not an inevitable consequence of transportation. Many animals arrive at their destination without overt signs of welfare issues, are healthy and in good condition. There are economic pressures on the industry to avoid mortality, morbidity, poor meat quality, weight loss, food safety, and biosecurity issues that can sometimes be associated with transportation. Therefore, there is some degree of synergy between the standards required for optimal productivity and those for animal welfare. That said, if transport is not undertaken well, animals can experience pain, fear, distress, fatigue, and prolonged hunger and thirst. These adverse effects can occur when occasional issues arise. However, there are commercial pressures on the manner in which animals are routinely transported, and these can affect the quality of the journey. There are costs associated with optimising some welfare conditions, for example, by careful handling and driving, providing environmental control, reducing stocking density, and providing feed, water, and rest at frequent intervals.

 DOI: 10.4324/9781003182351-13

Transportation reasons

The main reasons for transporting farmed animals are:

- Slaughter;
- Transfer between production units during stages of production;
- Sale at an auction market/saleyard;
- Transfer of breeding animals;
- Movement to areas with improved access to feed and/or water.

Most farm animals that are reared for meat and those that are slaughtered after they are culled following a period of breeding or production are transported from a farm to a place of slaughter. Many breeding animals have high value and are transported in superior conditions. The intensification of production has increased the number of times that many animals are transported. For example, animals can be born in one location, fattened/reared in one or more locations and then slaughtered in a specialised facility. Some journeys can be long and complex and can involve multiple stages and destinations, sometimes with export between countries.

Welfare concerns

During transport, animals can be exposed to a range of potential stressors, including food and water deprivation, thermal extremes, physical injuries, motion sickness, mixing with unfamiliar animals, close confinement, and novel experiences (Fisher et al., 2009; Schwartzkopf-Genswein et al., 2012). If a journey is undertaken according to best practice and in compliance with all aspects of any regulations designed to protect the welfare of animals in transit, many animal welfare problems would either not be expected to occur or would be minimised. A range of behavioural and physiological measurements have been used to assess the responses of animals to transportation. However, changes in some of these variables do not necessarily reflect reduced welfare. Other than monitoring behaviour, there are few animal-based welfare assessments that can be used during a commercial journey. The main animal-based assessments are made during and after unloading (Cockram, 2020). The welfare implications of transport are assessed from observations of behaviour, physiological and biochemical measurements, mortality, morbidity, injury, and carcass characteristics (EFSA, 2011, Table 11.1). The occurrence of some meat quality issues, such as Pale Soft Exudative and Dark-Firm-Dry/Dark cutting meat, is related in part to handling and transport (Adzitey and Nurul, 2011). However, they do not have the specificity to be used to assess animal welfare.

When considering the implications of journeys, it is important to consider each risk factor, their interactions and the multiple stages that can be part of a transport continuum. Whether welfare issues arise during transportation will depend upon the type of animal, their fitness for transport, the quality of the journey, the environmental conditions, and the associated handling and management of the animals.

The ability of animals to cope with handling and transport varies between:

- Animals that respond differently to periods of feed and water deprivation, e.g., between ruminants that have potentially larger stores of water and feed in their rumen than monogastrics that have a smaller stomach, and between unweaned and weaned animals;
- Animals that respond differently to heat stress due to differences in their effectiveness to lose heat via respiratory evaporative water loss and sweating, and to cold stress due to differences in their amounts of fat and coat insulation;

Table 11.1 Potential welfare outcomes of transportation, their main risk factors, and common methods used in their assessment

Outcome	*Main risk factors*	*Examples of methods of assessment*
Hyperthermia and heat stress	inadequate ventilation high stocking density high ambient temperature and humidity	behavioural and physiological responses (increased respiration rate, panting and sweating) raised body temperature mortality
Hypothermia and cold stress	low ambient temperature wet or diminished coat insulation air movement/wind	behavioural responses (huddling, shivering) lowered body temperature mortality frostbite
Injury	manual handling contact with structures animal interactions loss of stability due to vehicle acceleration and slippery floors	skin cuts and lacerations post-mortem bruising post-mortem bone fractures lameness and non-ambulatory conditions plasma creatine kinase activity
Fatigue	prolonged standing muscular exertion to brace against acceleration and to make frequent foot adjustments inadequate rest exhaustion of body energy reserves	behavioural responses (reduced responses to fear and other external stimuli, resting) non-ambulatory conditions blood and tissue measurements indicative of exhaustion of body energy reserves and accumulation of metabolites (lactate)
Hunger	prolonged periods without access to feed fasting before transport	behavioural responses (vocalisation, reduced latency to eat, increased feed consumption, investigation of non-food materials) blood and tissue measurements indicative of utilisation of body energy reserves: liver glycogen concentration, raised plasma concentrations of β-hydroxybutyrate and free fatty acids, and reduced blood glucose concentration (hypoglycaemia)
Thirst and dehydration	prolonged periods without access to drinking water high ambient temperature consumption of feed	increased drinking behavioural responses (reduced latency to drink) blood variables indicative of decreased water content: increased plasma osmolality and plasma total protein concentration reduced body weight, skin tenting sunken eyes
Stress	fear of humans and other animals, novelty, noise, acceleration and vibration, light and odours	behavioural responses raised concentrations of plasma glucocorticoids and catecholamines increased heart rate and heart rate variability.
Illness	mixing with animals from different sources exposure to pathogens from fomites immunosuppression poor air quality	behavioural responses clinical and post-mortem examinations and laboratory tests

- Animals that respond differently to changes in social grouping (due to differences between species and rearing conditions), e.g., some fight and use teeth or horns to injure each other;
- Animals that are different in size as this affects how they are manually handled and their centre of gravity affects their response to changes in acceleration;
- Animals in different physical condition due to their fitness as a result of injury, disease, weakness, genetic traits for productivity, and drugs given;
- Different rearing conditions as this can influence experience of exposure to stressors and humans;
- Different physiological states, such as lactation and pregnancy (Transportation Code of Practice Scientific Committee, 2018).

Species-specific information on handling and transportation can be found in Grandin (2019), and online guides to good practice are available, e.g., Spoolder (2019).

Transportation modes

Land transport

Road transport

The most common mode of animal transport is by road. Vehicles designed for the transportation of animals usually consist of an integrated cab and livestock compartment or an articulated vehicle consisting of a vehicle/tractor attached to one or more trailers. In less regulated areas, animals can be transported on open trailers, trucks, bicycles, and motorcycles. The livestock compartment normally consists of a chassis-mounted box that can be single tier or multi-tier, where animals are carried on several floors or decks. Internal ramps within a vehicle are often steep. Especially in a multi-decked vehicle, it is important that the deck height is sufficient for the animals to stand in their normal position without the deck coming into contact with any part of the animal, and there is adequate airflow over the animals. Many vehicles or trailers have sides that are enclosed with ventilation openings and a roof to provide protection from the external environment (preferably reflective and insulated). The walls can be thermally insulated, but in situations where protection from cold conditions is not required, the compartment can be more open, with sides consisting of a fence-type construction. For ease of cleaning and durability, most livestock vehicles are constructed of aluminium and steel. There should be no sharp edges or protrusions, and sheeting is used to provide a smooth surface to prevent bruising. Internal partitions across the width of the vehicle, adjustable for different sizes, numbers, and types of animal, are used to reduce movement due to vehicle acceleration and to separate groups. Vehicle exhaust stacks should be at least as tall as the roof, otherwise, diesel fumes can enter the livestock area. The floors can be constructed of pressed metal sheeting or a metal grid to provide a non-slip surface. In some countries, bedding materials such as sawdust, wood shavings, straw, and sand are used to absorb urine and faeces, provide traction to reduce slipping, comfort, and thermal insulation. When animals are transported in containers, the containers are normally stacked on a flat-bed trailer that is completely or partially open on the lateral sides of the vehicle.

Rail transport

Transport by rail (Bisschop, 1961; Sutton, 1961) is no longer a common mode of transportation and is restricted to specific locations within some countries. However, there is potential for the expansion of rail transport and a rail journey is likely to be conducted at a relatively constant speed with fewer stops than on a road journey (Woodhead et al., 2016). Rail transport can be used if handling facilities such as loading ramps are available at the railheads and there are direct links to the common destinations.

Walking

Droving/walking/trekking of livestock is a traditional mode of transportation and still occurs in some countries and in specific locations, such as during transhumance in mountain regions. However, walking is only used where the infrastructure for road transport is inadequate; the journeys are short or insufficient resources are available to use road transport. It is a slow method of transport and exposes the animals to risks such as accidents, predation, toxic plants, poor walking surfaces, infectious diseases, and environmental extremes. Arrangements need to be made for opportunities for grazing, watering, and overnight rest. Some animals can experience weight loss, muscular damage, injury, starvation, dehydration, and exhaustion.

Sea transport

Livestock are transported between countries on long journeys by sea in specialist vessels. As some journeys can last for days or weeks, arrangements are made for the daily care and management of the animals. Welfare issues can occur if sea and weather conditions are extreme. Sheep and cattle are transported on long sea journeys from Australia to the Middle East. Some sheep die during these journeys due to heat distress, starvation from failure to eat the pelleted food offered during the journey, and salmonellosis. The animals can also experience discomfort from the accumulation of ammonia and from the motion of the ship (Phillips and Santurtun, 2013). Many shorter journeys are conducted using roll-on-roll ferries where the livestock remain in the vehicle. Although there is forced ventilation below deck, this does not cause sufficient air movement within the vehicles. Extra adjustable ventilation openings are required, and for long journeys, the stocking density may need to be reduced (Watts, 1982).

Air transport

Air transportation is expensive and mainly restricted to high-value animals and day-old chicks. The International Air Transportation Association (IATA) (2021) have produced guidelines for the transport of animals by air. The animals are normally placed in containers. Their stability can be affected by turbulence, take-off, and landing. Air-conditioning units should be available at departure and on arrival (Watts, 1982; Le, 2012; Collins et al., 2020).

Preparation before transport

Fasting prior to transport for slaughter is practised to reduce faecal contamination and stomach and intestinal distension that can pose a risk of inadvertent puncture/rupture during evisceration (Hogan et al., 2007). Fasting pigs and broilers before transport can also reduce the risk of mortality during journeys in warm weather (Averos et al., 2008; Caffrey et al., 2017). Vaccinating and pre-weaning calves before they are transported on a long journey reduces mortality and morbidity (especially respiratory disease) after arrival (Earley et al., 2017).

Fitness for transport and compromised animals

The fitness of animals for their intended journey must be assessed before loading and many countries have regulations that define when animals are considered to be unfit for the intended journey. Examples of animals that are unfit for transport include those that are sick, injured, weak, disabled, or fatigued; those that are unable to stand unaided and bear weight on each leg; those in poor body condition, are pregnant and likely to give birth, neonatal animals and those with a condition that indicates that they cannot be transported without suffering (OIE, 2018). In some cases, animals arrive unfit at their destination because their health deteriorated during the journey, and in others, the animals may have had pathology that was not readily apparent before loading. Issues can arise when different stakeholders have different views on the criteria that make an animal unfit for transport (Dahl-Pedersen et al., 2018).

Compromised animals with major pathology are likely to experience pain, systemic illness, and have reduced physiological function, e.g., to respond to changes in environmental temperature and undertake physical movement (walking on and off the vehicle and maintaining stability). In addition, their physical condition (e.g., low body condition score, weakness, and chronic disease) can increase their vulnerability to injury from handling and to extended periods without feed, water, and rest. Animals that are not in good health are more likely to become fatigued, injured, non-ambulatory, or die during transport (Cockram, 2019).

Compromised and vulnerable animals require additional care, such as restricting the duration of the journey and ensuring that the animals are transported directly to their destination; individual loading and unloading without having to negotiate internal ramps; loaded last and unloaded first; segregated on the vehicle; provided with additional bedding; not transported in extreme thermal conditions, and given increased provision of feed and water. Lactating animals require drying-off before a journey; otherwise, they will need regular milking to avoid udder engorgement and discomfort. Of particular concern is the transport of cull animals that are sent for slaughter after a period of breeding or production of milk/eggs. These animals should be transported while they are still fit, or they should be killed on-farm for consumption or disposal (Cockram, 2021). If end-of-lay hens are transported to slaughter rather than killed on-farm, this can cause severe welfare issues. The layers are susceptible to bone fractures during handling and cold conditions as a consequence of reduced feathering (Newberry et al., 1999; Vecerkova et al., 2019).

Handling

Stress during handling and the risk of physical injury can be minimised by ensuring that the facilities are well designed and the handlers trained and supervised. Animals can be stressed by movement from their normal pen and mixing before or at the time of loading. Before loading, animals destined for slaughter may need to be segregated from the rest of the group, weighed, and their condition assessed. Some animals require identification, such as ear tagging or branding for traceability, and this can involve additional handling and discomfort. In most circumstances, it is preferable to segregate the following during handling and transportation: different species; significantly different sizes or ages; adult breeding males; sexually mature males from females; animals with horns from animals without horns; animals hostile to each other; and tied animals from untied animals. Animals that have been handled previously are easier to handle than those that have never been handled, and the temperament of the animal can affect the ease of handling. Tools such as panels, flags, and rattles can be used to encourage and direct the movement of the animals. However, animals should not be hit or kicked; pressure should not be applied to any particularly sensitive part of the body, and they should not be lifted or dragged by their extremities. Instruments that administer electric shocks (goads/prods) should be avoided as far as possible. Grandin (2020) has pioneered the use of handling methods that consider the herding and/or flocking behaviour of animals, movement of animals in small groups, flight zones and points of balance, field of view, depth perception, visual and auditory distractions, movement towards light, and avoidance of slippery floors. Performance standards can be established in which numerical scoring is used to evaluate the use of driving instruments and the percentage of animals slipping or falling.

Loading and unloading

The assembly/holding areas used before loading should be designed to provide protection from the weather, separate social groups, and, if the animals are kept for an extended period, opportunities for rest, feed, and water. The loading facilities should be designed, constructed, maintained, and operated to minimise the risk of injury and to facilitate movement of the animals. Larger animals walk onto the vehicle; some are lifted onto the vehicle; poultry and small mammals are normally loaded onto the vehicle in containers. Animals can be loaded using (a) a ramp that is integral to the vehicle or via an external ramp that can be longer and not as steep; (b) a loading bridge where the animals can walk onto the vehicle without having to walk up an incline; (c) an elevator or hydraulic lift that allows a group of animals to walk into a pen or the floor of the vehicle that is then raised to the relevant vehicle deck level; (d) manual catching and placement into a container that is then loaded onto the vehicle; or (e) manual catching and placement directly onto the vehicle. Ramps, bridges, gangways, and lifts should be non-slip and have sides, railings, or some other means of lateral protection. Ramps used for loading should have the minimum possible incline. Where the slope is steep, the ramp should be fitted with foot battens or steps.

Broiler chickens are usually caught manually from the barn floor then carried to a receptacle consisting of a crate or module placed either inside or outside of the barn. The container is carried out of the barn either manually or via a forklift truck. The manner in which the birds are carried and placed in the container affects the risk of injury and mortality (Cockram and Dulal, 2018). The stocking density in the container is adjusted according to the live weight of the birds and the thermal conditions. Mechanical catching using a machine to collect the birds from the floor and move them into a receptacle for loading onto a transport trailer is used in some countries. In some systems, a conveyor belt is used to load the birds.

Loading can sometimes be prolonged, and some vehicles have to wait before unloading at the destination. During this time, ventilation may be inadequate and external mechanical ventilation using banks of fans can be beneficial to reduce the risk of heat stress. On arrival at the destination, some animals may have become non-ambulatory because of injury, a metabolic condition or weakness. Large non-ambulatory animals should be killed on the vehicle and not unloaded.

Transportation effects

Stress

Transport involves exposure to many simultaneous stressors. Stress responses are affected by interactions between genetics, experience, and the manner in which animals are handled and transported (Fisher et al., 2009). Stressors during handling and transport include those that initiate fear, such as novel stimuli, unpredictable stimuli (such as noise, motion, and acceleration), proximity to humans, other animals (especially if in mixed social groups), handling systems that separate/isolate animals from other members of their group, moving animals too fast, exposure to heat and cold, and restraint. Some breeding practices have made certain strains more susceptible to stress, heat, and exercise (Grandin, 2021).

Injury

Injuries such as cuts, lacerations, bruising, fractures, dislocations, and trauma to existing lesions can occur during handling and transport. These can be caused by design flaws such as protrusions, slippery floors and moving animals too fast, resulting in slips and falls, physical force used by a handler, either directly or with a tool, aggressive interactions and mounting between animals (especially in mixed social groups), and by movement caused by vehicle acceleration resulting in loss of stability during the journey. Post-transport assessment of injuries to benchmark the percentages of animals with bruising, fractures, and dislocations provide an indication of the number and severity of physical insults sustained during handling and transportation.

Hunger and thirst

Animals can be exposed to long periods without feed and water, and this predisposes them to hunger and thirst. Feed and water in the rumen can provide ruminants with a source of energy and water, but other animals are more vulnerable, especially in hot or cold conditions. If feed and/or water is restricted, the animals do not eat and/or drink enough before transport, during a journey, or while at an intermediate stop, the cumulative effect is loss of weight, and some animals will be at risk of dehydration and some will show signs of significant mobilisation of body energy reserves.

Fatigue

Fatigue during transport may result from long periods of standing, muscular tension required to brace the body in response to vehicular movements, and frequent limb movements as a result of a loss of balance. Exhaustion can reduce the capacity of an animal to respond to vehicle movement and predispose to injury. Muscle fatigue can be associated with a depletion of muscle energy stores, such as glycogen, the accumulation of metabolites, and muscular damage. However, clear evidence of fatigue after a long journey in terms of muscle exertion and damage and a short latency to lie down is not always apparent (Fisher et al., 2009).

Infectious disease

Mixing animals from different sources increases the risk of contact with infected animals and/or materials. Vehicles and containers need to be cleaned and disinfected after every journey. There is an increased susceptibility to infection and disease if the animals are "stressed" by the journey with increased shedding of pathogens and immunosuppression. Prior vaccination, segregation, and quarantine of transported animals at the destination may be necessary. When animals are exported between countries, considerable care is required to follow biosecurity protocols and to complete any necessary testing, inspection, and documentation.

Mortality

Some animals are dead on arrival, and others die within a few days of a journey. The risk of mortality tends to be greatest in cull animals, poultry, and pigs. Risk factors for mortality in poultry are heat and cold stress, trauma, and disease. The mortality risk in pigs is affected by the genotype, fasting, and heat stress.

Thermoregulation

During transportation, animals can be exposed to thermal environments that can cause thermal stress, and sometimes it can exceed their thermoregulatory capacity, and they can experience hyperthermia and hypothermia. Animals can die from hyperthermia if the conditions are too hot and humid or from hypothermia if the conditions are too cold, the animals are wet and cold, or they have lost some of their coat insulation, and they are exposed to wind. The conditions inside a vehicle can be very different from those outside the vehicle. The internal environment is affected by the metabolic heat and moisture produced by the animals, the efficiency of the ventilation system, the external environment, vehicle movement and vehicle insulation. High stocking density, poor ventilation, exercise, stress, and movement from their established environment can predispose animals to thermal environments that exceed their capacity to maintain homeostasis. Major differences exist in the thermoneutral range of transported farmed animals and their ability to avoid thermal stress. Species vary in their surface area to body mass ratio and their consequential rate of heat exchange with their environment. Age, body condition, type of digestive system, duration of fasting, and presence of fat can affect the energy reserves available to increase heat production in cold conditions. The ability of animals to thermoregulate can also depend on the space available to move and adjust their posture. There are major differences between species in their relative abilities for evaporative heat loss via respiration and sweating. For example, pigs and poultry do not have sweat glands that increase sweating in response to raised temperatures.

Ideally, animals should be transported within their thermal comfort zone, i.e., the range of effective environmental temperatures where an animal is able to thermoregulate with the least behavioural and physiological effort by changing exposed body surface, tissue insulation (sensible heat loss), and latent (evaporative) heat loss without panting (EFSA, 2004). A temperature monitoring and recording system, as well as a warning system to alert the driver when temperatures in the animal compartments reach a maximum or minimum limit, is beneficial.

Heat stress

The greatest risk of heat stress occurs at high temperatures and high humidity. As the ambient temperature approaches body temperature, sensible cooling becomes less effective, and the animal relies increasingly on evaporative cooling. In response to heat, all farmed animals increase their respiratory rate and some, such as cattle and sheep, can increase heat loss by sweating. The ability of animals to lose heat via the evaporation of water is dependent on a temperature and vapour pressure gradient. As humidity increases, the effectiveness of evaporative cooling decreases. The temperature and humidity within a vehicle are dependent on the weather conditions, the number of animals within the vehicle and the efficiency of the ventilation to remove heat and moisture. A forecast of the temperature-humidity index can be used to assess the risk of heat stress during transportation.

Cold stress

During journeys in cold conditions, the animals (especially poultry and young pigs) require protection from cold external temperatures, and it is essential that the animals do not become wet or exposed to excessive air movement. In cold conditions, protection from excessive air movement and precipitation involves the use of protective barriers (e.g., screens, curtains, and tarpaulins) around part or all of the vehicle/trailer or the partial closure of ventilation openings by use of flaps and boards. Unfortunately, this reduces the ventilation, and the internal trailer temperature rises. In extremely cold conditions, this temperature rise can be beneficial in that it can raise the internal temperature above potentially lethal cold external temperatures. However, in a closed or partially closed ventilation configuration, internal thermal cores consisting of pockets of raised temperature and moisture from the animals can occur at one or more locations within the vehicle (Kettlewell et al., 1993; Mitchell and Kettlewell, 1998). Especially in poultry, if the internal temperature and humidity within areas in the core of the vehicle rises too high, some of the animals can experience hyperthermia, even though the external temperature is so low that it would otherwise have caused hypothermia (Mitchell and Kettlewell, 1998). In cold conditions, the air entering the vehicle through air inlets is at a low external temperature and might be accompanied by moisture and excessive air movement. The parts of the vehicle or vessel near air inlets and those on the sides of the vehicle close to the cold external temperature can expose some animals to a risk of hypothermia of sufficient severity to cause death. If an animal cannot move away from the side of the vehicle, it may also be susceptible to frostbite and freezing to metal surfaces. Adding extra bedding, such as straw, when the temperature is low can provide increased floor insulation, but wet bedding should be removed to avoid freezing.

Factors affecting journey quality

Ventilation

Natural ventilation

An efficient ventilation system removes the heat and moisture produced by the animals and replaces it with external air. It should be designed, constructed, and maintained in such a way that, at any time during the journey, whether the vehicle is stationary or moving, it is capable of maintaining a thermal environment that meets the animals' requirements. Most livestock vehicles are ventilated by natural/passive ventilation. Airflow is provided through apertures along the sides (and sometimes the roof) of the compartment, and this allows air exchange between the internal and external environment. Air movement can also provide some convective cooling. The control of natural ventilation is achieved by the driver opening and closing ventilation apertures while the vehicle is stationary. Most air movement is caused by external pressure changes produced by a moving vehicle. As the vehicle moves, air passing over the front edge of the container separates from the vehicle and creates an area of low pressure (suction). The airflow tends to re-attach along the length of the vehicle and enter through the rear openings. Air usually moves forward within the livestock area and leaves through ventilation apertures near the front of the vehicle. Airflow is dependent on vehicle speed, wind direction, vent area, and the degree of obstruction caused by the animals (Gilkeson et al., 2009). In climates where the risk of exposure to cold is low, the vent area can be greater. There should be sufficient space inside the compartment and at each of its levels to ensure that there is adequate airflow above the animals and between containers. When the vehicle is stationary, airflow within the vehicle is dependent on either wind or the stack effect. Convective airflow from the heat of the animals is responsible for thermal buoyancy (the stack effect). During stationary periods, the ventilation is often inadequate, and in hot weather, parking for prolonged periods in direct sunlight or close to obstructions to wind should be avoided. Parking vehicles at right angles to the wind direction is optimal. Effective ventilation is also important to maintain air quality. High concentrations of ammonia, particles, and microorganisms can accumulate during a journey and increase the risk of health and welfare issues.

Mechanical ventilation

As passive ventilation may not always provide consistent, effective ventilation, some vehicles have mechanical ventilation. Mechanical ventilation is especially useful for stationary periods. The direction of airflow created by the fans should be used to enhance the natural airflow within the vehicle caused by the forward movement of the vehicle. Extraction fans near the front and air inlet apertures near the rear of the vehicle provide the most effective arrangement. In case of mechanical failure, a fan ventilated vehicle should have the capability of opening sufficient side apertures to enable emergency natural ventilation (Kettlewell et al., 2001). When livestock are transported by sea in a vessel that depends on mechanical ventilation there should be a back-up system to prevent heat distress (Schultz-Altmann, 2008).

Motion, driving, and road surface

During transport, animals are exposed to acceleration from vehicle movement. Acceleration is the rate of change in the velocity of an object (whether it is forward acceleration or deceleration) and is affected by the balance of forces that act on an animal and its mass. Acceleration occurs in three axes (longitudinal, lateral, and vertical) and consists of vibrations and shocks (Gebresenbet et al., 2011). Shocks are short-duration, high-amplitude acceleration events that occur randomly and are produced in response to a driving event, such as braking, cornering, or running over a pothole. Random, high-magnitude acceleration events (shocks) pose the greatest risk of loss of postural stability (Tarrant, 1990). Vibration represents background acceleration. If the vibration is close to the whole-body resonant frequency, it is aversive, stressful, can reduce resting behaviour, and in pigs cause motion sickness (Santurtun and Phillips, 2015). The roughness, undulation, and curvature of roads, the type of vehicle suspension system, and the manner in which the vehicle is driven can affect the acceleration experienced by the animals. This can affect their stability and ability to rest (Cockram and Spence, 2012).

Stocking density

The stocking or loading density refers to the number or live weight of animals within a specified area of floor space. The space allowance can be quantified as the floor area per animal, but the relevant live weight range must be specified. Allometric equations ($kW^{2/3}$ where k is a constant and W represents live weight) are used to estimate the space that a stationary animal occupies as a consequence of its mass (Petherick and Phillips, 2009). A k value of at least 0.02 is required, but if all of the animals within a pen need to lie down simultaneously, a k value of at least 0.027 is required. During transport, animals need increased space to adopt postural changes to brace themselves while standing and to adjust their footing in response to acceleration. For long journeys, animals may need to lie down and sometimes to drink and feed onboard the vehicle. In this situation, extra space is required to provide access to troughs and drinkers so that the animals can attempt to eat or drink simultaneously. If insufficient space is provided, the animals can experience reduced stability, fatigue, bruising, and stress (Tarrant, 1990). If the vehicle is driven well on good quality roads, the animals benefit from plenty of space. Overcrowding must always be avoided, but in some situations where they are exposed to violent movement, there is some evidence that animals can benefit from mutual and lateral support to reduce their potential for movement and subsequent injury (Eldridge and Winfield, 1988; González et al., 2012). Increasing stocking density increases the number of animals in a container or vehicle and the amount of metabolic heat and moisture produced. Unless this extra metabolic heat and moisture can be effectively removed by ventilation, it can be detrimental at warmer temperatures and predispose to heat stress. Lowering stocking density reduces the risk of heat stress. At low temperatures, a high stocking that provides beneficial metabolic heat can reduce the risk of cold stress. In some situations, overcrowding can predispose to suffocation and sometimes it can obstruct ventilation openings.

Journey duration

Land transportation can last several hours or several days. Sea transportation can last several days or weeks and can include road journeys. Air transport may only last several hours but it also requires road journeys. The major concerns that have been expressed over the effects of long journeys on the welfare of animals are that animals are exposed to prolonged stress; excessive periods without feed and water; fatigue and lack of rest; an uncomfortable physical environment; and increased risk of injury, ill-health, or death (Transportation Code of Practice Scientific Committee, 2018). Some countries have regulations that restrict maximum journey duration (European Council, 2005). However, there is no consensus on specific maximum journey durations. A rationale for restricting specific journey durations can be made when it is clear that (a) aspects of welfare are adversely affected after a specific journey duration and thus stopping a journey before this occurs would help to minimise these adverse effects; (b) the animals are exposed to a continuous, aversive experience, and that restricting journey duration would minimise the duration of this experience; and (c) that the many risk factors associated with transportation that have the potential to adversely affect aspects of welfare cannot be mitigated by an improved journey quality, and the longer the journey, the greater the risk (Cockram, 2007). However, there is a strong argument that too much emphasis has been placed on journey duration, and greater focus should be placed on the quality of the journey. The quality of a journey will affect how the animals respond to a journey of long duration. If environmental conditions (including driving style, road conditions, vehicle design and operation, space allowance, thermal conditions, and ventilation), the fitness of the animals and the pre- and post-transport handling of the animals are optimal, it should be possible to transport certain types of animals over long distances without major welfare problems. If, however, there is widespread non-compliance with regulations or standards, together with inadequate enforcement or supervision, and optimal conditions are not provided, the argument for limiting journey durations is strengthened (Cockram, 2007).

Feed and water deprivation

Many farm animals are not provided with feed and water during a road journey. The period of feed and water deprivation is the sum of the journey duration without access to feed and water and the periods pre-and post-transport during which the animals are not given access to feed and/or water. On long journeys, some vehicles, especially in Europe, carry equipment and a quantity of appropriate feed and water to feed and water the animals during the journey. However, "rest" stops during long journeys need to be long enough for each animal to eat and drink. There is considerable debate over when journey duration should be limited by the physiological requirements of the animals for feed and water and to avoid severe hunger and thirst. Stopping a journey to unload the animals to provide a period of rest, feed, and water can increase the risk of stress, injury, and infectious disease.

Contingencies

Protocols and procedures are required to monitor and, when necessary, inspect the animals and essential equipment during a journey. For long sea journeys, a formal risk assessment and procedure for management of the risks is required (Stinson, 2008). A contingency plan is required to establish procedures for events such as delays, accidents, and compromised or unfit animals. Road accidents can occur due to the vehicle overturning, collisions, or mechanical failure. This can result in mortality, injury, or escape of animals onto the road (Miranda-de la Lama et al., 2011). During a journey, an efficient means of communication should be available, and navigation, tracking, and recording systems are valuable to provide a journey log and benchmarking to improve future journeys.

Conclusions

If the fitness of the animals and the quality of the journey conditions that they experience are optimal, farmed animals are likely to be stressed by the novelty of their environment, but they can be transported without suffering. However, there are numerous factors that can increase the risk of animals experiencing negative affective states as a consequence of transportation, and many welfare issues occur during routine transportation. The potential welfare issues associated with transport include stress, thermal and physical discomfort, pain, fatigue, sickness, hunger, and thirst. The quality of a journey is affected by factors such as vehicle design, stocking density, ventilation, the standard of driving and quality of the road, the journey duration, the environmental conditions, and the associated handling and management of the animals. The welfare issues are multifactorial, transport conditions are diverse and challenging, and different types of animals have distinct requirements. These issues are of such significance that they are regulated by detailed industry standards, codes of practice, and legislation.

References

Adzitey F and Nurul H, 2011. Pale soft exudative (PSE) and dark firm dry (DFD) meats: causes and measures to reduce these incidences-a mini review. *International Food Research Journal*, 18, pp. 11–20.

Averos X, Knowles TG, Brown SN, Warriss PD and Gosalvez LF, 2008. Factors affecting the mortality of pigs being transported to slaughter. *Veterinary Record*, 163, pp. 386–390.

Bisschop JHR, 1961. Transportation of animals by rail. *Journal of the South African Veterinary Association*, 32(2), pp. 235–268.

Caffrey NP, Dohoo IR and Cockram MS, 2017. Factors affecting mortality risk during transportation of broiler chickens for slaughter in Atlantic Canada. *Preventive veterinary medicine*, 147, pp. 199–208.

Cockram MS, 2007. Criteria and potential reasons for maximum journey times for farm animals destined for slaughter. *Applied Animal Behaviour Science*, 106(4), pp. 234–243.

Cockram MS, 2019. Fitness of animals for transport to slaughter. *The Canadian veterinary journal*, 60(4), pp. 423–429.

Cockram M, 2020. Condition of animals on arrival at the abattoir and their management during lairage. In: Grandin, T and Cockram M (ed) *The Slaughter of Farmed Animals: Practical Ways of Enhancing Animal Welfare*. CABI, Wallingford, UK, pp. 49–77.

Cockram MS, 2021. Invited Review: The welfare of cull dairy cows. *Applied Animal Science*, 37(3), pp. 334–352.

Cockram MS and Dulal KJ, 2018. Injury and mortality in broilers during handling and transport to slaughter. *Canadian Journal of Animal Science*, 98(3), pp. 416–432.

Cockram MS and Spence JY, 2012. The effects of driving events on the stability and resting behaviour of cattle, young calves and pigs. *Animal Welfare*, 21(3), pp. 403–417.

Collins T, Stockman C, Hampton JO and Barnes A, 2020. Identifying animal welfare impacts of livestock air transport. *Australian Veterinary Journal*, 98(5), pp. 197–199.

Dahl-Pedersen K, Foldager L, Herskin MS, Houe H and Thomsen PT, 2018. Lameness scoring and assessment of fitness for transport in dairy cows: Agreement among and between farmers, veterinarians and livestock drivers. *Research in Veterinary Science*, 119, pp. 162–166.

Earley B, Buckham Sporer K and Gupta S, 2017. Invited review: Relationship between cattle transport, immunity and respiratory disease. *Animal*, 11(3), pp. 486–492.
Eldridge GA and Winfield CG, 1988. The behaviour and bruising of cattle during transport at different space allowances. *Australian Journal of Experimental Agriculture*, 28(6), pp. 695–698.
European Council, 2005. Council Regulation (EC) No 1/2005 of 22 December 2004 on the protection of animals during transport and related operations and amending Directives 64/432/EEC and 93/119/EC and Regulation (EC) No 1255/9.
EFSA, European Food Safety Authority, 2004. Opinion of the Scientific Panel on Animal Health and Welfare (AHAW) on a request from the Commission related to standards for the microclimate inside animal road transport vehicles. *EFSA Journal*, 122, 1–25.
EFSA, European Food Safety Authority Panel on Animal Health and Welfare, (AHAW), 2011. Scientific Opinion Concerning the Welfare of Animals during Transport. *EFSA Journal*, 9(1).
Fisher AD, Colditz IG, Lee C and Ferguson DM, 2009. The influence of land transport on animal welfare in extensive farming systems. *Journal of Veterinary Behavior: Clinical Applications and Research*, 4(4), pp. 157–162.
Gebresenbet G, Aradom S, Bulitta FS and Hjerpe E, 2011. Vibration levels and frequencies on vehicle and animals during transport. *Biosystems Engineering*, 110(1), pp. 10–19.
Gilkeson CA, Thompson HM, Wilson MCT, Gaskell PH and Barnard RH, 2009. An experimental and computational study of the aerodynamic and passive ventilation characteristics of small livestock trailers. *Journal of Wind Engineering and Industrial Aerodynamics*, 97(9–10), pp. 415–425.
González LA, Schwartzkopf-Genswein K, Bryan M, Silasi R and Brown F, 2012. Relationships between transport conditions and welfare outcomes during commercial long haul transport of cattle in North America. *Journal of Animal Science*, 90(10), pp. 3640–3651.
Grandin T, 2019. *Livestock Handling and Transport*. CABI, Wallingford, UK.
Grandin T, 2020. Behavioural principles of stockmanship and abattoir facility design. In: Grandin, T and Cockram M. (ed) *The Slaughter of Farmed Animals: Practical Ways of Enhancing Animal Welfare*. CABI, Wallingford, UK, pp. 90–110.
Grandin T, 2021. Welfare during transport of livestock and poultry. In: Grandin T. (ed) *Improving Animal Welfare: A Practical Approach*. CABI, Wallingford, UK, pp 347–384.
Hogan JP, Petherick JC and Phillips CJC, 2007. The physiological and metabolic impacts on sheep and cattle of feed and water deprivation before and during transport. *Nutrition Research Reviews*, 20, pp. 17–28.
International Air Transport Association, (IATA), 2021. *Live Animals Regulations (LAR)*. www.iata.org
Kettlewell P, Mitchell M and Meehan A, 1993. The distribution of thermal loads within poultry transport vehicles. *Agricultural Engineer*, 48, pp. 26–29.
Kettlewell PJ, Hoxey RP, Hampson CJ, Green NR, Veale BM and Mitchell MA, 2001. Design and operation of a prototype mechanical ventilation system for livestock transport vehicles. *Journal of Agricultural Engineering Research*, 79(4), pp. 429–439.
Le L, 2012. Safe transport of live animal cargo. *AERO Magazine*, QTR_02.12, pp. 17–24.
Miranda-de la Lama, G. C., Sepúlveda WS, Villarroel M and María GA, 2011. Livestock vehicle accidents in Spain: Causes, consequences, and effects on animal welfare. *Journal of Applied Animal Welfare Science*, 14(2), pp. 109–123.
Mitchell MA and Kettlewell PJ, 1998. Physiological stress and welfare of broiler chickens in transit: Solutions not problems! *Poultry Science*, 77(12), pp. 1803–1814.
Newberry RC, Webster AB, Lewis NJ and Van Arnam C, 1999. Management of spent hens. *Journal of Applied Animal Welfare Science*, 2(1), pp. 13–29.
OIE, Office International des Epizooties, 2018. Transport of animals by land. *Terrestrial Animal Health Code*. OIE, Paris, France. https://www.oie.int.
Petherick JC and Phillips CJC, 2009. Space allowances for confined livestock and their determination from allometric principles. *Applied Animal Behaviour Science*, 117(1–2), pp. 1–12.
Phillips CJC and Santurtun E, 2013. The welfare of livestock transported by ship. *Veterinary Journal*, 196(3), pp. 309–314.
Santurtun E and Phillips CJC, 2015. The impact of vehicle motion during transport on animal welfare. *Research in Veterinary Science*, 100, pp. 303–308.
Schultz-Altmann A, 2008. Engineering and design of vessels for sea transport of animals: The Australian design regulations for livestock carriers. *Veterinaria italiana*, 44(1), pp. 247–258.
Schwartzkopf-Genswein K, Faucitano L, Dadgar S, Shand P, Gonzalez LA and Crowe TG, 2012. Road transport of cattle, swine and poultry in North America and its impact on animal welfare, carcass and meat quality: A review. *Meat Science*, 92(3), pp. 227–243.

Spoolder H, 2019. Animal transport guides. Wageningen Livestock Research, the Netherlands. http://www.animaltransportguides.eu.

Stinson PR, 2008. Two incidents that changed quality management in the Australian livestock export industry. *Veterinaria italiana*, 44(1), pp. 177–186.

Sutton GD, 1961. Transportation of animals by rail II. Transport of animals by rail. *Journal of the South African Veterinary Association*, 32(2), pp. 271–276.

Tarrant PV, 1990. Transportation of cattle by road. *Applied Animal Behaviour Science*, 28(1–2), pp. 153–170.

Transportation Code of Practice Scientific Committee, 2018. Code of practice for the care and handling of livestock and poultry: Transportation. Review of scientific research on priority welfare issues. NFACC, Lacombe, Canada. https://www.nfacc.ca/resources/codes-of-practice/transport/transportation_sc_report_mar2018.pdf.

Vecerkova L, Vecerek V and Voslarova E, 2019. Welfare of end-of-lay hens transported for slaughter: effects of ambient temperature, season, and transport distance on transport-related mortality. *Poultry Science*, 98(12), pp. 6217–6224.

Watts MET , 1982 .Bulk transportation of farm animals by air and vehicular ferries. In: Moss R. (ed) *Transport of Animals Intended for Breeding, Production and Slaughter.* Martinus Nijhoff Publishers, The Hague, the Netherlands, pp. 147–165.

Woodhead A, Nugent T, McDonald L and Rezazade F, 2016, *Quilpie to Brisbane moving cattle by rail freight, Australian Centre for Sustainable Business and Development.* University of Southern Queensland, Toowoomba, Queensland, Australia. https://eprints.usq.edu.au/33955/

12
SLAUGHTER, EUTHANASIA, AND DEPOPULATION

Temple Grandin

Introduction

When animals have to be killed for any reason, it is essential to either eliminate or reduce both pain and fear stress. The major emphasis for this chapter is the slaughter of terrestrial animals for food. The welfare of fish at slaughter is considered in Chapters 10, 16, and 17. The welfare implications of killing animals used for scientific and educational purposes are considered in Chapter 13. Some of the principles in this chapter also apply to euthanasia of sick or debilitated animals on the farm and mass depopulation of animals for disease control.

The public has become increasingly concerned about the welfare of the livestock that are raised for food. When I am talking to people who do not work in the meat industry, I am often asked if cattle and other animals know they are getting slaughtered. Early in my career, I was also looking for an answer to this question. I observed that the willingness of cattle to move through a race (chute) to the stunner was the same at both the slaughter house and at a feedlot, where they were vaccinated. If they knew they were going to die, the behaviour of the cattle should have been more agitated at the slaughter house.

At both the beef slaughter house and at the feedlot, the handling systems have many similarities. The cattle are moved from group holding pens (lairage) to a drive alley that leads to either the stun box or a restraining squeeze chute for vaccinations. In both places, there is a single file race (chute) where the cattle wait in line. To direct the cattle into the single file race, they are moved in small groups into a small pen called a crowd pen or forcing pen (Figure 12.1). The crowd pen is used to direct the cattle into the single file race. Figure 12.2 shows lairage pens that hold livestock after they are unloaded from the trucks.

After visiting many slaughter plants, ranches, and feedlots where large numbers of cattle were handled, I observed that visual distractions, which most people do not notice, would cause cattle and pigs to stop and refuse to move through a race to be restrained for either vaccination or stunning. Some examples of visual distractions are loose chains hanging down, seeing moving equipment up ahead, or reflections on wet surfaces (Grandin and Cockram, 2020). Sharp shadows cast by the sun on the floor of an alley have been associated with more cattle stopping and refusing to move compared to blurry shadows or no shadows (Willson et al., 2021). A large truck with a loud engine parked alongside the lairage (stockyards where livestock are held before slaughter) would also cause cattle to stop and be more difficult to move. This is likely to

DOI: 10.4324/9781003182351-14

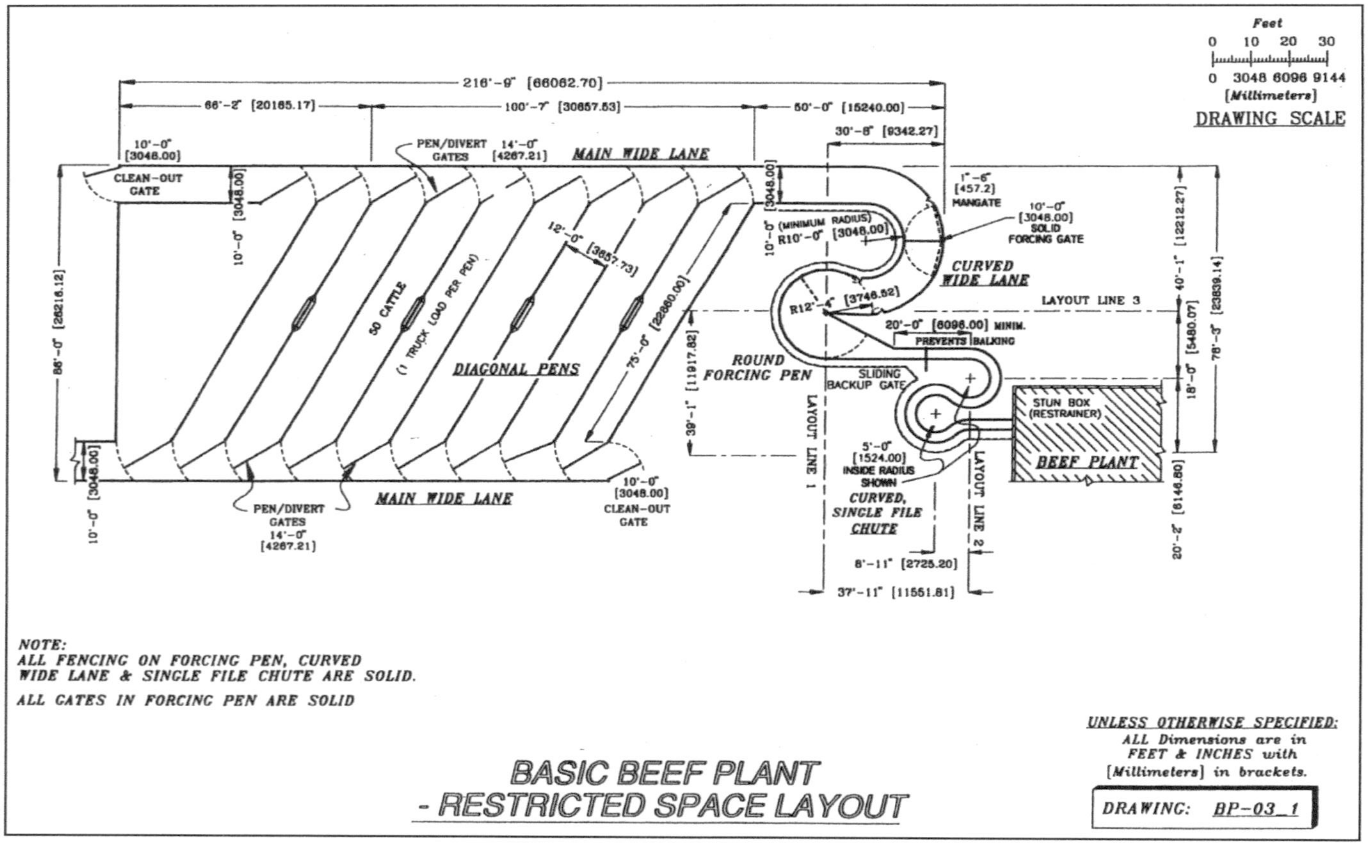

Figure 12.1 Layout for a large beef slaughter plant with diagonal holding (lairage) pens, a round forcing (crowd) pen and a curved single file chute (race). A major advantage of this layout is that all cattle traffic through the holding pen is one way. When the cattle move through the round crowd pen, they will enter the single file race easily because they think they are going back to where they came from.

Figure 12.2 Holding pens (lairage) for market weight pigs. This pen is stocked at the correct density and all the pigs have space to lay down without being on top of each other (photo credit: Temple Grandin).

result in the use of more aversive methods to move cattle, such as electric prods or tail twisting. However, sometimes simple solutions are available that do not entail causing pain to the animals. For example, the ease of moving cattle and pigs can be improved by adding a lamp to illuminate a dark race or stun box entrance. Cattle and pigs avoid entering dark places (Grandin and Cockram, 2020; Grandin, 2001). The addition of the lamp greatly reduces electric prod use because the animals are more willing to move.

Short-term physiological measures of stress

Stress in the abattoir caused by visual distractions, aversive handling methods, such as with the aid of electric prods, and unfamiliar environments are major welfare concerns and can also have a detrimental effect on meat quality. It is important to remove visual distractions because when animals stop moving, handlers are more likely to use the aversive handling methods, which include not only electric prods but also tail twisting. When pigs are jammed in a race or moved with an electric prod within five minutes of stunning, they are more likely to have poor meat quality. Their lactate levels are higher (Edwards, et al., 2010) and their meat is more likely to be Pale, Soft, and Exudative (known as PSE meat). This is a severe quality defect. In cattle, repeated shocks with an electric prod within a few minutes before stunning create problems with tough meat (Warner et al., 2007).

I have reviewed a number of studies and compared cortisol levels after handling on the farm for veterinary procedures with cortisol levels after stunning at a slaughter plant (Grandin, 1997). The levels ranged from high to low, but the range was similar in both places. Studies conducted after the publication of Grandin (1997) continue to confirm this.

Behavioural principles of livestock handling

It is really important for managers to train employees who work in abattoirs about the behavioural principles of livestock handling. When handlers understand behaviour, it will reduce stress because handling methods that cause stress will be reduced. The first principle is that a calm animal is much easier to handle compared to an agitated, frightened one. One sign of a highly stressed animal is that it has diarrhoea. It takes 20 to 30 minutes for all livestock to calm down if they become highly agitated. Handlers need to remain calm and must never yell at animals. Yelling at animals is much more stressful than normal conversation (Hemsworth et al., 2011).

Flight zone and point of balance

Handlers should be trained in three basic behavioural principles. First, a tame animal that is trained to lead has no flight zone (the animal's personal space which, when entered by a human, causes it to flee) and it can be touched by people. In cattle, pigs, and sheep that are not completely tame, the flight zone can vary from one metre to many metres. The animal will stop moving when the handler backs up and retreats out of the flight zone. If an animal rears up in a race, the handler should back up to remove themselves from the inside of the animal's flight zone.

The second important behavioural principle is that there is a point of balance at the animal's shoulder. An animal in a single file race will move forward when the handler is behind the point of balance. Further information is in Grandin and Cockram (2020) and Grandin (2021).

Moving small groups

One of the most common mistakes that people make when handling cattle and pigs is placing too many animals in the crowd pen that leads to the single file race (Yost et al., 2020). Calm, low-stress handling requires stock people to walk more to enable them to move small groups of animals from the holding pens to the crowd pen. There are species differences. Sheep can be moved in larger groups due to their intense following behaviour. The basic principle is that cattle, pigs, and goats should be moved in small separate bunches and sheep can be moved to the stunner in a continuous flow.

Driving aids

Flags, plastic paddles, and sort boards for pigs are all acceptable driving aids. An electric prod (goad) should never be a person's primary driving aid. Many slaughter plant managers only allow an electric prod to be used at the entrance to the stun box. Abusive methods of moving animals are prohibited by both legislation and voluntary industry guidelines (FSIS/USDA, 2020; NAMI, 2021; OIE, 2019). The European Food Safety Authority (EFSA) publishes scientific opinion reports for every species. These can be easily found online by searching for the EFSA report by species. Some examples of abusive driving methods that should never be used are poking sensitive areas of the animal, such as the eyes, ears, or anus. Other abusive methods for moving animals that must be avoided are dragging conscious animals, poking them with pointed sticks, beating or breaking tails. If an animal refuses to move into a stun box, a brief application of an electric prod would be preferable if it prevented the use of an abusive method. If many animals refuse to move, there may be a distraction that needs to be removed.

Problems caused by farm conditions

At the abattoir, some livestock are extremely difficult to handle in a low-stress manner due to on-farm factors. One problem is lame animals that have difficulty walking. Some dairy cows are in a really poor debilitated condition before they leave the farm (Edwards-Callaway et al., 2018). Young feedlot grain-fed cattle may also be lame and reluctant to move due to a variety of on-farm factors. Four factors that can cause increased lameness in these cattle are 1) raised for long periods on concrete floors (Magrin et al., 2020); 2) poor leg conformation due to excessive genetic selection for growth; 3) high-grain diets and a lack of roughage (Magrin et al., 2020); and 4) high doses of beta-agonists (Peterson et al., 2015). Beta-agonists are drugs that increase muscle mass. Cattle and pigs fed high doses are more likely to become lame or become difficult to move.

In all types of livestock, animals that have not experienced people walking among them on the farm may be difficult to move at the abattoir. They have never learned to move quietly away from a person walking through their pen or pasture. On intensive pig farms, people should walk through the fattening pens every day. Cattle that have been handled exclusively by riders on horseback may be dangerous to handle when they first encounter stock people walking at the abattoir. The flight distance is greatly increased because the horse and rider are perceived as familiar and safe, and the person walking is perceived as new and frightening.

Methods to render livestock unconscious

The animal welfare laws and industry guidelines in the United States, Europe, and many other countries require that animals are rendered unconscious and insensible to pain before slaughter procedures start (FSIS/USDA, 2020; OIE, 2019; Welfare Quality, 2009; NAMI, 2021). The European guidelines are published by EFSA. In most countries, these laws apply to all mammals and poultry. After the animal is rendered unconscious by an approved stunning method, the throat is cut to drain the blood ("exsanguination"). After bleeding, further procedures such as skinning and removal of the internal organs are performed.

Restraint for stunning

An animal has to be restrained so that a stunning method to render it unconscious can be correctly applied. Cattle, sheep, pigs, and other animals are usually held in either a single animal stun box or they ride on a conveyor restrainer (Grandin and Cockram, 2020). A single animal stun box is a small stall which can hold one animal. It should be narrow so that the animal cannot turn around. A non-slip floor is essential to prevent slipping and falling. Agitated behaviour and jumping around in a stun box are often caused by multiple little slips on a floor that is worn out and slick. Many stun boxes are equipped with a device to hold the head still for stunning (Figure 12.3). In large abattoirs, a conveyor restrainer where cattle, pigs, or sheep ride in a continuous single file line may be used. There are two types of conveyor restrainers: the V conveyor restrainer and the centre track conveyor that the animals straddle (also called a band, belly, or double rail restrainer). Research has shown that it is a low-stress method of restraint (Westervelt et al., 1976).

Captive bolt and gunshot

A well-maintained captive bolt gun, when correctly positioned, renders animals such as cattle, pigs, or sheep instantly unconscious by shooting a steel rod into the brain. It is propelled by

Figure 12.3 Single cattle stun box with a headholder to hold the head still for captive bolt stunning. This box must have a non-slip floor to prevent agitated behaviour due to slipping (photo credit: Temple Grandin).

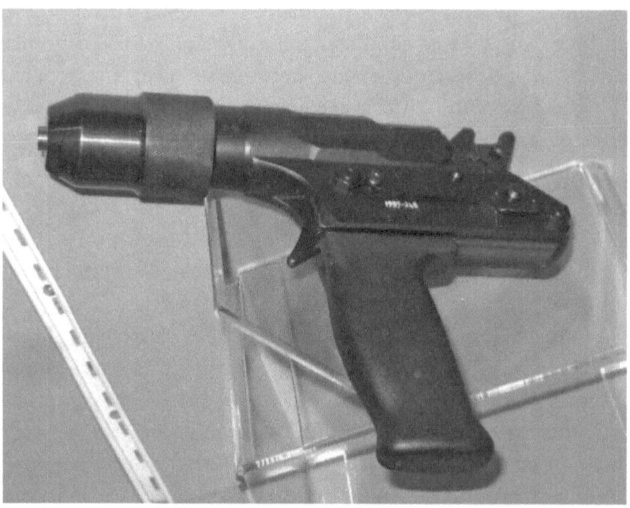

Figure 12.4 Pistol-style captive bolt stunner that uses a blank cartridge to propel the bolt. To insure maximum effectiveness, this tool requires three things: 1) careful cleaning and maintenance every day; 2) dry cartridges; 3) the correct cartridge size for the type of animal (photo credit: Temple Grandin).

either compressed air or a blank cartridge (Figure 12.4). The rod penetrates the brain of the animal. This device is called a captive bolt because the rod is retracted and then reset for the next animal. There are two types, the penetrating captive bolt, which was just described, and a non-penetrating mushroom head that does not penetrate the skull. A penetrating captive bolt that causes physical damage to the brain is more effective than a non-penetrating gun for large animals such as bulls (Gibson et al., 2019). Both types depend on a bolt being propelled at a high speed to produce the required concussive force to make the animal instantly unconscious. Research clearly shows that a well-maintained and accurately positioned penetrating captive bolt gun is effective (AVMA, 2016; Gibson et al., 2019). Gunshot with firearm using a free bullet is also an effective method of slaughter. It is used in some abattoirs. A major advantage of the captive bolt is safety. Captive bolt guns are also used for euthanasia of animals on the farm and mass depopulation for disease control.

Captive bolt guns are available in three designs, inline blank cartridge fired, pistol-type blank cartridge fired (Figure 12.4), and pneumatic. All types require careful maintenance to maintain effectiveness. Poor maintenance is a major cause of captive bolt failure. The blank cartridges used for propelling the bolt must be kept in a dry location. Damp cartridges will cause poor stunning because they will be less powerful. On pneumatic captive bolts, the air supply must be sufficient and set at the manufacturer's specifications.

Both captive bolt and gunshot are most effective when the shot is placed in the correct location on the bovine's forehead (Figure 12.5). Poor aim and shooting in the wrong location is another reason for failure of a captive bolt to render an animal instantly unconscious. Diagrams which show the correct location for shooting all species of animals are widely available (AVMA, 2016; Humane Slaughter Association in the UK, 2021; NAMI, 2021).

Electrical stunning

Electrical stunning induces instantaneous unconsciousness by passing a sufficient electrical current through an animal's brain to induce a grand mal epileptic seizure (Anil and McKinstry,

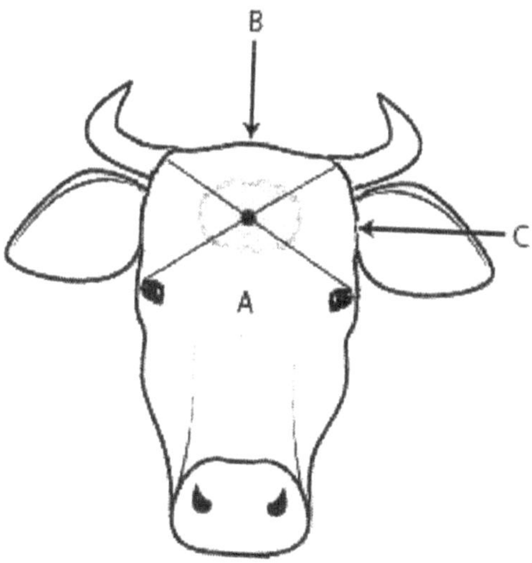

Figure 12.5 Correct location on the forehead for shooting cattle with a captive bolt gun. (diagram credit: State of Queensland, Australia, Creative Commons).

1998; AVMA, 2016; HSA, 2021). To achieve this, the electrodes have to be placed so the electric current will pass through the brain (Anil and McKinstry, 1998). These principles apply to all mammals, birds, and fish. Sufficient electrical amperage has to be passed through the head to produce the seizure. The recommended electrical parameters can be found in AVMA (2016) and HSA (2021). Some common causes of electric stunning failure are dehydrated animals that have been off water for long periods, low electrical amperage, and wrong electrode placement.

For mammals, there are three types of electrical stunning. They are 1) head-only reversible; 2) head to heart cardiac arrest non-reversible; and 3) sequential head and then heart non-reversible. When head-only stunning is used, the current is passed through the brain and it will temporarily make the animal unconscious. The animal may start to regain consciousness unless it is promptly bled within 10 to 15 seconds (Lambooij, 1982). When head to heart stunning is used, the current is simultaneously passed through the brain and the heart. This stops the heart and the animal will not recover (Figure 12.6). Many small abattoirs will use sequential head and heart stun: the stunning tong is first applied to the head to produce instantaneous unconsciousness and then re-applied to the chest to stop the heart (Vogel et al., 2010).

For poultry, there are two types of electrical stunning. They are head-only and water bath stunning. Water bath stunning is used in many large commercial poultry plants. In this system, the operator hangs the live chickens by their legs on shackles. They are then moved by an overhead conveyor to a trough filled with water where their heads are submerged. The electrical current flows through their heads to the shackle. Electrical stunning of poultry will make the birds instantly unconscious. Even when this system is working correctly, small numbers of birds may raise their heads and not be stunned. Suspending the fully conscious birds by their legs to present them to the water bath is highly stressful (Bedanova et al., 2007) (Figure 12.7).

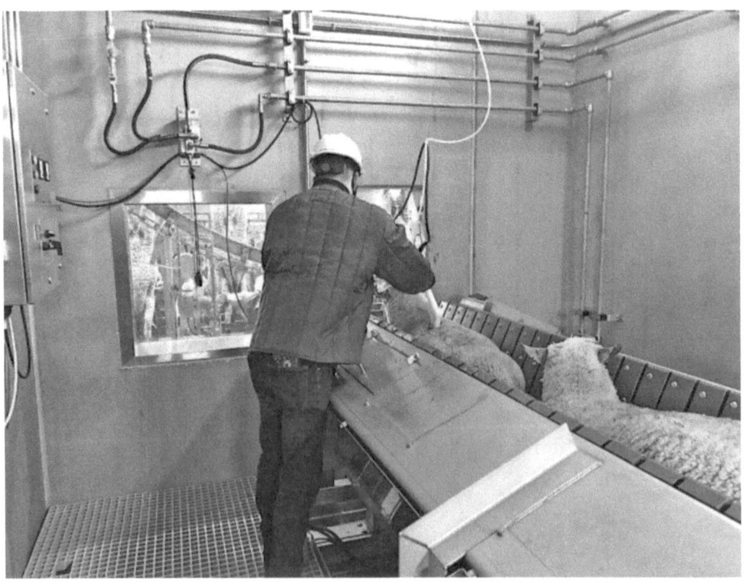

Figure 12.6 Head to back electrical stunner being used on sheep. This device simultaneously passes the electric current through the brain and the heart to produce both instantaneous unconsciousness and cardiac arrest. The sheep are riding to the stunning a V-conveyor restrainer (photo credit: Temple Grandin).

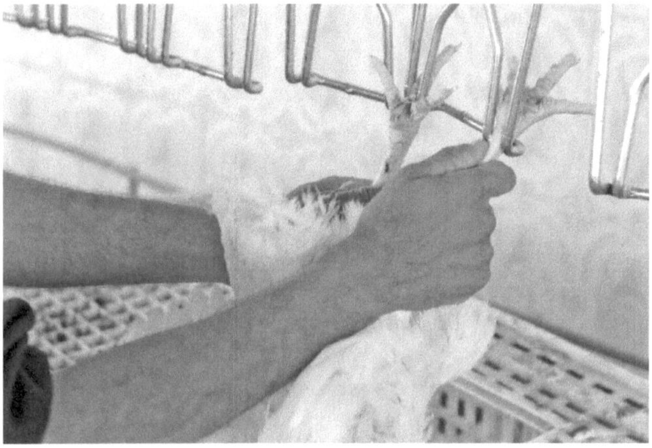

Figure 12.7 Shackling of live poultry to position them for water bath electrical stunning. The live birds are then moved by the conveyor to the water bath electrical stunner (photo credit: Daniel Schneider).

There have been some welfare controversies about the electrical settings for broiler chickens. High settings that will reliably induce cardiac arrest are preferred for welfare reasons. Lower settings that are less likely to damage meat are often used in the industry. There are electrical parameters that can be used that will provide both reliable induction of unconsciousness and better meat quality.

Controlled Atmospheric Stunning (CAS)

Containers of groups of either pigs or poultry are conveyed into a chamber containing CO_2. This is used commercially in many plants. The big advantage of this system from a welfare standpoint is that stressful handling practices, such as hanging live birds on the shackle line, are eliminated. For turkeys and chickens, the same containers that are used for transporting the birds from the farm are moved by a conveyor through the gas stunning system. Handling of individual live birds by plant employees is eliminated. This is a huge welfare advantage because the stress caused by hanging live birds on the shackles is eliminated. The birds enter the stunner in the transport containers and people almost never handle live birds. Unconscious birds are hung on the shackle line after they emerge from the stunner.

Another method that has been developed for broiler chickens is LAPS (Low Atmospheric Pressure Stunning). In this system, air is slowly removed from a chamber that contains travel containers filled with chickens. It is approved by the European Authority and in the United States for chickens. The air withdrawal cycle must comply with the specifications (see Grandin and Cockram, 2020). It must never be used for pigs due to severe ear damage and pain (McKeegan, 2020).

For pigs, the animals are walked from the lairage to the CO_2 chamber in small groups. Each group of pigs enters a gondola (elevator car) and it is submerged into an atmosphere of 80 to 90% CO_2 Since the single file race is eliminated, it is possible to completely eliminate electric prods and other aversive handling methods. For both birds and pigs, the live animal handling is low stress in these systems.

Welfare implications of CAS

When an animal or a bird is successfully stunned with either a captive bolt or electricity, the induction of unconsciousness is instantaneous. In a gas stunning system, the induction of unconsciousness is not instantaneous. The big welfare question is: what is the animal or bird experiencing before it becomes unconscious? Some of the welfare concerns are mucosal irritation and breathlessness. In broiler chickens, slowly increasing the levels of CO_2 as the travel containers pass through a series of stations causes only mild behavioural reactions (Gerritzen et al., 2013). If the CO_2 levels are raised too quickly, the birds may flap and attempt to escape. To prevent escape movements, the CO_2 level must be kept under 40% until the birds lose consciousness. It is my opinion that a CAS method that causes the animals or birds to attempt to escape is not acceptable. A book by Grandin and Cockram (2020) contains information from EFSA on behavioural and physiological measures for assessing suffering, pain, and distress. When slowly increasing CO_2 in the correct manner, there will be no escape reactions from broiler chickens. The best poultry systems are equipped with windows so that the bird reactions can always be observed. Some mild reactions such as gasping or head shaking may be a reasonable trade-off for using a system that eliminates highly stressful pre-slaughter handling, such as hanging live birds on shackles.

For pigs, the use of CO_2 has become more controversial from a welfare standpoint (see the scientific opinion of EFSA, 2020). There are similar EFSA scientific opinions for cattle, poultry, and sheep. I have observed pigs in 90% CO_2 and their reactions range from mild to squealing escape attempts when the pigs are obviously still conscious (Grandin and Cockram, 2020). There is a huge variation in the reactions of pigs ranging from acceptable to really bad. From years of observing CO_2 used to stun pigs, I have learned that genetic differences may explain why some pigs have a relatively peaceful induction and others are squealing and trying to escape from the gondolas. Various other gasses have been tried to replace CO_2 with mostly poor results (Terlouw, 2021). The pork industry has often been reluctant to do research because they may be afraid to bring more attention to the CO_2 problems in pigs. I am confident that this problem can be solved through the use of conventional breeding to remove genetic lines of pigs that have bad reactions.

Religious slaughter without stunning

When slaughter without stunning is performed by members of either the Jewish or the Muslim faith there are three major welfare concerns. They are 1) stressfulness of the methods of restraint used to hold the animal in position for the throat cut; 2) painfulness of the throat cut; and 3) the time required for the animal to become unconsciousness after the throat cut. Many countries have regulations in their Humane Slaughter laws that allow slaughter without stunning to protect religious freedom. The first step in improving welfare during slaughter without stunning is to eliminate highly stressful methods of restraint, such as suspending cattle or sheep by their hind legs. This is prohibited by the World Animal Health Organisation (OIE, 2019) and NAMI (2021). Cattle and sheep can be restrained either standing upright in a box, inverted onto their backs, or tilted on their sides. Restraining methods should be evaluated by the use of outcome-based measurables such as vocalisation and struggling. There is a serious welfare problem with the restraint equipment if a high percentage of cattle vocalise or struggle before the throat cut. In one study, reducing excessive pressure applied by a head restrainer reduced the percentage of cattle vocalising from 23% to 0% (Grandin, 2001). Excessive pressure applied by a restraint device is a common cause of cattle vocalisation (Bourquet et al., 2011). In every type of animal, struggling before the throat cut indicates a welfare problem that must be improved. Either the design or the operation of the restraint device must be corrected and for further guidance on restraint methods see Grandin and Cockram (2020).

Since sheep and goats are smaller, they can often be easily restrained by hand. Animal welfare issues during restraint may be greatest for large animals such as cattle.

Pain during the cut can be reduced by using an extremely sharp knife (Imian et al., 2021). The results of studies are mixed on the issue of pain during the cut (Grandin, 1994; Gibson et al., 2009). One study showed that the knife cut was painful and the other showed that it was not painful. This may be due to the type of knife that was used. Many studies have shown that cattle take longer to lose consciousness after the cut compared to sheep or goats. Numerous studies have shown that sheep will lose consciousness within 2 to 14 seconds and cattle will become unconscious in 17 to 85 seconds. The author has observed that when procedures become sloppy, cattle may remain fully conscious for several minutes. This is due to differences in the anatomy of the blood vessels. When slaughter without stunning is done skilfully, the time to loss of consciousness will be shorter (Gregory et al., 2010; Grandin and Cockram, 2020).

Some religious authorities will accept stunning either before the cut or immediately after the cut. In the Muslim faith, some religious authorities are concerned about when the animal dies (Fuseini, 2019). They want to ensure that a stunning method does not cause death by stopping the heart. When head-only electrical stunning is used, it does not cause death, because the animals or birds can completely recover (Sabow et al., 2017). When water bath stunning is used for halal chicken, the current setting is often sufficiently low to produce a reversible stun. Some Muslim religious authorities will accept the captive bolt method because the heart will continue to beat for several minutes (Vimini et al., 1983). Therefore, the animal dies from the knife cut to the throat, as required by the Muslim faith, and not from the stunning method (Fuseini, 2019).

I have previously argued that slaughter without stunning can be performed with an acceptable level of welfare if everything is done perfectly, but this requires constant attention from management (Grandin, 1994). It requires constant attention to exact details of the process. This level of process control does not occur in most plants. The use of stunning is strongly recommended to ensure the highest standard of animal welfare.

Assessing consciousness

A common mistake that many people make is thinking that an animal is still conscious after stunning when they see the legs kicking. Kicking can still occur in the carcass after the head is removed or the spinal cord is severed (Terlouw et al., 2015). This occurs because the neurological circuit that creates the reciprocal motion of walking is located in the middle of the back in the spinal cord (Grillner, 2011). There are three stages between full consciousness and brain death. These are fully conscious, a transition zone from between fully conscious and unconscious, and brain death (Terlouw et al., 2016; NAMI, 2021; AVMA, 2016; Grandin and Cockram, 2020). An animal that has been bled, *must* be completely brain dead and all signs of return to consciousness must be absent before invasive procedures are started such as skinning or leg removal. The following signs must be absent to ensure that the animal is brain dead (OIE, 2019; FSIS/USDA, 2020; AVMA, 2016; NAMI, 2021). These same criteria also apply to slaughter without stunning:

- Loss of posture (LOP) which is the ability to stand (fully conscious);
- Righting reflex – lifting up the head (fully conscious);
- Natural blinking that looks like the eye movements of live animals (fully conscious);
- Menace reflex – no blink reaction when a hand is quickly moved in front of the eye. The eye is not touched (fully conscious);
- Eyelashes respond to touch (unconscious but not completely brain dead);
- Corneal reflex – blink when the eye is touched (unconscious but not completely brain dead).

It is possible for an animal to be unconscious and have a weak corneal reflex (Vogel et al., 2010) or rhythmic breathing. At this point, unconsciousness has occurred prior to the onset of complete brain death. If a corneal reflex occurs, the animal must be immediately restunned because the process of returning to consciousness has started. For both mammals and poultry, a limp floppy head and neck is a good sign that the animal is unconscious. Sometimes an animal will have eye nystagmus (vibrating eyelid). This should not be confused with the natural blink of a fully conscious animal. If you are not sure what a natural blink looks like, the reader should observe live animals and birds in the lairage (stockyards). Nystagmus is most likely to occur in electrically stunned animals. If it occurs in an animal shot with a captive bolt, the animal should be immediately restunned. Unconscious animals or birds that are properly stunned with electricity will sometimes make gasping movements like a fish out of water. This should not be confused with rhythmic breathing, which is a sign of starting the process of returning to consciousness.

Depopulation for disease control or other emergencies

The standards for animal welfare in many industry and government guidelines for depopulation are less strict than for slaughter or euthanasia. The American Veterinary Medical Association (AVMA) has three separate standards for euthanasia, slaughter, and depopulation. Some methods of stunning animals that are used in slaughter plants are also used for mass depopulation for disease control, such as causing death inside an enclosure and increasing carbon dioxide concentration in the air. Others, such as foam filling of poultry houses are unique to mass depopulation. There are some extremely stressful and cruel methods that must never be used, such as drowning, burning alive, burying alive, and turning off the ventilation in a building and allowing the animals to die from heat stress. It is beyond the scope of this chapter to discuss all the advantages and disadvantages of different depopulation methods. Some of the worst situations where animal welfare is severely compromised can arise during mass depopulation of farm animals.

One factor that has to be considered is the mental stress on the people who have to kill hundreds of animals. This is especially a problem when people have to kill hundreds of healthy animals on a farm by individually shooting each one. If possible, one of the best methods for depopulation is to use the facilities in a slaughter plant.

Euthanasia and killing of surplus animals

Euthanasia of individual cattle, sheep on the farm that are sick or unwanted is usually carried out by shooting, which requires careful positioning of the bullet and attention to safety. The AVMA has guidelines for euthanasia. There are large numbers of unwanted male chicks that are usually killed by carbon dioxide gassing or maceration, although this is being phased out in some European countries. Similarly, many male calves are killed at just a few days of age. Sexing techniques are developed that could reduce this problem of surplus males. The use of injectable anaesthetics by a veterinarian is usually impractical for livestock in all situations except hobby farms. It is, however, widely used for euthanasia of companion animals. Laboratory rodents are often killed by cervical dislocation, although carbon dioxide gassing is increasingly common.

Managing an animal welfare program

In most large slaughter plants, there is a designated animal welfare officer who oversees handling and stunning. Their job is to ensure that the plant is in compliance with both their county's legislative codes and their company welfare requirements. It is also possible to evaluate many on-farm and transport welfare issues at the slaughter plant, such as animals with bruises, animals arriving dead, and animals arriving unable to walk (non-ambulatory). Other issues, such as lameness, poor body condition, and disease of the internal organs can also be easily assessed (Grandin and Cockram, 2020).

Recommendations for auditing and assessing welfare at slaughter

The trend in animal welfare auditing and assessment is to move away from input and engineering specifications toward animal-based outcome measures, or measurables (OIE, 2019). Instead of requiring a specific equipment design, the outcome of the use of the equipment is assessed. Below is an outline of the variables that should be assessed by using numerical scoring. This enables management to determine if procedures are improving or becoming worse. The use of numerical scoring also makes it possible to determine if an improvement such as adding a lamp to a dark race entrance or improved stunner maintenance has improved performance. The use of measurement is an essential component of continuous improvement programs.

Handling outcome-based measures for cattle, pigs, and sheep

- Percentage of animals that are moved without an electric prod (NAMI, 2021);
- Percentage of animals that do not slip or fall during handling (NAMI, 2021l Welfare Quality, 2009; OIE, 2019). Falling should be 1% or less of the animals. The major causes of falling are poor handling methods, slippery floors, or lame animals that have difficulty walking. Figure 12.8 shows a good, non-slip floor;
- Percentage of animals that move easily without stopping or turning back (Welfare Quality, 2009);

Figure 12.8 A good non-slip flooring surface that can be used for sheep or pigs. It is created by stamping the pattern of expanded metal mesh into the wet concrete. For cattle, a slightly deeper pattern is recommended (photo credit: Temple Grandin).

- Percentage of cattle or pigs that remain silent and do not bellow or squeal in the stun box, restrainer conveyor, or religious slaughter box. Vocalisation in cattle and pigs during handling and restraint is associated with physiological indicators of stress, such as higher lactate or cortisol levels (Dunn, 1990; Hemsworth et al., 2011; Edwards et al., 2010; Warriss et al., 1994). High percentages of pigs or cattle vocalising are usually associated with obvious aversive events such as electric prods, jamming in the race, excessive pressure from a restraint device or sharp edges (Grandin, 2001; Bourquet et al., 2011; Edwards et al., 2010). Vocalisation scoring cannot be used with sheep because they do not vocalise when they are hurt. In well-managed cattle operations, the percentage of cattle that bellow in the stun box or religious slaughter box should be under 5%;
- Percentage of cattle, sheep, or pigs with bruises. Assessment of bruising is covered in Grandin and Cockram (2020).

Handling outcome-based measures for poultry

- The percentage of birds with no broken wings should be under 1%. High percentages of broken wings are an indicator of either on-farm handling problems or damage caused by removing live birds from the travel containers for shackling. The birds should be scored with their feathers on to avoid confusing broken wings caused by handling with damage caused by the machinery that removes the feathers.
- Percentage of birds with no broken legs;
- Percentage of animals rendered unconscious with one application of the stunner (NAMI, 2021; AVMA, 2016);
- Percentage of electrically stunned animals or birds where the stunner is placed correctly to pass the electric current through the brain (NAMI, 2021; AVMA, 2016);
- Percentage of birds with no bruises. Count any red mark that is larger than 1 cm as a bruise. Poor handling practices are a major cause of bruises.

Stunning outcome-based measures for livestock and poultry

- Percentage of cattle and pigs that remain silent when an electric stunner is applied. If a bovine vocalises or a pig squeals, this is due to the stunner electrode being energised before it is in firm contact with the animal (NAMI, 2021);
- Percentage of animals that show no signs of return to consciousness (NAMI, 2021; AVMA, 2016);
- Religious slaughter without stunning. Record the time required for an animal or bird to lose consciousness. When good technique is used, 90% of the cattle should either lose the ability to stand or show eye rollback in 30 seconds. Sheep should lose consciousness in about half that time.

Conclusions

Both now and in the future, one of the biggest welfare issues I see is what I call biological system overload. The animal's or bird's biology is pushed so hard for meat production by either genetic selection, feed additives, or too much grain in the diet that it is barely functional. Cattle may become lame and have difficulty walking, heart failure, increased heat stress, or other serious problems (Grandin and Whiting, 2018). In my recent work with slaughter plants, the biggest welfare issue I observe is young cattle or pigs that are stiff and sore and they are reluctant to get up and walk. I call this bad that has become normal. People get so accustomed to seeing it that they do not notice it. For the early years of my career from the 1970s through the 1990s, these problems did exist in young, grain-fed cattle. To give the animals we raise for food a quality life that is worth living, these problems caused by pushing their biology too hard must be corrected. To have both sustainability and good welfare, we need to strive for optimum treatment of animals at all stages in their life, but particularly at the time of slaughter.

References

Anil MH and McKinstrey JL, 1998. Variation in electrical stunning tong placement and relative consequences in slaughter pigs. *Meat Science* 31, pp. 481–491.

AVMA, 2016. *Guidelines for the Humane Slaughter of Animals*, 2016 edn. Schaumberg, IL: American Veterinary Medical Association. avma.org/sites/default/of;es/resources/Humane-Slaughter-Guidelines.pdf (Accessed on August 17, 2021).

Bedanova I, Voslarova E, Chloupek P, Pistekova V, Suchy P, Blahova J, Dobsikova R and Vecerek V, 2007. Stress in broilers resulting from shackling. *Poultry Science*, 86, pp. 1065–1069.

Berg L and Raj M, 2015. A review of different stunning methods for poultry-animal welfare aspects. *Animals* 5, pp. 1207–1219, doi:10.3390/ani5040407.

Bourquet C, Deiss V, Cohen C, Tannugi EM and Terlouw C, 2011. Behavioral and physiological reactions of cattle in a commercial abattoir: Relationship with organization aspects of the abattoir and animal characteristics. *Meat Science,* 7, pp. 19–168.

Dunn CS, 1990. Stress reactions in cattle undergoing ritual slaughter using two methods of restraint. *Veterinary Record*, 125, pp. 522–525.

Edwards LN, Grandin, T, Engle TE, Porter SP, Ritter MJ, Sosnicki AA and Anderson DB, 2010. Using exsanguination blood lactate to assess the quality of pre-slaughter handling. *Meat Science,* 86, pp. 384–360.

Edwards-Callaway LN, Walker J and Tucker CB, 2018. Culling decisions and dairy cattle welfare during transport to slaughter in the United States, Frontiers in *Veterinary Medicine* 18. https://doi.org/10/3389/fvets.2018.00343.

EFSA, 2020. The welfare of pigs at slaughter. *EFSA Journal.* https://doi.org/10.2903/j.efsa.2020.6148 (Accessed on August 17, 2021).

FSIS/USDA, 2020. *Humane Slaughter of Livestock.* Washington, DC: USDA Food Safety Inspection Service. www.animalhandling.org/producers/regulations (Accessed on August 17, 2021).

Fuseini A, 2019. The brain unconsciousness and death: A critical appraisal with regard to halal meat production. *Animal Welfare*, 28(2), pp. 165–171 (7).

Gerritzen M, Reimert HGM, Hindle VA, Verhoeven MTW and Veerkamp WB, 2013. Multistage carbon dioxide gas stunning of broilers. *Poultry Science*, 92, pp. 41–60.

Gibson TJ, Johnson CB, Murrell JC, Hulls CM, Mitchinson SL, Stafford KJ, Johnstone AC and Mellor DJ, 2009. Electroencephalographic responses of halothane-anesthetized calves to neck cutting by central neck incision without stunning. *New Zealand Veterinary Journal,* 57, pp. 78–85.

Gibson TJ, Octavio Oliveira SE, Dalla Costa FA and Gregory NG, 2019. Electroencephalographic assessment of pneumatically powered penetrating and non-penetrating captive bolt stunning of bulls. *Meat Science,* 151, pp. 54–59.

Grandin T, 1994. euthanasia and slaughter of livestock. *Journal American Veterinary Medical Association,* 204, pp. 1354–1360.

Grandin T, 1997. Assessment of stress during handling and transport. *Journal of Animal Science*, 9, pp. 249–257.

Grandin T, 2001. Cattle vocalizations are associated with handling and equipment problems in beef slaughter plants. *Applied Animal Behavior Science,* 71, pp. 191–201.

Grandin T and Cockram M, 2020. *The Slaughter of Farmed Animals: Practical Ways of Enhancing Animal Welfare.* Wallingford, Oxfordshire, UK: CABI Publishing.

Grandin T and Whiting M, 2018. *Are We Pushing Animals to Their Biological Limits? Welfare and Ethical Implications.* Wallingford, Oxfordshire, UK: CABI International.

Grandin T, 2021. *Improving Animal Welfare: A Practical Approach.* Wallingford, Oxfordshire, UK: CABI Publishing.

Gregory NG, Fielding HR, von Wenzlawowicz M and von Hollen K, 2010. Time to collapse following slaughter without stunning. *Meat Science,* 85, pp. 66–69.

Grillner T, 2011. Human locomotor circuits conform. *Science*, 334, pp. 912–913.

Hemsworth PH, Rice M, Karlen MG, Calleja L, Barnett JL, Nash J and Coleman GJ, 2011. Human-animal interactions at abattoirs relationships between handling and animal stress in sheep and cattle. *Applied Animal Behavior Science,* 135, pp. 24–33.

HSA (Humane Slaughter Association), 2021. Wheathempstead, UK. www.hsa.org.uk publications/online-guides (Accessed on August 15, 2021).

Imian JC, Kaka U, Yong-Meng G, Idrus Z, Awad EA, Abubakar AA, Ahmad T, Nizamuddin HNQ and Sazili AQ, 2021. Effects of slaughter knife sharpness on blood biochemical and electroencephalogram change in cattle. *Animals*. https:doi.org/10.390/ani10040579.

Lambooij E, 1982. Electrical stunning of sheep. *Meat Science,* 6, pp. 123–135.

Magrin L, Brscic M, Armato L, Contiero B, Lotto A, Cozzi G and Gottardo F, 2020. Risk factors for claw disorders in intensively finished Charolais beef cattle. *Preventative Veterinary Medicine*, 175. https://doi.org/10.1016/j.prevetmed.2019.104864.

McKeegan DEF, Dec. 2020. LAPS is not a humane alternative to stunning pigs with carbon dioxide. *Meat Hygienist,* 180, pp. 20–22.

NAMI, 2021. *Recommended Animal Handling Guidelines and Audit Guide.* Washington, DC: North American Meat Institute. www.animalhandling.org (Accessed on August 17, 2021).

OIE, 2019. Chapter 7.5: Slaughter of animals. In *Terrestrial Animal Health Code*. Paris, France: World Organization of Animal Health. www.oie.int/what-we-do/standards/codes-and-manuels/terrestrial-code-online-access/

Peterson CM, Pilcher CM, Rothe HM, Marchant-Forde JN, Ritter MJ, Carr SN, Puls CL and Ellis M, 2015. Effect of feeding ractopamine hydrochloride on the growth performance and responses to handling and transport in heavy weight pigs. *Journal of Animal Science,* 93, pp. 1239–1249.

Sabow AB, Nakyinsige K, Adeyemi KD, Sazili AO, Johnson CB, Webster J and Farouk MM, 2017. High frequency stunning of ruminants: A review. *Meat Science,* 202, pp. 124–134.

Terlouw C., 2021. Stunning of pigs with different gas mixtures: Behavioral and physiological reactions. *Meat Science*, 175, 108452.

Terlouw C, Bourguet C, Deiss V and Mallet C, 2015. Origins of movements following stunning and during bleeding in cattle. *Meat Science,* 110, pp. 135–144.

Terlouw C, Bourguet C and Deiss V, 2016. Consciousness and unconsciousness and death in the context of slaughter, Part II evaluation methods. *Meat Science,* 118, pp. 147–156.

Vimini RJ, Field RA, Riley ML and Varnell TR, 1983. Effect of delayed bleeding after captive bolt stunning and heart activity and blood removal in cattle. *Journal of Animal Science,* 57, pp. 628–631.

Vogel KD, Badtram G, Claus JR, Grandin T, Turpin S, Weyker RE and Voogd E, 2010. Head only followed by cardiac arrest electrical stunning in an effective alternative to head only electrical stunning. *Journal of Animal Science,* 89, pp. 1412–1418.

Warner RD, Ferguson DM, Cottrell JJ and Knee BW, 2007. Acute stress induced by the preslaughter use of electric prodders causes tougher beef meat. *Australian Journal of Experimental Agriculture,* 47, pp. 782–788.

Warriss PD, Brown S, Adams SJM and Corlett IN, 1994. Relationship between subjective and objective assessments of stress at slaughter and meat quality in pigs. *Meat Science*, 38, pp. 329–340.

Welfare Quality Network, 2009. Assessment protocols. https://welfarequalitynetwork.net/en-us/reports/assessment-protocals/ (Accessed on August 15, 2021).

Westervelt RG, Kinsman D, Prince RP and Giger W, 1976. Physiological stress measurement during slaughter of calves and lambs. *Journal of Animal Science,* 42, pp. 833–834.

Willson D, Baier FS and Grandin T, 2021. An observation field study on the effects of changes in shadow contrasts and noise on cattle movement in a small abattoir. *Meat Science*, 179. https://doi.org/10.1016/jomeatsci.2021.108539.

Yost JK, Yates JW, Davis MP and Wilson ME, 2020. The Stockman's Scorecard, quantitative evaluation of beef cattle, Stockmanship. *Transactions Animal Science* 4(4) https://doi.org/10.1093/tax.txaa175.

PART III

Animal use for other purposes

13
SCIENTIFIC AND EDUCATIONAL ANIMAL USE

Andrew Knight

Introduction

Scientific animal use is one of the most controversial animal use issues. This controversy may stem from the fact that animals may be deliberately harmed during such use, and sometimes gravely so.

This chapter briefly reviews the history of scientific animal use, from ancient Greece to the present day. The multiplicity of animal welfare concerns created by such animal use are explored, including not only those associated with invasive procedures, but also with routine procedures, and laboratory environments. The justifications for such animal use are critically scrutinised – particularly, the key claim that such research is essential for the advancement of human healthcare. Systematic reviews of animal research within various fields are reviewed. These provide quantitative evidence of its limited utility in advancing human healthcare, and insights into the reasons for this. Replacement alternatives are reviewed, along with recommendations for the future of policy and practice relating to scientific animal research, in accordance with evidence and best practice.

Requirements for students to harm or kill animals during their education or training are also particularly controversial. Next, this chapter reviews the history and contemporary status of educational animal use. The animal welfare concerns associated with such use are explored, followed by a review of alternative teaching methods. The educational efficacy of the two methods has been studied within systematic reviews. These have clearly demonstrated that humane teaching methods usually produce learning outcomes as good or better than those achieved via harmful animal use. This evidence is then reviewed, and finally recommendations are provided for increasing the implementation of humane teaching methods within life and health sciences education.

Scientific animal use

Historical and contemporary scientific animal use

The first recorded scientific animal usage was steeped in controversy. Social taboos about dissecting human corpses greatly hampered the physicians of ancient Greece, during their anatomical and physiological studies (Von Staden and Von Staden, 1989), with the result that some

DOI: 10.4324/9781003182351-16

turned to the use of animals. Alcmaeon of Croton (6th–5th century BCE) and a few others even went as far as practising surgical and other invasive procedures on living animals (*vivisection*) (Court, 2005).

As scientific activity grew during the 17th century Renaissance, such experiments on living animals increased. Predating anaesthesia, some of these surgical investigations, demonstrations and experiments were infamously cruel. French philosopher René Descartes (1596–1650) prominently justified such practices, claiming that animals were merely mindless automata (Descartes 1989), whose cries were of no greater moral importance than the squeals of a poorly oiled machine. Such instrumental views of animal worth, and minimisation of their interests in living and avoiding suffering, continue to be used to attempt to justify invasive scientific and educational animal use to this very day.

Nevertheless, by the end of the 17th century, animal suffering within research and other social endeavours had become an increasingly prominent social concern. By the mid-1980s, animals were becoming broadly appreciated as beings with moral status and interests worthy of protection (Lairmore and Ilkiw, 2015), and campaigns against both scientific and educational animal use were increasing.

The most accurate evidence-based estimate of global laboratory animal use in recent times describes the year 2015. Global laboratory animal use for all purposes was estimated at 192 million (Taylor and Alvarez, 2019) – a 51% increase on the approximately 127 million animals used a decade previously, in 2005 (Knight, 2008, Taylor et al., 2008). Although very large, these totals nevertheless represent conservative estimates. Several animal categories are excluded, including advanced foetal developmental stages, and certain invertebrate species believed capable of suffering. Major drivers for the significant rise in overall numbers include greater production and use of genetically modified (GM) animals, and the implementation of large-scale chemical testing programmes in Europe and the US, following increasing concern about the possible toxicity of many chemicals produced in high volumes.

Animal welfare concerns

The magnitude and nature of animal welfare concern created by scientific animal use depends not only on the numbers of animals used, but also on the type of animals, and the procedures and environments they experience. More specifically, it depends on their sentience and other morally relevant characteristics, the level of invasiveness of scientific procedures, welfare impacts due to environmental, social or other circumstances, and the degree to which these are mitigated by strategies such as anaesthesia, *analgesia* (painkillers), and environmental enrichment.

Europe represents the largest region providing harmonised reporting between Member States, and at time of writing, the most recent EU reports described animal use from 2015 to 2017 (EC, 2020). In 2017, the main species used were mice (61%), fish (13%), rats (12%), and birds (6%), which together represented 92% of animals used. Similar proportions of these species are used internationally.

These animals are all higher vertebrates, with the neuroanatomical architecture and psychological capacities necessary to experience negative affective states such as pain, fear, and psychological distress. Some capacities for sentience and affective states may also exist in the small proportion of other animals, including some invertebrates, which are used.

A considerable array of stressors may cause significant stress and even fear, in laboratory animals. Relatively rarely, these may be associated with the capture of wild-sourced species such as primates, to supply breeding centres or research facilities. More commonly, stress may result from transportation, which may be prolonged for some animals, such as GM mouse strains available only from certain suppliers. Extremely commonly, laboratory housing and environments cause stress, as do laboratory procedures – both invasive, and more routine.

An

> *[i]nvasive procedure* [is] one interfering with bodily integrity, whether through puncture or incision, or insertion of an instrument or foreign material, as in surgical and some experimental procedures. Markedly invasive procedures include those resulting in death (whether or not the subjects are conscious), surgical procedures … , major physiological challenges, and the production of genetically modified animals.
>
> *(Knight, 2011)*

There is a widespread view within the laboratory animal community that procedures, including those resulting in death, do not harm an animal, providing the animal is killed humanely, e.g., whilst under anaesthesia (Webster 1994). This provides a very important "legitimisation" of the killing of many millions of laboratory animals annually. However, this is not consistent with modern understandings of animal welfare, which note that for an animal to experience good welfare it requires more than the absence (as far as reasonably possible) of negative states (which can be achieved through death). Good welfare also requires that animals have the opportunity to experience positive states. As we've noted elsewhere (Zemanova et al., 2021),

> Death permanently prevents such positive states, and indeed, the achievement of any other interests animals could seek to fulfill during the remainder of their lives (Kaldewaij, 2006, Yeates, 2010, Jensen, 2017). Accordingly, death is in fact one of the most profound harms that can be inflicted, barring exceptional cases such as genuine euthanasia of those faced with severe, ongoing suffering, with a poor prognosis for recovery.

This understanding of death is in accordance with both sound reasoning, and common sense. In contrast, lethal procedures (euphemistically termed "non-recovery"), are not considered "severe", or even "moderate" or "mild", in the severity classifications now required in the reporting of laboratory animal use within EU Member States, and some other nations. In 2017, EU animal uses were reported as severe (11%), moderate (32%), mild (51%), and non-recovery (6%) (EC, 2020). Figures for 2015–2016 were similar.

Moderate procedures are those "likely to experience short-term moderate pain, suffering or distress, or long-lasting mild pain, suffering or distress, as well as procedures that are likely to cause moderate impairment of the well-being or general condition of the animals" (Herrmann and Flecknell, 2018). Moderate, severe, and non-recovery procedures jointly accounted for almost half of all EU procedures in 2017.

This alone is concerning enough. However, it appears these figures markedly underestimate the harms experienced by laboratory animals. Herrmann and Flecknell (2018) published a systematic analysis of 684 surgical procedures within 506 animal research applications made to German competent authorities in 2010. They found that "researchers frequently underestimated the levels of pain, suffering, distress and lasting harm that were to be inflicted on the animals. Furthermore, the planned health monitoring strategies were generally flawed". Germany is a leading EU Member State, with one of the largest and most developed animal research sectors. Its laboratory animal practice standards are likely to be at least as good as those of most other nations. Accordingly, it is likely that systematic underestimation of procedural severity, and inadequate animal health monitoring, also occur in many, if not most, other nations conducting invasive animal research.

The impacts of moderate or severe procedures can be mitigated via appropriate provision of anaesthesia or analgesia. These should normally be provided for any procedures likely to result in significant pain or discomfort, including for surgical procedures, which are normally among the most painful. However, Herrmann and Flecknell (2019) found that postoperative analgesia was not proposed for 30% of the 684 surgical procedures they analysed. In 10% of cases, animals were to be provided with analgesics if investigators considered this necessary; however, the use of recognised or validated pain assessment tools to detect pain were lacking. Where analgesia was proposed it was often suboptimal. Optimal techniques, such as *multimodal analgesia* (the concurrent use of multiple analgesics), were virtually never used, to alleviate postoperative pain. Once again, Germany is a leader within the animal research sector, and these disturbing results indicate that suboptimal analgesic provision is probably widespread within animal research internationally.

It is well understood that invasive procedures may cause stress to animals. Less commonly appreciated is that routine laboratory procedures may also cause stress. However, numerous studies have demonstrated that routine procedures such as handling associated with cage-cleaning, blood sampling, and gavaging, can cause profound, statistically significant distortions in physiological parameters, including serum concentrations of hormones such as stress hormones, glucose, and various cardiovascular parameters (Balcombe et al., 2004). *Gavaging* is the insertion of a tube within the oesophagus to allow the forced administration of test compounds orally, and is one of the most common routes by which animals are dosed during toxicity tests (Knight et al., 2006).

Laboratory housing and environments, even when enriched, remain very significantly deprived compared to the natural environments of laboratory animal species, with the diverse stimuli and cognitive challenges intrinsic to these (Balcombe, 2006). The chronic stress caused by long-term confinement within standardised, relatively barren laboratory environments, combined with stress caused by both routine and more invasive laboratory procedures, is often sufficient to result in marked behavioural indicators of stress. Examples include stereotypies, aggression, self-injurious behaviour, lethargy, and other abnormalities such as "floating limb syndrome". Lack of environmental stimulation when compared to natural environments also appears to result in cognitive deficits, such as decreased cerebrocortical thickness and weight, and impairments of memory and learning capacity (Balcombe et al., 2004, Balcombe, 2006, Baldwin and Bekoff, 2007).

Chronic stress causes not just psychological and behavioural effects, but also physiological effects, including immunosuppression. It can increase susceptibility to various pathologies. As well as creating significant animal welfare problems, acute and chronic stressors may distort a range of experimental outcomes, such as those dependent on accurate measurement of physiological, behavioural, or cognitive characteristics.

Accurate assessment of laboratory animal welfare nationally and internationally is frequently impeded by reporting deficits and inconsistencies, including important matters such as frequency of analgesic or anaesthetic use, their correlation with markedly invasive procedures, and the prevalence of environmental enrichment and socialisation opportunities. Additionally, we now understand that welfare impacts are cumulative over time (Honess and Wolfensohn, 2010). This warrants monitoring and reporting of welfare impacts over animals' lifetimes, and of consideration of historical, as well as contemporary, welfare impacts. However, such monitoring and reporting are also rare.

Human healthcare advancement

The greatest justification for the frequent and potentially severe welfare impacts experienced by laboratory animals are the societal benefits it is hoped will flow from such research. We now understand that most laboratory animal species are highly sentient, with intrinsic worth in their own right, independent of any potential benefit for humans. Accordingly, fulfilment of scientific curiosity alone – as occurs in fundamental research, when it is methodologically sound – cannot reasonably be considered adequate justification for the harms inflicted upon the many millions of animals used annually within scientific research.

"Translational and applied" research comprised 23% of all EU laboratory animal use in 2017 (EC, 2020). Most of this research is aimed at developing clinical interventions to combat human diseases. If this research were effective and efficient at achieving this goal, this would provide a much stronger justification for laboratory animal research. But is it?

Advocates of invasive animal research have regularly claimed such research is essential for preventing, curing, or alleviating human diseases (e.g., Festing, 2004), with their opponents making counter-claims (e.g., Greek and Greek, 2004). However, the most reliable, quantitative information about the utility of such research in advancing human healthcare, comes from systematic reviews. A *systematic review* is

> a review of a clearly formulated question that uses systematic and explicit methods to identify, select, and critically appraise relevant research and to collect and analyse data from the studies that are included in the review. Statistical methods (meta-analysis) may or may not be used to analyse and summarise the results of the included studies.
>
> *(Moher et al., 2009)*

Many systematic reviews of animal experiments within various research fields have now examined their utility for advancing human healthcare. Among 20 relevant published systematic reviews located by this author during a previous survey, animal models demonstrated significant potential to contribute towards clinical interventions that were efficacious in human patients, in only two cases, one of which was contentious due to a small sample size. This was despite some of these systematic reviews focusing on those animal experiments most likely to provide human benefit. These included experiments approved by ethics committees on the basis of specific claims that medical advances were likely to result from the animal research; very highly cited animal experiments published in leading scientific journals; and chimpanzee experiments, given that chimpanzees are the species most generally predictive of human outcomes, because they're genetically most similar to humans (Knight, 2011).

Seven additional systematic reviews demonstrated poor reliability of animal models in predicting human toxicological outcomes, including carcinogenicity and teratogenicity – the propensity to cause cancer and birth defects, respectively. These are the toxicities of greatest public health concern (Knight, 2011). Since then, many additional systematic reviews have yielded similar results (Knight, 2019). To date, no published systematic reviews in any healthcare fields appear to have yielded contrary results – that invasive animal research is an effective and efficient tool for the advancement of human healthcare.

Limitations of animal models

The poor rates of translation of animal outcomes into human patients and consumers are due both to the animal models themselves, and to the manner in which they are used. Animals differ from humans in multiple relevant ways. Differences in absorption, distribution, metabolism, and elimination pathways or rates, affect *toxico- or pharmaco-kinetics* (i.e., bodily distribution of test compounds). *Toxico- and pharmaco-dynamics* (mechanisms of action and biological effects) may also differ. Jointly these may alter organ systems affected, and the nature and magnitude of those effects (Knight, 2011).

Human predictivity is further compromised by the experimental protocols used. Young animals, of single strains and sexes, lacking in biological variably and concurrent human risk factors, such as common comorbidities, become even less likely to predict outcomes of human patients, consumers, or workers (Knight, 2011).

Many toxicity tests also use *maximum tolerated doses* (above which dose increases become impossible, due to acute, toxicity-related effects), as well as chronic dosing. These factors do maximise sensitivity to toxins. However, these doses can also overwhelm physiological defences that are effective at environmentally realistic doses. As a result, many compounds that would not normally result in toxicity, are falsely indicated as toxic in animal tests, seriously undermining the reliability of any positive results. Human routes of exposure (e.g., inhaled) may also differ from those used in animals, requiring extrapolation between routes of exposure, introducing further uncertainty (Knight, 2011).

And as noted previously, laboratory animals experience stress both chronic and acute, resulting from laboratory environments and procedures. These stressors can alter physiological, hormonal, and immune status, and even behavioural repertoires and cognitive capacities, in ways that may be unpredictable (Balcombe et al., 2004, Balcombe, 2006, Baldwin and Bekoff, 2007).

Methodological quality of animal studies

Additionally, a sizeable body of systematic reviews have confirmed that significant methodological flaws are highly prevalent in most published animal experiments (e.g., Knight, 2019). To date, no systematic reviews have found that a majority of animal studies in any field, were of good methodological quality.

Bias of results occurs when factors systematically alter research outcomes. This may be conscious, but usually results from unconscious factors. Hooijmans et al. (2014) described ten types of bias with potential to influence animal research results. They grouped these into selection bias, performance bias, detection bias, attrition bias, reporting bias, and other sources of bias. Many of these flaws are highly prevalent within animal studies. Common examples include use of apparently arbitrary numbers of animals, rather than statistically justified and significant sample sizes. Failure to use randomisation during allocation to treatment and control groups, and blinding during outcomes assessment, are also common, as is lack of reporting of basic characteristics of animals used. Percie Du Sert et al. (2020) found that randomisation was reported in 30–40% of published animal studies, blinding in around 20%, sample size justification in < 10%, and all basic characteristics of animals used reported in < 10% of publications (Macleod et al., 2015, Avey et al., 2016, Leung et al., 2018).

Across a diversity of fields, studies that incorporate the fewest measures to minimise sources of bias, have also reported the greatest treatment effect sizes (e.g., Crossley et al., 2008, Vesterinen et al., 2010). Accordingly, we can conclude that such apparent increases in effect size, are not real, but are *artefacts*, resulting from flaws in experimental design, conduct, or reporting.

In response to such problems, in 2010 Kilkenny and colleagues proposed the *Animal Research: Reporting of In Vivo Experiments* (ARRIVE) guidelines. These comprised a checklist of 20 items, designed to minimise such flaws by ensuring animal research publications include basic information on animal numbers and characteristics, housing and husbandry conditions, and experimental, statistical, and analytical methods employed. Steps to reduce bias were prominent, such as randomisation, blinding, statistical justifications of sample sizes, reporting of exclusion criteria, and of investigator conflicts of interest. Several similar guidelines have been published, but the ARRIVE guidelines are most prominent. Despite their very widespread publication and endorsement by research journals, major funding agencies, and biomedical research organisations, multiple studies have demonstrated that compliance with the ARRIVE guidelines remains poor (Macleod et al., 2015, Avey et al., 2016, Leung et al., 2018, Percie du Sert et al., 2020). In response, the guidelines have been simplified into "essential" and "recommended" checklists in ARRIVE 2.0 (Percie du Sert et al., 2020). It remains to be seen whether this will improve compliance.

3Rs alternatives

Given that animal models are so unreliably predictive of humans, what alternatives might be used instead? Famously proposed by Russell and Burch in 1959 (e.g., USDA, 2015), the *3Rs* are the:

1. *Replacement* of animal use with non-animal alternatives, wherever possible;
2. *Reduction* of animal numbers to the minimum possible;
3. *Refinement* of animal use, to avoid or minimise animal pain, distress, or other adverse effects suffered at any time during the animals' lives, and to enhance well-being (Buchanan-Smith et al., 2005).

Compliance with these 3Rs is universally considered fundamental to good laboratory animal practice. As stated by Russell and Burch, "Refinement is never enough, and we should always seek further reduction and if possible replacement ... replacement is always a satisfactory answer".

I've previously reviewed 3Rs alternatives in detail (Knight, 2011). Replacement alternatives include mechanisms to enhance sharing and assessment of existing data, physicochemical evaluation of test compounds, and computerised modelling of their effects. Advanced tissue cultures include *immortalised cell lines* (which continue to differentiate indefinitely), *stem cells* (which can differentiate into other cell types), and *organotypic cultures* (three-dimensional cell cultures that retain features of the original organ). Tests using bacterial, yeast, protozoal, mammalian, or human cell cultures exist for numerous toxic and other endpoints. Human *hepatocyte* (liver cell) cultures and metabolic activation systems may allow identification of metabolic pathways (which break down test compounds), and of resultant compounds produced. "Human on a chip" systems connect cell cultures from different organs via microfluidic systems that mimic the circulatory system, allowing assessment of organ–organ interaction. Microarray technology can allow genetic expression profiling of toxins, greatly speeding up their detection, well prior to more invasive endpoints. Surrogate human tissues, e.g., harvested during surgery or childbirth, advanced imaging modalities, and human epidemiological, sociological, and psychological studies, may all increase understanding of illness *aetiology* (causation) and *pathogenesis* (development). Finally, human clinical trials may be enhanced in various ways to increase safety for volunteers, and predictivity for diverse patient populations. As I've noted previously (Knight, 2011), "Non-animal investigative methods cannot, of course, provide answers to all questions about humans, particularly given present technological limitations. However, the same is certainly true of animal models, which have a more limited capacity for further development". Additionally, when human tissues or volunteers are used, these methods may generate faster, cheaper results, that yield superior insights into human biochemical processes, and that are ultimately more reliably predictive for human patients, consumers, and workers.

Recommendations for scientific animal use

As we've described elsewhere (De Boo and Knight, 2008), a multifaceted strategy is warranted to increase the implementation of 3Rs principles, improve the welfare of laboratory animals, and improve the methodological quality of animal research.

Compliance must become mandatory, with 3Rs principles, the ARRIVE guidelines, and other best practice standards, during the design, conduct, and reporting of animal experiments. Such standards should cover animal sourcing, housing, handling, environmental enrichment, socialisation opportunities, appropriate use of anaesthetics and analgesics, and of refinement modalities such as non-invasive or *humane endpoints* (the latter being the humane killing of animals early within terminal protocols). Compliance with a range of measures designed to minimise bias and ensure methodological quality, must also become mandatory. Compliance should be necessary for securing ethical approval and research funding; for licensing of researchers, facilities, and experimental protocols; and for publication of subsequent results.

To enable animal researchers and technicians to meet the necessary standards, regular training in 3Rs methodologies, and in the design, conduct, and reporting of animal research, should be universally compulsory. The widespread lack of attention to replacement methods (in favour of refinement methods) must be rectified.

Greater efforts must also be made to publish negative results. Studies that fail to show a treatment effect are generally less likely to be published, as they're considered less noteworthy. The subsequent exclusion of negative results from systematic reviews that aim to consider all published evidence concerning test treatments leads to overestimations of treatment efficacy, and partly explains the widespread failures in human patients of treatments apparently effective in animals.

To date, compliance with such best practice standards by the animal research community has been demonstrably poor (Leung et al., 2018, Percie du Sert et al., 2020). To achieve the substantial improvements for both laboratory animal welfare, and human predictivity, that are so urgently needed, widespread change is needed. This would require a willingness and commitment to very significant change, from researchers and their professional associations, regulators, licensing bodies, ethical review committees, funding bodies, and scientific journals.

Educational animal use

Historical and contemporary educational animal use

Animals have also been, and still are, widely used within life and health sciences education. Students dissect dead animals within biology and anatomy courses, and conduct invasive procedures on living animals in subjects such as physiology, biochemistry, pharmacology, and parasitology. Animals are frequently killed prior to, or at the end of, such procedures.

In many veterinary schools, animals have been used to teach surgical and clinical procedures, including invasive procedures such as resuscitation. Surgical procedures have progressed historically from multiple survival procedures on individual animals, to terminal procedures, with animals killed at the completion of the surgery, usually via anaesthetic overdose. Such terminal surgical procedures have been common within many countries, with a notable exception being the UK, where students instead gain surgical experience through closely supervised externships and internships. Alternatives to terminal surgeries, such as cadavers and inanimate models, have become increasingly common in countries such as the US and Canada (Bauer 1993). Nevertheless, Bauer reported that 27% of veterinary schools were still utilising terminal surgeries, with 69% using terminal exercises in "small" animals (typically, dogs and cats), and 20% using them in "large" animals (typically, large agricultural species).

In 2001 this author completed the veterinary surgical program at Western Australia's Murdoch University. At that time, terminal animal use within surgery and other subjects was routine within most Australian veterinary schools. However, student-led campaigns by this author and others resulted in the introduction of alternatives in all Australian veterinary schools. In 2000 terminal animal use was completely abolished at the University of Sydney, and by 2005, the first students had graduated from all four established Australian veterinary schools without participating in terminal surgical training. By 2012 terminal animal use was uncommon, and was expected to cease entirely within a few years. Similar developments have occurred at numerous other veterinary schools internationally, usually also driven by student-led campaigns.

Educational animal use has been estimated at 1–10% of total numbers of animals used globally for scientific purposes (Akbarsha et al., 2013). Applied to the estimated 192 million animals used in 2015 (Taylor and Alvarex, 2019), this equates to some 2–19 million animals used annually, worldwide. Among those EU Member States which reported data, from 2014 to 2018, total EU educational animal use was 124,000–172,000 (Zemanova et al., 2021). However, these estimates are very conservative, because animals killed for cadaver use (as is very common within anatomy or biology courses) are usually excluded from reported figures.

Animal welfare concerns

Not all educational animal use is harmful to animals, of course. Non-harmful uses of animals may include handling and physical examination of domesticated species, and observational studies of wild, free-living, or sanctuary animals, where animals are not stressed by excessive use or human presence.

However, many of the animals used in demonstration experiments do suffer significant welfare impacts, including during initial sourcing and transportation, the involuntary disruption of their social networks, confinement within relatively unenriched environments, as well as during the experiments themselves (Knight, 2011). Clinical skills training may be stressful for animals, particularly when invasive procedures are being demonstrated or practiced. Many animals used within education are also killed, whether for sourcing of cadavers or body parts for anatomy, physiology or biochemistry laboratories, or at the end of terminal physiology demonstration experiments, or practice surgeries. As discussed previously, viewpoints are common that humane killing of animals does not constitute a harm to them, and this viewpoint serves to legitimise large-scale animal killing for educational purposes. However, a more reasoned and critical consideration of the impacts of killing clearly reveals it to be one of the most profound harms that can be inflicted on healthy animals.

Alternatives to educational animal use

Many humane alternatives have been developed for harmful educational animal use, and successfully implemented within life and health sciences curricula internationally. I've described these in detail elsewhere (Knight, 2012). They include computer simulations and videos of professionally performed dissections (*prosections*) and experiments, non-invasive self-experimentation, *ethically-sourced cadavers* (from animals who have died naturally, or in accidents, and been donated for teaching purposes), anatomical specimens preserved using several different methods, models, mannequins and surgical simulators, and supervised clinical experiences.

Alternative surgical training

Humane training of surgical practical skills should comprise three main stages. First, students should practice basic skills such as instrument handling and suturing, and should refine their motor skills, using knot-tying boards, plastic organs, and other models. Second, they should participate in simulated surgery, using ethically sourced cadavers. Third, students should observe, assist with, and then finally perform, beneficial surgery on real patients. The latter should be conducted under close one-to-one supervision, similarly to the training of human surgeons (Knight, 2011).

Spaying and castrating cats and dogs are some of the most common procedures veterinary students will later need to perform in clinical practice, and shelter animal neutering programs are a very popular way for veterinary students to gain surgical experience. In the Shelter Medicine Program at Mississippi State University, for example, fourth year students averaged 65 sterilisation surgeries in two weeks (Shivley et al., 2018), demonstrating the high volume of surgical experience these programmes can provide, when compared to other forms of surgical training. Additionally, neutered shelter animals are more likely to be adopted, decreasing pet overpopulation due to uncontrolled breeding, and delivering important animal welfare benefits.

Educational efficacy of alternatives

Despite the successful implementation of such teaching alternatives within numerous courses worldwide, harmful animal use persists within many others. Why does such harmful animal use persist? The answers may be revealed through systematic analysis of the summary reports of scientific and educational animal use that EU nations are required to publish annually. Our analysis of reported summaries from 18 EU and EEA Member States during 2017–2019 (Zemanova et al., 2021) revealed that the two main reasons why some educators felt animal use remained necessary, were 1) the necessity of using a living animal for "proper" learning; and 2) the perceived lack of an adequate alternative.

However, in 2021 Zemanova and I published a systematic review of published studies which compared learning outcomes achieved by humane teaching methods with those achieved through harmful animal use. Such studies are often conducted by educators, when trialling a new teaching method, and subsequently published. Fifty such studies were published from 1968–2020, primarily from the US, UK, and Canada. Humane teaching methods produced learning outcomes that were superior (30%), equivalent (60%), or inferior (10%) to those produced by traditional harmful animal use (Figure 13.1). This is the most comprehensive systematic review published to date within this field, with studies covering all educational levels and disciplines in which animals are used, and its results are clear. Accordingly, it may be concluded that the preference of some educators for harmful animal use is not evidence-based; indeed, it is contrary to the best available evidence in this field.

The demonstrably superior educational efficacy of humane alternatives in 30% of relevant published educational studies may be due to certain advantages offered by alternatives. Unlike animals, many simulators accurately replicate key elements of humans, allowing human medical students to practise clinical skills procedures. Simulated procedures in any species may generally be repeated or otherwise customised to individual learner needs. Repeated practice results in superior skill retention (Andreatta et al., 2015). Live animal laboratories are also very time- and resource-intensive, with the majority requiring an entire morning or afternoon to set up, prepare and stabilise animals, conduct procedures, recover or euthanase animals, and clean and pack

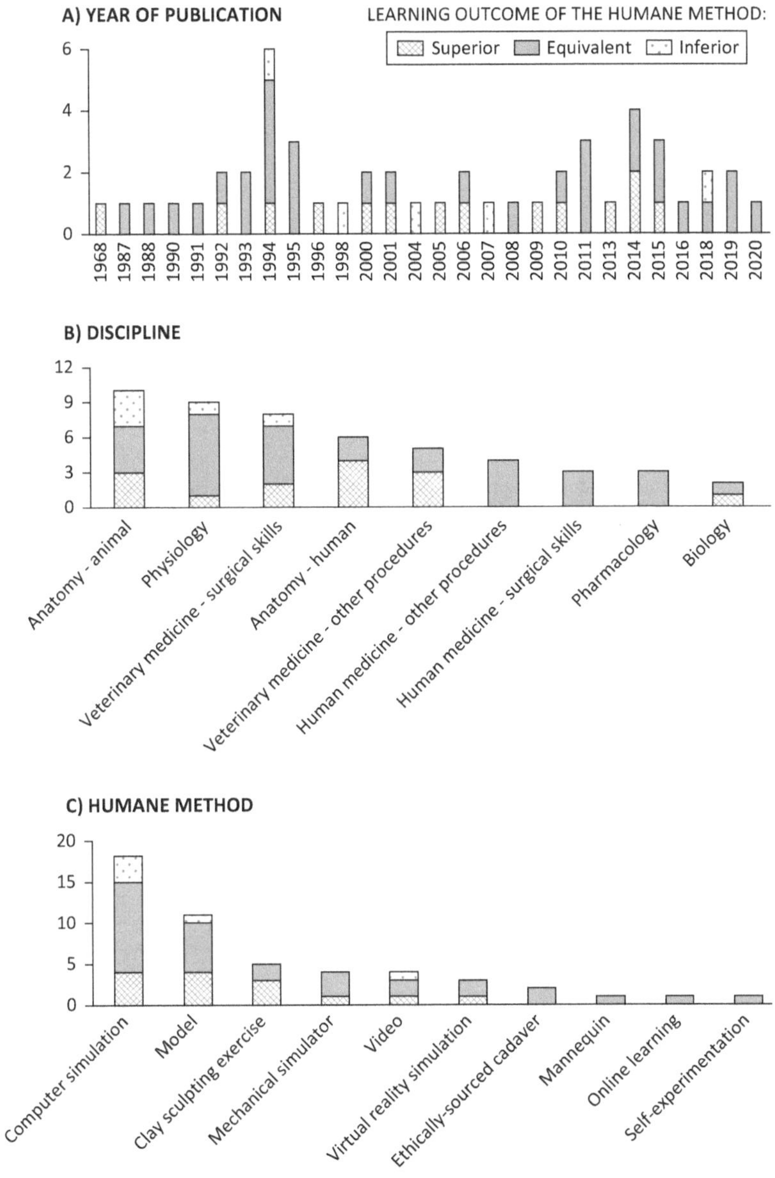

Figure 13.1 Number of studies comparing learning outcomes of humane teaching method and harmful animal use: (A) from 1968 to 2020, (B) grouped by discipline, and (C) by humane method used. Note: years with zero publications are not included in (A). Adapted from: Zemanova and Knight, 2021.

away. Humane alternatives frequently offer significant savings in both time and costs (Leonard 1992), freeing student and staff time, space and financial resources, for other learning or academic activities. Accordingly, as we noted (Zemanova and Knight, 2021), "wide-spread implementation of humane teaching methods would not only preserve learning outcomes, but may in fact be beneficial for animals, students, educators, and institutions".

Recommendations for educational animal use

Clearly, remaining harmful animal use within life and health sciences courses worldwide should be replaced with humane teaching alternatives, as soon as possible. We recently recommended several steps to facilitate the appropriate implementation of humane teaching methods (Zemanova and Knight, 2021):

(1) the training of life and health sciences educators should be designed to increase their awareness about the efficacy of humane teaching methods,
(2) exchange of information and best practice strategies among universities should be encouraged,
(3) there needs to be more financial support from governmental and international institutions to universities for implementing alternatives, as well as for non-profit organizations that are distributing information about humane teaching methods (e.g. InterNiche, Animalearn), and
(4) more stringent enforcement of legislation requiring alternatives to animal use, is necessary.

Additionally, those universities offering courses in which harmful animal use continues to persist, should implement policies committing to providing alternatives for students (or staff) who conscientiously object to participating in harmful animal use. I've previously provided detailed guidance on this matter, including examples of such conscientious objection policies, and the jurisprudential (legal) bases for their implementation (Knight, 2014).

Detailed information about curricular animal use and related conscientious objection policies should also be publicised to all students, well in advance of such animal use, via university handbooks, curricular and course guides. Such information should also be circulated to teaching faculty, along with guidelines about the assessment of conscientious objection claims, and the necessary provision of alternative teaching or assessment activities.

Conclusions

Scientific and educational animal use is particularly controversial. With 192 million animals conservatively estimated as being used globally for scientific and educational animal purposes in 2015, the numbers are large, even if significantly smaller than some other fields in which animals are impacted by humans. Animal research is also one of the very few fields in which animal suffering may be knowingly or deliberately inflicted, as well as severe. Requiring students to engage in harmful or lethal use of animals is also fraught with controversy. Social unease in both domains has been reflected by abundant campaigns and lawsuits, and decreasing public support over time for such uses of animals (Funk and Rainie, 2015).

A paradigm change is clearly warranted concerning scientific and educational animal use. Instead of uncritically assuming human benefits, we must subject such use to much more rigorous and critical evaluation, consistent with common legislative requirements for researchers and ethics committees to conduct harm–benefit analyses, before proposing or approving such work (e.g., EU, 2010). Systematic reviews have clearly indicated that most animal research does not yield hoped-for human healthcare benefits, and have also identified multiple reasons for this. Systematic reviews have also clearly demonstrated that learning outcomes achieved by humane teaching methods, are normally as good or better than those achieved through harmful animal use. Modern understanding of the animals used for scientific and educational purposes has clearly demonstrated the existence within such species of a range of morally relevant characteristics. It is clear these animals have lives and interests that matter profoundly to them. And it is incumbent on us to respect these, if we aspire to act as moral agents.

Accordingly, when conducting the harm–benefit analyses required both by good ethics, and commonly, by legislation, it is not normally reasonable to conclude that benefits accruing for human patients, consumers, industry workers, or students, or even for those motivated by simple scientific curiosity or profit, exceed the harms incurred by the animals used.

Accordingly, a range of measures are clearly warranted, to increase compliance with 3Rs methods, across the domains of scientific and educational animal use. Where animal use persists, measures are also warranted to improve the methodological quality of animal research. Where scientific and educational animal use fails to meet the harm–benefit standards expected by society, and frequently required by legislation, such animal use should cease. Resources consumed could then be redirected into more justifiable, and potentially more promising, research and teaching modalities.

References

Akbarsha MA, Zeeshan M and Meenekumari KJ, 2013. Alternatives to animals in education, research and risk assessment: an overview with special reference to Indian context. *ALTEX Proceedings*, 2, pp. 5–19.

Andreatta PB, Klotz JJ, Dooley-Hash SL, Hauptman JG, Biddinger B, House JB, et al. 2015. Performance-based comparison of neonatal intubation training outcomes. *Advances in Neonatal Care*, 15(1), pp. 56–64.

Avey MT, Moher D, Sullivan KJ, Fergusson D, Griffin G, Grimshaw JM, et al., 2016. The devil is in the details: incomplete reporting in preclinical animal research. *PLoS One*, 11(11), e0166733.

Balcombe JP, 2006. Laboratory environments and rodents' behavioural needs: a review. *Laboratory Animals*, 40(3), pp. 217–235.

Balcombe JP, Barnard ND and Sandusky C, 2004. Laboratory routines cause animal stress. *Journal of the American Association for Laboratory Animal Science*, 43(6), pp. 42–51.

Baldwin A and Bekoff M, 2007. Too stressed to work. *New Scientist*, 194(2606), p. 24.

Bauer MS, 1993. A survey of the use of live animals, cadavers, inanimate models, and computers in teaching veterinary surgery. *Journal of the American Veterinary Medical Association*, 203(7), pp. 1047–1051.

Buchanan-Smith HM, Rennie A, Vitale A, Pollo S, Prescott MJ and Morton DB, 2005. Harmonising the definition of refinement. *Animal Welfare*, 14(4), pp. 379–384.

Court WE, 2005. Pharmacy in the ancient world to 1100 AD. In Anderson S, ed., *Making Medicines: A Brief History of Pharmacy and Pharmaceuticals*. London: Pharmaceutical Press, pp. 21–36.

Crossley NA, Sena E, Goehler J, Horn J, van der Worp B, Bath PM, et al., 2008. Empirical evidence of bias in the design of experimental stroke studies: a metaepidemiologic approach. *Stroke*, 39(3), pp. 929–934.

De Boo J and Knight A, 2008. Increasing the implementation of alternatives to laboratory animal use. *Alternatives to Animal Testing and Experimentation*, 13(3), pp. 109–117.

Descartes R, 1989. *Animal Rights and Human Obligations*, Regan T and Singer, P, eds. Upper Saddle River, NJ: Prentice Hall, pp. 13–19.

European Commission (EC), 2020. *Report from the Commission to the European Parliament and the Council. 2019 Report on the Statistics on the Use of Animals for Scientific Purposes in the Member States of the European Union in 2015–2017*. Brussels: EC.

European Union (EU), 2010. Directive 2010/63/EU of the European Parliament and of the Council of 22 September 2010 on the protection of animals used for scientific purposes. *Official Journal of the EU*, 276, pp. 33–79.

Festing MF, 2004. Is the use of animals in biomedical research still necessary in 2002? Unfortunately, "yes". *Alternatives to Laboratory Animals*, 32(1_suppl), pp. 733–739.

Funk C and Rainie L, 2015. Chapter 7: Opinion about the use of animals in research. In Pew Research Center (Ed.) *Americans, Politics and Science Issues*. Washington, DC: Pew Research Center, pp. 141–144.

https://www.pewresearch.org/internet/wp-content/uploads/sites/9/2015/07/2015-07-01_science-and-politics_FINAL-1.pdf.

Greek R and Greek JS, 2004. Are animals necessary in 2002? reply to Dr Michael Festing's book review of sacred cows and golden geese: the human cost of experiments on animals. *Alternatives to Laboratory Animals*, 32(1_suppl) suppl 1B, pp. 741–746.

Herrmann K and Flecknell P, 2018. Severity classification of surgical procedures and application of health monitoring strategies in animal research proposals: a retrospective review. *Alternatives to Laboratory Animals*, 46(5), pp. 273–289.

Herrmann K and Flecknell P, 2019. Retrospective review of anesthetic and analgesic regimens used in animal research proposals. *ALTEX*, 36, pp. 65–80.

Honess P and Wolfensohn S, 2010. The extended welfare assessment grid: a matrix for the assessment of welfare and cumulative suffering in experimental animals. *Alternatives to Laboratory Animals*, 38(3), pp. 205–212.

Hooijmans CR, Rovers MM, De Vries RB, Leenaars M, Ritskes-Hoitinga M and Langendam MW, 2014. SYRCLE's risk of bias tool for animal studies. *BMC Medical Research Methodology*, 14, p. 43.

Jensen KK, 2017. How should death be taken into account in welfare assessments? *Journal of Agricultural and Environmental Ethics*, 30(5), pp. 615–623.

Kaldewaij F, 2006. Animals and the harm of death. In Kaiser M and Lien ME, eds., *Ethics and the Politics of Food*. Oslo, Norway: Congress of the European Society for Agricultural and Food Ethics, pp. 528–532.

Knight A, 2008. Estimates of worldwide laboratory animal use. *Alternatives to Laboratory Animals*, 36(5), pp. 494–496.

Knight A, 2011. *The Costs and Benefits of Animal Experiments*. Houndmills, Basingstoke, UK: Palgrave Macmillan.

Knight A, 2012. The potential of humane teaching methods within veterinary and other biomedical education. *ALTEX Proceedings*, 1, pp. 365–375.

Knight A, 2014. Conscientious objection to harmful animal use within veterinary and other biomedical education. *Animals*, 4(1), pp. 16–34.

Knight A, 2019. Critically evaluating animal research. In Herrmann K and Jayne K, eds, *Animal Experimentation: Working Towards a Paradigm Change*. Leiden, The Netherlands: Brill, pp. 321–340.

Knight A, Bailey J and Balcombe J, 2006. Animal carcinogenicity studies: 2. Obstacles to extrapolation of data to humans. *Alternatives to Laboratory Animals*, 34(1), pp. 29–38.

Lairmore MD and Ilkiw J, 2015. Animals used in research and education, 1966–2016: evolving attitudes, policies, and relationships. *Journal of Veterinary Medical Education*, 42(5), pp. 425–440.

Leonard WH, 1992. A comparison of student performance following instruction by interactive videodisc versus conventional laboratory. *Journal of Research in Science Teaching*, 29(1), pp. 93–102.

Leung V, Rousseau-Blass F, Beauchamp G and Pang DS, 2018. ARRIVE has not ARRIVEd: support for the ARRIVE (animal research: reporting of in vivo experiments) guidelines does not improve the reporting quality of papers in animal welfare, analgesia or anesthesia. *PloS One*, 13(5), e0197882.

Macleod MR, Lawson McLean A, Kyriakopoulou A, Serghiou S, de Wilde A, Sherratt N, et al., 2015. Risk of bias in reports of in vivo research: a focus for improvement. *PLoS Biology*, 13(10), e1002273.

Moher D, Liberati A, Tetzlaff J, Altman DG and Prisma Group, 2009. Preferred reporting items for systematic reviews and meta-analyses: the PRISMA statement. *PLoS Medicine*, 6(7), e1000097.

Percie du Sert N, Hurst V, Ahluwalia A, Alam S, Avey MT, Baker M, et al., 2020. The ARRIVE guidelines 2.0: updated guidelines for reporting animal research. *PLoS Biology*, 18(7), p. e3000410. https://doi.org/10.1371/journal.pbio.3000411.

Shivley JM, Brookshire WC, Bushby PA and Woodruff KA, 2018. Clinically prepared veterinary students: enhancing veterinary student hands-on experiences and supporting hospital caseload using shelter medicine program. *Frontiers in Veterinary Science*, 5, p. 95.

Taylor K and Alvarez LR, 2019. An estimate of the number of animals used for scientific purposes worldwide in 2015. *Alternatives to Laboratory Animals*, 47(5–6), pp. 196–213.

Taylor K, Gordon N, Langley G and Higgins W, 2008. Estimates for worldwide laboratory animal use in 2005. *Alternatives to Laboratory Animals*, 36(3), pp. 327–342.

US Department of Agriculture, Animal and Plant Health Inspection Service, Animal Care (USDA), 2015. Policy #12: consideration of alternatives to painful/distressful procedures. In USDA (Ed.) *Animal Care Policy Manual*. Riverdale, MD: USDA.

Vesterinen HM, Sena ES, Ffrench-Constant C, Williams A, Chandran S and Macleod MR, 2010. Improving the translational hit of experimental treatments in multiple sclerosis. *Multiple Sclerosis Journal*, 16(9), pp. 1044–1055.

Von Staden H and Chalcedonius H, 1989. *Herophilus: The Art of Medicine in Early Alexandria: Edition, Translation And Essays*. Cambridge, UK: Cambridge University Press.

Webster J, 1994. *Animal Welfare: A Cool Eye Towards Eden*. Oxford, UK: Blackwell Publishing.

Yeates JW, 2010. Death is a welfare issue. *Journal of Agricultural and Environmental Ethics*, 23(3), pp. 229–241.

Zemanova MA and Knight A, 2021. The educational efficacy of humane teaching methods: a systematic review of the evidence. *Animals*, 11(1), p. 114.

Zemanova MA, Knight A and Lybæk S, 2021. Educational use of animals in Europe reveals reluctance to implement alternatives. *ALTEX – Alternatives to Animal Experimentation*, 38(3), pp. 490–506. doi: 10.14573/altex.2011111.

14
ANIMALS IN ENTERTAINMENT

David A Fennell and Sarah Coose

Introduction

The focus of this chapter is on the broad spectrum of animal uses in entertainment, thus the level of detail that may be examined for each category is limited. In some cases, further details may be found in other chapters of this book. There are two main categories to consider in such use. The first and most abundant is tourism, while the second includes film, television, and advertising. One can easily find cases around the world where the use of animals for entertainment has resulted in compromised animal welfare leading to suffering and often death. Many of these uses take place in captive settings where animals must perform daily under conditions of negative reinforcement and deprivation. Others are too frequently spontaneous such as the case of the baby dolphin that found its way to the shore of a popular tourist destination in Argentina, only to be passed around for numerous selfies until the animal ultimately died (O'Neil, 2016). There is ongoing discussion in the literature regarding the ethical implications of making animals perform at all. Those who support animal use point to the behavioural enrichment benefits that come from animal performances, while those who reject such use claim that it goes against an individual animal's agency and dignity (Keulartz and Bovenkerk, 2016).

For the purposes of this chapter, entertainment is defined as the diversion and/or enjoyment experienced by an audience from viewing an animal performance, with "performance" defined as an animal exhibiting a behaviour for an audience (Brando, 2016). Thus, animal welfare as it applies to entertainment encompasses meeting the needs of an animal who performs for an audience. According to the UK Animal Welfare Act, 2006, these needs include *physical health*, which includes protection from "pain, suffering, injury and disease"; *mental and emotional well-being*, which can be indicated by the display of typical or "normal behaviour patterns"; and *sufficient access to resources*, such as a "suitable environment and diet" as well as "any need it has to be housed with, or apart from, other animals" (Animal Welfare Act, 2006).

Tourism

Even though tourism has been a focus of research for half a century (far longer in practice), consideration of the welfare of animals used in tourism traces back only to the turn of the century when the first articles emerged (Fennell 2000). It was not until more than a decade later, however,

DOI: 10.4324/9781003182351-17

that more intensive treatment of the topic of animal ethics started to emerge on the theoretical landscape of animal ethics in tourism, culminating in a series of comprehensive publications (Fennell 2012; Markwell 2015; Carr and Broom, 2018; Kline 2018; Rickly and Kline, 2021).

If the interests of animals are taken into consideration at all in tourism, animal welfare is the perspective adopted. Indeed, a climate that considers moral issues only in self-interest is rampant in the industry. More protectionist perspectives like animal rights and newer conceptions of animal use based on a posthumanist approach (Thomsen et al., 2021) question the use of animals for pleasure and profit (Fennell, 2012). The animal rights organisation, PeTA, has recently reported that over 50 global tour operators have removed elephant riding tourism from their itineraries, prompting PeTA to claim that we are "winning" the war against this type of animal use (PeTA, 2021b). At the macro scale, research is pushing the world's largest tourism organisation, the UNWTO (World Tourism Organization), to be even more responsible and sustainable by recommending that its Global Code of Ethics include an 11th Article (the Code presently has 10 Articles) on the welfare consideration of animals used in tourism (Fennell, 2013). At present the Global Code of Ethics has little to say about the interests of millions of animals drawn into the tourism industry.

The main categories of research and practice in tourism include wildlife viewing and ecotourism, animals as captives, animals forced into competition, and animals pursued for sport and subsistence (Fennell, 2012). Indeed, animal work for humans in tourism (Rickly and Kline, 2021) comes in many different forms. These practices often vary according to setting, mode of engagement, animals' state, and mediators (Cohen, 2012):

1. *Fully natural settings*, such as wilderness, jungles or deserts, which are unframed, and in which animals are in no way restrained.
2. *Semi-natural settings*, such as national parks and wildlife sanctuaries, which are regulated and bounded to various degrees by the authorities, thus separated from the flow of ordinary life; but the animals remain unrestrained within their context.
3. *Semi-contrived settings*, such as zoos, aquariums and animal theme parks, in which at least nominally wild but captured, animals are kept in framed, contrived surroundings; while some simulate the animals' natural habitats, in others the animals are confined to narrow, restraining habitation, which significantly constrain their ability to reproduce their natural behaviour patterns.
4. *Fully contrived settings*, such as establishments featuring animal performances and shows, in which captured animals, though they might remain wild, are trained, tamed or humanized to varying degrees, mostly to enact behaviors which are not part of their natural repertoire.

(Cohen, 2012, pp. 194–195)

Wildlife ecotourism

The viewing of charismatic megafauna (along with other aspects of the natural world like plants and geological formations) has become an important economic driver in the tourism industry. Countries rich in fauna have been able to compete with other international destinations for a share of the domestic and international market because of these natural resources and assets. But these forms of tourism are also important because of the focus on conservation and sustainable development, learning, and the ethical planning, development, and management of these natural features (Fennell, 2014). While wildlife tourism can be non-consumptive (wildlife viewing), it can also be consumptive (hunting and fishing). Ecotourism, in contrast, is solely non-consumptive. The problem with many wildlife tourism venues is that they are not subject to national and international regulations, so their regulation and policy development is contingent upon what tourists find acceptable or not. As long as tourists do not leave poor reviews of these establishments, there is no perceived need to change how animals are handled and presented. Establishing standards according to welfare, conservation, and proper governance in ways that tourists can use and understand is of clear importance if the lives of animals are to be improved.

Wildlife viewing

Perhaps the purest form of ecotourism is wildlife viewing in backcountry and wilderness areas which supports the existence of wildlife in fully natural settings, completely unframed and unrestricted (Cohen, 2009), where wildlife live freely and are not subject to manipulation and control (Fennell, 2013). The frequency of this type of recreation is low, likely due to the numerous accessibility challenges arising from the necessary absence of infrastructure, such as roads (Whittaker 1997; Nettles et al., 2022). Recreationists who seek this type of tourism frequently do so in the pursuit of novelty, solitude, or a deeper harmony with nature and can often be found floating the rivers, on horseback, hiking, climbing, or hunting (Whittaker, 1997; Nettles et al., 2022).

Animals in fully natural settings often lead very harsh and difficult lives. Humans tend to romanticise the idea of "the wild" and it should be noted that simply because an animal is "free", it does not necessarily follow that the animal is living under good welfare conditions (Mehrkam and Fad, 2020). Threats to physical health include infectious diseases, injuries, infections, and parasites, among others (Atuman et al., 2019; Mehrkam and Fad, 2020). Lack of veterinary care means that wild animals suffering from medical conditions often die prematurely compared to their non-wild counterparts and these afflictions can sometimes be painful (Tidiere et al., 2016). In the case of mental and emotional well-being, of course, the advantage of a fully natural setting is that an individual animal is allowed to operate under its own agency free to engage in natural behaviours in pursuit of mental and emotional well-being. Wild animals also encounter a variety of natural environmental stressors including climate variations, predator–prey interactions, territorial defence, and food scarcity which can lead to both acute and chronic stress (Dickens and Romero, 2013; Atuman et al., 2019). In terms of sufficient access to resources, wild animals have the freedom to roam large distances across their range in hunt of resources; however, especially during winter or dry seasons when resources are naturally scarce, it can be difficult to find sufficient food and water (Atuman et al., 2019). Wildlife viewers should consider the range of major welfare challenges wild animals face in spite of commonly idealised depictions of what life in "the wild" is like. The argument can also be made that those in wildlife management should consider current welfare conditions of individual animals in their management areas and seek out measures to improve these conditions where possible (e.g. trapping animals to administer vaccines and veterinary care, providing and stocking supplementary feeding stations, translocation from dangerous or overpopulated areas). Of course, the fiscal ramifications of such measures generate a significant obstacle to their implementation.

Animals as captives

Animals are captured or bred for presentation to the public in a number of different ways, which underscore the pleasure and profit motivations of tourists and operators. Various degrees of confinement and freedom are emphasised in these venues, along with use for entertainment and conservation and education (Shackley, 1996). Examples include safari parks, zoos, circuses, aquaria, sanctuaries, butterfly parks, and crocodile farms. Wildlife tourism attractions, zoos and aquaria, sanctuaries, and circuses are discussed briefly, below.

Wildlife tourism attractions (WTAs)

Several WTAs occur in wild uncontrived settings, but many also occur in captive settings. WTAs have been defined by Moorhouse et al. (2015) as non-zoo, non-hunting attractions that offer opportunities for tourists to interact with specific taxa of non-domestic animals, either in captive or wild settings, many of which claim benefits for wildlife which they do not deliver. There are several examples of these attractions including venues that allow for direct interactions (e.g. touching, feeding, taking selfies) of tigers, lions, and dolphins in captivity, trekking to observe gorillas, visiting civet coffee farms, viewing rehabilitated or rescued animals (e.g. orangutan sanctuaries), or watching wildlife-based shows (such as "snake charming") (Moorhouse et al., 2015). These authors found that 24 types of wildlife tourist attractions collectively impacted the welfare status of 230,000–550,000 individual animals, and that 120,000–340,000 animals were maintained in WTAs likely to reduce their species' conservation status (Moorhouse et al., 2015).

Zoos and aquaria

Zoos as a form of tourism are particularly interesting and controversial in regard to how they have "rebranded" themselves over the years. The evolution of zoos began as private collections and menageries owned by the wealthy as symbols of status, purely for human amusement and later transitioned into public spaces for recreation and profit (Rabb, 2004). However, with the rise of animal rights and welfare voices, zoos began to move away from these antiquated backgrounds and started placing heavy emphasis on enrichment, research, conservation, and education in their mission statements, with the main argument being that zoos act as a vehicle for conservation both *in situ* by raising funds to support projects for wild animals and *ex situ* through captive breeding programs (Patrick et al., 2007; Iossa et al., 2009; Zimmermann, 2010). Individual zoos have been successful at this to varying degrees with some zoos contributing a large proportion of their income towards conservation initiatives. The Species Survival Program (SSP) and European Endangered Species Program (EEP) oversee captive breeding in zoos, select mates for animals based on genetic compatibility, and fund the transfer of these animals to appropriate facilities (Conway, 2011). The welfare concerns associated with zoos and aquaria are explored in depth in Chapter 15.

Sanctuaries

Sanctuaries are a semi-contrived setting that, at first glance, appear quite similar to zoos. Animals are kept in enclosures simulating a natural environment, similar husbandry techniques are used, and sometimes there are even animals on display for tourists. However, in contrast to zoos, the purpose of a sanctuary is not to keep animals captive but to hold them temporarily until such a time as they can be rehabilitated and safely released. Some animals may be held indefinitely due to complications that would preclude their survival in the wild. Many sanctuary models operate mixed-access facilities in which there is a side open to ecotourists that holds such animals indefinitely and a rehabilitation side, closed to the public in which animals can recover in privacy (Thomsen et al., 2021). There are also pre-release enclosures that are meant to simulate a natural environment as closely as possible in order to ensure an animal is ready for release after time spent in an artificial environment for medical rehabilitation (Thomsen et al., 2021). Interactions with locals who call sanctuaries to report injured or orphaned wildlife can expose the community to pro-wildlife ideas that may be in contrast with historically negative cultural attitudes (Thomsen et al., 2021).

Not all sanctuaries are created equal, however. There are serious welfare concerns associated with such enterprises that market themselves as "green" or "conservation-minded". While many sanctuaries have remarkable missions and truly transformative impacts on their communities and local ecosystems, others may simply act out of self-interest, by putting on a facade for visitors and appearing to be legitimate from the "frontstage" viewing areas, while neglecting animal welfare in the "backstage" operation of the facility (Moorhouse et al., 2015; Thomsen et al., 2021).

Circuses

In contrast to zoos, circuses have not fared well under the modern cultural transition towards animal rights and posthumanist ideals. Due to the inherent fully contrived, artificial, and spectacle-centric nature of a circus environment, circuses have not been able to "rebrand" themselves in the same way as zoos.

The nomadic nature of travelling circuses adds an additional layer of difficulty when it comes to maintaining proper animal welfare standards – especially when it comes to giving animals adequate space. In a travelling-circus environment, the frequent upheaval makes compactness necessary for ease of storage and movement, which is frequently at odds with the needs and interests of animals. Both transit enclosures and exercise enclosures in circuses are far smaller than what is required for the same animals in zoo enclosures (Iossa et al., 2009). Animals in circuses are frequently housed either alone or in groups smaller than those of their wild counterparts. Additionally, animals are often exchanged among circuses, uprooting any previous social bonds animals may have formed with each other (Iossa et al., 2009). There are a number of stressors associated with transport including "forced movement, human handling, noise, cage motion, and confinement" (Iossa et al., 2009).

When it comes to the performances themselves, animals run into further welfare issues. Training in circuses often utilises positive and negative reinforcement with a focus on reducing levels of fear and anxiety. However, the quality of training depends on the skill of the trainer (Iossa et al., 2009). Acute stress driven by the stimuli of circus performances may contribute to a number of medical conditions, defensive and escape responses, and stereotypical behaviours in different species (Iossa et al., 2009). Examples of medical conditions driven by circus stimuli include gastroenteritis in tigers, which can develop as a result of noise exposure and septicaemia infection in Indian pythons, which can develop as a result of light exposure (Iossa et al., 2009). Primates, bears, and ungulates display defensive and escape behaviours in the presence of human crowds (Iossa et al., 2009). Stereotypical behaviours such as pacing increase in tigers and elephants in the time leading up to a performance, suggesting either anticipation or anxiety (Iossa et al., 2009). Despite these issues, enforcing legislation to protect animal welfare in circuses can often prove challenging as lack of clarity in definitions of words such as "domesticated animal, a wild species, a travelling circus, a mobile zoo, and performance" often lead to discrepancies (Harris and Pickett, 2016). A working theoretical framework for animal welfare as well as a consensus on the precision of language used in animal welfare literature could contribute to clarifying present ambiguities.

Sport and subsistence

Animal sport is inherently fully contrived, artificial, and spectacle-based. Furthermore, it is almost always laced with an undercurrent of innate violence which some have attributed to historically gendered behaviours related to the demonstration of masculinity and virility (Atkinson and Young, 2005; Kalof, 2014). This violence is coined by Atkinson and Young as "sports-related violence" (SRV) justified in the minds of participants on the basis of a set of historical and or sociological norms (Atkinson and Young, 2005; Kalof, 2014; Ahluwalia, 2016).

One theory on why sport, especially combat-based and bloodsport, has persisted so resiliently for thousands of years is that it serves as an "identity prop" in which males assert their masculinity and virility within their social groups (Kalof, 2014). Throughout history and across cultures, man's inherent power, aggression, and control within the context of bloodsport has been woven into the symbolism of literary tradition, depicting parallels between hunting of animals and the acquiring of a female mate (Kalof, 2014). This chapter will examine four types of sport in which animals are used: *combat*, such as bullfighting or dogfighting, where either a human and animal or two animals fight each other until there is a winner; *hunting and angling*, in which wild or feral animals are pursued and harvested; *racing*, such as greyhound racing or horse racing in which animals are pitted against each other in a competition of speed to see who can reach a certain point first; and *rodeo*, in which livestock animals are used for a variety of events showcasing various demonstrations of cowboy skill.

Combat

Combat sports include human-on-animal events such as bullfighting, in which a non-human animal and human fight in a direct contest of strength or skill. Combat also includes animal-on-animal events, in which two non-human animals are pitted against each other to fight until there is a winner (e.g. dogfighting, cockfighting). A brief overview of both sets of practices is included.

Bullfighting is steeped in tradition in countries like Spain and Mexico, where upwards of 40,000 bulls are killed yearly in this bloodsport (Hall and Brown, 2006). Critics argue that the bulls are placed at a great disadvantage before the event through beatings, laxatives, drugs, vision impairment, and shaved horns (PeTA, 2021a). This cultural practice involves several acts (as in a theatre play) and the use of weapons to dispatch the bull, described as elegant, beautiful, and tragic (McCormick, 1997), with the event taking approximately 20–25 minutes. Marvin's (1994) description of the meaning of the bullfight underscores the line between humans and nature, with domination, manipulation, and control of nature in the elevation of human agency and cultural advancement as a necessary ingredient. Inherent in the practice is representation of masculinity, played out through expressions of sexual potency, independence, assertiveness, and strong will (Marvin, 1994). The bull is killed slowly to represent the process of moving from a wild and exotic state to domestication. Studies indicate that the cultural significance of the bullfight is changing. Spaniards, for example, are recognising that the practice is a serious form of animal cruelty and sanctioned abuse, with entertainment geared more for tourists than locals (Bailey, 2007).

Dogfighting evolved out of the use of dogs for hunting companions and personal protectors. For instance, hounds have historically been pitted against other animals in various hunting disciplines including foxhunting and hare coursing, while mastiffs were at one time trained to protect their owners by fostering aggression in mock fights against "bait" animals like bears and bulls (Atkinson and Young, 2005; Kalof, 2014). Training for fights includes a variety of methods that aim to strengthen and prepare a dog in a number of different ways

(Kalof, 2014). Dogs are run on both non-baited treadmills and baited "Catmills" or "Jennys" to increase cardiovascular fitness and endurance. Sometimes a flirt pole will be used where a dog chases a lure attached to a handheld pole. Spring poles or jump poles strengthen the jaw muscles and back legs. Chains and weights build neck and upper body strength. Drugs, vitamins, and supplements are used to condition and/or incite fighting behaviours. The fights are violent and end when one dog kills the other or they stop fighting. Injuries are often crudely tended to, often using easily obtained items like superglue and staples in lieu of proper veterinary care (League Against Cruel Sports) "Losing dogs, especially at gang-run fights, are often shot, set on fire, tied to train tracks or left to die in abandoned buildings as 'punishment'" (Hageman, 2004).

Much like dogfighting, cockfighting has persisted for thousands of years and likely sprang up around the same time that chickens were first domesticated around 3,000 BC (Forsyth, 1996). In fact, it's frequently claimed to be the oldest sport in existence (Darden and Worden, 1996). Even though it too is illegal in most places, a healthy and lucrative network of underground cockfighters still thrives in defiance of the laws, especially in the US South (Maunula, 2007). Birds are carefully bred and well cared for and offered free-range living conditions and special diets with high quality food – even regularly massaged (Maunula, 2007). Even during the course of the fights, great care is shown towards the birds. After a time fighting, handlers will take breaks where they untangle the birds, retreat to separate corners of the ring and care for their birds. Handlers might sponge off the birds' heads, give them a drink of water, stroke their backs or breathe on their necks to warm them, or even place their bird's beak in their mouth to suck up obstructions in the birds' throats (Worden and Darden, 1992). The fight ends when one bird wins and the other dies. If a fight goes on for too long, they may be moved to a secondary location with a smaller "drag pit" where the fight may drag on for hours until one bird prevails or the handlers step in and kill them (Worden and Darden, 1992; Forsyth, 1996).

While many look down upon the addition of additional accessories, some cockfighting circles may employ gaffs and/or knives in their cockfights (Forsyth, 1996). Gaffs are sharp 1-inch to 2.5-inch curved steel spikes resembling ice picks that replace a cock's natural spur (Forsyth, 1996). Knives, which are more extreme and gaining in popularity, are 1-inch to 3-inch-long steel blades that are attached like bayonets to one leg of a cock (Forsyth, 1996). These fights are often over in a matter of seconds due to the lethality of the blade and are more likely to come down to luck rather than the actual fighting ability of the birds (Forsyth, 1996).

Hunting and fishing

The debate on the ethical legitimacy of hunting and fishing is voluminous. Staunch advocates like the rock star Ted Nugent, argue that hunting is a natural evolutionary right of humans, as top of the food chain, to use nature's bounty for their own purposes (Bauer and Herr, 2004). Critics contend that there is no need to hunt because we can obtain protein in any number of different ways through a vegetarian or vegan lifestyle – notwithstanding the ecological and sustainable benefits of not having to rear livestock for meat consumption (see Chapter 23). If the gorilla as one of the most powerful terrestrial animals can reach formidable strength through a vegan lifestyle, surely humans can gain all the protein they need through similar consumptive practices.

Increasingly, hunters have had to justify hunting through the use of concepts and terms like fair chase, sustainable, and conservation, instead of "sport", when seeking to make it more socially acceptable. Yet although labels and approaches have changed in the support of hunting, critics argue that hunting is an act of violence, and the game lacks symmetry. For instance, technology has advanced to the point where there is an unfair advantage for the humans that use it. Additionally, because the freedom to participate in the activity (i.e., sport involves two willing combatants) is solely one-sided, despite fair-chase rules, the activity can hardly be considered a sport and the animal becomes simply a recreational resource to satisfy our pleasures and desires (Scruton, 2002).

The arguments for and against fishing are not exactly the same, although the intent – to pursue an animal, is. Hunting involves the consumptive use of an animal, i.e., the removal of the animal from its environment. Fishing can be consumptive, but also non-consumptive through catch and release. But as some philosophers argue, fishing for food and subsistence is acceptable, whilst fishing for sport and tournaments and catch-and release are immoral because the welfare of the animal is compromised simply for pleasure and entertainment (Balon, 2000). The welfare concerns associated with hunting and fishing are explored in depth in Chapter 16.

Racing

The two species most commonly used in animal racing are dogs and horses. Dog racing in particular is disturbing due to alarmingly high injury and fatality rates. According to Atkinson and Young (2005), Lines in greyhound racing begin to blur between sport and bloodsport, as

> estimates published by the Greyhound Protection League suggest that nearly 30,000 young greyhounds are killed in North America every year when they are no longer able to win or "place". Approximately 5,000–7,000 farm puppies are "culled" annually, and more simply "go missing" without being registered to an owner.
>
> *(Atkinson and Young, 2005: 336–337)*

Dogs deemed unsuitable for racing are frequently killed. These include young puppies and older dogs that have lost their ability to place well in competitions (Atkinson and Young, 2005).

Dogs are also raced in events such as the Iditarod dog mushing contest – the 1,049-mile race between Willow and Nome Alaska that takes between 9 and 14 days to complete with temperatures from –34 to +1°C (Stafford, 2008). Humans and dogs are viewed as athletes in these competitions, but the choice to participate is solely human. An account of the 2020 Iditarod by PeTA (2020) describes the significant dog welfare violations that were committed in the name of competition. Examples include vomiting of dogs, fighting between dogs, frostbite, twisted intestines, and pneumonia. Common additional injuries include injuries to the pads, web or nail beds of the foot, injuries of the carpal joint and tendons of the foreleg, hypothermia, diarrhoea, and dehydration, and gastric ulcers – which can lead to sudden death (Stafford 2008). Due to these risks, up to a third of dogs who participate in the Iditarod do not finish (Stafford 2008).

Over 150 dogs have died since the inception of the Iditarod in 1973; many more dogs have endured the conditions described above. Dog "culling", as noted above, was observed at Howling Dog Tours, a dog sled tourism operation in British Columbia, which killed dozens of sled dogs after overestimating the increase in business they would receive as a result of the Paralympic Winter Games being hosted in Vancouver, BC (Fennell and Sheppard, 2011).

Horse racing is far less lethal than dog racing with only about two fatalities per 1,000 starts in the US (Werner, 2021). In fact, "99.86% of flat racing starts at the US racetracks participating in the Equine Injury Database were completed without a fatality" (Werner, 2021). Between 2009 and 2014, 80% of racehorse fatalities were the result of a fracture (Georgopoulos and Parkin, 2017). Fractures can be prevented by having horses race on a flat synthetic surface as opposed to natural substrates more prone to unevenness like turf or dirt (Arthur, 2010). Between races, racehorses are extremely well cared for, in some ways. According to Mundy (2000) in a paper on Equine welfare in racing for the JAVMA Animal Welfare Forum, "This care includes assigned grooms, around-the-clock monitoring, regular professional health care, individualized training programs, and excellent husbandry, all at considerable expense to these horses' owners" (Mundy, 2000). However, racehorses may also experience extended confinement and social isolation. Their financial value can make owners reluctant to risk injury or disease risk increasing with greater access to outdoor paddocks and social groupings. The welfare concerns associated with horse racing are also explored in Chapter 19.

Rodeo

Rodeo began as a way for cattlemen to demonstrate the skills necessary for their day-to-day life by featuring skills used in everyday husbandry practices, such as the ability to rope and restrain a calf for branding (Furman, 2001; Rizzuto et al., 2020). However, in contrast to common husbandry practices, the entertainment value of rodeos depends on the distress and misbehaviour of the animals, as more violent and exaggerated behaviours from the animals increase the challenge for the participants and thus add to the drama of the spectacle (Franzky, 2005; Ahluwalia, 2014; Rizzuto et al., 2020).

Horse Disciplines include events such as bare back riding and saddle bronc riding, in which a flank strap is placed and tightened on a horse's sensitive flanks to induce bucking (Franzky, 2005; Petition 2014/53 of Ahluwalia, 2014; RNZSPCA Submission on Rodeos for the Primary Production Select Committee, 2016). These horse disciplines also often employ the use of spurs to give signals and cue certain behaviors in horses (Franzky, 2005). However when these spurs are wheeled, sharpened, or applied forcefully to areas such as the sides of the neck, they can pose significant welfare concerns (ibid). While bucking is a natural behaviour for horses that can be playful, these playful displays of bucking are usually accompanied by relaxed facial expressions, while defensive bucking behaviours are accompanied by stressed facial expressions (Franzky, 2005). Horses that have learned that successfully throwing off their rider results in the immediate release of the flank strap, are visibly more relaxed before and after their performance, than horses that have not yet learned this (Franzky, 2005). The fact that horses are able to learn and anticipate how their behaviour will affect the stimuli applied to them indicates that it may be possible to condition a horse through training alone to buck – either without the continued use of a flank strap or perhaps without a flank strap at all (Franzky, 2005).

Cow disciplines include bull riding, steer wrestling, and team roping. Flank straps used in bull riding are often tightened around the urethra, which may result in increased pain for bulls in contrast to horses, but normally the flank strap loosens on its own during the performance (Franzky, 2005). Cattle prods are also used to administer an electric shock to encourage bulls and calves to move into the ring (Ahluwalia, 2014). Aggressiveness in bulls is desirable in that it contributes to the drama between man and animal. However, research shows that subtle changes in handling during pre-performance, such as how many handlers are in a given area, where they stand, and the cues they use have the potential to greatly improve the experience of the bulls without altering the bull's performance (Goldhawk et al., 2016). Calf-related events such as calf roping also pose significant welfare concerns as they result in a significant number of injuries and acute stress responses (RNZSPCA, 2016; Rizzuto et al., 2020).

Film, television, and advertising

Film and television

In the US, there are no direct laws that have been passed to protect the welfare of animals used in the film and television industry. Rizzo (2012) points out that the film industry still must abide by the "exhibitors" category of the Animal Welfare Act and are also not allowed to use animals that are threatened or endangered under the Endangered Species Act. Some states, such as California, have laws that criminalise the filming of animal cruelty. The organisation acting as a watchdog for the ethical treatment of animals used in the film industry is the American Humane Association (AHA), which has long been empowered by the Motion Picture Association of America to eliminate abusive practices. "Humane", according to the AHA, is "marked by an emphasis on humanistic values and concerns; characterized by kindness, mercy or compassion" (American Humane Association, 2015, p. 7). The AHA's certification program "No Animals Were Harmed" outlines several basic principles for the safe use of animals in filmed media (American Humane Association, 2015, p. 6).

An important component of the work of the AHA is the development of species-specific guidelines, suggesting that welfare for animals cannot be generalised. Separate categories of guidelines are included for dogs, domestic cats, birds, fish, insects and arachnids, horses and livestock, exotic/captive wildlife, primates, reptiles, amphibians, and wildlife. This precludes oversights of the unique needs of individual animals on film sets and ensures that these needs are properly met. Of course, meeting these AHA guidelines costs money and the burden is on producers to source and manage animals within a film's budgetary limits. This poses a particular challenge to television programs, as they typically run on smaller budgets than feature films and are likely to want to use animals more than once throughout the course of a program (Wilkins 1981). In many cases, the use of live animals is no longer necessary, as computer-generated imagery technology and special effects methods have advanced to such a degree of realism. Such technology is especially useful for depicting violence and recreations of historical events involving the suffering of animals, for example the use of puppetry in the film "*War Horse*" (Tait, 2016).

Documentaries allow the opportunity to observe animals in a fully "natural" setting as opposed to a contrived film set. However, even in this context, there are welfare concerns regarding the filming process and the disruption of animals (Mills, 2010). In some cases, documentaries don't even capture true wild behaviours and instead contrive situations for increased drama. The Centre for Active Animal and Nature Protection created a documentary in 1981 called "Cruel Camera" on the cruel treatment of animals in movies and documentaries (Cory, 1986). A classic example is Mutual of Omaha's "*Wild Kingdom*", starring Merlin Perkins, that aired between 1968 and 1971. Cory (1986) illustrates one of many different strategies designed to get animals to cooperate: "How do you get an alligator to attack a water moccasin? Tie a string to the water moccasin's tail; throw him out and reel him in. Eventually, the alligator will attack the water moccasin out of sheer boredom". Not all documentaries exploit animals in this way, however, and technology such as filming equipment that can capture footage from high altitudes in helicopters and planes, as well as hidden cameras that can capture footage without a camera operator, can reduce invasiveness to wildlife to a minimum. Even then, some animal rights and posthumanist voices call for filmmakers to consider an animal's innate desire to not be seen and to entertain the idea of an animal's right to privacy (Mills, 2010).

Advertising

Advertisements are a reflection of the values implicit in society. A case in point is the Boost Mobile commercial (Boost Mobile, 2009). Two pigs are dining in an upscale restaurant, with one pig commenting, "I like a nice ham. Do you think that's wrong? We're just enjoying the flavours of a fallen friend". The commercial generated debate from those that hated it on the basis of its disregard for pigs, to those who loved it based on its humour. In the end, Boost Mobile appears to have emerged victorious because the advertisement got people talking about their product.

Another example is an advertisement run by Nike that "showed two dogs lunging at one another attempting to fight". The company spokesperson denied that the ad was about dog-fighting at all but rather about "the compelling need to win, to beat your opponent and win at all odds … People have to understand the youth culture we cater to … Our market is the urban, edgy, hip-hop culture" (cited in Gibson, 2005, note 14; Kalof, 2014). Dogfighting is also glorified in the advertisements of many clothing and toy manufacturers aimed at this demographic (Kalof, 2014).

The British Veterinary Association (BVA, 2018) is assuming an active role in protecting the interests and welfare of pets based on their framework for good practice and responsibility in advertisement. The BVA advocates five main welfare needs for pets, as follows:

- Suitable environment: pets used in advertising should be shown to be living in environments that meet their physical, social, and behavioural needs;
- Suitable diet: pets used in advertising should be shown to be eating proportionate amounts of a nutritionally balanced diet and/or around appropriate food for their species;
- Behavioural needs: pets used in advertising should be exhibiting, or shown to have the potential to exhibit, normal behaviours for their species or breed type;
- Social needs: pets used in advertising should be shown to be housed and interacting with, or apart from, other animals appropriate to their species;
- Protection from pain, disease, and suffering: pets used in advertising should be protected/free from pain, disease, and suffering.

Conclusion

This chapter has shown the broad spectrum of animal uses for purposes of entertainment and pleasure as well as for profit. New research is questioning the conventional contractarianist mindset in tourism research and practice by offering new perspectives on welfare, posthumanism, and other theoretical domains, although welfare still dominates these discussions and practices. Those who continue to participate as operators and tourists in many events and attractions (e.g. bullfighting), must be prepared to endure social costs as changing values and priorities disrupt the historically perceived legitimacy of these practices. We argue that tourism will need to more formally develop policy and regulations that place the interests of animals on a much higher level as tourists become better educated on the welfare, conservation, and governance issues that continue to plague the industry. As the world's largest industry, tourism must be prepared to invest considerable resources into animal welfare changes in keeping pace with sectors such as scientific animal use and intensive animal farming, where the critical lens has been used more liberally, and, in some ways, to greater effect.

References

Ahluwalia S, 2014. Petition 2014/53 of Shanti Ahluwalia on behalf of SAFE, SPCA, and Farmwatch. Petition 2014/53 of Shanti Ahluwalia on behalf of SAFE, SPCA, and Farmwatch.

Ahluwalia S, 2016. *Submission Calling for a Rodeo Ban*. Kelburn, Wellington, NZ: SAFE.

American Humane Association, 2015. 'No Animals Were Harmed': A certification program of American Humane Association. *Guidelines for the Safe Use of Animals in Filmed Media*. https://www.americanhumane.org/app/uploads/2016/08/Guidelines2015-WEB-Revised-110315-1.pdf (Accessed: 23 June 2021).

Animal Welfare Act 2006, 2015. https://www.animallaw.info/sites/default/files/UKAnimalWelfareAct2006.pdf (Accessed: 23 June 2021).

Arthur RM, 2010. Comparison of racing fatality rates on dirt, synthetic, and turf at four California racetracks. *AAEP Proceedings*, 56, pp. 405–408.

Atkinson M and Young K, 2005. Reservoir dogs: Greyhound racing, mimesis and sports-related violence. *International Review for the Society of Sport*, 40(3), pp. 335–356.

Atuman YJ, Kudi CA, Abdu P and Abubakar A, 2019. Prevalence of parasites of wildlife in Yankari game reserve and Sumu wildlife park in Bauchi State, Nigeria. *Sokoto Journal of Veterinary Sciences*, *17*(4), pp. 70–79.

Bailey C, 2007. We are what we eat: Feminist vegetarianism and the reproduction of racial identity. *Hypatia*, *22*(2), pp. 39–59.

Balon EK, 2000. Defending fishes against recreational fishing: an old problem to be solved in the new millennium. *Environmental Biology of Fishes*, *57*(1), pp. 1–8.

Bauer J and Herr A, 2004. *Hunting and Fishing Tourism. Wildlife Tourism: Impacts, Management and Planning*. Altona Vic: Common Ground Publishing, pp. 57–77.

Boost Mobile, 2009. Boost mobile UNwrong'D. https://www.facebook.com/boostmobile/videos/1100730952681/?extid=SEO---- (Accessed: 29 June 2021).

Brando S, 2016. Wild animals in entertainment. In Bovenkerk B, Keulartz J, Eds, *Animal Ethics in the Age of Humans, The International Library of Environmental, Agricultural and Food Ethics*, vol 23. Cham, Switzerland: Springer International Publishing AG, pp. 295–318.

British Veterinary Association, 2018. *Pets in Advertising: A Social Concern*. London, UK: British Veterinary Association.

Carr N and Broom DM, 2018. *Tourism and Animal Welfare*. Wallingford, UK: CABI.

Cohen E, 2009. The wild and the humanized: Animals in Thai tourism. *Anatolia*, 20(1), pp. 100–118.

Cohen E, 2012. Tiger tourism: from shooting to petting. *Tourism Recreation Research*, *37*(3), pp. 193–204.

Conway WG, 2011. Buying time for wild animals with zoos. *Zoo Biology*, 30, pp. 1–8.

Cory J, 1986. 'Cruel camera,' about animal abuse. *New York Times Section C*, p. 18, New York, March 24, 1986. https://www.nytimes.com/1986/03/24/movies/cruel-camera-about-animal-abuse.html (Accessed: 25 July, 2021).

Darden, DK and Worden SK, 1996. Marketing deviance: The selling of cockfighting. *Society and Animals*, 4(2), pp. 211–231.

Fennell DA, 2000. Tourism and applied ethics. *Tourism Recreation Research*, 25(1), pp. 59–70.

Fennell DA, 2012. *Tourism and Animal Ethics*. London: Routledge.

Fennell DA, 2013. Contesting the zoo as a setting for ecotourism, and the design of a first principle. *Journal of Ecotourism*, 12(1), pp. 1–14.

Fennell DA, 2014. Exploring the boundaries of a new moral order for tourism's global code of ethics. *Journal of Sustainable Tourism*, 22(7), pp. 983–996.

Fennell DA and Sheppard VA, 2011. Another legacy for Canada's 2010 Olympic and Paralympic Winter Games: Applying an ethical lens to the post-games' sled dog cull. *Journal of Ecotourism* 10(3), pp. 197–213.

Forsyth CJ, 1996. A pecking disorder: Cockfighting in Louisiana. *International Review of Modern Sociology*, pp. 15–25.

Franzky A, 2005. *Expert Opinion Regarding Rodeo Events in the Federal Republic of Germany from a Legal, Ethological and Ethical Perspective*. Rolfsen, Germany: Registered Association of Veterinarians for Animal Protection

Furman JW, 2001. Rodeo Cattle's many performances. *Journal of the American Veterinary Medical Association*, 219, pp. 1394–1397.

Gibson H, 2005. *Dog Fighting Detailed Discussion*. East Lansing, MI: Animal Legal and Historical Center: Michigan State University College of Law

Goldhawk C, Bond G, Grandin T and Pajor E, 2016. Behaviour of bucking bulls prior to rodeo performances and relation to rodeo and human activities. *Applied Animal Behaviour Science, 181*, pp. 63–69.

Hageman W, 2004. A child, a pup a blood sport. *Chicago Tribune*. [Online] 11th June. https://www.chicagotribune.com/news/ct-xpm-2004-05-11-0405110252-story.html, (Accessed: 30 June 2021).

Hall DR and Brown F, 2006. *Tourism and Welfare: Ethics, Responsibility and Sustained Well-Being*. Cabi.

Harris S and Pickett H, 2016. The welfare of wild animals in traveling circuses. https://www.academia.edu/28471968/The_welfare_of_wild_animals_in_travelling_circuses (Accessed: 23 June 2021).

Iossa G, Soulsbury CD and Harris S, 2009. Are wild animals suited to a travelling circus life. *Animal Welfare*, *18*(2), pp. 129–140.

Kalof L, 2014. Animal blood sport: A ritual display of masculinity and sexual virility. *Sociology of Sport Journal*, 31(4), pp. 438–454.

Keulartz J and Bovenkerk B, 2016. Changing relationships with non-human animals in the anthropocene: An introduction. In Bovenkerk B and Keulartz J, Eds, *Animal Ethics in the Age of Humans, The*

International Library of Environmental, Agricultural and Food Ethics, vol 23. Cham, Switzerland: Springer International Publishing AG, pp. 1–24.

Kline C, 2018. *Animals, Food, and Tourism*. London: Routledge.

Markwell K, 2015. *Birds, Beasts and Tourists: Human-animal Relations in Tourism*. Clevedon, UK: Channel View.

Marvin G, 1994. *Bullfight*. Champaign, IL: University of Illinois Press.

Maunula M, 2007. Of chickens and man: Cockfighting and equality in the South. *Southern Cultures*, 13(4), pp. 76–85.

McCormick J, 1997. The bullfight gentrified. *Society*, *34*, pp. 48–50.

Mehrkam LR and Fad O, 2020. Animal welfare science and "a life worth living" for wild and captive elephants. *Animal Sentience*, *5*(28), p. 10.

Mills B, 2010. Television wildlife documentaries and animals' "right to privacy." *Continuum: Journal of Media and Cultural Studies* 24(2), pp. 193–202.

Moorehouse T Dahlsjö CAL Baker SE D'Cruze NCD and Macdonald DW, 2015. The customer isn't always right: Conservation and animal welfare implications of the increasing demand for wildlife tourism. *PLoS ONE*, 10(10), e0138939. DOI: 10.1371/ journal.pone.0138939.

Mundy GD, 2000. Racing. *Journal of the American Veterinary Medical Association*, 216, pp. 1243–1246.

Nettles JM, Brownlee MTJ, Sharp RL and Verbos RI, 2022. The utilization distribution: Wildlife research methods as a tool for understanding visitor use in remote parks and protected areas. *Human Dimensions of Wildlife*, 27(2), pp. 151–163, DOI: 10.1080/10871209.2021.1885766

O'Neil L, 2016. Dolphin calf dies after tourists use it for selfies. https://www.cbc.ca/news/trending/dolphin-calf-dies-after-tourists-use-it-for-selfies-1.3456188 (Accessed: 30 June 2021).

Patrick PG, Tunnicliffe SD, Matthews CE and Ayers DF, 2007. Mission statements of AZA-accredited zoos: Do they say what we think they say? *International Zoo News* 54(2), pp. 90–98.

PeTA, 2020. The 2020 Iditarod is over, but PeTA's fight for dogs continues. https://www.peta.org/blog/2020-iditarod-death-race-coverage/ (Accessed: 23 July 2021).

PeTA, 2021a. Bullfighting. https://www.peta.org/issues/animals-in-entertainment/cruel-sports/bullfighting/ (Accessed: 23 July 2021).

PeTA, 2021b. We're winning: More than 50 travel companies pull elephant rides. https://www.peta.org/blog/winning-travel-companies-pull-elephant-rides/ (Accessed: 23 July 2021).

Rabb GB, 2004. The evolution of zoos from menageries to centers of conservation and caring. *Curator* 47(3), pp. 237–246.

Rickly J. and Kline C, (Eds) 2021. *Exploring Non-human Work in Tourism: From Beasts of Burden to Animal Ambassadors*. Berlin: De Gruyter.

Rizzo V, 2012. Detailed discussion of the legal protections of animals in filmed media. *Animal Legal & Historical Center*. https://www.animallaw.info/article/detailed-discussion-legal-protections-animals-filmed-media (Accessed: July 26, 2021).

Rizzuto S, Evans D, Wilson B and McGreevy P, 2020. Exploring the use of a Qualitative Behavioural Assessment approach to assess emotional state of calves in rodeos. *Animals*, 10(1), 113.

Royal New Zealand Society for the Prevention of Cruelty to Animals Inc., 2016. *Submission on Rodeos for the Primary Production Select Committee*. NZSPCA.

Scruton R, 2002. Ethics and welfare: The case of hunting. *Philosophy*, 77(4), pp. 543–564.

Shackley M, 1996. *Wildlife Tourism*. London: International Thomson Press.

Stafford K, 2008. Welfare of working and sport dogs. In World Small Animal Veterinary Association World Congress Proceedings, August, Dublin.

Tait P, 2016. *Fighting Nature: Traveling Menageries, Animal Acts, and War Shows*. NSW, Australia: Sydney University Press.

Thomsen B, Thomsen J, Copeland K, Coose S, Arnold E, Bryan H, Prokop K, Cullen K, Vaughn C, Rodriguez B, Muha R, Arnold N, Winger H and Chalich G, 2021. Multispecies livelihoods: A post-humanist approach to wildlife ecotourism that promotes animal ethics. *Journal of Sustainable Tourism*. Advance online publication. DOI: 10.1080/09669582.2021.1942893.

Tidiere M, Gaillard JM, Berger V, Muller DWH, Lackey LB, Gimenez O, Clauss M and Lemaitre JF, 2016. Comparative analyses of longevity and senescence reveal variable survival benefits of living in zoos across mammals. *Scientific Reports*, 6, p. 36361. DOI: 10.1038/srep36361.

Werner K, 2021. The Jockey Club releases data from the equine injury database for 2020. *The Jockey Club*, March 29, 2021. http://jockeyclub.com/Default.asp?section=Resources&area=10&story=1258 (Accessed: July 22, 2021).

Whittaker D, 1997. Capacity norms on bear viewing platforms. *Human Dimensions of Wildlife*, 2(2), pp. 37–49.

Wilkins DB, 1981. Animals in film and television. *International Journal for the Study of Animal Problems*, 2(6), pp. 284–287.

Worden S and Darden D, 1992. Knives and gaffs: Definitions in the deviant world of cockfighting. *Deviant Behavior*, 13(3), pp. 271–289.

Zimmermann A, 2010. The role of zoos in contributing to *in situ* conservation. In Kleiman DG, Thompson KV and Baer CK, Eds, *Wild Mammals in Captivity: Principals and Techniques for Zoo Management*, vol 23. Chicago: University of Chicago Press, pp. 281–287.

15
ZOOS AND AQUARIA

Terry L Maple and Bonnie M Perdue

Introduction

Can zoos and aquariums be humane? Throughout the world, many people visit zoos and aquariums and leave with a negative impression. Since the goals of zoological institutions are to educate, conserve, study, and safely exhibit wildlife, often in an entertaining way, zoo professionals and their consultants spend millions of dollars designing facilities that are aesthetically pleasing and functionally stimulating for the animals that reside there. One of the most salient questions that must be answered is whether zoos and aquariums can be humane in their treatment of a diversity of animals. Because they have historically been compared to prisons and mental hospitals (Sommer, 2008), the denizens of zoos and aquariums must be confined in ways that encourage the expression of their natural behaviour patterns. To meet this requirement, zoological institutions must be evidence-based, and their standards and practices have to be beyond reproach. Maple and Perdue (2013) have identified institutions in this category as "empirical zoos". Elite modern zoos provide innovative operating standards and best practices for achieving animal welfare.

Animal welfare generally refers to the basic conditions necessary for the maintenance of the physical and psychological well-being of animals. Adopted from livestock regulations in the UK, a gold standard for animal welfare was first proposed by the Brambell Committee (Brambell et al., 1965). For example, freedom from hunger and thirst; freedom from discomfort; freedom from pain, injury, or disease; freedom to express normal behaviour; and freedom from fear or distress. Current standards emphasise positive rather than negative outcomes following Mellor's construct of the "five domains" (Mellor et al., 2020). While it is likely that many zoo species are still coping with captivity, a growing number of superior naturalistic exhibits have demonstrated that many zoo animals are indeed thriving. Thriving is the goal of optimal animal welfare (Maple and Perdue, 2013).

The keys to optimal animal welfare in zoos are environmental, social, and developmental variables. Historically, zoo and aquarium animals may have suffered in facilities that were inadequate and restrictive. Small cubic cages left few degrees of freedom for animals to behave normally. Raising animals as singletons or in inappropriate social groups induced idiosyncratic deprivation acts and stereotypies. Normal social development requires mother-rearing, peer experience, and sufficient space to comfortably socialise with others. We know how to design and build zoos that

 DOI: 10.4324/9781003182351-18

contribute to psychological wellbeing but far too many zoos worldwide have failed to achieve the upgraded standards and practices demanded by the science of animal welfare. Substandard roadside zoos, most of them lacking accreditation by regional associations, far outnumber the good ones. Institutions such as these are responsible for widespread suffering of captive animals.

The present chapter aims to succinctly summarise the long history of zoos in relation to animal welfare and some of the key figures historically involved in these developments. This review will necessarily be limited and will not include every relevant event or development, but we argue that continuing to teach and understand the historical context and developments will benefit decisions to be made in the future. We will then consider areas that are likely to be of increasing importance in the future of zoos.

A brief history of zoos in relation to animal welfare

Like zoo animals, occupants of mental hospitals were once a source of entertainment for a curious public (e.g., MacKinnon, 2009). With time, mental hospitals began to provide the humane treatment that their patients required. Higher standards and better mental health practices were the psychiatric equivalent of animal welfare reform. Unlike zoos, however, mental hospitals continued to operate within the limitations of hard architecture. If mental hospitals had elected to soften their facilities, they might have provided for more access to natural landscapes, including exposure to botanical gardens, where patients could enjoy the surroundings or work with plants as therapy. During the 1970s, governments chose to close these facilities rather than reform them, leading to an explosion of homelessness in American cities. Substandard zoos have been closed by governments or irate citizens, but the animals cannot be released. They must be transferred to better facilities or sanctuaries to improve their quality of life. A better solution is to reimagine and rebuild zoos so they enhance psychological wellbeing. North American examples of institutional relocation and renewal include the Indianapolis Zoo, Zoo Miami, Fresno Zoo, and scores of aquariums in cities such as Long Beach, California, the Georgia Aquarium in Atlanta, and the Tennessee Aquarium in Chattanooga.

Fewer zoos are rebuilt from scratch, but aquariums and aquatic parks have proliferated throughout the world, especially in Asia. Unfortunately, many of these facilities follow the amusement park model and fail to support normative animal welfare standards. With the priority of entertainment, the treatment of dolphins, orcas, and whales are a major problem within some Asian aquariums. SeaWorld Corporation was once revered for its commitment to professional animal training and enrichment, but many visitors have been disappointed by the company's failure to advance the quality of life for these huge, complex marine mammals. In our opinion, no whale tank on earth is fully adequate for the enlightened exhibition of a free-swimming, migrating whale or dolphin (Maple, 2016) but human contact through training does provide some enriching moments if carried out properly.

Key historical figures

Robert M Yerkes

The standards and practices of zoos have been generated by a variety of scientists from different backgrounds and professions. Robert M Yerkes was widely recognised as the foremost expert on the behaviour and management of great apes. Although the field of animal welfare did not yet exist when Yerkes' most important books were published (for example, *The Great Apes*, 1929), he understood that he could not study these complicated animals unless he could first keep them alive and well. One of the core principles of the labs was the requirement that staff would observe every animal every day to evaluate their individual health and wellbeing. Yerkes also pioneered the idea that animals should be occupied to avoid boredom. He concluded that they enjoyed manipulating objects so he introduced items as toys or tools for their amusement. His own interest in the mental potential of apes led him to conduct experiments to test their aptitude. In *The Great Apes*, he and his wife, Ada Watterson, provided a comprehensive breakdown of comparisons of chimpanzee, gorilla, and orangutan intelligence based on his studies and the research of colleagues and students. Today, this type of research is characterised as comparative cognition. He and his Yale research associates managed apes with enlightened protocols that influenced management and husbandry in zoos throughout the world. Robert M Yerkes could be considered one of the founding fathers of the science of zoo animal welfare.

Hediger and the founding of zoo biology

While Yerkes was a leading contributor to the literature of North American comparative psychology, Prof. Heini Hediger was building a reputation in Europe as an applied ethologist and a highly innovative director of zoological gardens. At various points in his long career, he was director of each of the leading zoos in Switzerland, in Basel, Bern, and Zurich. In Zurich he also served as Professor of Ethology at the University of Zurich where he mentored some of Europe's most distinguished students of the emerging science of zoo biology. One of his students was the distinguished primatologist Hans Kummer, who carried out studies of hamadryas baboons at Zoo Zurich and field investigations of geladas and hamadryas species in Ethiopia. Hediger wrote many important books and papers and is arguably the most prolific of all European zoo directors. *Wild Animals in Captivity* (1950) was the book with the earliest and greatest impact on principles of animal welfare. The English translation is still in print. Hediger also influenced the field of environmental psychology with his observations of personal space in captive wildlife.

Robert Sommer

Robert Sommer – a leading scholar in this field – acknowledged Hediger's contributions when he published his iconic reference book *Personal Space* (Sommer, 1974). Sommer recognised that environmental psychology was based on research by anthropologists, psychologists, and ethologists who studied both animals and people. For example, Dutch ethologist Niko Tinbergen articulated a theory of human autism after a career studying birds. Konrad Lorenz (1963) theorised on the origins of human aggression based on his ethological studies of birds and reef fish. Both scientists shared with Karl Von Frisch, the Nobel Prize in Physiology and Medicine in 1973, at a time when the impact of ethology had peaked.

Influenced by Hediger's observation that zoo exhibits based on 'the cube' were essentially un-biological, Sommer identified key features of zoos that qualified as 'hard architecture'. He contrasted these prototypical features with their antithesis, 'soft architecture'. A parallel movement in landscape architecture produced the school of design known as landscape immersion, where naturalistic exhibits merged the visitors into the landscape, creating the illusion that the animal and the visitor were in the exhibit together. Naturalistic features were enriching to animals, while the perception of naturalism rendered the exhibit more acceptable to the visitor. In an essay published in *Natural History* (1972), Sommer asserted that zoos were teaching all the wrong ideas about animals. Living in restricted settings, socially deprived animals exhibited abnormal behaviour patterns and failed to socialise or reproduce normally. These flawed exhibits were examples of the educational deficiencies of traditional zoos.

In an early contribution to the journal *Zoo Biology*, Jon Coe (1985) affirmed that the perception of naturalistic design made the zoo experience more realistic. In this paper he argued that by presenting animals on a higher plane than the visitor, the animal is placed in a dominant rather than a subordinate position. He also suggested that exhibits that surrounded the animals with perimeter viewers were detrimental to their well-being. Surrounding and subordinating zoo animals befitted the category of hard architecture. Soft architecture, by contrast, offered both safe and elevated distance from people. This idea derived from Hediger's experience in safely managing a variety of species to prevent the flight or fight response. The ideas of Hediger and Sommer are fundamental to the science of animal welfare. Today's best design firms generate ideas from a storehouse of knowledge in the literature of animal behaviour.

John B Calhoun: The chaos of crowding

John B Calhoun (1966) engineered experimental rodent cities to examine the ontogeny of psychopathology. His 'mouse universe' was a project supported by the US National Institutes of Mental Health. Although the enclosure provided an abundance of food, water, and nesting materials, the colony abruptly ceased growing at day 600 of the experiment. In addition, there was a complete breakdown of behaviour and social structure. Females failed to reproduce and males adopted a solitary life style. Calhoun's research typifies psychological studies of crowding through animal models. In a study of a chimpanzee facility in both summer and winter (crowded) conditions, Nieuwenhuijsen and deWaal (1982) discovered that apes coped with crowding without an increase in aggression. However, a similar study in South African zoos (Duncan et al., 2013) reported increases in abnormal behaviour in response to crowding. Abnormal behaviour may be a common coping response in chimpanzees when they experience crowding. Because they are capable of coping, they are a more appropriate model for human behaviour than rats and mice.

Harry F Harlow

A student of CP Stone at Stanford, Harlow joined the faculty and spent his entire career at the University of Wisconsin. Because the psychology department had no animal facilities, he and his students began their research on monkeys at the tiny Vilas Park Zoo. Many of the monkeys were exhibited on islands, so his early work was largely observational. His first graduate student was Abraham Maslow who later gained fame as the founder of humanistic psychology. Towards the end of his training at Wisconsin, Maslow and Harlow collaborated on studies of monkey dominance and sexual behaviour at the Central Park Zoo in New York.

Harlow's greatest contribution to the science of animal welfare was his benchmark research on social deprivation (Harlow et al., 1965). In a series of experiments, Harlow and his collaborators demonstrated that social deprivation and isolation led to abnormal behaviours such as catatonic, self-injurious, and hyper-aggressive responses in social situations, and inadequate social, reproductive, and parental behaviour. Motherless monkeys, raised on mechanical surrogates, were incapable of establishing normal social relationships. While Harlow was vilified by animal rights groups for his cruel experiments, his findings greatly benefitted zoo primates as he identified the variables that controlled socialisation. Zoos of this period were populated by psychologically damaged individuals, created by neglect and ignorance. The first author (TLM) used this knowledge in making decisions about the future direction of Zoo Atlanta during its reformation period.

Examples of key institutional developments

A landscape revolution at Woodland Park Zoo

The first zoo in North America designed on principles of landscape immersion was Woodland Park. The design team of landscape architects at Jones & Jones company, led by lead architect, Jon Coe, and zoo director David Hancocks – also a trained architect, exposed their entire collection to simulations of natural habitats. The zoo's gorilla exhibit was revolutionary in its impact on more traditional zoo architecture. This exhibit was heavily planted with botanicals available to the animals. They were encouraged to climb in large shade trees within the exhibit. The former exhibit was barren with crude paintings of trees on the concrete walls of the cage. Opening in 1977, Woodland Park marked the beginning of the era of radical naturalistic design in zoos throughout the world. In 2000, Hancocks published a scathing critique of traditional zoos in his book *A Different Nature*. Pulling no punches, Hancocks upset many of his colleagues in the zoo world by identifying substandard exhibits and their deleterious effect on animal welfare. He was also critical of the approach taken by Professor Hal Markowitz (1982) who was employed by two zoological parks before he joined the faculty of San Francisco State University. Markowitz is an iconic contributor to the animal welfare literature having applied operant techniques to engineer manipulanda to occupy and enrich the lives of many zoo animals, including gibbons, mandrills, servals, polar bears, and elephants. Hancocks observed that naturalistic settings were disturbed by the mechanical tools that Markowitz deployed. At one point, Markowitz experimented with applications of operant devices in landscapes such as the Hilo (Hawaii) zoo. This debate between naturalistic design and mechanistic enrichment was not resolved until Forthman-Quick (1984) published an essay reviewing how operant technology in applied behaviour analysis and environmental design could work together. The difference between the two schools of thought is that Hancocks softened traditional zoo architecture with landscape innovations whereas Markowitz softened zoos by changing the behaviour of the resident animals. Applied behaviour analysis in zoos, pioneered by Markowitz, is an essential set of procedures and methodology supporting the scientific study of animal welfare. Although operant labs declined in universities for many years, zoos and aquariums have rediscovered the value of training for husbandry and medical intervention (Maple and Segura, 2015).

Zoo Atlanta's iconic ape

Gorilla exhibits led the way in reforming standards and practices to enhance animal welfare in zoos. Soon after Woodland Park's reformation, Zoo Atlanta built the first facilities to exhibit a population of gorillas. The Ford African Rain Forest exhibit displayed contiguous breeding units from animals loaned to the zoo from the Yerkes National Primate Research Center. Because the solitary male Willie B was a high priority in Atlanta, the addition of 13 lowland gorillas made it possible to provide for his re-socialisation after 27 years in isolation (Maple, 2021). By housing breeding units, the animals were provided with the strongest form of enrichment; a species-appropriate social group. From introductions to surplus females, Willie B began a socialisation process in 1989, culminating in five successive offspring. Because he was born in the wild, his genes were an important contribution to the genetic history of the species survival program of the Association of Zoos and Aquariums. The Bronx Zoo in New York and Chicago's Lincoln Park Zoo quickly followed with the construction of superior gorilla exhibits that also served as research centres. The research leader at Lincoln Park, Steve Ross, and his research team, have contributed dozens of publications to advance the science of animal welfare for non-human primates. A similar design revolution has powered many new ape exhibits in Europe, including some outstanding examples of chimpanzee facilities in Scotland, Holland, and Germany. At Arnhem's Burger Zoo in Holland, the chimpanzee group is so large that it is considered to be a colony. Large collections of chimpanzees have also been established in Los Angeles and Detroit. Exhibits such as these support the design principle that social opportunity is the most important variable in naturalistic design.

Experience with captive Gorilla groups has produced a series of discoveries about best practices. Gartland et al. (2021) demonstrated that overnight separation protects bachelor gorillas from wounding by peers. Most zoos separate silverback males, but allow females to sleep with offspring. The investigators determined that separation was unnecessary for the first six years of bachelor group formation. Because so many zoos exhibit bachelor groups, sleeping arrangements require great care and planning once the animals reach puberty.

Current and future areas of importance

As illustrated in the previous section review, many issues relating to animal welfare have a long history. Despite great progress in many respects, there continue to be critically important areas that zoos will face moving forward. Here we describe some of these issues that zoos should proactively address bringing into account the history, new technologies, funding, shifting public opinions, and more.

Psychological constructs

A new construct that expands the scope of animal welfare, is wellness (Maple and Bocian, 2013; Maple, 2019). Like any psychological construct, the method through which one defines and measures a construct is of critical importance and will be benefited by close associations between zoos and academics studying such issues. Wellness is multifaceted and is the equivalent of optimal animal welfare. When wellness is achieved, animals tend to thrive rather than merely cope with captivity. Coping is widely acknowledged as an acceptable outcome, but thriving is a much higher standard. Although wellness is ubiquitous in the human potential literature, the first use of the term in animal studies was a publication by Fritz and Howell (1993). Wellness programs in zoos combine wellness research and wellness services.

In a recent publication, Prinzing (2021) identified two approaches to the study of wellbeing. The first is largely practised by philosophers who emphasise the importance of normative theory. By contrast, social scientists carry out research to explore the causes and consequences of wellbeing. As an alternative to these two approaches, Prinzing proposed a conceptual engineering strategy to bring the two together whereby normative theorising, empirical investigation, and conceptual revision articulate optimal concepts of psychological wellbeing. Since wellbeing is a construct that applies to both animals and human beings, a unified theory would provide an explanatory framework for a large amount of comparative data. Dimensions of psychological wellbeing are essential components of animal welfare. Prinzing concluded his argument by asserting that wellbeing is normative. Theories of wellbeing will suggest how to measure the construct provided there is sufficient precision. The best way to identify the phenomenon of wellbeing, he argued, is to design and deploy it. Animal welfare standards depend on a valid definition of psychological wellbeing.

The most effective way to guarantee innovation is to encourage scientific partnerships between academic and zoological institutions. An atmosphere of debate, discussion, constructive criticism, and replication is necessary for a healthy dialogue about quality of life and animal welfare science. Because zoos and aquariums prefer to invest in service personnel, they are prepared to intervene to upgrade welfare deficiencies rather than spend their limited funding on research personnel. The most successful combinations of research and services require partnerships with universities. A good working example of such partnerships is the Leipzig model, where the zoo serves as a living laboratory for great ape research and exhibition supported by significant funding from the Max Planck Institute (Tomasello and Call, 1994). Standards dictated by a centralised authority tend to suppress new ideas, and therefore should be avoided, in our view.

Taxonomic bias

Burghardt (2013) and others (Robbins and Margulis, 2016) have argued that there is a distinct bias in the zoo animal welfare literature favouring mammals over birds, fish, and reptiles. Research is skewed towards social mammals, especially non-human primates. Among mammals there are also large gaps that ignore charismatic species such as rhinos and hippos. Hediger noted that rhinos need certain substrate features to keep their horns sharp. Rhinos in barren exhibits often mutilate their horns. Hippos may be one of the most neglected of the African mammals in zoos as they are invariably exhibited without riverine water sources that would normally facilitate their social tendencies. In North America, only Disney's Animal Kingdom has simulated an appropriate riverine habitat for hippos while other zoos provide small, substandard dump and fill pools. Blowers et al. (2012) studied space use in female hippos in Disney's naturalistic, riverine exhibit. The Disney animals preferred certain sections of the river that met their depth requirements. Apparently, hippos conserve energy by resting in shallower waters. Tennant et al. (2018) suggested that most of the world's captive hippos suffer poor welfare. Normally active at night, both rhinos and hippos are typically confined in hard night quarters without access to grazing opportunities. Two closely related species where night access to the outdoors is becoming more common are Asian and African elephants. Singapore Open Zoo has experimented with night exhibition for a variety of species including elephants in order to provide entertainment for guests who purchase tickets to dine there at night. Night safari experiences are proliferating in North American zoos with warm-weather climates like San Diego. For an animal like the Asiatic fishing cat, access to the outdoors at night is clearly enriching and entertaining.

To demonstrate the power of this effect, it is necessary to perform objective post occupancy evaluations (POEs). This methodology is initiated with unobtrusive observations of behaviour at night. Brockett et al. (1999) found that elephants could be permitted contact with peers in the night house when the standard protocol called for protective chaining. Elephants prefer to be socially active at night. POEs can be conducted on the animals and the human visitors. Ogden et al. (1990) and Ogden et al. (1993) examined the behavioural effects of new, naturalistic habitats at Zoo Atlanta and the San Diego Zoo respectively. Combining access to the outdoors 24/7 with devices that provide more control for zoo animals is the new pathway to optimal animal welfare. For example, zoo animals trained for this purpose can control ambient light, sounds, access to water and food, and images of their keepers working on their behalf. Markowitz pioneered this type of self-control when he provided an automatic car wash system that elephants could operate to take showers on demand.

Taxa that suffer in zoos

In addition to a bias concerning which species are studied, there are likely differences in how different taxa respond to or are treated in captivity. Warwick et al. (2019) observed that captive snakes are commonly confined in small enclosures with dimensions that prevent occupants from adopting straight line body postures. Captive snakes may be the only vertebrates where management policy commonly involves deprivation of the welfare need to freely extend their bodies to natural full length. The investigators concluded that future policies for snake husbandry will require a paradigm shift from an erroneous belief system to acknowledge the greater spatial needs of reptiles. This problem is exacerbated with larger snakes such as the anaconda, pythons, and other constrictors. Since pythons are trainable, it should be possible to design contingencies of reinforcement to shape movement in these largely sedentary animals. This issue should be addressed in the accreditation standards of regional associations.

The great size and mobility of elephants and orcas present a challenge to institutions that attempt to confine them for exhibition. Animal rights groups in North America have repeatedly attempted to translocate suffering elephants from dilapidated zoos to spacious sanctuaries. These protests in some cases produced better outcomes when the targeted zoos re-imagined and rebuilt superior exhibits. Renaissance exhibits include zoos in Atlanta, Cleveland, Dallas, Denver, San Diego, and the Smithsonian National Zoo. Dublin Zoo in Ireland is considered to be a model for enlightened management of elephants. In each case, the science of animal welfare has informed a functional, landscape immersive effect to benefit elephants. Several conferences organised by stakeholders preceded design innovations. Persuasive, evidence-based essays have appeared in Forthman et al. (2009) and Wemmer and Christen (2008). In their chapter in Forthman et al. (2009), Maple et al. compared elephant welfare to the history of non-human primate welfare.

Public criticism of orca exhibits and shows at SeaWorld led to a prohibition of trainers entering the water with orcas (see Anderson et al., 2016, for a review). Business losses associated with the lack of public confidence have damaged the financial standing of the company, and the effects reverberated throughout the industry. Although Asian aquatic parks continue to import large specimens of whale sharks, orcas, and beluga whales, animal rights groups are monitoring the mortality and morbidity data for these facilities. Asian aquariums have been the last to acknowledge the ethical lapses that have decimated wild populations. Expeditions to capture whales are stressful for the entire group under attack. Injuries are common and some animals are killed. The connection between animal welfare and conservation has never been more evident. Associations and Governments are becoming more vigilant in their efforts to protect wild populations from exploitation by unethical aquarium operators. Several major transactions involving North American aquariums have been declined by responsible government agencies. Thankfully, it is becoming quite difficult to acquire a beluga, orca, or whale shark.

Vertical space

Some taxa might be particularly challenging to stimulate in a captive environment given natural history characteristics such as a high degree of arboreality. In 1995, the Smithsonian National Zoo experimented with expansions into vertical space with their O-Line travel system for orangutans. The apes are able to brachiate 50 feet above the heads of visitors and move away from their exhibits to adjacent platforms. This system has operated successfully for three decades (Molotsky, 1995). There are now many other zoos that have installed vertical pathways for apes and other species. The Jacksonville Zoo built an exhibit that encourages tigers to travel away from their exhibit through arboreal pathways. This exhibit was created to fulfil wellness requirements dictated by the zoo's commitment to enhancements in animal welfare standards and practices. The exhibit won AZA's 2018 competition for significant achievement in design. Travel systems have also been installed in the new African Primate Exhibit for bonobos, gorillas, and mandrills. The Philadelphia Zoo has designed extensive travel systems for a variety of animals. These pathways are utilised throughout the day and sometimes during night safaris. The discovery of leopards in an elevated travel tube is eerily similar to the behaviour of leopards in African forests. Given the investments that support expansions of useable space in zoos, Browning and Maple (2019) presented a formula for calculating usable, spatial volumes in complex exhibits. It is no longer sufficient to measure exhibits by area. Calculations in cubic feet are more descriptive.

Global perspectives on welfare

The importance of good animal welfare is not universally accepted. There is resistance that must be overcome by dialogue and intervention. Compliance is the norm in North America and Europe, but zoological facilities in some Asian nations tend to be focused on entertainment rather than conservation and animal welfare. Although the giant panda is revered in China and wherever it has been exhibited, other species, both domestic and wild, have not been well treated in China (Watts and Han, 2010). The abuse of Asiatic black bears in service for their bile (a traditional Chinese medicine) is deplorable. Li (2021) suggested that Chinese exhibitions of giant pandas at the Chengdu breeding centre is too labour-intensive. Like other places that practise animal tourism, cubs in Chengdu are produced *en masse* and utilised as revenue (Li, 2021). In the context of aquariums, the elicit capture of beluga whales, dolphins, orcas, and whale sharks continues to populate the many aquatic facilities that have proliferated around the world. Advocating for global changes requires a degree of cultural sensitivity. This is especially true because many of the practices that can be criticised as oppositional to animal welfare were recently practised in a Western context as well, but advocating for such change is important in an increasingly interconnected and global world.

Enrichment in many forms

Training animals may also yield an enriching effect. Melfi (2013) concluded that training is essential for some species but for others it should be avoided. Of five hypotheses tested by this investigator, only access to learning opportunities were strongly supported by the data. Melfi classified training as a husbandry technique that needed further evaluation on a cost–benefit basis. By contrast, practitioners of applied behaviour analysis agree that training is, in fact, enriching. Fernandez and Timberlake (2019) utilised foraging devices to encourage species-specific feeding behaviour and eliminate stereotypies in walruses. In this case, training and enrichment was the right combination to change undesirable behaviour. According to Fernandez (personal communication), animal training has the benefit of giving animals the ability to choose if they will be involved in any procedure. Trained animals in a zoo have choice and control over their environment. By this definition, training is certainly enriching and liberating.

Another avenue for continued exploration is how technology can be used to serve as enrichment for animals and improve welfare. Non-human primates are innately interested in manipulating objects. Wilson (1983) reported that great apes were influenced by objects in their immediate environment that were moveable. For many years, Washburn (2015) and his colleagues have deployed joysticks for monkey and ape communication and enrichment. The animals are easily trained on these manipulanda. Similarly, orangutans at the Smithsonian National Zoo have mastered the use of occupational devices such as iPads and apps (Raghavendran, 2013). In Queensland, Australia, Julia Hoy has pioneered enrichment and husbandry systems that can dispense food, toys, and medicine depending on the needs of microchipped animals. The technology is able to recognise individual animals by their unique imbedded microchip (Hoy et al, 2010). Computer work stations can be installed in most exhibits if they can resist the strength of animals such as elephants, gorillas, and chimpanzees. Perdue (2016) developed a computerised touchscreen system for use with sun bears, and other research suggests that observing animals engaged with cognitive tasks might be beneficial to the visitor experience (Perdue et al., 2012). A work station at the Jacksonville Zoo and Gardens is connected to a large artificial Kapok tree branded as the Wellness Tree. Data are continuously collected from a population of bonobos, gorillas, and mandrills in a mixed species exhibit (Maple, 2019). Computer technology represents a profound breakthrough in animal welfare science. It is adaptable to almost any species. Australian zoos in Melbourne and Sydney have led the way in advancing computerised solutions to confinement. However, these exciting possibilities must also be met with adequate support and staffing for equipment maintenance and upkeep. Further, zoos should consider the potential negative effects that an overload of technology might have on zoo visitors who might expect a zoo visit to be an escape from the technology of everyday life and a visit to a natural setting.

Conclusions

We have presented a number of current and future issues, though far from a complete list, that merit deep consideration in regards to zoo animal welfare. The recommendations laid out here are that the efforts of zoos should remain deeply rooted in evidence-based practice. In addition, zoo staff should proactively seek opportunities to openly educate the general public about the zoo's perspective on the complex issue of animal welfare, and to broadly advocate for evidence-based policies and processes.

Evidence-based standards and practices

Returning briefly to a historical perspective, which remains relevant today, Hediger (1950) asserted that research is always the last priority in the zoo. As an esteemed Professor of Ethology at the University of Zurich, Hediger's commitment to research conflicted with the priorities of his governing board. With meagre resources, he still mentored many young zoo biologists through their doctoral degrees and careers carried out in zoos, universities, and field settings. Hans Kummer (1995) is the best example of Hediger's mentorship. After a career in research, Kummer replaced Hediger on the faculty. He described his early research on Hamadryas baboons at Zoo Zurich in his 1995 book published by Princeton University Press. His observations and experiments at the zoo informed his field work on this species in Ethiopia. He was a distinguished contributor to primatology who also worked at the Delta Primate Research Center in the United States. His primer on primate social behaviour (Kummer, 1971) had great impact on a generation of zoo biologists. Oddly, when we met with him in Zurich in 1988, he convened a seminar with his students to argue that it was impossible to conduct good research in zoos. His many publications did not support this claim, so we can only conclude he was using the occasion as a heuristic device. The baboons at Zoo Zurich were living in a simulated colony and enjoyed optimal welfare. The Zoo Zurich habitat provided the perfect setting to construct an ethogram for this species and prepare his students for their participation in demanding field studies.

Because animal welfare is now considered an equivalent priority to conservation and education, and animal welfare has been built on a foundation of research, scientific zoo biology must be regarded as a necessary prerequisite to the achievement of well-being. For this reason alone, accrediting bodies encourage the recruitment of scientific colleagues to dedicated staff positions or in scientific partnerships. A few zoos at their inception developed a reputation for their scientific programs, for example, London Zoological Society, founded in 1826, and the Smithsonian

National Zoo, created by Congress in 1989, were both originally dedicated to conservation and science (Maple and Bashaw, 2010).

Importance of accreditation, legislation, advocacy

Zoo leaders, curators, scientists, and educators deserve much of the credit for meeting the higher operating standards supported by the public. Associations and governments promulgated written guidelines that originated in the United Kingdom and the United States and were subsequently adopted by other zoological institutions throughout the world. One of the most powerful instruments for ensuring compliance is the process of accreditation. North American zoos and aquariums must apply for reaccreditation every five years when they are evaluated by their peers. The Association of Zoos and Aquariums, representing North American zoological institutions, has one of the strongest accreditation programs in the world. Unfortunately, some inferior programs are competing with AZA to enrol institutions with major deficiencies. Some AZA institutions have elected to support these alternatives rather than resist them. If accreditation in these alternative programs is easy to achieve, zoo animals will pay a heavy price. Accreditation should be challenging and it should evolve to meet the very highest standards and best professional practices. The quality of the accreditation program is evident by the lowest common denominator of those institutions enrolled. We have taken the position that zoos and aquariums accredited by AZA should not accept simultaneous accreditation by lesser accrediting bodies. Only the gold standard of AZA and other credible associations (e.g. EAZA; WAZA) can ensure compliance with acceptable norms.

Final words

To the extent possible, efforts to address the issues raised in this chapter should take a broad historical context, with policies and accreditation processes being developed in conjunction with empirical research, and appealing to the broad perspectives of the general public. These are all challenging – but increasingly critical – tasks for zoos to pursue now and in the future.

References

Anderson R, Waayers R, and Knight A, 2016. Orca behavior and subsequent aggression associated with oceanarium confinement. *Animals*, 6(8), pp. 1–16.

Blowers TW, Waterman JM, Kuhar CW, and Bettinger, TL, 2012. Female Nile hippopotamus space use in a naturalistic exhibit. *Zoo Biology*, 31, pp. 129–136.

Brambell FWR and Technical Committee to Enquire into the Welfare of Animals kept under Intensive Livestock Husbandry Systems, 1965. *Report of the Technical Committee to Enquire into the Welfare of Animals kept under Intensive Livestock Husbandry Systems*. London: HM Stationery Office.

Brockett R, Stoinski TS, Black J, Markowitz T, and Maple TL, 1999. Nocturnal behavior in a group of unchained African elephants. *Zoo Biology*, 18(2), pp. 101–109.

Browning H and Maple TL, 2019. Developing a metric of useable space for zoo exhibits. *Frontiers in Psychology*, 10, pp. 1–11.

Burghardt G, 2013. Environmental enrichment and cognitive complexity in reptiles and amphibians: concepts, review, and implications for captive populations. *Applied Animal Welfare Science*, 147, pp. 286–298.

Calhoun JB, 1966. The role of space in animal sociology. *Journal of Social Issues*, 22, pp. 46–59. DOI:10.1111/j.1540-4560.1966.tb00548.x

Coe JC, 1985. Design and perception: making the zoo experience real. *Zoo Biology*, 4, pp. 2197–208.

Duncan LM, Jones MA, van Lierop M, and Pillay N, 2013. Chimpanzees use multiple strategies to limit aggression and stress during spatial density changes. *Applied Animal Behaviour Science*, 147(1–2), pp. 159–171.

Fernandez E and Timberlake W, 2019. Foraging devices as enrichment in captive walruses (*Odobenus rosmarus*). *Behavioural Processes*, 168, p. 103943. DOI:10.1016/j.beproc.2019.103943
Forthman D, Kane LF, Hancocks D, and Waldau PF (Eds), 2009. *An Elephant in the Room: The Science and Well-being of Elephants in Captivity*. North Grafton, MA: Tufts University.
Forthman-Quick D, 1984. An integrative approach to environmental engineering in zoos. *Zoo Biology*, 3, pp. 29–43.
Fritz J and Howell SM, 1993. Psychological wellness for captive chimpanzees: an evaluative program. *Humane Innovations and Alternatives*, 7, pp. 426–434.
Gartland KN, Carigan J, and White FJ, 2021. Preliminary relationship between overnight separation and wounding in bachelor groups of western lowland gorillas (Gorilla gorilla gorilla). *Applied Animal Behaviour Science*, 241, p. 105388.
Harlow HF, Dodsworth RO, and Harlow MK, 1965. Total social isolation in monkeys. *Proceedings of the National Academy of Sciences of the United States of America*, 54(1), p. 90.
Hediger, H, 1950. *Wild Animals in Captivity*. London: Butterworths.
Hoy JM, Murray PJ, and Tribe A, 2010. Thirty years later: enrichment practices for captive mammals. *Zoo Biology*, 29(3), pp. 303–316.h
Kummer H, 1971. *Primate Societies*. New York: Routledge.
Kummer H, 1995. *In Quest of the Sacred Baboon*. Princeton: Princeton University Press.
Li PJ, 2021. *Animal Welfare in China*. Sydney, Australia: Sydney University Press.
Lorenz K, 1963, *On Aggression*. London: Methuen.
MacKinnon D, 2009. 'Amusements are provided': asylum entertainment and recreation in Australia and New Zealand c. 1860–c. 1945. In *Permeable Walls*. Leiden, Netherlands: Brill, pp. 267–288.
Maple TL, 2019. Zoo. In Vonk, J and Shackelford, K, Eds, *Encyclopedia of Animal Cognition and Behavior*. Cham: Springer-Verlag. DOI:10.1007/978-3-319-47829-6_389-1.
Maple TL, 2021. *Atlanta's Iconic Ape: The Story of Willie B*. Fernandina Beach: Palmetto/Red Leaf Press.
Maple TL and Bashaw M, 2010. Trends in zoo research. Chapter 24 in Kleiman D et al., Eds, *Wild Animals in Captivity*, Vol. 2. Chicago: University of Chicago Press, pp. 288–298.
Maple TL and Bocian D, 2013. Wellness as welfare. *Zoo Biology*, 32(4), pp. 364–365.
Maple TL and Perdue BM, 2013. *Zoo Animal Welfare*. Berlin/Heidelberg: Springer-Verlag.
Maple TL and Segura V, 2015. Advancing behavioral analysis in zoos and aquariums. *The Behavior Analyst*, 38(1), pp. 77–91.
Markowitz H, 1982. *Behavioral Enrichment in the Zoo*. New York: Van Nostrand Reinhold.
Melfi V, 2013. Is training zoo animals enriching? *Applied Animal Behaviour Science*, 147, pp. 299–305.
Mellor DJ, Beausoleil NJ, Littlewood KE, McLean AN, McGreevy PD, Jones B, and Wilkins C, 2020. The 2020 five domains model: including human–animal interactions in assessments of animal welfare. *Animals*, 10(10), p. 1870.
Molotsky I, 1995. National zoo puts six orangutans to work in a high-wire act. *New York Times, August, 21*, p. A7.
Nieuwenhuijsen K and de Waal FBM, 1982. Effects of spatial crowding on social behavior in a chimpanzee colony. *Zoo Biology*, 1(1), pp. 5–28.
Ogden JJ, Finlay TW, and Maple TL, 1990. Gorilla adaptations to naturalistic environments. *Zoo Biology*, 9(2), pp. 107–121.
Ogden JJ, Lindburg DG, and Maple TL, 1993. Preferences for structural environmental features in captive lowland gorillas. *Zoo Biology*, 12(4), pp. 381–396.
Perdue BM, 2016. The effect of computerized testing on sun bear behavior and enrichment preferences. *Behavioral Sciences*, 6(4), p. 19.
Perdue BM, Clay AW, Gaalema DE, Maple TL, and Stoinski TS, 2012. Technology at the zoo: the influence of a touchscreen computer on orangutans and zoo visitors. *Zoo Biology*, 31(1), pp. 27–39.
Prinzing M, 2021. How to study well-being: a proposal for the integration of philosophy with science. *Review of General Psychology*, 25(2), pp. 152–162. DOI:10.1177/10892680211002443.
Raghavendran B, 2013. At National Zoo, even orangutans have latest high-tech gadgets. *Washington Times*, February 1, https://article.wn.com/view/2013/02/01/At_National_Zoo_even_orangutans_have_latest_hightech_gadgets_o/ (accessed 01 Nov. 2021).
Robbins L and Margulis SW, 2016. Music for the birds: effects of auditory enrichment on captive bird species. *Zoo Biology*, pp. 1–6. DOI:10.1002/zoo.21260.
Sommer R, 1972. What do we learn at the zoo? *Natural History*, 81(7), pp. 26–29, pp. 84–85.
Sommer R, 1974. *Tight Spaces*. Englewood Cliffs: Prentice-Hall.

Sommer R, 2008. The development of the zoological garden and the mental hospital. *American Journal of Orthopsychiatry*, 78, pp. 378–382.

Tennant K, Morris M, Segura V, Denninger-Snyder K, Bocian D, Lee GH, and Maple TL, 2018. Achieving optimal welfare for the Nile hippopotamus in North American zoos and aquariums. *Behavioural Processes*, 156, pp. 51–57.

Tomasello M and Call J, 1994. Social cognition in monkeys and apes. *American Journal of Physical Anthropology*, 37(S19), pp. 272–305.

Warwick C, Arena P, and Steedman C, 2019. Spatial considerations for captive snakes. *Journal of Veterinary Behavior*, 30, pp. 37–48. DOI:10.1016/j.jveb.2018.12.006.

Washburn DA, 2015. The four C's of psychological wellbeing: lessons from three decades of computer-based environmental enrichment. *Animal Behavior and Cognition*, 2(3), pp. 218–232. DOI:10.12966/abc.08.02.2015.

Watts J and Han Y, 2010. Chinese zoo closed amid tiger starvation investigation. *Guardian*, 17 March. https://www.theguardian.com/world/2010/mar/17/chinese-zoo-tiger (accessed 01 Nov. 2021).

Wemmer C and Christen CA, 2008. *Elephants and Ethics: Toward a Morality of Coexistence.* Baltimore: Johns Hopkins University Press.

Wilson PJ, 1983. *Man, the Promising Primate: The Conditions of Human Evolution.* Yale University Press.

16

HUNTING, FISHING, AND WHALING

Laetitia Nunny and Mark P Simmonds

Introduction

Hunting, trapping, fishing, and whaling can all impact wild animal welfare. They are undertaken for a variety of reasons including:

- For food and other products, such as fur, either as part of a subsistence or commercial hunt;
- To stop wild animals from competing with humans for a resource (competition may be proven or perceived);
- To control wild animals around livestock/crops;
- To prevent carnivores from posing a physical threat;
- As a form of population control or culling, for example pest or invasive species control or to prevent the spread of disease;
- For recreation/sport;
- For cultural reasons (Hampton et al., 2016; Hampton and Hyndman, 2019; Feber et al., 2020; Nunny, 2020).

When assessing the welfare of an animal during a hunting, trapping, fishing, or whaling event, the time to death (TTD) can be measured but it is generally considered more appropriate to measure the time to insensibility or time to irreversible unconsciousness (TIU) (Nunny, 2020). Animals that are killed in a manner that causes them to lose consciousness instantaneously so that they become insensible to pain and distress, therefore, do not experience any negative welfare after loss of consciousness (Broom, 1999). If the animal does not experience poor welfare just before the start of the killing procedure, such a death can be considered 'humane'. Sharp and Saunders (2011) recommend an assessment method based on Mellor and Reid's Five Domains Model, which can be used to assess the effects on welfare of a hunting or trapping method by considering the severity of any poor welfare before death and the duration of the poor welfare as well as the killing method itself (if death is the ultimate goal and, in some cases, e.g. catch-and-release fishing, it may not be).

DOI: 10.4324/9781003182351-19

Hunting and trapping

We hunt animals on land, on ice, at sea, and in the air. We hunt them from solid ground and from moving vehicles including helicopters, motor vehicles, and boats. We use firearms, harpoons, bows and arrows, or we catch them in traps or nets which either kill them or hold them until we can despatch them. We pursue them with dogs or birds of prey and we lure them with decoys. Depending on the species being hunted and the country where the hunt is taking place, legislation and codes of practice will differ and with them the methods used. Here, some methods and their welfare implications are described.

Hunting methods

Firearms, including centrefire and rimfire rifles, shotguns, and pistols, are one of the most widely used methods for culling wildlife and for commercial harvesting (Hampton et al., 2021). Wildlife regulations may specify which bullet weights and calibres should be used for targeting particular species, but these can vary between countries, for example see Nunny et al. (2018) for the range of firearms and ammunition used throughout Europe to kill seals. Other factors which can differ and which can impact welfare outcomes include the distances from which animals are shot and the area of the animal's body that is targeted (Hampton et al., 2021). These disparities may affect the welfare outcome for the individual animal. An assessment of feral camel (*Camelus dromedarius*) management in Australia, for example, determined that when camels were shot in the chest the outcome was less humane than when they were shot in the head (Hampton et al., 2016).

Bullets kill either by causing trauma to the central nervous system which leads to irreversible unconsciousness, or by causing fatal haemorrhage, when major blood vessels are impacted, or when major organs are lethally damaged (Stokke et al., 2018; Hampton et al., 2021). Most bullets used in hunting are designed to expand upon impact (Stokke et al., 2018) and although kinetic energy (determined by velocity and mass of a bullet) influences the likelihood of killing or injuring an animal, it is how a bullet behaves, including fragmentation, that can be more relevant for determining animal welfare outcomes (Hampton et al., 2021).

It is argued that killing an animal with a firearm allows it to carry out normal behaviours up until the moment of death, meaning that it has the potential to experience positive welfare until the last moment (Gamborg et al., 2020). When deer are undisturbed before being shot and killed instantly they have lower cortisol levels (a stress indicator) than deer killed after trauma (Gentsch et al., 2018).

Trauma could include pursuit of the animal before the shot is taken. Hunts which involve chasing an animal, on horseback and/or with dogs, mean that a target animal may experience negative welfare for hours or even days depending on how long it is pursued (Jones and Draper, 2018). Bateson and Bradshaw (1997) found that red deer (*Cervus elaphus*) hunted with dogs were chased for an average of 19 km and for a mean duration of 3.12 hours. These long hunts led to several physiological outcomes including high concentrations of cortisol and the authors concluded that red deer are not well-adapted to cope with the activity levels required during a hunt with hounds, and that long hunts lead to extreme exhaustion.

A report carried out for the UK government found that stalking was preferable to chasing a deer with dogs in terms of animal welfare (Burns et al., 2000). The same report also found that when a fox is caught and killed by hounds it "seriously compromises the welfare of the fox". The digging out of foxes from their dens was also considered to seriously impact welfare as the fox has no escape and the process can take a considerable time. The UK has subsequently banned hunting with dogs (Hunting Act, 2004) but the practice is still common in other countries, e.g.

in Australia hunting dogs are used to track and restrain feral pigs (*Sus scrofa*), which are then killed with a large knife (Orr et al., 2019).

Other hunting methods may be species-specific such as the use of a club or hakapik (a club with a metal ferrule which has a spike on one side) for killing young harp seals (*Pagophilus groenlandicus*) (Daoust and Caraguel, 2012). The animal is struck on the head with the club or blunt part of the hakapik, checked to make sure that it has been properly stunned by palpating the skull to ensure that it has been crushed, and then bled out to ensure death. Although this method of killing has been controversial, the sealer is, in theory, able to check the seal's state of consciousness instantly, whereas for seals which are shot from a distance on the ice there may be some delay before they are checked. Checking is even harder if the seal is shot in water, and Daoust and Caraguel (2012) found that shooting a seal in water meant a 30% risk of a poor welfare outcome, whereas for a seal shot on the ice there was only a 2.6% chance. The location of the hunt may, therefore, be as important as the method used in some circumstances.

Hunting methods which are considered traditional, or which are undertaken by indigenous people, are often not scrutinised scientifically for their impacts on animal welfare (Hampton and Hyndman, 2019). No animal-based welfare studies have been published regarding dugong (*Dugong dugon*) hunting in Australia, for example, despite concerns about welfare outcomes, because one of the killing methods includes drowning (Hampton and Hyndman, 2019) – which is generally considered to have severely negative welfare consequences (Ludders et al., 1999).

Trapping

Worldwide, tens of millions of animals are trapped legally each year and an unknown number are trapped illegally (Iossa et al., 2007). Killing traps aim to render the animal unconscious and kill it, whereas restraining traps hold the animal until the trapper checks the trap and kills or releases it. As many traps were developed by trappers to capture furbearers, the main aim of the trap was to capture the animal without damaging the pelt. Welfare was at best a secondary consideration.

Leghold traps are prohibited in many countries including throughout the European Union (European Communities, 1991) but are legal in Canada and many states of USA (Iossa et al., 2007). In Australia they may be combined with toxins which are applied to the trap jaws to kill the animals once trapped (Meek et al., 2019). This practice can have negative welfare implications depending on the toxin used. Strychnine, for example, intoxicates the animal while it is still conscious and is considered an inhumane poison due to the severe welfare impacts (Sharp and Saunders, 2011). It has been banned in many countries, e.g. New Zealand and the UK.

The 1999 Agreement on International Humane Trapping Standards (AIHTS) banned steel-jawed leghold traps in the European Union, Canada and Russia (European Communities, 1998). The AIHTS have been criticised for not reflecting the latest trapping technology, for omitting commonly used traps and commonly trapped species, and for not including assessments to ensure that animal welfare is properly protected (Proulx et al., 2020). The TIUs given in the AIHTS are, in some cases, considered to be too long and could be reduced for many species if new technology and materials were implemented.

The International Organization for Standardization (ISO) standards 10990-4 1999 ("Methods for testing killing-trap systems used on land or underwater") can be used to evaluate traps for animal welfare, capture efficiency, selectivity, and user safety (ISO, 1999). Tests are carried out with anaesthetised animals, although they also recommend testing on conscious animals as the effects of trap forces can vary. Few traps have actually been tested according to ISO standards and how traps perform in experimental testing circumstances can be different to how they perform in the field (Iossa et al., 2007).

Due to their indiscriminate nature, cable snares are prohibited in many places including most central African nations, but Noss (1997) found that they were regularly in use in the Central African Republic and that foot snares had clear welfare consequences. Over one-third of animals caught escaped from the snare injured, often with a severed limb. The animals that did not escape would fight to free themselves, often breaking the captured limb and dying of shock, blood loss, exhaustion, and starvation.

How a trap impacts welfare will, in part, depend on how often the trap is checked by the hunter. The legal requirement for checking leg-hold jaw traps used for trapping dingoes (*Canis lupus dingo*) and wild dogs (*Canis lupus familiaris*) in Victoria, Australia, is only every three days, whereas the international recommendation for checking soft-jaw traps in research projects is daily, and many researchers check more frequently (Hampton and Hyndman, 2019). There are no legal requirements regarding trap and snare checking times in Canada, and in the USA times vary according to state, often exceeding 24 hours, leading to the recommendation that all killing traps, including those certified as 'humane', should be monitored frequently, ideally every 12 hours (Proulx and Rodtka, 2019).

Trophy hunting

Trophy hunting often involves the use of methods which are less likely to result in a quick death for the target animal, e.g. bows and arrows, and many trophy hunters are not expert shots (Jones and Draper, 2018). As part of the animal's body (often the head) will be retained as a trophy, this area is not targeted to avoid damaging it, meaning that the weapon and/or target area is not necessarily chosen with the aim of minimising negative welfare consequences. The killing of Cecil the Zimbabwean lion (*Panthera leo*) by a trophy hunter in 2015 sparked intense public interest because he was part of an ongoing research project and had been lured out of a national park to an area where hunting was permitted. Cecil was initially shot and wounded with a bow and arrow before being killed many hours later.

The practice of canned hunting, where animals are bred to be shot by trophy hunters, raises several concerns including the poor conditions in which they are kept (Feber et al., 2020) and the hand-rearing of cubs so that the mothers will be ready to breed again in a short period of time (Jones and Draper, 2018). Tourists are sometimes allowed to interact with these cubs – an activity which also has negative welfare consequences (Hunter et al., 2012).

When animals are bred and raised to be hunted, their welfare at other points of their lives needs to be considered (not just the moment when they are hunted). In the UK, over 40 million pheasants (*Phasianus colchicus*) are bred and released to be hunted each year (Feber et al., 2020). Although welfare guidelines for their rearing are available (DEFRA, 2010), these birds are susceptible to starvation, disease, predation, and roadkill (Feber et al., 2020).

Wounding

If an animal is not killed outright but is wounded by a shot, for example, then its welfare will be negatively impacted to some degree and for some duration. A seriously wounded animal could die relatively quickly (after a few minutes or hours) whereas an animal with a less serious wound could live for several days or weeks (Fox et al., 2005) in pain and with difficulties carrying out necessary behaviours such as foraging. The animal may also suffer sickness, discomfort, and psychological effects. Welfare may be very poor before the animal finally succumbs (Broom, 1999). It is, therefore, important that hunters are able to determine when an animal has been wounded, so that they can locate it and humanely kill it. Stokke et al. (2018) proposed a way to assess animal welfare outcomes in the field during the hunting of terrestrial mammals based on body mass and flight distance. Such wounding thresholds may be affected by factors such as terrain, weather conditions and animal stress levels.

Aebischer et al. (2014) assessed wounding of deer shot with rifles. Of 2,281 first shots, 4.5% resulted in a clean miss, 88.8% resulted in the deer being killed instantly and 6.7% wounded the deer. Of the wounded animals, 81.7% were killed with a subsequent shot and 18.3% escaped. The authors, therefore, estimated wounding rates of 1–12%, although a worst-case scenario (where apparently missed animals were actually wounded) would give wounding rates of 3–17%. Bow hunting has been associated with high rates of wounding and increases the risk of a slow and painful death (Gamborg et al., 2020). In two white-tailed deer (*Odocoileus virginianus*) hunts in the USA, there was an 18% wounding rate for animals shot with modern compound bows or crossbows (Pedersen et al., 2008) and a 50% wounding rate for those shot using traditional archery equipment (recurve and longbows) (Ditchkoff et al., 1998).

Hunter skill level

Wounding and other negative welfare outcomes tend to be more common in recreational hunts, whereas professional culling techniques usually have very low nonfatal wounding rates (Hampton and Hyndman, 2019). However, Caudell et al. (2009) found that many people involved in wildlife management using firearms are not adequately trained.

Accuracy (how closely a projectile strikes to a target) and precision (the closeness of shots to each other) are important and can be affected by the stability of the shooting platform and the hunter's position (Hampton et al., 2021), and a number of best practice recommendations have been made, including regular hunter training (for example by Aesbischer et al., 2014). In Denmark and Norway, crippling rates of pink-footed geese (*Anser brachyrhynchus*) were reduced thanks to awareness campaigns and appropriate hunter training (Clausen et al., 2017).

Effects on non-target animals

To prevent young animals from being orphaned and starving, close seasons are imposed at particular times of year for certain species (Nunny, 2020). However, there are many examples of animals not being protected, e.g. in England and Wales there is currently no close season for hares (*Lepus europaeus*) meaning that young hares (leverets) can starve to death if their mothers are killed (Butterworth et al., 2017a). A close season is being considered by the British government (DEFRA, 2021).

The removal of an individual animal from a social group or population can have welfare consequences for other animals. For example, the killing of a male lion may lead to other males replacing him, often with the accompanying infanticide of cubs (Whitman et al., 2004), whilst the removal of a matriarch African elephant (*Loxodonta africana*) can have consequences for the rest of her group because she influences their social knowledge (McComb et al., 2001).

The welfare of domestic animals used in hunts also deserves consideration. In Australia, there is a lack of information regarding the health and welfare of hunting dogs (Orr et al., 2019). Many are trained using aversive training techniques such as electric shock collars which can increase anxiety and aggression, reduce motivation, and create other conditions (Australian Veterinary Association, 2014). Orr et al. (2019) expressed concern over the health of hunting dogs due to their poor living conditions and because they are exposed to diseases and parasites, injuries sustained during the hunt and increased risk of heat exhaustion, poisoning, being hit by vehicles, snake bite, accidental shooting, and dehydration.

Humane alternatives

If the aim of the hunting or trapping is to control a pest or predator, then there are many non-lethal alternatives available such as enclosures, livestock guarding, the use of deterrents and repellents, or translocation (see Nunny, 2020 for a discussion). Fertility control reduces or eliminates the need for lethal control and may have some welfare benefits by averting the physical risks and energetic costs of pregnancy, birth, and lactation for female animals (Gamborg et al., 2020). However, it may present some welfare challenges itself, for example, when fertility control drugs are administered to deer, the animals may experience fear and pain, and, depending on the drug given, behaviours such as rutting and mating may be suppressed or unnaturally extended. These behavioural changes could lead to negative welfare or, at least, prevent the possibility of positive welfare being experienced (if mating and raising offspring are considered to be rewarding experiences).

Fishing

Do fish feel pain?

Browman et al. (2019) discuss which aquatic animals experience pain and recommend that fish welfare assessments should be based on indicators of stress, health status and behaviour specific to the situation and species. Some authors, e.g. Rose et al. (2014) have concluded that from the behavioural and neurobiological evidence, fish have limited responses to nociceptive stimuli and are unlikely to experience pain. However, other experts assert that there is evidence that fish have nociceptive systems similar to those of mammals and that the behavioural and physiological changes they exhibit following potentially painful events are indicative of a pain response (Sneddon, 2020). Fish will avoid areas where they have experienced pain demonstrating cognitive engagement including learning and memory, and an emotional response (e.g. fear) to the negative stimuli (Vila Pouca and Brown, 2017; Sneddon, 2020). The painful event may be so consuming that they cease to exhibit normal fear or antipredator responses (Sneddon, 2020).

Recreational fishing

Although some fish are harvested and killed during recreational fishing, many more are released during catch-and-release angling (Cooke and Sneddon, 2007). Fish may be released because they are not the target species, because they are undesirable in some way (sex, size, food value) or because there is a regulation in place to preserve resources. There has been a lot of debate in recent years about the consequences of catch-and-release and it is generally recognised that all caught fish will experience some level of injury and stress (Brownscombe et al., 2017).

When a fish is caught with a hook the extent of injury or tissue damage will depend on where on the body the injury is and the type of fishing gear used (Cooke and Sneddon, 2007). Most fish are hooked in the jaw/mouth area, subsequently impacting respiration (ventilation), foraging and feeding, reproduction (e.g. in the case of mouth brooding), or social interactions. Eye injuries are also common, taking place during hooking and handling. Impaired vision has been linked to mortality and may affect behaviours such as foraging, predator avoidance and finding a mate.

Landing is a key moment regarding welfare outcomes as fish can be exposed to air and can be injured by the landing gear, handling by the angler, or from contact with other surfaces (e.g. boat, shore) (Brownscombe et al., 2017). Bluegill (*Lepomis macrochirus*) that were kept in a landing net for 30 seconds had increased fin abrasion compared to fish that were angled and held out of the water but which were not netted (Barthel et al., 2003). Scale and mucous loss also increased for netted fish, as did mortality rates post-treatment.

Lure type, size, and the number of hooks used all influence fish injury and handling time for dehooking indicating that lure choice is important for ensuring improved animal welfare outcomes (Clarke et al., 2021). Dehooking time depends on the number of hooks and the location of the hook on the fish's body. Hooks left in place in a released fish may subsequently be expelled or may remain embedded in the oesophagus, gut or throat leading to lesions and infections (Cooke and Sneddon, 2007). Further research is needed so that advice can be given to anglers about whether to remove or leave hooks.

Angler experience, knowledge, and skill can all impact fish welfare outcomes. When an angler is unprepared in terms of tools (including hook type, landing gear, unhooking tools) and knowledge then the fish is more likely to experience stress and injury (Brownscombe et al., 2017).

Killing methods

If a fish is bleeding, injured, deeply hooked, hooked in a vital organ, severely exhausted, or if the hook cannot be removed without causing significant damage, then experts recommend that the fish should be killed using a method which minimises stress and suffering prior to unconsciousness (Davie and Kopf, 2006). Davie and Kopf (2006) recommend percussive stunning of the cranium to cause immediate death or unconsciousness, followed by pithing, exsanguination, or decapitation to prevent the fish from recovering consciousness. They recommend bleeding-out for active fish such as *Salmonidae* which require high concentrations of oxygen, and suggest that pithing (only after stunning) may be more appropriate for less active fish, e.g. catfish (*Ictaluridae*), which can survive for longer with poor blood flow to the brain. Pithing (or spiking) involves inserting a rod into the cranial cavity to destroy the central nervous tissues. Bleeding-out without prior stunning can lead to fish showing aversive behaviour and results in a slow death (Robb and Kestin, 2002).

Other killing methods which are considered unacceptable on welfare grounds include hypothermia (through the use of ice slurries) and asphyxia (death caused by lack of oxygen) (Davie and Kopf, 2006). Fish placed in ice slurries or exposed to air die from hypoxia (oxygen deficiency). In higher ambient air temperatures, fish die more quickly and with increased stress levels, whereas fish in ice slurries die more slowly but with a lower physiological stress response. Hypoxic-tolerant fish will die more slowly than active fish species. Due to the slow TTD or TIU, ice slurries and asphyxia are not considered appropriate killing methods (Davie and Kopf, 2006).

Subsistence fishing

Subsistence fishers around the world use a number of different fishing methods including hook and line, purse seine nets, gillnets, longlines, dredging, traps, pots, spears or harpoons, and even dynamite (FAO, 2015). Different fishing methods have different welfare outcomes. Not all can be reviewed here but it is noted, for example, that gillnets can cause scale, skin, and fin injuries, stress, and asphyxiation, whilst fish caught on hooks will have injuries to their mouth, throat, or gut (Veldhuizen et al., 2018; Brown and Dorey, 2019). Fish caught at depth may exhibit pressure injuries, and greater capture depth and longer fishing duration results in more injuries across a range of gear types (Veldhuizen et al., 2018). The use of explosives not only has welfare consequences but severe environmental impacts as all fish and many other organisms in a 15–20 m radius are killed or injured and habitat, e.g. coral reef, is completely destroyed (Slade and Kalangahe, 2015).

Diggles et al. (2011) assert that for many subsistence fishers it would be financially, logistically, and even culturally impossible to adapt traditional fishing methods to take animal welfare into account.

Bycatch

Bycatch, whereby non-target species are caught incidentally by fishers, is recognised as a major problem in commercial fisheries but it also takes place in subsistence or artisanal fisheries and often at alarmingly high levels. For example, artisanal fisheries based at the port of Salaverry, Peru take approximately 2,412 small cetaceans as bycatch each year (Mangel et al., 2010), as well as leatherback turtles (*Dermochelys coriacea*) which are either released alive or are landed for human consumption (Alfaro-Shigueto et al., 2007).

The welfare consequences for bycaught animals may be severe with most bycaught dolphins asphyxiating and experiencing a period of extreme stress before they succumb (Dolman and Moore, 2017). Animals which escape or are released from nets may be injured and/or effected by stress leading to behavioural changes and physiological costs which can impact their welfare and survival. Bycaught loggerhead turtles (*Caretta caretta*) have been recorded as suffering from decompression sickness evidenced by gas embolisation (GE) after being caught in trawls and gillnets at depth (García-Párraga et al., 2014). Turtles with GE that are released alive, may die within hours or days of release. Non-target fish species taken as bycatch experience trauma whilst trapped in fishing gear or during hauling (Metcalfe, 2009). If the fish are released alive, their subsequent welfare and survival will depend on injuries and stress sustained when they were bycaught.

Whaling

Since 1990, people in at least 114 countries have taken one or more of at least 87 marine mammal species (Robards and Reeves, 2011). This is principally to provide food, although arguments about killing fish-eating mammals as a form of pest control prevail in many places. Japan has the largest-scale targeted takes and also takes the highest diversity of species (32).

Table 16.1 provides a summary of some of the better known cetacean hunts, outlining the species taken and the wide range of methods deployed. An idea of scale is provided by the figures for 2019 where available. Here we focus on two types of cetacean hunting that have been subject to critical scrutiny.

The International Whaling Commission (IWC) currently calculates and endorses quotas for member nations under its aboriginal subsistence whaling category and this form of whaling continued after the IWC's international moratorium on commercial whaling came into force in 1986. Norway took out a legal objection to the moratorium and, hence, legally whales "under objection". The IWC's 'Whale Killing Methods and Welfare Issues Working Group' considers welfare, including information related to whale hunting methods, and last met in 2018 (IWC, 2018). In recent years, several whaling nations have preferred to provide relevant information to the North Atlantic Marine Mammal Commission (NAMMCO) instead. NAMMCO's Committee on Hunting Methods was established in 1994 and it also hosts other relevant groups and publishes their reports on its website.

Table 16.1 The better-described cetacean hunts of recent years

Country	*Species*	*Killing methods*	*Notes*	*2019 take*
Canada	Bowhead whale, narwhal, beluga, harbour porpoise, white-beaked dolphin, Atlantic white-sided dolphin.	Darting gun and penthrite grenades for the bowheads. Rifles for the smaller species.	Ongoing hunts.	4 bowheads taken. Some 1,300 others.
Faroe Islands	Long-finned pilot whale, bottlenose dolphin, white-beaked dolphin and Atlantic white-sided dolphin.	Drive hunting: animals are herded into the shallows and then killed with sharp instruments.	Ongoing hunts.	682 long-finned pilot whales and 10 Atlantic white-sided dolphins
Greenland	Narwhal, beluga, common minke whale,* fin whale,* humpback whale,* bowhead whale,* harbour porpoise, long-finned pilot whale, Atlantic white-sided dolphin, white beaked dolphin, orca, northern bottlenose whale.	Penthrite grenades as primary and secondary method for the larger species and penthrite grenades as primary with high calibre rifles as the secondary killing method for the common minke whale, apart from the "collective" hunt for common minke whales which uses non-explosive harpoons and rifles. Other species are mainly taken with rifles of various calibre. Narwhal and beluga are hunted by harpoon from qayaqs and with rifles from small boats. In some places in northern and eastern Greenland, narwhal and beluga are also captured with nets.	Ongoing hunts.	536 narwhals, 265 beluga, 171 common minke whales, 8 fin whales, 4 humpback whales, 2,569 harbour porpoises, 285 long-finned pilot whales, 126 Atlantic white sided/ white-beaked dolphins, 31 orcas, 8 northern bottlenose whales.
Iceland	Common minke and fin whales.	Penthrite harpoon plus secondary killing methods when necessary.	No whaling in 2019 or 2020.	0

(Continued)

Table 16.1 (Continued)

Country	*Species*	*Killing methods*	*Notes*	*2019 take*
Japan	Sei whale, common minke whale, Bryde's whale.	Penthrite harpoon plus secondary killing methods as required.	Japan left the IWC in 2019 and after this has, to date, restricted its cetacean takes to its exclusive economic zone.	25 sei whales, 187 Bryde's whales and 123 common minkes taken.
	Baird's beaked whale, short-finned pilot whale ('northern' and 'southern' 'forms'), Risso's dolphin, false killer whale, striped dolphin, bottlenose dolphin, pantropical spotted dolphin, Pacific white-sided dolphin, Dall's porpoise ('Dalli type' and 'Truei-type'), rough-toothed dolphin, melon-headed whale.	'Small-type, whaling' (i.e. smaller vessels with smaller harpoon canons), hand-harpoon, drive hunting.		1,911 cetaceans were reported taken across all these species. 818 of these were Dall's porpoises (Truei-type) all of which were taken with hand harpoons. Some cetaceans are also taken in net traps. Whilst this is termed 'bycatch'; they appear to be a predictable component of the catch and are commercially sold.
Norway	Common minke whale.	Penthrite harpoon plus secondary killing methods when necessary.	Ongoing hunt.	429 taken including 2 lost.
Russian Federation	Gray whale,* beluga, harbour porpoise, long-finned pilot whale, southern bottlenose whale, Baird's beaked whale, orca.	For gray whales: Automatic weapons.	Ongoing hunts.	138 grays and 1 bowhead (taken by mistake). 2 lost grays. Less than 100/year other species.

(Continued)

Table 16.1 (Continued)

Country	*Species*	*Killing methods*	*Notes*	*2019 take*
St Vincent and the Grenadines	Humpback whales,* Atlantic spotted dolphin, short-finned pilot whale, spinner dolphin, Fraser's dolphin, orca, false killer whale, Risso's dolphin, rough-toothed dolphin, melon-headed whale, dwarf sperm whale, pygmy killer whale, common bottlenose dolphin, Gervais' beaked whale, Clymene dolphin, striped dolphin, pantropical spotted dolphin.	Harpoons.	Ongoing hunts.	3 humpbacks taken. More than 500 from other species taken annually.
USA	Bowhead whale,* beluga also taken.	For Bowheads: Penthrite projectile (darting gun) alone; the penthrite projectile (darting gun) with a black powder shoulder gun as a backup; or black powder (darting gun) with a black powder shoulder gun as a backup.	Ongoing hunts.	36 Bowhead taken. 6 struck and lost. Figures for belugas not found.

* indicates that these takes are classified as Aboriginal Subsistence Whaling and are granted a quota by the IWC.
Sources: NAMMCO, 2011, 2015; Altherr and Hodgins, 2018; IWC 2018; Minamikawa, 2020; IWC, 2021; Simmonds et al., 2021; Whaling.fo, 2021.

The hunting and killing weapons used for cetaceans (alone or in combination) are (NAMMCO, 2011, 2015; Øen, 2021):

- Cold harpoons (i.e. non-explosive) delivered by harpoon gun or by hand;
- Explosive grenades delivered by harpoon gun or darting gun (only used on large cetaceans);
- Firearms – rifles and various types of ammunition;
- Lances/spears and knives;
- Nets.

Vessels are used in many hunts to get close enough to strike or shoot and, in some hunts, to drive animals to shore. Cetaceans also need to be secured, so that they do not escape or sink once struck. This explains the prevalence of various types of harpoons with barbs or toggling claws that embed in the flesh of the animals and which are typically attached to an appropriately robust rope, allowing the animal to be winched in.

The three key factors related to the humaneness of cetacean hunting are: firstly, the actions leading up to the animals being killed, particularly whether they are chased or confined, both of which can cause stress. Secondly, the actual killing method deployed linked to the time taken until the animal is deemed to have died or be insensible prior to death. TTD and Instantaneous Death Rates (IDR) are key terms in the assessments. The rapid subduing of the animal is also an issue for the hunters' safety, given that these large animals may thrash around when struggling to escape or in their death throes. A further reason for trying to quickly render the animal immobile is that this reduces its opportunity to escape. This links to the third key attribute, the issue of "struck and lost", which typically refers to animals that have been hit with a harpoon or shot but then lost and a similar concept applies to animals that have been otherwise captured (e.g. driven ashore as part of a drive hunt) but were then either released or escaped.

Commercial take of large whales

The chase: large-scale whaling in Norway, Iceland, and Japan in recent years has been conducted from dedicated whaling vessels or modified multi-purpose fishing vessels, all of which use a harpoon tipped with a penthrite explosive-containing grenade, which is fired from a harpoon canon.

In the Norwegian hunt, there is reported to be no chase (Øen, 2021). The boat approaches the whale and manoeuvres to strike it from the side, rather than from behind or the front, which both offer smaller targets. By contrast, Japan's large whale hunting both chases the whale (apparently using sonar) and, also, tends to fire the harpoon from behind (NAMMCO, 2015). Both factors are bound to increase TTD and decrease IDR.

The kill: determining death or insensibility in mammals that can hold their breath for considerable periods of time is challenging. Additionally, reflex movements, such as the thrashing of the tail or flukes, may be exhibited for several minutes after loss of consciousness or even death. The grenade explodes when the harpoon has penetrated about 70 cm inside the whale and, in the Norwegian hunt, rifles of calibres .375 and .458 aimed at the whale's brain are used as a secondary killing method when the animal has not been killed outright (Øen, 2021). Norwegian gunners have to pass annual shooting tests with the harpoon canon and backup rifle and, from 2006 onward, the hunt has been monitored by an electronic trip recorder and "spot controls" in harbours. Studies in Norway have shown that the range, the size of the whale, and the angle of the shot relative to the animal's long axis all strongly influenced survival time.

The TTD data available from recent commercial hunts are in most cases far from comprehensive and, as differing methods may have been used to gain and analyse them, comparisons between hunts are difficult and may not be valid. However, here are some examples: in the Norwegian hunt, TTD data for 271 common minke whales (*Balaenoptera acutorostrata*) taken in 2011 and 2012 showed an IDR of 81.9% with an average TTD of 60 s (Øen, 2021). The median TTD for the 49 whales that were not recorded as instantly dead was 300 s, and one that had only been wounded was reshot after 20–25 minutes. Similarly, in a sample of Iceland's minke whale hunt, 9 (69%) of 13 whales taken in 2014 and 2015 were reported instantly dead (NAMMCO, 2015). The median survival time for the 4 whales that did not die instantly was 4 minutes, with one recorded as surviving for 13 minutes. In a sample from Iceland's fin whale (*Balaenoptera physalus*) hunt in 2014, 42 whales (84%) were recorded as instantaneously killed. The others (8) were re-shot with penthrite grenades, and their median survival time was 8 minutes, with the longest TTD being 15 minutes.

Data reported from Japan from 2010-2015 for sei whales (*Balaenoptera borealis*) show TTD of 2–4 minutes and an IDR of 48–60% depending on the year (NAMMCO, 2015). Data for Bryde's whales (*Balaenoptera brydei*) appear broadly similar. Information is not provided about the whales that did not instantaneously die.

Struck and lost: recent data on the very sensitive issue of recent struck and loss rates appears scant, but some examples are noted in Table 16.1.

Drive hunting in Taiji, Japan

The chase: the animals are herded by a coordinated group of fishing vessels deploying noise created by fishermen banging on trumpet-shaped metal poles. This generates powerful acoustic signals (greater than 170 dB) inducing flight, escape and avoidance behaviours (Vail et al., 2020). The primary sense of cetaceans is hearing and the deployment of loud noise is likely to be highly stressful.

The kill: The animals are driven into a bay which is then closed off with a net. Previously they were killed by lances that were thrown at them but, in 2000/2001, trials were started with a "spinal lance". This is a narrow metal blade inserted posterior to the blow hole and intended to sever the spinal cord and key blood vessels (NAMMCO, 2015; Butterworth et al., 2017b). This method is based on that used in the Faroe Islands and now seems to be the main killing method. In 2009, apparently because of adverse publicity linked to the reporting of clouds of blood in the water, the Taiji hunters modified their methods. A narrower lance/rod was used to minimise the area of the wound and a wooden plug was then inserted into the wound. This approach was criticised by the NAMMCO Expert Group which "emphasised that the process of bleeding out animals is part of the killing process and that it is a widely accepted principle both from an animal welfare point of view and from the point of view of meat quality". Butterworth et al. (2017b) analysed this method, concluding that whilst it would cause damage to the vertebral blood vessels and the complex nets of blood vessels around the head, leading to significant haemorrhage, it would not lead to a rapid death in a large mammal of this type. The method causes paraplegia and death through trauma and gradual blood loss. An animal that is paralysed would be less of a hazard to the hunters.

Once contained at the shoreline, the dolphins are sometimes held up to five days before they are either selected for slaughter or, in the case of a small minority, sold to dolphinaria, or they are released. Those to be killed are roped around their tails and towed backwards to the killing area. Vail et al. (2020) concluded that the prolonged and strenuous chase and herding, capture, and restraining of the dolphins in Taiji can result in acute stress and injury.

Lost and released animals: individuals, including calves, are sometimes released, probably so as not to violate quotas but this process involves their rough handling further to being stressed and exhausted from the chase. Calves are unlikely to survive and this may also be true of injured and heavily stressed adults.

In many instances, these highly social and intelligent animals will be aware of the herding, captivity and killing of the other members of their groups. Butterworth et al. (2017b) concluded "from a scientific, humane and ethical perspective, the treatment of dolphins in drive hunts sharply contradicts current animal welfare standards employed in most modern and technologically advanced societies". They also noted that 'termination of movement' is not a credible measure of death for a mammal, highlighting that in this instance it could be the result of severance of the spinal cord.

Consideration of these examples of cetacean hunting clearly shows effort to improve the efficiency and humaneness of whale hunting in Norway. Arguably the biggest step in this being when cold (non-explosive) harpoons were replaced with grenade harpoons in the 1980s, when IDR went from 17% to 44.8% (Øen, 2021). Other improvements followed but many would still argue that even the relative efficiency of the hunt in Norway still does not compare well with the methods used to kill farmed animals in abattoirs and clearly animals that are not killed outright or, even worse, struck and lost may have prolonged and painful deaths. Japan's whaling activities seem less transparent; its large whale takes are less efficient and its drive hunting raises highly significant welfare concerns.

The hunts focused on here are only a small part of the full picture. Approximately 100,000 small cetaceans (all the toothed cetaceans bar the sperm whale *Physeter macrocephalus*) are intentionally killed each year worldwide (Altherr and Hodgins, 2018). In most cases, these hunts are unregulated, or even illegal and poorly documented.

Conclusions

Compared to farmed animals, companion animals and animals used in research, there have been relatively few independent studies assessing how animal welfare is impacted in hunting, fishing, and whaling scenarios. Many welfare issues are identified here and the killing of cetaceans, in particular, is worrying, noting that many hunts are not documented at all or only poorly. Some authors, e.g. Nunny (2020); Proulx et al. (2020), have called for the development of appropriate and widely accepted animal welfare assessment approaches which can be used in predator control, trapping, and other hunting scenarios.

Acknowledgements

Our thanks to the various experts with whom we have discussed these matters over the years. The views presented are our own and may not reflect the views of any institutions that we are or have been associated with.

References

Aebischer NJ, Wheatley CJ and Rose HR, 2014. Factors associated with shooting accuracy and wounding rate of four managed wild deer species in the UK, based on anonymous field records from deer stalkers. *PLoS ONE*, 9(10), p. e109698. doi: 10.1371/journal.pone.0109698.

Alfaro-Shigueto J, Dutton PH and Van Bressem M-F, 2007. Interactions between leatherback turtles and Peruvian artisanal fisheries. *Chelonian Conservation and Biology*, 6(1), pp. 129–134.

Altherr S and Hodgins N, 2018. Small cetaceans big problems: A global review of the impacts of hunting on small whales, dolphins and porpoises. *Published by Pro Wildlife, Animal Welfare Institute and Whale and Dolphin Conservation*. Available at: https://awionline.org/sites/default/files/press_release/files/AWI-ML-Small-Cetaceans-Report.pdf.

Australian Veterinary Association, 2014. Use of behaviour-modifying collars on dogs. Available at: https://www.ava.com.au/policy/613-use-behaviour-modifying-collars-dogs.

Bateson P and Bradshaw EL, 1997. Physiological effects of hunting red deer (*Cervus elaphus*) *Proceedings of the Royal Society of London B*, 264, pp. 1707–1714.

Barthel BL, Cooke SJ, Suski CD, et al., 2003. Effects of landing net mesh type on injury and mortality in a freshwater recreational fishery. *Fisheries Research*, 63(2), pp. 275–282. doi: 10.1016/S0165-7836(03)00059-6.

Broom DM, 1999. The welfare of vertebrate pests in relation to their management. In Cowan PD and Feare CJ, eds. *Advances in Vertebrate Pest Management*. Fürth, Germany: Filander Verlag, pp. 309–329.

Browman HI, Cooke SJ, Cowx IG, et al., 2019. Welfare of aquatic animals: Where things are, where they are going, and what it means for research, aquaculture, recreational angling, and commercial fishing. *ICES Journal of Marine Science*, 76(1), pp. 82–92. doi: 10.1093/icesjms/fsy067.

Brown C and Dorey C, 2019. Pain and emotion in fishes: Fish welfare implications for fisheries and aquaculture. *Animal Studies Journal*, 8(2), pp. 175–201.

Brownscombe JW, Danylchuk AJ, Chapman JM, et al. 2017. Best practices for catch-and-release recreational fisheries: Angling tools and tactics. *Fisheries Research*, 186(3), pp. 693–705. doi: 10.1016/j.fishres.2016.04.018.

Burns T, Edwards V, Marsh J, et al. 2000. *Report of the Committee of Inquiry into Hunting with Dogs in England and Wales*. Available at: https://www.gov.uk/government/publications/report-of-committee-of-inquiry-into-hunting-with-dogs-in-england-wales.

Butterworth A, Turner KME and Jennings N, 2017a. Minimising orphaning in the brown hare *Lepus europaeus* in England and Wales: Should a close season be introduced? *Wildlife Biology*, 2017(1), pp. 1–7. doi: 10.2981/wlb.00279.

Butterworth A, Reiss D, Brakes P, et al., 2017b. Welfare issues associated with small toothed whale hunts: An example, the 'Drive Hunt' in Taiji, Japan. In: Butterworth A. (ed.) Marine Mammal Welfare. *Animal Welfare*, 17, pp. 91–110. doi: 10.1007/978-3-319-46994-2_6.

Caudell JN, West BC, Griffin B et al., 2009. Fostering greater professionalism with firearms in the wildlife arena. In Boulanger JR (Ed.). *Proceedings of the 13th WDM Conference*, pp. 95–99. Lincoln, NE: Internet Center for Wildlife Damage Management, University of Nebraska. https://digitalcommons.usu.edu/cgi/viewcontent.cgi?article=1230&context=wdmconference.

Clarke SH, Brownscombe JW, Nowell L, et al., 2021. Do angler experience and fishing lure characteristics influence welfare outcomes for largemouth bass? *Fisheries Research*, 233, p. 105756. doi: 10.1016/j.fishres.2020.105756.

Clausen KK, Holm TE, Haugaard L, et al., 2017. Crippling ratio: A novel approach to assess hunting-induced wounding of wild animals. *Ecological Indicators*, 80, pp. 242–246.

Cooke SJ and Sneddon LU, 2007. Animal welfare perspectives on recreational angling. *Applied Animal Behaviour Science*, 104, pp. 176–198.

Daoust P-Y and Caraguel C, 2012. The Canadian harp seal hunt: Observations on the effectiveness of procedures to avoid poor animal welfare outcomes. *Animal Welfare*, 21, pp. 445–455. doi: 10.7120/09627286.21.4.445.

Davie PS and Kopf RK, 2006. Physiology, behaviour and welfare of fish during recreational fishing and after release. *New Zealand Veterinary Journal*, 54(4), pp. 161–172.

DEFRA, 2010. Code of practice for the welfare of gamebirds reared for sporting purposes. Department for Environment, Food and Rural Affairs. Available at: https://assets.publishing.service.gov.uk/government/uploads/system/uploads/attachment_data/file/69379/pb13356-game-birds-100720.pdf.

DEFRA, 2021. Policy paper. Action plan for animal welfare. Published 12 May 2021. Available at: https://www.gov.uk/government/publications/action-plan-for-animal-welfare/action-plan-for-animal-welfare.

Diggles BK, Cooke SJ, Rose JD et al., 2011. Ecology and welfare of aquatic animals in wild capture fisheries. *Reviews in Fish Biology and Fisheries*, 21, pp. 739–765. doi: 10.1007/s11160-011-9206-x.

Ditchkoff SS, Welch ER, Lochmiller RL, et al., 1998. Wounding rates of white-tailed deer with traditional archery equipment. *Proceedings of the* Annual *Conference Southeastern Association of Fish and Wildlife Agencies*, 52, pp. 244–248.

Dolman SJ and Moore MJ, 2017. Welfare implications of cetacean bycatch and entanglements. In Butterworth A, ed., *Marine Mammal Welfare. Animal Welfare*, 17, pp. 41–65. doi: 10.1007/978-3-319-46994-2_4.

European Communities, 1991. Council Regulation (EEC) 3254/91 of 4 November 1991 prohibiting the use of leghold traps in the Community and the introduction into the Community of pelts and manufactured goods of certain wild animal species originating in countries which catch them by means of leghold traps or trapping methods which do not meet international trapping standards. Available at: https://eur-lex.europa.eu/legal-content/EN/TXT/?uri=CELEX:31991R3254.

European Communities, 1998. Agreement on international humane trapping standards between the European Community, Canada and the Russian Federation. *Official Journal of the European Communities*. 14.2.98. Available at: https://eur-lex.europa.eu/legal-content/EN/TXT/PDF/?uri=CELEX:21998A0214%2802%29.

FAO, 2015. Artisanal fisheries. *The Fish Project*. Available at: http://www.fao.org/family-farming/detail/en/c/335263/.

Feber RE, Johnson PJ and Macdonald DW, 2020. Shooting pheasants for sport: What does the death of Cecil tell us? *People and Nature*, 2, pp. 82–95. doi: 10.1002/pan3.10068.

Fox NC, Blay N, Greenwood AG, et al., 2005. Wounding rates in shooting foxes (*Vulpes vulpes*). *Animal Welfare*, 14, pp. 93–102.
Gamborg, C, Sandøe, P and Palmer, C (2020) Ethical management of wildlife. Lethal versus nonlethal control of white-tailed deer. *Conservation Science and Practice*, 2, p. e171. doi: 10.1111/csp2.171.
García-Párraga D, Crespo-Picazo JL, Bernaldo de Quirós Y, et al., 2014. Decompression sickness ('the bends') in sea turtles. *Diseases of Aquatic Organisms*, 111, pp. 191–205. doi: 10.3354/dao02790.
Gentsch RP, Kjellander P and Röken BO, 2018. Cortisol response of wild ungulates to trauma situations: Hunting is not necessarily the worst stressor. *European Journal of Wildlife Research*, 64, p. 11. doi: 10.1007/s10344-018-1171-4.
Hampton JO and Hyndman TH, 2019. Underaddressed animal-welfare issues in conservation. *Conservation Biology*, *33*(4), pp. 803–811.
Hampton JO, Jones B, Perry AL, et al., 2016. Integrating animal welfare into wild herbivore management: Lessons from the Australian Feral Camel Management Project. *The Rangeland Journal*, 38, pp. 163–171. doi: 10.1071/RJ15079.
Hampton JO, Arnemo JM, Barnsley R, et al., 2021. Animal welfare testing for shooting and darting free-ranging wildlife: A review and recommendations. *Wildlife Research*, 48(7), pp. 577–589. doi: 10.1071/WR20107.
Hunter LB, White P, Henschel P, et al., 2012. Walking with lions: Why there is no role for captive-origin lions *Panthera leo* in species restoration. *Oryx*, 47(1), pp. 19–24. doi: 10.1017/S0030605312000695.
Hunting Act 2004. Available at: https://www.legislation.gov.uk/ukpga/2004/37/contents.
Iossa G, Soulsbury CD and Harris S, 2007. Mammal trapping: A review of animal welfare standards of killing and restraining traps. *Animal Welfare*, 16, pp. 335–352.
ISO, 1999. *Animal (Mammal) Traps: Part 4: Methods for Testing Killing-trap Systems Used on Land or Underwater.* Geneva: International Organization for Standardization (ISO) 10990–4 1999.
IWC, 2018. *Report of the Whale Killing Methods and Welfare Issues Working Group IWC/67/04.* Available at: https://iwc.int/welfare.
IWC, 2021. Total catches. https://iwc.int/total-catches. Last visited 24/6/2021.
Jones M and Draper C, 2018. Trophy hunting and animal welfare. In Butterworth A, ed. *Animal Welfare in a Changing World.* Wallingford, UK: CAB International, pp. 46–56.
Ludders, JW, Schmidt RH, Dein FJ, et al., 1999. Drowning is not euthanasia. *Wildlife Society Bulletin*, 27, pp. 666–670.
Mangel JC, Alfaro-Shigueto J, Van Waerebeek K, et al., 2010. Small cetacean captures in Peruvian artisanal fisheries: High despite protective legislation. *Biological Conservation*, 143(1), pp. 136–143. doi: 10.1016/j.biocon.2009.09.017.
McComb K, Moss C, Durant SM, et al., 2001. Matriarchs as repositories of social knowledge in African elephants. *Science*, 292, pp. 491–494. doi: 10.1126/science.1057895.
Meek PD, Brown SC, Wishart J, et al., 2019. Efficacy of lethal-trap devices to improve the welfare of trapped wild dogs. *Wildlife Research*, 46, pp. 89–95.
Metcalfe JD, 2009. Welfare in wild-capture marine fisheries. *Journal of Fish Biology*, 75, pp. 2855–2861. doi: 10.1111/j.1095-8649.2009.02462.x.
Minamikawa S, 2020. *Japan's Scientific Progress Report on Small Cetaceans in the Fiscal Year 2019 (April 2019 to March 2020), with Statistical Data for the Calendar Year 2019.* Japan ProgRep. SM/2020_rev1.
NAMMCO, 2011. Expert Group Meeting to Assess the Hunting Methods for Small Cetaceans, 15–17 November 2011, Copenhagen, Denmark. Norway, NAMMCO. Available at: https://nammco.no/topics/expert-group-meetings/.
NAMMCO, 2015. Expert Group Meeting on Assessing TTD Data from Large Whale Hunts, 4–6 November 2015, Copenhagen, Denmark. Norway, NAMMCO. Available at: https://nammco.no/topics/expert-group-meetings/.
Noss AJ, 1997. The impacts of cable snare hunting on wildlife populations in the forests of the Central African Republic. *Conservation Biology*, 12(2), pp. 390–398.
Nunny L, 2020. Animal welfare in predator control: Lessons from land and sea. How the management of terrestrial and marine mammals impacts wild animal welfare in human-wildlife conflict scenarios in Europe. *Animals*, 10(2), p. 218. doi: 10.3390/ani10020218.
Nunny L, Simmonds MP and Butterworth A, 2018. A review of seal killing practice in Europe: Implications for animal welfare. *Marine Policy*, 98, pp. 121–132. doi: 10.1016/j.marpol.2018.08.013.
Øen EO, 2021. Animal welfare in the conduct of whaling: A review of the research and developments to improve animal welfare in the minke whale hunt in Norway 1981–2005. In Kishigami N, ed. *World*

Whaling: Historical and Contemporary Studies. Senri Ethnological Studies, Vol. 104, pp. 287–318. https://books.google.co.uk/books/about/World_Whaling.html?id=le9wzgEACAAJ&redir_esc=y.

Orr B, Malik R, Norris J, et al., 2019. The welfare of pig-hunting dogs in Australia. *Animals*, 9, p. 853. doi: 10.3390/ani9100853.

Pedersen MA, Berry SM and Bossart JC, 2008. Wounding rates of white-tailed deer with modern archery equipment. *Proceedings of the Annual Conference Southeastern Association of Fish and Wildlife Agencies*, 62, pp. 31–34.

Proulx G and Rodtka D, 2019. Killing traps and snares in North America: The need for stricter checking time periods. *Animals*, 9(8), p. 570. doi: 10.3390/ani9080570.

Proulx G, Cattet M, Serfass TL, et al., 2020. Updating the AIHTS trapping standards to improve animal welfare and capture efficiency and selectivity. *Animals*, 10, p. 1262. doi: 10.3390/ani10081262.

Robards MD and Reeves RR, 2011. The global extent and character of marine mammal consumption by humans: 1970–2009. *Biological Conservation*, 144(12), pp. 2770–2786. doi: 10.1016/j.biocon.2011.07.034.

Robb DHF and Kestin SC, 2002. Methods used to kill fish: Field observations and literature reviewed. *Animal Welfare*, 11, pp. 269–282.

Rose JD, Arlinghaus R, Cooke SJ, et al., 2014. Can fish really feel pain? *Fish and Fisheries*, 15, pp. 97–133.

Sharp T and Saunders G, 2011. *A Model for Assessing the Relative Humaneness of Pest Animal Control Methods*. 2nd edn. Canberra, ACT: Australian Government Department of Agriculture, Fisheries and Forestry.

Simmonds MP, McLellan F, Entrup N et al., 2021. Whaling in Europe: An ongoing welfare and conservation concern. In Nunny L, ed. *Under Pressure: The need to Protect Whales and Dolphins in European waters*. Switzerland: OceanCare, pp. 62–76.

Slade LM and Kalangahe B, 2015. Dynamite fishing in Tanzania. *Marine Pollution Bulletin*, 101(2), pp. 491–496. doi: 10.1016/ j.marpolbul.2015.08.025.

Sneddon LU, 2020. Can fish experience pain? In Kristiansen T, Fernö A, Pavlidis M et al., eds. *The Welfare of Fish. Animal Welfare*, Vol. 20. Cham, Switzerland: Springer. doi: 10.1007/978-3-030-41675-1_10.

Stokke S, Arnemo JM, Brainerd S, et al., 2018. Defining animal welfare standards in hunting: Body mass determines thresholds for incapacitation time and flight distance. *Science Reports*, 8, p. 13786.

Vail CS, Reiss D, Brakes P, et al. 2020. Potential welfare impacts of chase and capture of small cetaceans during drive hunts in Japan. *Journal of Applied Animal Welfare Science*, 23(2), pp. 193–208. doi: 10.1080/10888705.2019.1574576.

Veldhuizen LJL, Berentsen PBM, de Boer IJM, et al., 2018. Fish welfare in capture fisheries: A review of injuries and mortality. *Fisheries Research*, 204, pp. 41–48. doi: 10.1016/j.fishres.2018.02.001.

Vila Pouca C and Brown C, 2017. Contemporary topics in fish cognition and behaviour. *Current Opinion in Behavioral Sciences*, 16, pp. 46–52. doi: 10.1016/j.cobeha.2017.03.002.

Whaling.fo, 2021. Whales and Whaling in the Faroe Islands, Catches 2000–2021. Available at: https://www.whaling.fo/en/regulated/450-years-of-statistics/catches/.

Whitman K, Starfield A, Quadling H et al., 2004. Sustainable trophy hunting of African lions. *Nature*, 428, pp. 175–178. doi: 10.1038/nature02395.

17
COMMERCIAL FISHERIES

Moira Harris

Introduction

Worldwide, approximately one trillion wild fish are captured every year, outnumbering any farmed animal. Table 17.1 illustrates that numbers of wild-caught fish far exceed those of the most popularly produced species slaughtered for meat – poultry and farmed fish (fishcount.org .uk, 2016; Food and Agriculture Organization of the United Nations, 2021). In the past, the capacity of fish to experience pain and suffering has been overlooked but there is now extensive scientific evidence that finfish are sentient (capable of experiencing pain and distress as well as positive emotions; e.g. see EFSA, 2009a). Public concern and consumer awareness about fish have traditionally been less than for terrestrial farmed species. However, recent research across Europe (Eurogroup for Animals/Compassion in World Farming, 2018) showed that a majority of consumers understand fish are sentient, think their welfare should be better protected, and want to use welfare as a guide in their purchasing choices. In this survey, 89% of people said they believe that humane slaughter is important or essential for good fish welfare.

Compared to farmed fish, those in the wild may enjoy the majority of their lives in relatively natural conditions but, by contrast, the end of life is commonly extremely unpleasant and often prolonged. A variety of methods are used to extract fish from water, each of which has associated threats to their welfare (described in Table 17.2). Outlined in this chapter are the many potential problems that can befall fish during the period between initial contact with fishing gear, through catching, handling, and finally death. These welfare issues can be quite different from those associated with fish farming. In particular, the vast majority of wild-caught fish are killed without any effective means of pre-slaughter stunning and are thus fully conscious immediately before their death.

The situation is not without hope, however. The chapter incorporates recommendations for improving the welfare of wild-caught fish, covering fishing methods, handling, stunning and slaughter, and concludes that following these recommendations would result in significant mitigation of many current issues.

Hazards to fish welfare during capture

Capture of fish and landing, during which they are brought onboard the fishing vessel or onto land, can involve multiple different threats to fish welfare, which are summarised below.

 DOI: 10.4324/9781003182351-20

Table 17.1 Number of animals slaughtered every year globally for meat

Species or animal group	*Number*
Fish	Wild: at least 787 billion (up to 2.3 trillion) Farmed: at least 51 billion (up to 167 billion)
Chickens and other poultry	86.5 billion
Pigs	2.2 billion
Rabbits	1.5 billion
Sheep and goats	1.3 billion
Cattle	340 million

Source: adapted from fishcount.org.uk (2016); Food and Agriculture Organization of the United Nations (2021); Waley et al. (2021).

Duration of capture

The time between first contact with fishing gear and hauling onboard a vessel or onto land can last from minutes to several hours. Capture involves physiological changes, forcing fish to combine aerobic and anaerobic activity, and results in the depletion of energy stores, osmoregulatory changes, pH disruption, and accumulation of metabolites such as lactate. Fish may die due to a direct cause such as asphyxiation, or a combination of events such as crowding, injury, hypoxia, or exhaustion (Chopin et al., 1996). The longer the capture process, the more risk of undesirable effects, and the higher the levels of physiological distress.

Crowding density

The presence of a high density of fish in a given space during capture (crowding) forces fish into direct physical contact with each other and/or with fishing gear, potentially resulting in injury, asphyxiation, and elevated stress levels (Raby et al., 2015). It can lead to hypoxia if respiration is restricted either because the operculum (the bony structure protecting the gills) cannot move and/or due to depletion of oxygen in the water (Raby et al., 2015). Sea bass and sea bream that had been overcrowded displayed vigorous movements for several minutes before death, suggesting high levels of stress, and further increasing oxygen needs due to their vigorous movements (Robb and Kestin, 2002). Negative consequences of overcrowding have been reported in lingcod, sablefish, walleye pollock, Pacific halibut, sardines, and salmon.

Physical injuries

Some physical injuries to fish are intentional, such as hook and line injuries, and capture can also cause unintentional injuries (Raby et al., 2015) including scale loss, fin damage, dermal lesions, haemorrhages, damage to gills and eyes, and puncture wounds. These injuries are caused by excessive crowding or by interaction with fishing gear, the vessel, or its crew. Fish skin has pain receptors (Sneddon et al., 2003) and nerve fibres which means that injuries hurt fish. Epidermal injuries can further compromise welfare by disrupting osmotic balance or increasing vulnerability to infection (Noble et al., 2018).

Most commonly, lesions occur on a fish's dorsal surface and flanks, but deeper lesions can penetrate to the ribs or internal organs (Bottari et al., 2003). Trauma to gills can profoundly affect both health and welfare (Noble et al., 2011). In fish that were captured and

Table 17.2 Fishing methods used in EEA fisheries, their target species, and impacts on fish welfare

Method	*Description*	*Target species*	*Impact on fish welfare*
1. Trawling			
Pelagic or mid-water trawl	• Trawl net towed from vessel's bow or stern in mid-water • Tow times vary from a few minutes to a few hours depending on density of target species and size and power of vessel	• Generally a single pelagic species (e.g. mackerel, herring) with small bycatch (e.g. whiting, bass)	• Exhaustion, injury, asphyxiation, and crushing during towing and hauling • Barotrauma and thermal shock associated with greater depths
Beam trawl	• Trawl net towed on seabed, held open by a wooden or steel beam • Beam towed behind vessel and tow times vary from a few minutes to a few hours depending on density of target species and size and power of vessel	• Mainly flatfish and demersal species (e.g. plaice, sole, cod)	• Exhaustion, injury, asphyxiation and crushing during towing and hauling • Barotrauma and thermal shock associated with greater depths • Large catches of non-target species are common • Can have a significant impact on seabed fauna
Bottom trawl	• Large trawl net towed from vessel's bow or stern, on or near seabed, held open by pair of trawl doors • Tow times vary from a few minutes to a few hours depending on density of target species and size and power of vessel	• Demersal species (e.g. cod, sole, plaice, rays, anglerfish, bass, whiting)	• As for beam trawl
Dredge	• Rigid structure, consisting of a frame and toothed bar, with a collecting bag, towed along seabed to target shellfish • One or more dredges (up to 22 per side) towed on either side of vessel	• Shellfish, particularly scallops	• Shellfish come to the surface alive as this is often a requirement for sale • Non-target species may be injured or suffocated
2. Seine nets	• Large net used to surround a shoal of fish. Bottom of the net is then drawn together to enclose them • Headrope carrying floats is used to keep the net on the surface • Net has rings along its lower edge through which a cable is passed, forming a bowl-like shape and preventing fish from escaping downwards	• Pelagic species for Danish seine and demersal species for Scottish seine	• Crowding and then crushing when fish are lifted onto the deck • Large species may gaffed • Barotrauma and thermal shock associated with greater depths

(*Continued*)

Table 17.2 (Continued)

Method	*Description*	*Target species*	*Impact on fish welfare*
3. Hanging nets			
Drift nets	• Net suspended from buoys in the water and drifts anywhere between seabed and surface • Nets are either attached at one end to the vessel or left to drift and be recovered later • Fish become entangled when the mesh is caught behind their gills • Soak time is generally a few hours	• Mainly pelagic species (e.g. mackerel, herring) but can be set to drift along the seabed in sandy areas to catch prawns	• Suffocation, injury, exhaustion, depredation • Barotrauma and thermal shock associated with greater depths
Fixed and set nets	• Net suspended in the water, either hanging from buoys to drift or fixed to anchored poles, anywhere between seabed and surface • A gill net is a single wall of netting whereas a trammel/tangle net is a wall of small, fine mesh between two outer layers of rope • Fish become entangled when the mesh is caught behind their gills • Soak times vary from one tidal cycle to several days	• Demersal species (e.g. cod, hake, flatfish, monkfish, turbot, rays)	• As for drift nets
4. Hook and line			
Longline	• Left anchored or drifting with numerous baited hooks • The main line is made of light rope or heavy nylon monofilament and may be many kilometres long	• Can be rigged for demersal or pelagic species	• Fish may swallow the bait (deep hooking); being unhooked can result in gut and throat damage • Injuries from use of gaff hooks to bring fish onboard • May swallow bait and remain hooked underwater for several hours or days
Pole and line	• Single or multiple hooked rod and reel set-ups using live or dead bait, or artificial lures and feathers • Can also include trolling (towing baited lines behind a moving vessel) • In handlining, trolling and jigging the fisher is in physical contact with the line and reacts when a fish bites the bait	• Demersal species (e.g. mackerel, bass, cod, pollock)	• Fish may swallow the bait (deep hooking); being unhooked can result in gut and throat damage • Live bait is held in small containers until suddenly introduced to a new water environment and to a feeding frenzy

(Continued)

Table 17.2 (Continued)

Method	*Description*	*Target species*	*Impact on fish welfare*
5. Pots and traps	• Pots, creels and other fish traps are structures where fish are guided through funnels that encourage entry but limit escape • Traps differ in shape, size and material • Can be set singly on the seabed or in strings • Usually baited and can be left overnight or for several days	• Shellfish (e.g. nephrops, lobster, crab, whelk) • Trap fisheries for wrasse for use in salmon farms	• Depredation • Shellfish and some non-target species are trapped for several days and are usually captured alive • Main welfare impacts are on non-target species that are trapped, and on capture of the bait species

Source: adapted from Waley et al. (2021).

then released, gill trauma increased post-release mortality in Atlantic salmon (Mäkinen et al., 2000) and southern flounder (Smith and Scharf, 2011). A review of 85 published articles (Veldhuizen et al., 2018) found that scale, skin and fin injuries occurred more frequently in trawls, seines, gill nets, and traps than in capture involving hooks. Pressure injuries occurred with all gear types and mortality was higher in trawls and seines than with gill nets, hooks, or traps.

Some fishing gear aims to cause injury to fish by piercing parts of the body with a hook. Hooking occurs mostly in the jaw (Davis, 2002), tongue, gills, or eye but escaped fish have also been found with hooks in their oesophagus and stomach. Fish that are injured before being discarded are more likely to die, due to damage to skin, gill and muscle tissues, or secondary infections (Kojima et al., 2004). Fin damage may have a negative effect on movement and postural control, potentially affecting future welfare and survival.

Depredation

Many fishing methods involve long periods of constraint within fishing gear. As a result, fish can be incapable of any escape or defensive reaction, and so are vulnerable to predation. Other fish, marine mammals and seabirds specifically target fishing activities where it is likely that captured fish will be easy prey.

Thermal shock

Fish can be exposed to abrupt temperature increases during capture as water temperatures change rapidly at different depths. Exposure to warmer water increased heart rate and mortality in lingcod (Olla et al., 1997), and elevated water temperatures in sablefish led to increased mortality within 48 hours (Davis et al., 2001). Removing fish from water in freezing temperatures can cause immediate damage to wet soft tissues such as gills and eyes. Acute physiological responses occur when fish are brought into ambient air temperatures and then exposed to low temperatures as part of a chilling process or freezing medium.

Barotrauma

Fishing at depth can lead to decompression injuries as fish are hauled to the surface. Expanding gas can accumulate inside the organs, resulting in pressure-related injuries known as barotrauma, which can present as internal organ haemorrhage, organ distension, and organ rupture (Pribyl et al., 2011). Rupture of the swim bladder causes gases to escape into the abdominal cavity, distending it. In more severe cases, distension can cause eversion (turning inside out) of the stomach and gut. Externally visible pressure injuries can include protrusion of the gut or swim bladder from the mouth or anus, bulging of the eye and air trapped behind the cornea or under the skin (Mason and Lowe, 2008).

Exhaustion

Response to exhaustive exercise varies among fish species but in all cases the stress response from excess physical activity causes an increase in metabolites and measurable ion imbalances. In spring chinook salmon, researchers observed an initial flight response followed by struggles of decreasing magnitude due to exhaustion (Lindsay et al., 2004). When the extent of exercise stress is so great that consequent physiological stress response overcomes the fish's ability to cope, metabolic acidosis occurs which may lead to death. Swimming exhaustion and fatigue deaths have been observed in a variety of different capture methods and for different species.

Asphyxiation

Fish extract oxygen from water through the fine membranes (lamellae) of their gills and distribute it via the blood to the bodily organs. Some fish can obtain oxygen from air, either to supplement gill respiration or because they are obligate air breathers that need to access the surface to breathe. However, the gill lamellae can only function efficiently if water keeps moving across them from front to back. When the gill filaments are in contact with air, they stick to one another and collapse. As well as transporting oxygen, blood picks up carbon dioxide (CO_2) from cells and transports it back to the gills to be released. If this gas exchange is compromised, the fish asphyxiates. When this happens, several physiological systems are affected, and the fish suffers severely.

The main cause of asphyxiation in capture fisheries is air exposure (Ferguson and Tufts, 1992), but it also takes place when respiration is restricted, either because the operculum cannot move or due to water oxygen depletion. All of these causes have the same end result of acute anoxia, the severe loss of oxygen supply (Raby et al., 2015). Acute asphyxia results in an irreversible loss of consciousness and is considered to be one of the most stressful killing methods (Bagni et al., 2007).

Hazards to fish welfare after capture

After fish have been captured, they are usually subjected to handling followed by slaughter using one of a variety of methods. Many fish die before being brought onboard during subsequent handling, although it is very difficult to obtain numbers of these pre-slaughter casualties. Overcrowding with crushing and oxygen depletion, decompression, exhaustion, and long exposure to air are the main causes of death before the designated slaughter process. Welfare hazards associated with handling and slaughter are described below.

Handling

Handling fish in air is inherently stressful, and additionally so if the fish have been overcrowded, crushed, decompressed, or exhausted. Handling times can vary considerably with different fishing methods and vessel design and the longer the fish are handled, the greater the threat to their welfare (Davis, 2002). Handling out of water stresses fish in combined ways. Fish suffer simultaneously from the effects of direct handling and from deprivation of oxygen, with these events occurring during critical periods of physiological stress and heavy physical exertion. A study of Pacific salmon showed that lowest mortality was caused by a maximum handling time of ten seconds in air and three minutes in water (Patterson et al., 2016).

The response to handling extends the suite of acute stress response reactions initiated during capture, and this complex feedback is species-specific and dependent on the duration and nature of the stressor. During this stage, additional physiological disturbances as a result of exhaustive exercise may cause death. The effects of handling are magnified by the fact that it occurs in conjunction with air exposure and temperature increase.

Hauling onboard

Lifting fish out of the water is a critical handling step. When lifting nets full of fish from the water, the pressure can cause physical injuries, crushing, and hypoxia. Removing fish from fishing gear roughly can disrupt the mucous coat of the fish and cause scale loss and abrasions. Using gloves to handle fish can make injuries worse, and fish may be dropped.

When fish (especially smaller ones) are entangled in nets, fishers tend to pull them from the net rather than pushing them through, which causes further injuries and observable stress reactions (Veneranta et al., 2017). Fine twines and monofilament nets cause greater injury on de-netting. For seine fishing, coho salmon removed using traditional ramping (hauling the net onboard) had higher mortality and displayed higher stress than those removed by brailing (removing fish from the net still in the water; Farrell et al., 2000).

Some vessels that operate with trawl or seine nets use hydraulic fish lifting devices. Typically, a vacuum pump lifts the water, bringing the fish with it. Physiological responses to pumping and external injuries have been observed, but properly designed and operated fish pumps have the potential to be less stressful than alternative methods for hauling fish.

Hauling larger fish usually requires extra equipment and physical intervention. Common tools used to control large fish are nets, or a gaff consisting of a handle with a sharply pointed hook. The fisher places the point of the gaff deep inside the fish to support its weight when it is brought out of water. It is usual to gaff the fish in the gill operculum which allows it to be hauled and controlled without damaging the flesh, but this practice causes additional severe injuries including significant bleeding and may lead to exsanguination if death does not intervene (Davie and Kopf, 2006).

Onboard sorting

De-hooking inflicts extra injury on the fish. Hook removal methods vary from careful, manual removal to de-hooking devices or automated hook removal. Studies evaluating different hook removal methods found that the most common reason to consider halibut bycatch in poor condition, leading to greater mortality, was injuries sustained while being removed from the hook (Kaimmer and Trumble, 1998).

Sorting fish on deck can have a cumulative negative impact on their welfare and can cause physical damage due to throwing or movement using gaffs and picks, fish falling on the deck, and from other careless actions. Equipment such as sorting tables and conveyor belts may have sharp protrusions and design features that allow fish to become stuck. Large catches and longer and high-density tows and nets can increase sorting times, exposing fish to air and increased temperatures.

Slaughter

In many cases, no specific killing method is used and death results incidentally during capture and processing. Most specific slaughter methods are not preceded by stunning, and therefore can be described as inhumane. Several methods of more humane fish slaughter also exist, whereby killing is preceded by stunning that renders fish unconscious and insensible to pain, or the stunning method also causes fish to die.

Effectiveness of the various methods can be evaluated through indicators of the state of insensibility achieved until death occurs – however, identifying this state and differentiating it from the moment of death is difficult. Immobilisation may be misinterpreted as the absence of consciousness and, conversely, some fish species exhibit post-mortem reflexes that may be interpreted as them still being alive.

The World Organisation for Animal Health (2019) and European Food Safety Authority give some indicators for effective stunning of farmed fish, such as the immediate loss of body and respiratory movements, loss of visual evoked response (VER) resulting from brain dysfunction, incapacity to respond to light flashes directed at the eye, and loss of vestibule-ocular reflex as determined by the absence of eye rolling. This is confirmed by Kestin et al. (2002) who concluded that, in a range of species, behaviours such as swimming, response to stimuli like handling, and clinical reflexes – like eye rolling or breathing – indicate a state of awareness and the capacity to experience suffering.

Asphyxiation in air

Wild fish commonly undergo asphyxiation onboard until they die. The time taken to die from asphyxiation depends on the species, the exposure time, and the temperature. In general, when exposed to higher temperatures, most fish die more quickly due to increased metabolic rates and higher oxygen demand (Robb and Kestin, 2002). This is not a quick process: sea bream left to die in air lost self-initiated responses after four minutes, took around seven minutes to lose response to stimuli, and 14 minutes to cease reflexes (Kestin et al., 2002).

Concentrations of stress indicator variables such as plasma cortisol and glucose in Senegal sole after asphyxiation were significantly higher than resting values (Ribas et al., 2007). Asphyxiation in air is considered to be a killing method that causes a maximal stress response, violent attempts to escape, and aversive reactions with associated extreme physical activity (Robb and Kestin, 2002; Ribas et al., 2007; EFSA, 2009a).

Live chilling and death in ice slurry

Hypothermia is used to kill some fish species. They are placed in chilled water or water-ice slurry causing a temperature differential of up to 30°C. This induces cold shock which simultaneously chills, sedates and eventually kills them by asphyxia (Tanck et al., 2000; EFSA, 2009a). Initially, carp exposed to chilled water appeared comfortable and exhibited normal swimming activity; however, abnormal behaviours suggesting aversion followed (Rahmanifarah et al., 2011). The hypothermia effect on sea bream resulted in immobilisation before unconsciousness (van de Vis et al., 2003).

Cold shock causes progressive muscle paralysis which makes changes in behaviour difficult to assess. Sublethal physiological and behavioural consequences of cold shock stress include severe disruption of the fish's metabolic rate, movements, and behaviour, and as oxygen consumption is also impaired, it succumbs to hypoxia and becomes immobilised (Hovda and Linley, 2000). Live chilling before slaughter resulted in significantly increased blood levels of cortisol and lactate, indicating pre-slaughter stress. In Atlantic salmon, the muscle pH also fell, indicating that metabolic changes and consequent acidosis were occurring (Skjervold et al., 2001).

The hypothermic effect is induced more quickly when fish live in warmer waters, since the effectiveness of the process depends on the temperature difference between the ice slurry bath and the fish's usual habitat. When fish live in cold waters, their physiology is cold-adapted and they will be more likely to die from anoxia in the chilled water than from cold shock.

Sedation and loss of consciousness due to chilling is reversible if the fish is transferred back into its normal water conditions. Studies (Skjervold et al., 2001; Lambooij et al., 2006, Roth et al., 2007) have demonstrated chilled fish showing signs of consciousness when removed from the chilling tank. Immediate stress responses such as squirming and thrashing when fish were gilled and gutted, after being chilled, were also observed. Therefore, live chilling is an unsuitable method of stunning fish before slaughter as it does not induce insensibility.

Exsanguination

During death by exsanguination, blood is drained by cutting the major blood vessels. Methods of bleeding vary between species and can involve a throat cut, gill cut, or pectoral cut. All procedures consist of inserting a sharp knife and severing major blood vessels and the gills are often cut because they are heavily vascularised and readily accessible due to their external bodily location. Exsanguination often takes place without stunning and in some cases, non-stunned fish may also be subject to direct evisceration (removal of their internal organs). Following cutting of the blood vessels, fish struggle vigorously, initially due to being restrained, handled, and exposed to air. Tail flapping and head shaking were observed to last for about 30 seconds after gill cutting in salmon. VERs are present for up to seven minutes (van de Vis et al., 2003). Exsanguination without stunning appears to cause a maximal aversive stress response, but with more rapid loss of consciousness than asphyxiation.

Fish killed by methods that do not result in immediate insensibility, such as exsanguination without prior stunning, lose their response to stimuli and reflexes progressively over a prolonged period. Turbot took over 15 minutes to lose responses to stimuli (Morzel et al., 2002), and struggled and experienced the highest stress levels at slaughter, taking longer than an hour to cease ventilation movements or muscle activity (Ruff et al., 2002).

Movements slowly decrease, the fish loses consciousness as a direct result of exsanguination, and finally succumbs to anoxia due to ischaemia (a restriction in blood supply to tissues that results in a shortage of oxygen). Differences in the number of vessels severed and effectiveness of cutting cause variation in the bleeding and onset of unconsciousness, as determined in Atlantic salmon and turbot (Morzel et al., 2002). Differences in temperature also affect the time to lose brain function.

Decapitation

Decapitation consists of the complete separation of the head from the rest of the body. In the EU it is typically used for eels but it may be used elsewhere for larger fish. Loss of consciousness is not immediate or even quick: eel heads have shown signs of life for up to eight hours (Verheijen and Flight, 1997). On average, EEG tests showed that decapitated eels took more than ten minutes to demonstrate loss of VERs (van de Vis et al., 2003).

Carbon dioxide saturation

Fish can be rendered insensible by replacing oxygen with carbon dioxide. This is a relatively common method used in aquaculture where it has been mechanised and applied to fish on a batch basis. By contrast, it does not appear to be currently used in wild capture fisheries. Saturating water with CO_2 creates an acidic and oxygen-deficient environment that places fish in a narcotic state (EFSA, 2004). CO_2 immobilises the fish; however, there can be a delayed loss of consciousness which may result in them being slaughtered before becoming insensible. If used for prolonged periods, this technique can potentially cause death by acute hypoxia.

In response to CO_2 narcosis, fish express strong escape behaviours with aversive initial flight reactions. Vigorous head and tail shaking for up to nine minutes has been described in salmon. Similarly, in carp, strenuous activity was observed with fish keeping their mouths and gill covers closed, followed by collisions due to vigorous swimming (Rahmanifarah et al., 2011). Some fish, such as eels and sturgeon, appear to be more resistant to CO_2 saturation and were reported to show escape and aversive behaviour for more than an hour. Activity during exposure to CO_2 can lead to scale diffusion, increased mucus secretion, and gill haemorrhaging (Marx et al., 1997; Robb and Kestin, 2002; Roth et al., 2002).

The combined effects of live chilling and moderate CO_2 narcosis have been tested in Atlantic salmon and reported to be superior to narcosis alone (Erikson et al., 2006). In this case, live-chilled, gas-exposed fish may present limited reactions simply as a result of cold immobilisation, which is not enough to induce loss of brain function. Industry codes and guidance notes recommend sustaining fish in CO_2 saturated water for up to ten minutes before slaughter, which in effect means that fish are exsanguinated or gutted while still conscious (Yue, 2008). As it is also probable that CO_2 causes acute discomfort or pain, it is not considered an acceptable method of stunning fish before slaughter, either with or without simultaneous chilling.

Electrical stunning

Electricity can be used to render fish insensible by electrical stunning or kill them by electrocution. The principle of electrical stunning is to pass sufficient current through the brain, stimulating higher nerve centres to cause their dysfunction. This may be conducted dry, where fish are passed over an electrified surface out of water, semi-dry, where an electrical current is applied directly into the fish, or in water. In wild capture fisheries, few fish are stunned using electrical methods, compared to their extensive use in aquaculture. According to the OIE, electrical stunning/killing methods have been declared as a humane killing procedure for some species of farmed fish (World Organisation for Animal Health, 2019), but no advice has been given regarding capture fisheries.

The current that is passed through the fish's brain causes it to go through an epileptic-like fit involving seizures. Behaviour of fish during these phases varies between species, and increased intensity and duration of application can cause physical injuries. The effectiveness and duration of unconsciousness depends on the intensity of the current and the length of time it is applied, with death occurring if application is prolonged. If fish are not killed by the electrical process, they can recover consciousness gradually. Various studies have led to the formulation of some minimum requirements for effective electrical stunning (Kestin et al., 1995). Insufficient current, voltage, or duration can lead to immobilisation and unsuccessful stunning.

Percussive stunning/killing

In percussive stunning, the fish is removed from the water and restrained before a blow is delivered to its head via a club, hammer, or semi-automatic percussive stunning device (EFSA, 2009b). When a heavy blow is delivered correctly over the brain, cranial pressure massively increases causing disruption of normal electrical activity. Percussive stunning in Atlantic salmon was found to cause cerebral concussion leading to seizures and instant reduction or loss of consciousness. Induced brain haemorrhage may then impair the blood flow, ultimately leading to death (EFSA, 2004; Lambooij et al., 2010).

However, if the blow is inaccurate or not forceful enough, fish do not lose consciousness but may not display normal behaviours, can become immobile without losing consciousness, or sustain injuries. For some species, increasing percussive force to bring about instant insensibility also caused broken jaws and burst eyes (Roth et al., 2007). Other species such as sea bream are unsuited to percussive killing due to their anatomy (van de Vis et al., 2003). Percussive stunning also requires air exposure and individual handling of fish. In wild capture fisheries, this method could only be used for manual stunning of high-value fish, in low volumes.

Spiking

Spiking involves the insertion of a spike through the fish's skull to destroy the brain, also known as pithing. It requires individual handling and restraint of fish and can be done manually or by using a pneumatically operated pistol. It is most commonly used for larger fish such as tuna and salmon and, when spiked correctly, immediate brain death occurs (Poli et al., 2005). Perforation of the hindbrain produces an instant tonic reaction of a few flaps of the tail and minor muscle tremors before all motion stops. There is also immediate loss of VERs and electroencephalographic signs.

Modifications to spiking can include captive needle stunning and ikijime, a method originating in Japan, which follows spiking with insertion of a flexible pithing rod or wire along the spinal cord. Captive needle stunning involves pneumatically firing a needle into the brain, followed by injection of compressed air. In African catfish, inserting compressed air into the brain provoked slow muscle contractions for a few seconds and they subsequently demonstrated no reaction to painful stimuli, either through behavioural observation or EEG (Lambooij et al., 2002). If fish brains are small or the spike misses the target area, fish may not be effectively stunned and will suffer until death occurs.

Additional welfare concerns related to commercial fishing

In addition to procedures that directly involve fish being intentionally caught, commercial fishing is also associated with other activities that affect the welfare of fish and marine animals. Three of these are described in the following.

Discards and bycatch

Discards are the part of the catch that is returned to the sea, either dead or alive. Discarding may occur because fish are too small, or due to economic factors such as insufficient market demand, or due to fishing rules. Discards may include one or many species, and they can be thrown away on purpose, or fall through fishing gear by accident (FAO, 2011).

Discarding of live fish, which may be injured, carries a range of consequences for their survival and well-being. Releasing live unwanted caught fish can expose them to additional stress associated with onboard handling, air exposure and physical injuries. Some fish will die due to their experience of having been fished (known as fishing-induced mortality). The welfare of discarded fish is highly compromised throughout the process of harvest, capture, handling and release (Campana et al., 2009).

Discarding of fish may therefore result in 'hidden mortality'. Post-release mortality rates have been recorded for some species and some fishing methods – e.g. in skate caught with bottom trawlers, the overall short-term survival was 55% (Enever et al., 2009) and only 21% of those with poor health status survived being caught. Longline-caught Atlantic cod's short-term mortality rate varied from 0 to 69% (Milliken et al., 2009) and their survival rate was affected by depth, temperature, and de-hooking.

Sea turtles may be caught as bycatch during bottom trawling for shrimp, and as they are air breathers they will drown if they are unable to return to the surface. Turtle excluder devices (TEDs), incorporating a grid within the trawl net to prevent larger animals passing to the back of the net, and an escape opening, can be used to reduce turtle bycatch. Debris from the water can reduce the efficiency of TEDs and larger turtles may be too big to fit through escape hatches. As TEDs can reduce the efficiency of a catch, fishers may circumvent them by tying the escape opening shut.

Mutilations

Declawing is a procedure where one or both of a crab's claws are removed by hand before it is returned to the water. Crabs can regenerate lost limbs after a period of time, however larger crabs probably will not live long enough to regenerate their claws, which, together with newer knowledge about pain in crustaceans (e.g. Magee and Elwood, 2016), and the ethics of declawing, suggests the practice probably should not be carried out.

Claw removal can facilitate the storage and transport of crab meat, eliminate cannibalism within storage tanks, and make crabs easier to handle. In a study using commercial techniques, 47% of Florida stone crabs that had both claws removed died afterwards, as did 28% of single claw amputees, and around 75% of these deaths occurred within 24 hours of declawing (Davis et al., 1978). Declawing affects the ability of a crab to feed, leads to lower levels of activity, and to difficulty in attracting mates (Davis et al., 1978; McCambridge et al., 2016).

Consumer demand for shark fins has led to the practice of removing the fins of live sharks (shark finning) and returning them to the ocean. The fins, which are of much higher economic value than the rest of the body, are sold as ingredients for shark fin soup and traditional cures, particularly in China. After their fins have been removed, sharks are unable to swim effectively and will die, either by suffocation or being eaten by predators. In an attempt to prevent this practice, many countries have introduced legislation stating that, where sharks are fished, their fins must arrive back on land attached to their bodies. The 'fins naturally attached' policy bans shark finning by EU vessels (Humane Society International, 2013), and by late 2021, the UK was proposing introducing a ban on the import of detached shark fins and shark fin products (gov.uk, 2021).

Ghost fishing

Ghost fishing is a term applied to fishing gear (nets, traps or hook and line) that has been lost or discarded by fishers. Ghost gear is often made of plastic and other long-lasting materials (non-biodegradable gear is predicted to persist in the marine environment for up to 600 years; Global Ghost Gear Initiative, 2018). Ghost gear can continue to passively catch fish and other marine life, so has an ongoing negative impact on animal welfare. It may inflict physical injury or cause asphyxiation or depredation, and predators attracted by the captured prey as well as other non-target species can become trapped. Ghost fishing is an unrecorded source of fish mortality and there are concerns regarding its impact on sustainability. The Global Ghost Gear Initiative (Global Ghost Gear Initiative, 2020), comprising representatives of industry, governments, and animal protection groups, collaborates to address the problem of ghost gear worldwide.

Recommendations to mitigate welfare hazards

Fishing methods recommendations

Recommendations common across fishing methods

- The capture period should be minimised, minimising the tow duration during trawling and trolling, the soak time for gill and trammel nets, the deployment and drying up time of seine nets, and the time between checking of pots;
- Training should be used to increase the skills and knowledge of fishers on using fishing equipment, and on proper handling and slaughter of fish;
- Suffering and injury to fish should be minimised. Methods, handling practices, and equipment should be designed and manufactured with this goal in mind;
- Softer materials and knotless net construction should be preferred in all nets;
- The capture depth, ascent rate during hauling and towing speed should be minimised;
- Maximum target catch volumes per haul should be established, in relation to gear capacity, alongside a plan to reduce volumes if these are regularly over target.

Recommendations for specific fishing methods

Trawl

- The cod end and wings of trawl nets should be designed to reduce injuries;
- Fish should be brought onboard using fish pumps instead of by hauling;
- Catches so large that the net funnels are overwhelmed and selectivity fails, and where compression in the cod end is excessive, should be avoided;
- Bottom trawling, and especially beam trawling, should be prohibited.

Seine nets

- Fish should be crowded in steps and to the minimum density necessary. Maximal stress response should be avoided. Drying up time should be as short as possible;
- Fish should be brought onboard using fish pumps instead of brail nets.

Hanging nets

- Thicker twines should be used in place of fine twines and monofilaments.

Hook and line

- Barbless hooks and circle hooks (circular shaped hooks with a sharply curved back, found to do less damage than J-hooks) rather than J-style hooks should be used when possible;
- Live bait should not be used, including for chumming and for baiting hooks;
- Hook removal should be carried out by hand and with the appropriate training;
- Hooks should not be torn from fish.

Recommendations for fish handling between capture and slaughter

- Time spent out of water between capture and slaughter should be minimised;
- Live fish should be handled gently and in water;
- Fish should be brought onboard using fish pumps instead of by hauling trawl nets or using brail nets. Where this is not possible, the number of fish in the brail net should be limited and nets should preferably be fully lined to lift water with the fish or at least the sides of the net should be lined to reduce abrasion injury;
- Handling equipment and procedures should be organised to avoid throwing fish, moving them with gaffs or picks, fish falling on the deck, getting caught on equipment or being injured by equipment;
- Equipment coming into contact with fish should be kept moist;
- The use of gaff hooks should be minimised and avoided when possible, and must always be followed immediately by a killing procedure;
- No body part should be removed from a live animal, with the exception of the decapitation of stunned fish;
- Practices causing thermal shock to live, non-stunned fish should be eliminated;
- Fish species that are not used to captivity should not be held alive unless their welfare needs are met by the holding systems;
- Fish held and/or transported alive after capture should be held, transported, and killed in line with regulations and best practices applicable in aquaculture.

Slaughter recommendations

The most urgent need to improve welfare in wild capture fisheries is to further develop and implement humane slaughter practices. An effective stunning method followed by a suitable killing method, or a killing method that results in immediate loss of sensibility, should be applied as soon as possible after capture. Training and experience are essential for operating stunning equipment effectively, and especially for carrying out manual stunning and killing procedures.

- Out-of-water electrical stunners should be further implemented, and the technology adapted to other fish species;
- In-water electrical stunning technology should be further developed for wild capture fishing vessels;
- Manual percussive stunning should be used more, especially in small-scale fisheries;
- After stunning, a killing method must be applied. With large fish, this will typically be exsanguination or decapitation. With smaller fish, putting them quickly on ice will usually result in death before sensibility is recovered;
- Spiking is an immediate killing method that should be used with large fish, and with smaller fish that are handled individually.

Conclusions

It is clear that the practices used in the catching, processing, and killing of wild-caught fish lead to many threats to the welfare of the fish involved, as well as to other marine species. While it is not unusual for food production to involve compromises to animal welfare, most farmed animals, including fish, are routinely stunned before slaughter; however, stunning is used rarely in wild-caught fish and this lack of a procedure commonly agreed to minimise pain and suffering constitutes a major threat to their welfare. Additionally, killing methods range from relatively quick to much slower and more painful and fish routinely experience suffering during catching, handling, and processing.

It is possible to mitigate these welfare hazards, however, and following the recommendations above would facilitate significant improvements to the current situation. Several companies have successfully incorporated stunning into commercial fishing operations, demonstrating that this is both possible and practical. The Dutch flatfish trawling company Ekofish uses a conveyor belt and electric dry stunner onboard its vessels to render North Sea plaice, lemon sole, turbot, and brill unconscious within one second (Ekofish Group, 2021). The Alaska-based Blue North fishing company uses traditional hook-and-line fishery to catch cod singly via a 'moon pool' in the centre of the boat; fish are individually brought aboard directly from the water and stunned using a semi-dry automatic stunning table, immediately followed by manual bleeding (Humane Harvest Initiative, 2021).

Ultimately, to meet consumer demand for higher welfare fish products – and to continue to raise awareness of the importance and relevance of fish welfare – product labelling should include clear information that allows consumers to make welfare-based purchase decisions. Finally, a concerted effort is required from the fishery sector and regulators to implement meaningful improvements that will not only increase the welfare of fish, but also ensure issues related to fisheries management – such as bycatch and ghost fishing – are tackled in a comprehensive way.

References

Bagni M, Civitareale C, Priori A, Ballerini A, Finoia M, Brambilla G and Marino G 2007. Pre-slaughter crowding stress and killing procedures affecting quality and welfare in Sea Bass (*Dicentrarchus labrax*) and Sea Bream (*Sparus aurata*). *Aquaculture*, 263, pp. 52–60.

Bottari T, Greco S and Pamebianco A, 2003. Trolling lesions: Incidence in some fish species and preliminary statistical evaluations. *Veterinary Research Communications*, 27(Suppl. 1), pp. 285–288.

Campana SE, Joyce W and Manning MJ, 2009. Bycatch and discard mortality in commercially caught blue sharks Prionace glauca assessed using archival satellite pop-up tags. *Marine Ecology Progress Series*, 387, pp. 241–253.

Chopin FS, Arimoto T and Inoue Y, 1996. A comparison of the stress response and mortality of sea bream *Pagrus major* captured by hook and line and trammel net. *Fisheries Research*, 28, pp. 277–289.

Davie PS and Kopf RK, 2006. Physiology, behavior, and welfare of fish during recreational fishing and after release. *New Zealand Veterinary Journal*, 54, pp. 161–172.

Davis GE, Baughman DS, Chapman JD, MacArthur D and Pierce AC, 1978. *Mortality Associated With Declawing Stone Crabs, Menippe Mercenaria*. Homestead, FL: US National Park Service. Report T-522.

Davis MW, 2002. Key principles for understanding fish bycatch discard mortality. *Canadian Journal of Fisheries and Aquatic Sciences*, 59, pp. 1834–1843.

Davis MW, Olla BL and Schreck CB, 2001. A stress induced by hooking, net towing, elevated sea water temperature and air on sablefish: Lack of concordance between mortality and physiological measures of stress. *Journal of Fish Biology*, 58, pp. 1–15.

EFSA, 2004. Opinion of the Scientific Panel on Animal Health and Welfare on a request from the Commission related to welfare aspects of the main systems of stunning and killing the main commercial species of animals. *EFSA Journal*, 45, pp. 1–29.

EFSA, 2009a. General approach to fish welfare and to the concept of sentience in fish: Scientific opinion of the Panel on Animal Health and Welfare. *EFSA Journal*, 954, pp. 1–27.

EFSA, 2009b. Scientific opinion of the panel on animal health and welfare on a request from the European Commission on species-specific welfare aspects of the main systems of stunning and killing of farmed rainbow trout. *EFSA Journal*, 1013, pp. 1–55.

Ekofish Group, 2021. Our innovations. https://www.ekofishgroup.nl/en/our-innovations [accessed 6 November 2021].

Enever R, Catchpole TL, Ellis JR and Grant A, 2009. The survival of skates (Rajidae) caught by demersal trawlers fishing in UK waters. *Fisheries Research*, 97, pp. 72–76.

Erikson U, Hultmann L and Steen JE, 2006. Live chilling of Atlantic salmon (*Salmo salar*) combined with mild carbon dioxide anaesthesia: I. Establishing a method for large-scale processing of farmed fish. *Aquaculture*, 252, pp. 183–198.

Eurogroup for Animals/Compassion in World Farming, 2018. Fish welfare survey. https://comresglobal.com/polls/eurogroup-for-animals-ciwf-fish-welfare-survey/ [accessed 6 November 2021].

FAO, 2011. *Legislative and Regulatory Options for Animal Welfare*. Rome: Food and Agriculture Organization of the United Nations.

Farrell AP, Gallaugher PE, Clarke WC, Delury NC, Kreiberg H, Parkhouse W and Routledge R, 2000. Physiological status of coho salmon (*Oncorhynchus kisutch*) captured in commercial nonretention fisheries. *Canadian Journal of Fisheries and Aquatic Sciences*, 57, pp. 1668–1678.

Ferguson RA and Tufts BL, 1992. Physiological effects of brief air exposure in exhaustively exercised rainbow trout (Oncorhynchus mykiss): implications for "catch and release" fisheries. *Canadian Journal of Fisheries and Aquatic Sciences, 49*(6), pp. 1157–1162.

fishcount.org.uk, 2016. Fishcount estimates of numbers of individuals killed in (FAO) reported fishery production. http://fishcount.org.uk/studydatascreens/2016/fishcount_estimates_list.php [accessed 6 November 2021].

Food and Agriculture Organization of the United Nations, 2021. FAOSTAT. https://www.fao.org/faostat/en/#data/QL [accessed 15 November 2021].

Global Ghost Gear Initiative (GGGI), 2018. Annual report. https://static1.squarespace.com/static/5b987b8689c172e29293593f/t/5d0aada97196de000156ade8/1560980929408/LO-RES2.GGGI+annual+review.2018.spreads.pdf [accessed 6 November 2021].

Global Ghost Gear Initiative (GGGI), 2020. Protecting our oceans and the life within them. https://www.ghostgear.org/ [accessed 6 November 2021].

gov.uk, 2021. Government to introduce world-leading ban on shark fin trade. https://www.gov.uk/government/news/government-to-introduce-world-leading-ban-on-shark-fin-trade [accessed 15 November 2021].

Hovda J and Linley TJ, 2000. The potential application of hypothermia for anaesthesia in adult Pacific salmon. *North American Journal of Aquaculture*, 62, pp. 67–72.

Humane Harvest Initiative, 2021. The influence of humane harvesting on fish quality, nutritional value and safety. http://bluenorth.com/home/#/humane-harvest/editorial [accessed 6 November 2021].

Humane Society International, 2013. Shark finning and the European Union. https://www.hsi.org/news-media/shark_finning_europe/ [accessed 6 November 2021].

Kaimmer SM and Trumble RJ, 1998. Injury, condition, and mortality of Pacific halibut following careful release of Pacific cod and sablefish longline fisheries. *Fisheries Research*, 38, pp. 131–144.

Kestin S, Wotton S and Adams S, 1995. The effect of carbon dioxide, concussion or electrical stunning of rainbow trout (Oncorhynchus mykiss) on fish welfare. Abstract of poster. Aquaculture Europe '95, Trondheim, Norway, 9th - 12th August 1995. *European Aquaculture Society, Special Publication*, 23, p. 380.

Kestin S, van de Vis J and Robb D, 2002. Protocol for assessing brain function in fish and the effectiveness of methods used to stun and kill them. *The Veterinary Record*, 150, pp. 302–307.

Kojima T, Ishii M, Kobayashi M and Shimizu M, 2004. Blood parameters and electrocardiogram in squeezed fish simulating the effect of net damage and recovery. *Fisheries Science*, 70, pp. 860–868.

Lambooij E, van de Vis H, Kloosterboer RJ and Gerritzen M, 2002. Neural, physiological and behavioural observations after head-only electrical and captive needle stunning of African catfish (Clarias gariepinus). In: International Congress of Meat Science and Technology: ICoMST, Rome.

Lambooij E, Kloosterboer RJ, Gerritzen M and van de Vis H, 2006. Assessment of electrical stunning in fresh water of African Catfish (*Clarias gariepinus*) and chilling in ice water for loss of consciousness and sensibility. *Aquaculture*, 254, pp. 1–4.

Lambooij E, Grimsbø E, van de Vis H, Reimert H, Nortvedt R and Roth B, 2010. Percussion and electrical stunning of Atlantic salmon (*Salmo salar*) after dewatering and subsequent effect on brain and heart activities. *Aquaculture*, 300, pp. 107–112.

Lindsay RB, Schroeder RK, Kenaston KR, Toman RM and Buckman MA, 2004. Hooking mortality by anatomical location and its use in estimating mortality of spring Chinook salmon caught and released in a river sport fishery. *North American Journal of Fish Management*, 24, pp. 367–378.

Magee B and Elwood RW, 2016. Trade-offs between predator avoidance and electric shock avoidance in hermit crabs demonstrate a non-reflexive response to noxious stimuli consistent with prediction of pain. *Behavioural Processes*, 130, pp. 31–35.

Mäkinen ST, Niemelä E, Moen K and Lindström R, 2000. Effects of gill net and rod-and-reel capture on upstream migration of atlantic salmon (Salmo salar L.) following radio tagging. *Fisheries Research*, 45, pp. 117–127.

Marx H, Brunner B and Weinzierl W, 1997. Methods of stunning freshwater fish: impact on meat quality and aspects of animal welfare. *Zeitschrift für Lebensmittel und Untersuchung Forschning A*, 204, pp. 282–286.

Mason JE and Lowe C, 2008. The effects of barotrauma on the catch-and-release survival of Southern Californian near shore and shelf rockfish (*Scarpaenidae Sebastes spp.*). *Canadian Journal of Fisheries and Aquatic Sciences*, 65, pp. 1286–1296.

McCambridge C, Dick JTA and Elwood RW, 2016 Effects of autonomy compared to manual declawing on contests between males for females in the edible crab Cancer pagarus: Implications for fishery practice and animal welfare. *Journal of Shellfish Research,* 35, pp. 1037–1044.

Milliken H, Farrington M, Rudolph T and Sanderson M, 2009. Survival of discarded sublegal Atlantic cod in the Northwest Atlantic demersal longline fishery. *North American Journal of Fisheries Management*, 29, pp. 985–995.

Morzel M, Sohier S and van de Vis JW, 2002. Evaluation of slaughtering methods of turbots with respect to animal protection and flesh quality. *Journal of the Science of Food and Agriculture*, 82, pp. 19–28.

Noble C, Cañon Jones H, Damsgård B, Flood M Midling K, Roque A, Sæther B and Cottee S, 2011. Injuries and deformities in fish: their potential impacts upon aquacultural production and welfare. *Fish Physiology and Biochemistry*, 38, pp. 61–83.

Noble C, Gismervik K, Iversen MH, Kolarevic J, Nilsson J, Stien LH and Turnbull J, 2018. *Welfare Indicators For Farmed Atlantic Salmon: Tools For Assessing Fish Welfare.* Tromsø: Nofima.

Olla BL, Davis MW and Schreck CB, 1997. Effects of simulated trawling on sablefish and walleye Pollock: The role of light intensity, net velocity and towing duration. *Journal of Fish Biology*, 50, pp. 1181–1194.

Patterson DA, Robinson KA, Lennox RJ, Nettles TL, Donaldson LA, Eliason EJ, Raby GD, Chapman JM, Cook KV, Donaldson MR, Bass AL, Drenner SM, Reid AJ, Cooke SJ and Hinch SG, 2016. *Review and Evaluation of Fishing-Related Incidental Mortality for Pacific Salmon*. Nanaimo: Fisheries and Oceans Canada.

Poli B, Parisi G, Scappini F and Zampacavallo G, 2005. Fish welfare quality as affected by pre-slaughter and slaughter management. *Aquaculture International*, 13, pp. 29–49.

Pribyl A, Kent M, Parker SJ and Schreck C, 2011. The response to forced decompression in six species of Pacific rock fish. *Transactions of the American Fisheries Societies Society*, 140, pp. 374–383.

Raby GD, Wilson SM, Patterson DA, Hinch SG, Clark TD, Farrell AP and Cooke SJ, 2015. A physiological comparison of three techniques for reviving sockeye salmon exposed to a severe capture stressor during upriver migration. *Conservation Physiology*, 3, pp. 32.

Rahmanifarah K, Shabanpour B and Sattari A, 2011. Effects of clove oil on behavior and flesh quality of Common Carp (*Cyprinus carpio* L.) in comparison with pre-slaughter CO2 stunning, chilling and asphyxia. *Turkish Journal of Fisheries and Aquatic Sciences*, 11, pp. 135–143.

Ribas L, Flos R, Reig L, Mackenzie S, Barton BA and Tort L, 2007. Comparison of methods for anaesthetizing Senegal sole (*Solea senegalensis*) before slaughter. *Aquaculture,* 269, pp. 250–258.

Robb DHF and Kestin SC, 2002. Methods used to kill fish: field observations and literature reviewed. *Animal Welfare*, 11, pp. 269–282.

Roth B, Moeller D, Veland JO, Imsland A and Slinde E, 2002. The effect of stunning methods on *rigor mortis* and texture properties of Atlantic salmon (*Salmo salar*). *Journal of Food Science*, 67, pp. 1462–1466.

Roth B, Imsland A, Gunnarsson S, Foss Atle and Schelvis-Smit R, 2007. Slaughter quality and rigor contraction in farmed turbot (Scophthalmus maximus); a comparison between different stunning methods. *Aquaculture*, 272, pp. 754–761.

Ruff N, FitzGerald RD and Cross TF, 2002. Slaughtering method and dietary alpha-tocopheryl acetate supplementation affect rigor mortis and fillet shelf-life of turbot *Scophthalmus maximus* L. *Aquaculture Research*, 33, pp. 703–714.

Skjervold P, Olav Fjæra S, Braarød Østby P and Einen O, 2001. Live-chilling and crowding stress before slaughter of Atlantic salmon (Salmo salar). *Aquaculture*, 192, pp. 265–280.

Smith WE and Scharf FS, 2011. Post release survival of sublegal Southern Flounder captured in a commercial gill-net fishery. *North American Journal of Fisheries Management*, 31, pp. 445–454.

Sneddon LU, Braithwaite VA and Gentle MJ, 2003. Do fishes have nociceptors? Evidence for the evolution of a vertebrate sensory system. *Proceedings of the Royal Society of London B: Biology*, 270, pp. 1115–1121.

Tanck M, Booms G, Eding E, Wendelaar Bonga S and Komen J, 2000. Cold shocks: a stressor for common carp. *Journal of Fish Biology*, 57, pp. 881–894.

van de Vis H, Kestin S, Robb D, Oehlenschlager J, Lambooij B, Munkner W, Kuhlmann H, Kloosterboer K, Tejada M, Huidobro A, Ottera H, Roth B, Sørensen NK, Akse L, Byrne H and Nesvadba P, 2003. Is humane slaughter of fish possible for industry? *Aquaculture Research*, 34, pp. 211–220.

Veldhuizen LJL, Berentsen PBM, de Boer IJM, van de Vis JW and Bokkers EAM, 2018. Fish welfare in capture fisheries: A review of injuries and mortality. *Fisheries Research*, 204, pp. 41–48.

Veneranta L, Pakarinen T, Jokikokko E, Kallio-Nyberg I and Harjunpää H, 2017. Mortality of Baltic sea trout (*Salmo trutta*) after release from gillnets. *Journal of Applied Ichthyology*, 34, pp. 49–57.

Verheijen FJ and Flight WFG, 1997. Decapitation and brining: experimental tests show that after these commercial methods for slaughtering eel *Anguilla anguilla* (*L.*), death is not instantaneous. *Aquaculture Research*, 28, pp. 361–366.

Waley D, Harris M, Goulding I, Correira M and Carpenter G, 2021. *Catching Up: Fish Welfare In Wild Capture Fisheries*. Brussels: Eurogroup for Animals.

World Organisation for Animal Health (OIE), 2019. *Aquatic Animal Health Code: Section 7, Welfare of Farmed Fish*. https://www.oie.int/en/what-we-do/standards/codes-and-manuals/aquatic-code-online-access/?id=169&L=1&htmfile=titre_1.7.htm [accessed 6 November 2021].

Yue S, 2008. *The Welfare of Farmed Fish at Slaughter*. 2008. HSUS Reports 3. http://animalstudiesrepository.org/hsus_reps_impacts_on_animals/3. [accessed 6 November 2021].

PART IV

Species-specific concerns

18

CANINES AND FELINES

Heather Bacon

Introduction

Cats and dog have provided humans with companionship, pest control, labour, sensory and emotional support, and protection, for thousands of years. Whilst our unique and often emotional relationships with cats and dogs may be incredibly important to a wide variety of people, it's also important that our affection does not cloud our ability to objectively assess and provide for their species-specific and individual needs. This chapter will outline some of the welfare challenges faced by cats and dogs around the world, and has been divided into two primary themes: issues relating to physically or behaviourally restricted cats and dogs, and issues relating to non-restricted cats and dogs. Restricted cats and dogs are those typically considered as 'owned' in the Western world – these are cats and dogs living under permanent or long-term human guardianship, who are often well provided for physically but who may be behaviourally much more restricted and limited to a home/garden/kennel, than free-roaming dogs. I have not selected the term 'owned' to represent these cats and dogs, as in many communities in parts of Europe, Asia, Central/South America, and Africa, it is typical for cats and dogs to move freely outdoors, unrestricted, even when they are owned or fed and cared for by a human guardian. Whilst there is overlap between these groups, and for practical purposes they should be considered as a fluid, contiguous group where individual animals transition from one group to another, their welfare issues are often related to the level of human intervention they receive. Because of this, I consider welfare issues faced by each of these two groups before moving to more general welfare problems faced by cats and dogs, regardless of their roles in society.

Restricted cats and dogs

Restricted cats and dogs are those that we may typically consider as pets, working or sporting animals. They spend the majority of their time with their activities managed and limited by their human caregivers. However, even within this grouping there is variation in their level of restriction and experiences. For example, pet cats may be kept indoor only, or with access to both the indoors and outdoors, or occasionally outdoors only. These different levels of restriction present different behavioural and physical opportunities and risks and therefore generate different welfare problems. Cats kept indoors are often more behaviourally restricted, and thus may experi-

DOI: 10.4324/9781003182351-22

ence negative emotional states such as boredom or frustration. They may also be less active and more prone to weight gain and health problems associated with obesity. Cats with free-choice access to the outdoors, e.g., through a cat flap, are likely have good behavioural welfare as choice and control are important to welfare. However, they – along with entirely outdoor cats – may be at increased risk of trauma, social conflict, and road traffic injuries. By physically restricting cats and dogs we restrict our pets in many ways, for example we choose what they eat, where they sleep, when they toilet, exercise, and socialise. Such high-level restriction is entirely normal in many human societies, but can create behavioural problems when the behavioural and physical needs of cats and dogs are unmet, and this may result in negative emotional experiences and the development of problem behaviours. In many cases we also restrict reproductive activity in our pets. Whilst this has many benefits, including reduced overpopulation problems and burden on animal shelters, it also confers some health benefits. In some cases, it may also create new health and welfare problems. Whilst the spay/neutering of cats is generally recommended from the age of about 16 weeks, with no adverse effects reported, the sterilisation of male dogs and bitches is more complex, with very variable breed-specific risks that are only recently being explored (Hart et al., 2020a-b). Thus, in order to ensure good animal welfare, the restriction of behaviours and physical attributes of pet cats and dogs should be considered on an individual basis.

Sourcing

An animal's future behaviour, health, and welfare will be influenced both by its genetic make-up, and by its early life experiences. Temperament traits such as confidence and anxiety have been shown to have a level of heritability and may influence later behaviour and so breeding from parents that are temperamentally suitable is important. Additionally, both the mother's experience whilst pregnant and the perinatal environment may influence the later behaviours of the offspring. For example, stressful experiences during pregnancy have been shown to influence the behaviour and responses of the offspring to adverse experiences. For example, young animals born to stressed mothers are more likely to have increased sensitivity to pain, anxiety behaviours, and are more likely to develop abnormal behaviours (Braastad, 1998, Latham and Mason, 2008). It is for these reasons that obtaining a puppy or kitten from a reputable breeder raising animals bred from temperamentally stable parents, in a stimulating but not overtly fear-inducing environment (Rooney Clark et al., 2016), is important. Factors such as unpredictable handling, transport, and fear-provoking should be avoided because the experiences of dogs during their first year of life is crucial in determining their later behaviour and temperament (Foyer, Bjällerhag et al., 2014, Wauthier and Williams, 2018). In some regions (e.g., Europe, parts of the US), there is a large trade in puppies (which may be international), bred in intensive conditions. Such puppies may be at particular risk of such fear-inducing factors, as well as increased risks of transmissible diseases and parasites. The focus within the 'puppy mills' in which many are bred, is often on cost-minimisation, rather than optimal provision of preventative healthcare such as vaccination, parasiticides, or appropriate behavioural development and socialisation.

There are many commercial production systems including breeders and retailers that may supply puppies and kittens to meet a variety of commercial preferences, e.g., extreme conformation, particular aesthetics, etc. Puppies produced from commercial establishments have been shown to have poorer health and behavioural outcomes than those raised in the home (McMillan, 2017). Whilst shelters are a reputable and ethical way of sourcing a puppy or kitten, it is important that adequate socialisation opportunities and complex environments are provided.

Aversive experiences during early life such as pain, stress, or a lack of complexity will affect development and influence later behaviour. Due to the neuroplasticity of the neonatal nervous system, pain experiences at a young age can have lifelong impacts on an individual's behaviour and responses to future experiences (Schwaller and Fitzgerald, 2014). The socialisation window is a developmental period where young animals are able to learn appropriate social behaviours and become accustomed to new experiences. It occurs at approximately 4–16 weeks of age in puppies and 2–7 weeks of age in kittens. For puppies this is a period that would be affected by the pain experience associated with tail docking or ear cropping, potentially reducing exploratory and social behaviours and impeding appropriate social development. Inflicting deliberate unpleasant experiences such as cosmetic procedures like tail docking or ear cropping during this neonatal period has also been shown to influence pain sensitivity throughout life (Mellor, 2018, Reyes-Sotelo et al., 2020). Neonates have similar, if not increased, sensitivity to pain compared with adults (Noonan et al., 1996a, Fitzgerald and Beggs, 2001, Cameron et al., 2014) and an early pain experience may sensitise neonates to subsequent pain (Clark et al., 2014).

Convenience surgical procedures

A variety of elective surgical procedures may negatively impact the welfare of pet cats and dogs. Appendages such as tails, bony dew claws, and toes are often functional with complex anatomy including bone, nerves, muscle, and connective tissue. Surgical amputation of the tail in puppies results in severe pain (Noonan et al., 1996a), as does onychectomy (declawing) in cats. Transection of the nerves in the tail may result in neuroma formation. Neuromas are swellings of transected nerve endings. These may be associated with neuropathic pain or abnormalities which may include numbness, tingling, hypersensitivity, and actual pain, all of which can result in chronically poor welfare and self-injurious behaviour such as chewing or biting at animals' own bodies. Neuroma formation and pain-related behaviour, including severe self-injury, are documented in dogs after docking (Gross and Carr, 1990). In humans chronic pain associated with amputation occurs in 30–50% of cases (Kehlet et al., n.d.). The development of neuropathic pain can be prevented only by the administration of appropriate analgesic and anaesthetic drugs, including specifically ketamine (Wagner et al. 2002, Goldberg, 2017, Tsui and Chu, 2017) plus a local anaesthetic, but in one study only 10% of veterinarians used anaesthetics or analgesics in conjunction with tail docking of puppies (Noonan et al., 1996b). Only 0.59% of the total dog population visiting a veterinary practice are affected by tail injury, and only one in five of these tail injuries resulted in amputation, meaning that 0.118% (one in a thousand) dogs visiting a veterinary practice require therapeutic tail amputation (Cameron et al., 2014), with appropriate analgesia and anaesthesia delivered to minimise long-term welfare consequences. This risk should be balanced against the risk of acute pain in 100% of puppies that experience docking, and the potential chronic post-surgical pain in docked puppies (Bain, 2020). There are also no known benefits to ear cropping in dogs as it has not been shown to reduce the prevalence of ear infection or injury (Bain, 2020). In cats onychectomy is associated with significant pain (Wilson and Pascoe, 2016), increased lameness and house-soiling (potentially due to the discomfort of stepping on litter), indicative of chronic pain states (Tobias, 1994, Gerard et al., 2016).

Social communication in dogs relies on proper observation of tail signalling and ear positioning, suggesting that tail docking and ear cropping may impair social communication in dogs (Leaver and Reimchen, 2008), and this may contribute to behavioural problems. Similarly in cats, locomotion, climbing, and scratching are important elements of the normal feline behavioural repertoire. Frustration of behaviours has been suggested to be detrimental to emotional well-being and welfare (Hargrave, 2015). Behavioural problems have been shown to be significant and common indicators of welfare problems for pet dogs and cats (Rioja-Lang et al., 2019, Rioja-Lang et al., 2020, Yu et al., 2021). The lack of welfare benefits and the likely welfare detriments of cosmetic procedures mean they are ethically difficult to justify, and in all cases, veterinarians should consider their duty to their patient's welfare.

Breeding

Selective breeding of pet cats and dogs has resulted in a number of genetic and conformational disorders that negatively impact welfare. Some conformational disorders are rooted in genetics, e.g., the conformational changes seen in brachycephalic breeds are a result of selection that causes the expression of genes associated with pathology (Mansour et al., 2018), and so these groups of disorders are not mutually exclusive. Selected examples of these are shown in Table 18.1 (examples adapted from (Universities Federation for Animal Welfare, 2021)).

Nutritional management

Both dogs and cats have specific nutritional needs and require carefully balanced diets formulated for their individual species' needs. Cats are obligate carnivores and have an absolute requirement for taurine, an amino acid that may be derived from animal or synthetic sources. Alternatively, despite being classed as 'carnivores' dogs do not have an obligate requirement for meat, and the dietary preferences of domestic dogs are considerably different to their wild ancestors. Dogs on average choose to consume most of their calories from fat and then carbohydrate, with protein sources least preferred. Conversely cats choose to consume most of their calories from carbohydrate and then protein sources (Hall et al., 2018). These dietary preferences are interesting in the context of recent trends for bones and raw food (BARF) diets which often focus on high-protein ingredients. Such diets claim to be more 'natural', but are often modelled on the dietary preferences of ancestor species such as wolves, rather than those of domesticated dogs and cats. Additionally, such diets raise animal welfare and public health concerns as they increase the shedding of pathogenic bacteria from pet animals, and have been associated with infectious disease outbreaks in people and pets (Davies et al., 2019, O'Halloran et al., 2021). Cats fed fish-based raw diets may be at risk of thiaminase deficiency.

Similarly, a recent trend for 'grain-free' diets has been associated with the development of heart disease that improves upon conversion to a more typical commercial diet, in pet dog populations (Freid et al., 2021). Increasingly, obesity of pet dogs and cats is a significant welfare problem associated with a range of health problems as well as impacting on behavioural health and overall quality of life. Managing obesity in pet cats and dogs is challenging as it is a multifactorial and social issue, similarly to obesity in children.

Veterinary visits and medical considerations

Visits to the veterinary clinic are often stressful for a pet cat or dog as they may include a stressful or uncomfortable journey, a waiting room exposed to strange animals' sights and sounds, and unpleasant or painful experiences, e.g., vaccinations, microchipping, neutering. These aversive experiences may result in a learned fear response which may be a problem for future visits. Fear of the veterinary clinic may result in the cat or dog being forcefully restrained or muzzled which exacerbates the stressful experience. This is not only unpleasant for the animal but may be a reason for owners not to seek timely veterinary attention, and thus may result in delayed treatment or poor owner compliance with treatment. These negative experiences can be mitigated by veterinary professionals spending time "building the welfare bank account" (Fisher, 2015). To build the welfare bank account, owners should be advised to bring their pet to the

Table 18.1 Selected examples of genetic or conformational disorders in dogs and cats

Dogs			
Genetic disorders	*Impacts on animal welfare*	*Conformational disorders*	*Impacts on animal welfare*
Bedlington Terrier: Copper storage hepatopathy	Variable, ranging from mild in the early stages of disease to a severe reduction in the quality of life in the later stages, mainly due to animals feeling unwell	French bulldogs, English bulldogs, Pugs, etc.:Brachycephalic Obstructive airway syndrome	A syndrome of multiple conformational pathologies arising as a result of selection for brachycephalic conformation. May comprise, narrowed nares (nostrils), soft palate obstruction, enlarged tonsils, and everted laryngeal ventricles.
Dalmation, hyperuricosuria	Hyperuricosuria results in the formation of urate stones. Welfare issues range from moderate cystitis to complete obstruction of the urinary tract causing severe pain	Pekinese, Pug, French bulldog, Brachycephalic ocular syndrome	A syndrome associated with brachycephalic conformation (e.g., exophthalmos (bulging eyes), macropalpebral fissure (an excessively wide opening of the eyelids compared to the size of the eye) and lagophthalmia (inability to close the eyelids completely), distichiasis (abnormally placed eyelashes that rub on the eye), and poor tear production), resulting in lesions to the eyelid, conjunctiva and cornea

Cats			
Genetic disorders	*Impacts to animal welfare*	*Conformational disorders*	*Impacts to animal welfare*
Devon rex, Sphynx: Hereditary myopathy	Generalised muscle weakness caused by an abnormality in the transmission of signals from nerves to muscles. Affected cats may show signs of muscle weakness or be unable to walk and exercise normally. They may show signs such as muscle tremors, fatigue and collapse, and may be at risk of choking	Scottish Fold: Osteochondrodysplasia	Osteochondrodysplasia is a developmental abnormality of the cartilage and causes the characteristic ear fold as well as multiple joint pathologies. Affected cats may have significantly abnormal conformation including short wide limbs and a short, inflexible tail. They may display lameness, swollen carpal and tarsal joints, have an abnormal gait, and may be reluctant to move and jump. Severely affected individuals become crippled and unable to walk

(Continued)

Table 18.1 (Continued)

Cats			
Genetic disorders	*Impacts on animal welfare*	*Conformational disorders*	*Impacts on animal welfare*
British Shorthair, Maine Coon: Hypertrophic cardiomyopathy	Hypertrophic cardiomyopathy occurs when the thickness of the heart wall increases to an abnormal level. This thickening impedes normal heart contraction as the thickened wall does not function properly, resulting in heart failure. Heart failure is progressive and worsens over time, eventually leading to death due to the heart failing to pump blood adequately around the body or from the effects of a thromboembolus, or more slowly after progressive complications caused by poor circulation and fluid build-up in the chest and lungs affecting breathing	Manx: Manx syndrome/spina bifida	The genetic mutation that causes Manx (tail-less) cats also frequently causes severe disease consequences because of its effects on the development of the spine and spinal cord. The various forms of spina bifida which commonly occur lead to complications which can cause partial paralysis, prevent normal behaviours, cause incontinence, and lead to painful infections

veterinary clinic regularly, even if they have no medical concerns. This allows gentle exposure of the cat or dog to the staff and environment (weight or development checks provide the ideal opportunity for this type of interaction). Veterinary staff can pet and stroke the pet, if this is something he or she enjoys, and offer treats. Such activities start to develop a positive foundation and trusting relationship which may then offset any anxiety the pet has when actual treatments are needed. Many animals can cope with one stressor or trigger, but when several stressors are 'stacked' on top of each other, defensive aggression may occur. For example, a cat being forced into a carry box, transported in the car, placed next to a dog in the waiting room, then handled by the vet will experience multiple significant stressors in a short period and may well have exceeded its ability to cope even before the vet handles it, and the fairly benign experience of handling may be the 'final straw' for an animal that is already 'trigger-stacked'. For these reasons, considering the pet's experience when designing clinics and taking steps to minimise stressors (separate cat and dog waiting areas, pheromone diffusers – which can exert a calming effect, cat box shelving to place cats high up, etc.) can all help to improve the welfare of cat and dog patients.

Similarly, it is important that all members of the veterinary team are familiar with cat and dog behaviour and low-stress handling techniques, in order to ensure that a pet's visit is not more stressful than needed. Both cats and dogs will display behaviours such as lip-licking and gaze-aversion that indicate they are stressed. These behaviours will escalate into threat-distancing behaviours include moving away, growling and aggression. The behavioural response to stress in dogs is often referred to as the "ladder of aggression" (Shepherd, 2009). By participating in schemes such as the International Society for Feline Medicine (ISFM) feline friendly practice scheme or 'fear free' training programmes, veterinary staff can ensure that clinical stressors are minimised, and animal welfare is protected within the clinical environment.

Another way that veterinary professionals can safeguard animal welfare is by minimising the cat or dog's experiences of pain. Pain is a complex phenomenon and each individual animal will experience pain differently. Regardless of species, pain is best managed early and proactively as it has a significant impact on animal welfare. It is much more difficult to control pain once it is well established than it is to manage pain before it becomes severe. Preventing pain should always be the aim of the analgesia plan – surgical pain is 100% predictable and therefore a good analgesic plan should be in place for every patient, prior to undertaking any surgical procedure. Pre-emptive analgesia is the treatment of pain using analgesic drugs before the introduction of a potentially painful stimulus (i.e., surgery). Pre-emptive analgesia reduces the nociceptive input to the spinal cord, thus reducing peripheral and central sensitisation. This then reduces peri- and post-operative pain and hyperalgesia (increased pain perception). Similarly chronic pain due to degenerative conditions should be proactively managed and not simply considered as an unavoidable consequence of 'old age'. Dental disease and osteoarthritis are common sources of chronic pain in pet animals and may create significant welfare problems if left untreated. Pain from such degenerative conditions may be caused by inflammation (inflammatory pain), tissue damage (nociceptive pain), or nerve damage (neuropathic pain). A pet dog or cat may actually experience more than one type of pain concurrently, and thus multi-modal analgesia may be required to adequately manage the different types of pain. Different classes of drugs act at different sites along the pain pathway and thus can often be safely used in combination to provide optimal analgesia. Additionally because pain is the result of central processing by the nervous system, mental stimulation, environmental enrichment, and positive emotional experiences may play a role in mitigating the pain experience, so owners should also be engaged in this process.

Quality of Life (QoL) issues

Providing good standards of veterinary hospice and palliative care to terminally ill cats and dogs is currently hindered by an inadequate evidence base to guide veterinarians. The American Animal Hospital Association/International Association for Animal Hospice and Palliative Care (AAHA/IAAHPC) *End-of-Life Care Guidelines for Dogs and Cats* produced in 2016 provide guidance on central issues including client communication and patient care. Various ethical perspectives on animal death and euthanasia have been described from "death is not a welfare issue" (Webster, 1995) and the pragmatic view that euthanasia literally means killing painlessly (regardless of the reason for doing so), to "death is a welfare issue" as the potential for future suffering does not universally justify pre-emptive euthanasia (Yeates, 2010) (see Chapter 13 for a fuller discussion of this issue). From a welfare science perspective, we are generally less concerned with quantity of an animal's life and more concerned with the quality of that life and this remains the focus of animal welfare science research. The definition of, and decision-making around the acceptability of euthanasia, is primarily studied in the field of applied animal ethics, e.g. (Persson et al., 2020). In reality however, disentangling a pragmatic and pre-emptive approach to euthanasia, from the associated moral stress and emotional toll on the client and veterinary professionals, is challenging.

Working and service dogs and cats

The contribution that working dogs make to human society is undeniable – they provide assistance to people with disabilities, guard and herd livestock, protect and serve in the armed forces and police, ensure border security, and detect illegally trafficked drugs and other products. Even cats may provide services to human communities in terms of pest control, and as therapy animals in schools and residential settings. But how often do we consider the welfare of cats and dogs used in these ways? The life of a working dog varies enormously depending on the dog itself and the role it is expected to perform. For example, it may be considered that a working border collie engaged in shepherding work and field trials, and living with other dogs on a farm, is able to fulfil many aspects of its strongly motivated behavioural drives. In this scenario, work likely results in positive behavioural and emotional experiences. However, in some situations, such dogs may have primarily instrumental value, depending on the labour they provide, and as such a lack of comprehensive veterinary care or nutritional support could negatively impact on welfare (Littlewood and Mellor, 2016). Dogs trained to sniff out explosives or people, such as those used in war zones or natural disasters, may experience similar positive experiences through their work and the opportunity to satiate important behavioural drives. However, environmental stressors including extreme temperatures, low humidity, and the wearing of body armour may incur additional unpleasant physical experiences. On top of these welfare considerations, there is an additional ethical dilemma around the use of 'innocent' animals in such dangerous roles, where the risk of injury or death is apparent and the benefits to the dog are less obvious. Interestingly, we tend to be less ethically concerned with the use of support/assistance dogs for people with disabilities. In some situations it may be that such dogs actually lead lives of fairly significant behavioural restriction whilst working for long periods. In such cases it is important that the dogs get 'time off' to relax, play, and exercise. For visiting therapy dogs a working visit time of one hour is often suggested to ensure that dogs are not overly restricted or stressed, although in one survey of USA therapy dogs, around half of organisations surveyed did not offer any guidance on visit duration (Serpell et al., 2020).

Unrestricted cats and dogs

Cats and dogs roam freely in many parts of the world. Free-roaming offers animals significant behavioural choice, control, and complexity – the lack of which can cause significant behavioural and welfare problems in restricted pet animals, but it also generates risks including increased risk of physical trauma and illness For example, free-roaming animals may be more at risk of infectious disease such as transmissible venereal tumours, increased risk of non-infectious disease, e.g., from toxin exposure or injury due to fights or road traffic accidents, and increased risk of environmental stressors, e.g., extremes of temperature, extremes of resource availability, etc. In many situations cats and dogs free-roam but still rely on humans for resource provision either directly, for example a human guardian providing food or shelter, or indirectly through scavenging garbage or sheltering in human-constructed dwellings or under vehicles. Animals may enter the free-roaming population from birth, or be relinquished, abandoned, or lost. Cats and dogs that are abandoned or lost often end up in the free-roaming population for at least a period of time. They may remain there or alternatively may be caught and removed from that population. Their welfare will depend on their experiences. Free-roaming dog and cat populations may be managed for multiple reasons (Table 18.2).

Management of dog and cat populations

There are multiple reasons to manage dog and cat populations (Table 18.2).

Two primary methods of stray animal population control are recommended by the European Convention for the Protection of Pet Animals of 1988 (Council of Europe, 1987): (1) catch and removal methods, and (2) Catch-Neuter-Return methods.

Catch and removal

Catch and removal methods are usually unsuccessful at managing dog populations in the long term. The two primary methods used are catch and cull, or catch and remove to a shelter. Neither approach addresses the underlying cause of dog overpopulation but simply mitigates to some extent the symptoms of dog or cat overpopulation. This approach is therefore usually unsuccessful.

Table 18.2 Why manage dog and cat populations?

To protect public health	*To address dog and cat suffering*
Bites	Malnutrition and dehydration
Road traffic accidents	Illness and injury
Fouling	Fear and distress (both from competition between dogs/cats but also aggression from people, including inhumane methods of dog/cat control)
Noise	Poor human–animal relationships
Rabies cases	
Nuisance complaints	
Poor human–animal relationships	

Catch and cull

Culling is the removal of all or part of a dog or cat population. Culling is ineffective in population control as stray dog or cat numbers are often supported by the breeding of owned dogs and cats, or by migration from other areas. Culling raises a variety of animal welfare and ethical problems. Enforcing extreme levels of population reduction by culling is logistically impractical and often ethically unacceptable due to the severe welfare consequences of typical culling methods (blunt force trauma, gassing, poison). Non-targeted culling will often remove far more healthy and friendly dogs and cats than sick or aggressive ones. This may inadvertently select for a more aggressive population, thus potentially increasing conflict with humans or other animals.

Removal to a shelter

Sheltering and or 'rescue' of free-roaming cats and dogs is often well-intentioned, but it is resource-heavy and may inadvertently create welfare problems as well as potentially 'encouraging' relinquishment by providing an easy disposal option for owners. Shelters should aim to provide secure and suitable environments for the maintenance of physical and psychological well-being of stray animals, until a suitable permanent home can be found.

Shelters should not provide a long-term lifestyle for dogs or cats as it may be very difficult to deliver good welfare in the shelter environment, e.g.:

- Shelter dogs need 20–25 minutes per day of exercise and positive human contact to reduce stress (Menor-Campos et al., 2011);
- Dogs in long-term kennelling may suffer damage to their hearing due to noise exposure (Scheifele et al., 2012);
- Cats may struggle with the change to a confined environment (Jongman, 2007);
- Cats may become highly stressed when housed in proximity to other cats (Ottway and Hawkins, 2003).

Strategic removal/individual euthanasia

Strategic removal from a population may be necessary for dogs or cats suffering physical or behavioural conditions that adversely affect their welfare or pose a threat to humans, e.g., predatory aggression in dogs. Euthanasia for health or welfare reasons may be used to remove individuals but is unlikely to influence the overall population. In many cases shelters may utilise euthanasia as a management tool to ensure that limited resources are targeted most effectively towards dogs and cats with the greatest chances of rehoming. In others, shelters may have a no-kill policy and refuse to engage in euthanasia. Neither approach is universally right or wrong but should be appraised depending on the welfare impacts such strategies have on the cats and dogs. It may be posited that euthanasia (a 'good death') does not in itself generate any welfare problems as long as it is performed humanely, however the killing of potentially healthy pet animals is a source of ethical discomfort for many people, even though the killing of healthy farm animals for food often is not. Regardless of the approach taken, shelters should ensure they are able to provide good welfare for the animals they house, with access to adequate staff and resources to ensure their key welfare needs (see Chapter 3) are provided.

Catch-Neuter-Return (CNR)

There are a variety of CNR methods available. Well-planned CNR is the most successful long-term approach to managing dog populations, though modelling indicates that its efficacy in cat populations may vary. Pet dog and cat reproductive control must also be considered. Female-focused desexing is more efficient for population control and should be the focus of all canine and feline neutering programmes.

Surgical sterilisation (desexing)

All surgery requires excellent asepsis, analgesia, and good surgical technique; thus appropriate veterinary training is essential. Desexed dogs are generally healthier and live longer than non-desexed dogs. Appropriately supervised desexing of street dogs may provide valuable veterinary student practical experience, and standards of medicine and surgery practised should be equivalent to those applied for owned pet dogs. If such standards cannot be achieved then the ethics of the sterilisation programme should be re-examined – it is not acceptable to neuter cats and dogs simply for practice or as part of 'feel-good voluntourism' (Ryan et al., 2019). Even with good standards of care, cats and dogs will experience moderate to significant welfare problems including capture, transport, kennelling, surgery, and often inadequate analgesia or anaesthesia (Bacon et al., 2017, Bacon et al., 2019).

Medical contraception

Medical options are available to control reproduction in cats and dogs, but many are as yet not validated in terms of permanency or side effects. Medical contraception of female dogs and cats is non-permanent (progesterones) and may increase the risk of significant side effects (pyometra – uterine infection, etc.).

Medical sterilisation of male dogs (intra-epididymal CaCl or Zinc Gluconate) has had promising results but has not yet been robustly evaluated, and side effects do occur. 'Zeuterin' (zinc gluconate) is FDA-approved in the USA, but a Chilean study showed side effects including infection and pain requiring sedation and aseptic preparation of the area. 'Suprelorin' (deslorelin) acts as a GnRH agonist and provides approximately six months of contraception to male dogs. It is unlicensed in female dogs but has been trialled in free-roaming females with some success. As yet medical contraceptives are not reliable enough to form the basis of canine or feline population control, but may provide useful adjunctive therapy.

Potential solutions – the population management 'toolkit'

In addition to reproductive control as described above, stray dog and cat populations are often reliant on human-produced resources to maintain their populations. Thus, ancillary strategies can help to support population management.

Responsible pet ownership

Many pet owners do not recognise the intrinsic link between their own pets and more general dog or cat overpopulation. Indiscriminate breeding of pet dogs and cats generates puppies and kittens which may end up in shelters or free-roaming on the street, as there are limited lifelong homes available for them. Responsible pet ownership reduces human–animal conflict and disease transmission, through appropriate reproductive control, training, and vaccination of pet cats and dogs.

Public education in cat and dog behaviour

Problem behaviours are one of the leading causes of pet cat and dog abandonment or relinquishment. Children under 15 years of age are most likely to be bitten by dogs, and the majority of bites occur in the family home or at a family member or friend's home, by a known dog. Effective parental supervision of child–dog and child–cat interactions, and effective education of schoolchildren and the general public on the subject of safe dog–human interactions, are essential to reduce the risk of pet animal bites. There is no evidence that breed-specific or 'dangerous dog' legislation has any impact on improving public safety. Instead, a 'deed not breed' approach should be taken with a focus on behaviour rather than breed type. As pet dog or cat bites or scratches may lead to pets being abandoned or relinquished, preventing aggression can help to reduce stray dog and cat numbers, as well as improving the welfare of cats and dogs in the home, as owners use more informed and appropriate behavioural management techniques.

General welfare concerns

In addition to the potential welfare problems described above, there are a number of overarching welfare problems that may impact cats and dogs, regardless of their role in society and their level of restriction.

Neglect

In general dogs, and to a lesser extent cats, are dependent upon humans to provide them with resources and opportunities (nutrition, shelter, veterinary medicine, exercise, companionship, mental stimulation, behavioural opportunities) to ensure they enjoy a good quality of life. Neglect may be described as the state of being uncared for, and we may apply this description to dogs and cats that do not have appropriate provisions and care. Whilst we may generally consider unrestricted cats and dogs to be more at risk of neglect, as their relationships with their caregivers is often more tenuous and may even be absent entirely, other categories of cats and dogs may also experience neglect. For example, dogs valued instrumentally, e.g., working dogs, racing dogs, etc., may find themselves neglected or abandoned if they are no longer able to perform their expected tasks.

Even much-loved pet cats and dogs may be neglected as neglect of welfare needs may occur even with good intentions. An obvious example is that of much-loved pet dogs and indoor housecats that are under-exercised and do not receive adequate opportunities for social behaviour or mental stimulation – such dogs and cats may be bored and frustrated, even though their owner loves them. Cats living a semi-restricted lifestyle (indoor–outdoor) that find their needs unmet may even choose to move into another home that provides an improved quality of life, regardless of the desires of the original owner. In some situations the intense emotional connection that some people feel for cats or dogs may even create welfare problems. Hoarding of pets is recognised as a mental health problem and psychiatric disorder, and may require a multi-agency approach including mental health, social services, animal welfare, fire, law enforcement, and environmental health services. Hoarding can create significant animal welfare problems through a combination of a lack of resources for the cats and dogs involved, plus a denial by the hoarder that there is a problem, and persistence of the hoarding behaviour. In such cases there are often also public health concerns due to poor sanitation and waste management, and pest infestation.

Abuse

Abuse is a more active process than neglect, involving intentional actions which detrimentally affect the welfare of cats and dogs. Examples include physical or sexual injuries or deliberate cruelty aimed at depriving the animal of its needs (or the shooting, injuring, suffocating, burning, or scaring of cats and dogs, or their use in zoophilic sexual practices). There is also a connection between abuse and neglect of animals, and that of people, as the emotional, physical, and sexual abuse of people and animals may be committed by the same group of perpetrators (Arkow, 2019, Fitzgerald et al., 2020).

Three worrying behaviours in childhood that may indicate later antisocial behaviours have been suggested to be: animal cruelty, fire setting, and bed wetting. These are known as the 'MacDonald triad'. Whilst the 'triad' of behaviours has been challenged as a reliable indicator of later violence or aggression, there is a clear link between animal cruelty in childhood and abusive behaviour as an adult (Parfitt and Alleyne, 2020). Conversely, research also indicates that positive human–animal relationships may help with the development of empathy, compassion, and prosocial behaviours in children (Hawkins et al., 2017). This highlights the importance of recognising the connection between human and animal abuse, and the opportunity for promoting evidence-based rehabilitation strategies that may positively impact the lives of both people and animals.

Meat trade

The consumption of dogs and cats for meat is not a geographically limited practice, but is generally confined to certain cultural groups and communities within cultures (Li et al., 2017). Dog meat consumption may affect a range of 'types' of dog including free-roaming, pet, stray, and crossbreeds (Dugnoille, 2018). The animals affected may be farmed, stolen pets, or taken from the street. They are often subjected to long-distance transport in cramped, unsanitary conditions (including across national borders), may be force-fed rice porridge to artificially increase their slaughter weight, and then slaughtered without stunning using a range of methodologies, including blunt-force trauma, immolation (burning), skinning, or throat-cutting. Such experiences confer obvious welfare problems including fear, stress, exposure to infectious diseases, physical and thermal discomfort, pain, and physical and psychological trauma.

Conclusions: love is not enough

This chapter has outlined some of the many challenges faced by cats and dogs, whether managed under human care or roaming freely within human society, as well as exploring some overarching issues and the connection between human and animal welfare ('one welfare'). One key consideration is that animals with whom we have an emotional connection, such as our companion animals, are sometimes less well scrutinised and protected than animals in other industries, e.g., laboratory research or livestock animals. There may often be an assumption that our empathy and care towards cats and dogs confers an enhanced level of welfare upon them, but in reality, it may be that our affection for cats and dogs may actually limit our ability to objectively assess their welfare. We often assume that our good intentions will translate into good welfare outcomes. In many situations, however, this is not the case, and so separating out our emotional response to cat and dog welfare issues, and evaluating their welfare objectively, is an important first step in ensuring that we really do provide good welfare for the cats and dogs with whom we share our lives.

References

Arkow P, 2019. Chapter 17: The "dark side" of the human-animal bond. In Kogan L and Blazina C, eds. *Clinician's Guide to Treating Companion Animal Issues*. Cambridge, MA: Academic Press.

Bacon H, Vancia V, Walters H and Waran N, 2017. Canine trap-neuter-return: A critical review of potential welfare issues. *Animal Welfare*, 26, pp. 281–292.

Bacon HJ, Walters H, Vancia V and Waran N, 2019. The recognition of canine pain behaviours, and potentially hazardous Catch-Neuter-Return practices by animal care professionals. *Animal Welfare*, 28, pp. 299–306.

Bain M, 2020. Surgical and behavioral relationships with welfare. *Frontiers in Veterinary Science,* 519, p. 7.

Braastad BO, 1998. Effects of prenatal stress on behaviour of offspring of laboratory and farmed mammals. *Applied Animal Behaviour Science*, 61, pp. 159–180.

Cameron N, Lederer R, Bennett D and Parkin T, 2014. The prevalence of tail injuries in working and non-working breed dogs visiting veterinary practices in Scotland. *Veterinary Record*, 174, p. 450.

Clark C, Murrell J, Fernyhough M, O'rourke, T and Mendl M, 2014. Long-term and trans-generational effects of neonatal experience on sheep behaviour. *Biology Letters,* 10(7), p.20140273.

Council of Europe, 1987. European convention for the protection of pet animals. Strasbourg, 13.XI.1987. Europe. http://conventions.coe.int/Treaty/en/Treaties/Html/125.htm [accessed on 31st October 2021].

Davies RH, Lawes JR and Wales AD, 2019. Raw diets for dogs and cats: A review, with particular reference to microbiological hazards. *Journal of Small Animal Practice*, 60, pp. 329–339.

Dugnoille J, 2018. To eat or not to eat companion dogs: Symbolic value of dog meat and human–dog companionship in contemporary South Korea. *Food, Culture & Society*, 21, pp. 214–232.

Fisher S, 2015. Bank on it. In Dog's Trust, ed. International Companion Animal Welfare Conference, Porto, Portugal.

Fitzgerald M and Beggs S, 2001. The Neurobiology of pain: Developmental aspects. *The Neuroscientist*, 7, pp. 246–257.

Fitzgerald AJ, Barrett BJ, Gray A and Cheung CH, 2020. The connection between animal abuse, emotional abuse, and financial abuse in intimate relationships: Evidence from a nationally representative sample of the general public. *Journal of Interpersonal Violence*. 37(5–6). p. 0886260520939197. doi: 10.1177/0886260520939197.

Foyer P, Bjällerhag N, Wilsson E and Jensen P, 2014. Behaviour and experiences of dogs during the first year of life predict the outcome in a later temperament test. *Applied Animal Behaviour Science*, 155, pp. 93–100.

Freid KJ, Freeman LM, Rush JE, Cunningham SM, Davis MS, Karlin ET and Yang VK, 2021. Retrospective study of dilated cardiomyopathy in dogs. *Journal of Veterinary Internal Medicine*, 35, pp. 58–67.

Gerard AF, Larson M, Baldwin CJ and Petersen C, 2016. Telephone survey to investigate relationships between onychectomy or onychectomy technique and house soiling in cats. *Journal of the American Veterinary Medical Association*, 249, pp. 638–643.

Goldberg ME, 2017. A look at chronic pain in dogs. *Veterinary Nursing Journal*, 32, pp. 37–44.

Gross TL and Carr SH, 1990. Amputation neuroma of docked tails in dogs. *Veterinary Pathology*, 27, pp. 61–62.

Hall JA, Vondran JC, Vanchina MA and Jewell DE, 2018. When fed foods with similar palatability, healthy adult dogs and cats choose different macronutrient compositions. *Journal of Experimental Biology*, 221(14), p. jeb173450.

Hargrave C, 2015. Anxiety, fear, frustration and stress in cats and dogs: Implications for the welfare of companion animals and practice finances. *Companion Animal*, 20, pp. 136–141.

Hart BL, Hart LA, Thigpen AP and Willits NH, 2020a. Assisting decision-making on age of neutering for 35 breeds of dogs: Associated joint disorders, cancers, and urinary incontinence. *Frontiers in Veterinary Science*, 7. p. 388.

Hart BL, Hart LA, Thigpen AP and Willits NH, 2020b. Assisting decision-making on age of neutering for mixed breed dogs of five weight categories: Associated joint disorders and cancers. *Frontiers in Veterinary Science*, 7. pp. 472.

Hawkins RD, Williams JM and Animals S. S. F. T. P. O. C. T., 2017. Childhood attachment to pets: Associations between pet attachment, attitudes to animals, compassion, and humane behaviour. *International Journal of Environmental Research and Public Health*, 14, p. 490.

Jongman EC, 2007. Adaptation of domestic cats to confinement. *Journal of Veterinary Behavior*, 2, pp. 193–196.

Kehlet H, Jensen TS and Woolf CJ, n.d. Persistent postsurgical pain: Risk factors and prevention. *The Lancet*, 367, pp. 1618–1625.

Latham NR and Mason GJ, 2008. Maternal deprivation and the development of stereotypic behaviour. *Applied Animal Behaviour Science*, 110, pp. 84–108.
Leaver SDA and Reimchen TE, 2008. Behavioural responses of Canis familiaris to different tail lengths of a remotely-controlled life-size dog replica. *Behaviour*, 145, pp. 377–390.
Li PJ, Sun J and Yu D, 2017. Dog "meat" consumption in china: A survey of the controversial eating habit in two cities. *Society & Animals*, 25, pp. 513–532.
Littlewood KE and Mellor DJ, 2016. Changes in the welfare of an injured working farm dog assessed using the five domains model. *Animals*, 6, p. 58.
Mansour TA, Lucot K, Konopelski SE, Dickinson PJ, Sturges BK, Vernau KL, Choi S, Stern JA, Thomasy SM, Döring S, Verstraete FJM, Johnson EG, York D, Rebhun RB, Ho H-YH, Brown CT and Bannasch DL, 2018. Whole genome variant association across 100 dogs identifies a frame shift mutation in DISHEVELLED 2 which contributes to Robinow-like syndrome in Bulldogs and related screw tail dog breeds. *PLOS Genetics*, 14, p. e1007850.
Mcmillan FD, 2017. Behavioral and psychological outcomes for dogs sold as puppies through pet stores and/or born in commercial breeding establishments: Current knowledge and putative causes. *Journal of Veterinary Behavior*, 19, pp. 14–26.
Mellor DJ, 2018. Tail docking of canine puppies: Reassessment of the tail's role in communication, the acute pain caused by docking and interpretation of behavioural responses. *Animals*, 8, p. 82.
Menor-Campos DJ, Molleda-Carbonell JM and López-Rodríguez R, 2011. Effects of exercise and human contact on animal welfare in a dog shelter. *Veterinary Record*, 169, pp. 388–388.
Noonan GJ, Rand JS, Blackshaw JK and Priest J, 1996a. Behavioural observations of puppies undergoing tail docking. *Applied Animal Behaviour Science*, 49, pp. 335–342.
Noonan GJ, Rand JS, Blackshaw JK and Priest J, 1996b. Tail docking in dogs: A sample of attitudes of veterinarians and dog breeders in Queensland. *Australian Veterinary Journal*, 73, pp. 86–88.
O'halloran C, Tørnqvist-Johnsen C, Woods G, Mitchell J, Reed N, Burr P, Gascoyne-Binzi D, Wegg M, Beardall S, Hope J and Gunn-Moore D, 2021. Feline tuberculosis caused by Mycobacterium bovis infection of domestic UK cats associated with feeding a commercial raw food diet. *Transboundary and Emerging Diseases*, 68, pp. 2308–2320.
Ottway D and Hawkins D, 2003. Cat housing in rescue shelters: A welfare comparison between communal and discrete-unit housing. *Animal Welfare*, 12, pp. 173–189.
Parfitt CH and Alleyne E, 2020. Not the sum of its parts: A critical review of the MacDonald Triad. *Trauma, Violence, & Abuse*, 21, pp. 300–310.
Persson K, Selter F, Neitzke G and Kunzmann P, 2020. Philosophy of a "good death" in small animals and consequences for euthanasia in animal law and veterinary practice. *Animals*, 10, p. 124.
Reyes-Sotelo B, Mota-Rojas D, Martínez-Burnes J, Gómez J, Lezama K, González-Lozano M, Hernández-Ávalos I, Casas A, Herrera Y and Mora-Medina P, 2020. Tail docking in dogs: Behavioural, physiological and ethical aspects. *CAB Reviews*, 15(39). pp. 1–13.
Rioja-Lang F, Bacon H, Connor M and Dwyer CM, 2019. Determining priority welfare issues for cats in the United Kingdom using expert consensus. *Veterinary Record Open*, 6, p. e000365.
Rioja-Lang F, Bacon H, Connor M and Dwyer CM, 2020. Prioritisation of animal welfare issues in the UK using expert consensus. *Veterinary Record*, 187, pp. 490–490.
Rooney NJ, Clark CCA and Casey RA, 2016. Minimizing fear and anxiety in working dogs: A review. *Journal of Veterinary Behavior: Clinical Applications and Research*, 16, pp. 53–64.
Ryan S, Bacon H, Endenburg N, Hazel S, Jouppi R, Lee N, Seksel K and Takashima G, 2019. WSAVA animal welfare guidelines for veterinary practitioners and veterinary teams. *Journal of Small Animal Practice*, 60(5), p. 265.
Scheifele P, Martin D, Clark JG, Kemper D and Wells J, 2012. Effect of kennel noise on hearing in dogs. *American Journal of Veterinary Research*, 73, pp. 482–489.
Schwaller F and Fitzgerald M, 2014. The consequences of pain in early life: Injury-induced plasticity in developing pain pathways. *European Journal of Neuroscience*, 39, pp. 344–352.
Serpell JA, Kruger KA, Freeman LM, Griffin JA and Ng ZY, 2020. Current standards and practices within the therapy dog industry: Results of a representative survey of united states therapy dog organizations. *Frontiers in Veterinary Science*, 7, p. 35.
Shepherd K, 2009. Behavioural medicine as an integral part of veterinary practice. *BSAVA Manual of Canine and Feline Behavioural Medicine*. British Small Animal Veterinary Association, Gloucester, UK.

Tobias KS, 1994. Feline Onychectomy at a teaching institution: A retrospective study of 163 cases. *Veterinary Surgery*, 23, pp. 274–280.
Tsui PY and Chu MC, 2017. Ketamine: An old drug revitalized in pain medicine. *BJA Education*, 17, pp. 84–87.
Universities Federation for Animal Welfare, 2021. *Genetic Welfare Problems of Companion Animals* [Online]. Universities Federation for Animal Welfare. Available at: https://www.ufaw.org.uk/genetic-welfare-problems-intro/genetic-welfare-problems-of-companion-animals-intro [Accessed 31st October 2021].
Wagner AE, Walton JA, Hellyer PW, Gaynor JS and Mama KR, 2002. Use of low doses of ketamine administered by constant rate infusion as an adjunct for postoperative analgesia in dogs. *Journal of the American Veterinary Medical Association*, 221, pp. 72–75.
Wauthier LM and Williams JM, 2018. Using the mini C-BARQ to investigate the effects of puppy farming on dog behaviour. *Applied Animal Behaviour Science*, 206, pp. 75–86.
Webster J, 1995. *Animal Welfare: A Cool Eye Towards Eden*. Blackwell Science, Oxford, UK.
Wilson DV and Pascoe PJ, 2016. Pain and analgesia following onychectomy in cats: A systematic review. *Veterinary Anaesthesia and Analgesia*, 43, pp. 5–17.
Yeates JW, 2010. Death is a welfare issue. *Journal of Agricultural and Environmental Ethics*, 23, pp. 229–241.
Yu Y, Wilson B, Masters S, Van Rooy D and Mcgreevy PD, 2021. Mortality resulting from undesirable behaviours in dogs aged three years and under attending primary-care veterinary practices in Australia. *Animals*, 11, p. 493.

19
EQUINES

Sophie Hill

Introduction

Welfare concerns regarding equidae predominantly relate to training-, management-, and health-associated practices. Recent studies have determined the most important current issues for equine welfare are fear and stress involved in horse use; lack of owner knowledge of welfare needs of horses; inability of owners to recognise pain behaviour; obesity and inadequate feeding practices; lack of turnout and lack of social companions; inappropriate drug use and poor disease prevention; breeding practices including abrupt individual weaning and over-breeding; trading and unwanted horses; transportation; and delayed euthanasia decisions (DuBois et al., 2018a; McGreevy et al., 2018; Rioja-Lang et al., 2020).

Horses are often classified neither as companion nor livestock animals and will commonly pass through several different homes during their lifetimes. They are kept for leisure, racing, and other competitive roles, used in tourism, for draft purposes, in therapy, for the meat and skin trade, for conservation grazing, and some subsist as feral horse populations. These roles carry different risks for horse welfare.

Training methodology

The International Federation for Equestrian Sports, Fédération Equestre Internationale (FEI) is the international governing body of equestrian sports including dressage, combined driving, endurance, eventing, reining, show jumping, and vaulting. The FEI requires all those involved in international equestrian sports to adhere to the FEI's Code of Conduct for the Welfare of the Horse (FEI, 2013), "to acknowledge and accept that at all times the welfare of the horse must be paramount" and "must never be subordinated to competitive or commercial influences". This Code advises that all stages of preparation, training, on-site competition, and post-competition fate should be guided by welfare concerns (FEI, 2013). Events under these disciplines unaffiliated to FEI do not fall under FEI remit. The FEI does not govern horse racing which is regulated by International Federation of Horseracing Authorities (IFHA), or polo, which is regulated by the Federation of International Polo (FIP). However, a partnership exists between the FEI and FIP, and the International Horse Sports Confederation (IHSC) formally facilitates cooperation between the FEI and IFHA. The FEI, IFHA, and FIP

DOI: 10.4324/9781003182351-23

work in conjunction with the National Federation of each member country to regulate and govern equestrian disciplines. Despite steps taken to protect welfare of the horse by these and other organisations, horses suffer through many common and routine management and training practices.

There is growing awareness of arousal levels necessary for optimal learning and application of evidence-based ethical training approaches that can result in better welfare throughout a horse's life. For example, King et al. (2019) report on the successful application of the International Society for Equitation Science (ISES) Ten Training Principles (ISES, 2018) to early training of thoroughbreds, and describe horses better able to cope with the preparation for the next stages of their career, and increased chance of a career after racing. However, many common unethical and inhumane handling and training techniques are in use today. These range from use and abuse of training devices, to training which does not allow the musculoskeletal system to adapt to strains of the discipline. Dyson's (2021) ridden horse pain ethogram demonstrates behaviours which are at least ten times more likely in a horse with musculoskeletal pain, yet these same behaviours are often perceived as a sign of resistance and subsequently met with inappropriate responses.

Improper bits or improper use of bits can result in damage to the bars of the mouth, lacerations to the tongue or lip commissures, and compress tongue circulation. Aids such as spurs, whips, and side and draw reins are further tools of potential abuse. Analysis of the whip during racing has shown that the horses hit the most frequently, do not win (Arthur, 2011). British whipping rules recognise three legitimate uses of the whip; safety, correction, and encouragement. The British Horseracing Authority (BHA) has begun a consultation process on the use of the whip in the UK, with recommendations expected early 2022. Restrictive nose bands are associated with elevated physiological stress responses and increased prevalence of mouth injuries. They exert sustained restrictive pressures on sensitive tissues and may mask pain behaviours such as tongue movement and opening of the mouth, which may be performed to reduce oral pain resulting from bit pressure (ISES, 2019). Although virtually eliminated due to examination on exiting the ring under United States Equestrian Federation (USEF), United States Dressage Federation (USDF), and FEI regulations, protrusion of the tongue from the mouth – an example of perceived resistance – has in the past been managed by amputation of a portion of the front of the tongue, rather than addressing improper training techniques which resulted in pain or anxiety.

Excessive warm up length can mean that horses are unable to perform at their best in competition due to fatigue, and this can result in immediate corrective training which is likely to be ineffective or abusive. Excessive training or controlled exercise (e.g. lunging) is sometimes deliberately used to produce fatigue and compliance and give the appearance of calm behaviour desired in certain disciplines. Bleeding, marijuana, head tying high for several hours, placing a block of wood in a horse's mouth (so they are inclined to keep their mouth closed afterwards) are other practices which are used to result in a calm, quiet appearance in the show ring. Some trainers deliberately withhold water and feed to intentionally cause dehydration and depression.

The practice of tail blocking is commonplace in the Western performance horse. A substance – usually ethyl alcohol – is injected around the nerves that supply the muscle that elevate the tail, to block movement of the tail which is considered undesirable. The practice may affect faecal retention and elimination and may cause severe sloughing of tissue and infection. Horses are left unable to perform a normal behaviour. Tail swishing may be present as a sign of pain or conflict behaviour when ridden (Dyson, 2021), and the false appearance of a compliant horse can hide aversive training and riding techniques. American Quarter Horse Association (AQHA) rules prohibit tail blocking. The American Association of Equine Practitioners (AAEP) has also condemned this practice as unethical. However, horses with partial use of their tail continue to be successful in AQHA classes, which is a factor that drives continuation of the practice. The National Cutting Horse Association (NCHA), National Reined Cow Horse Association (NRCHA), and National Reining Horse Association (NRHA) have no restrictions on the practice of tail blocks.

Reining horses are commonly subject to a training practice termed 'fencing' whereby the horse is ridden at speed into a wall or fence to produce a desired effect of stopping with the hind legs under the body. Spinning is also common practice. Professional Rodeo Cowboys Association (PRCA) (2021) rules allow use of a livestock prod "when necessary".

Use of devices that inflict pain if a horse hits a jump, e.g. jump rails with carpet tacks, wires strung across the jump (sometimes electrified) that the horse cannot see, 'rapping' or 'poling', a practice of striking the horse's legs as they jump a fence in order to make them pick up their feet, are further welfare concerns. 'Chemical rapping' is a practice of sensitising lower legs with a skin irritant in an effort to make a horse less likely to knock jumps. Use of weighted or pressure boots in training to achieve hyper flexion of hind limbs on jump strides (to avoid knocking fences) may result in the horse 'tipping over' in a forward motion on landing. Trainers may use electric whips and spurs and 'excessive' use of conventional whips and spurs. Some horses trained in this manner may subsequently refuse to jump. When this happens the training may become even more abusive in attempts to force the horse to jump.

The effect of head and neck flexion on the welfare of dressage horses has received concern (ISES, 2015). FEI (2010b) re-defined "hyper-flexion/rollkur as flexion of the horse's neck achieved through aggressive force", as unacceptable.

High tail carriage, an arching neck and alert look, are desirable characteristics of the Arabian horse. Ginger – a mild chemical irritant – may be placed in the anus of the horse to encourage the tail to be held in a higher arch. 'Whip training' is a practice of hitting a horse in training with a whip until they poise with ears forward and a stretched, arched neck.

The Tennessee Walking Horse show industry desires an exaggerated high action forelimb step from horses. 'Soring' is the application of a caustic substance to the lower limb, or a mechanical device to the front feet, that result in pain such that the horse will alter their gait to relieve this pain. The practice of soring, although illegal, has continued unabated (Heird, 2011). The Prevent All Soring Tactics (PAST) Bill which was passed in the House of Representatives in 2019 aims to end the failed industry self-policing system, strengthen penalties, and ban the use of devices associated with soring not previously banned.

A common theme of these methods is the use of force and ignoring or underrecognition of underlying reasons for undesired behaviour. Unethical and inhumane training can lead to horses ultimately deemed unsuitable or dangerous and sent to sale, adding to the unwanted horse problem. While many of these practices are regulated at competition venues, their practice at home is more challenging to address, and public audiences have little awareness of the quality of the lives of the horses (Leitch, 2011).

Management and husbandry

Routine practices such as travel present welfare issues (McGreevy et al., 2018), such as sufficient space, ventilation, and restriction of head position, which may contribute to respiratory disease.

Horseman et al. (2017) identify management practices such as stabling for extended periods and unsuitable environments, as major areas identified by stakeholders for improvement. Many elite horses live in stables the vast majority of the time. Others, such as those used for carriage driving and pregnant mares' urine (PMU) production, are commonly held in tie stalls such that they are unable to turn around. Gastric ulceration, airway, and various musculoskeletal problems may be attributable to predominant stable confinement (Mitchell, 2011). Since welfare encompasses not just physical but also social and psychological aspects, effects of stabling on these aspects should also be considered. Horses evolved to be herd animals and confinement negatively effects social interaction and stimulation. Confinement can produce stress and alter physiological functions. Stereotypic behaviour such as weaving, cribbing, and wood chewing may be seen (Henderson, 2007). Time budgets have been used to compare management regimes with free-living horse behaviour. For example, a domesticated horse may be well nourished, but may spend a third of the normal time eating therefore suffer boredom, and be unable to locomote, which over time could lead to prolonged stress (Kiley-Worthington, 2011).

The International Society for Equitation Science 2018 conference illustrated that some countries now legislate aspects of housing, such as minimum turn out times. Group housing is increasingly used to improve social contact available to horses. A major barrier to more widespread adoption of this practice is the belief it increases injury risks (Randle and Waran, 2019). Social learning is limited by common practices such as forced weaning, youngsters in same-age groups, and isolation, which can lead to inappropriate behaviour and the assumption that horses are unable to live in groups (Kiley-Worthington, 2011). Subsequently, interactions are limited or heavily managed. Mating is commonly enforced by physical restraint with straps around the mare's legs and a twitch, justified to prevent injuries. However, socially educated equines allowed to court and cover freely have higher conception rates (McGreevy, 2012). Domestic breeding stallions are generally kept in isolation. In-hand breeding and semen collection are profoundly different from the normal ethogram (McGreevy, 2012). Preventing elements of stallion socio-sexual behaviour can contribute to handling difficulties.

Feeding

Genetics, life stage, environmental conditions, and level of exercise contribute to an individual's energy requirement and propensity to gain weight. In natural circumstances, horses will spend up to 18–20 hours foraging. Natural behaviour is to eat and move concurrently, and horses will rest from eating for no more than 4 hours at a time. Many domesticated horses and ponies spend much of the day in confinement housing (stables and small pens). Daily energy provisions often exceed requirements with resultant weight gain. Ponies turned out to pasture for 24 hours may consume up to 5% of body weight per day and up to 1% body weight in 3 hours of pasture turnout (Longland, 2013); 2–2.5% satisfies basic requirements. Additional problems with weight gain may occur because many horses and ponies are provided access to 'improved' pastures aimed at promoting growth and fattening sheep and cattle. This forage likely has much higher nutritional value when compared to forage available in non-domestic conditions (Geor and Harris, 2013). Risks of laminitis (inflammation of laminae inside the hoof), equine metabolic syndrome, insulin resistance, pedunculated lipomas, and heat and exercise intolerance are increased by obesity (Geor and Harris, 2013). Restriction of feed intake – though necessary in obese horses – can result in stress, increase risk of gastric ulceration, and hyperlipaemia, and practices to prolong feed intake time are recommended (Geor and Harris, 2013). Management of malnourished horses requires attention to their physical and thermal comfort, identification of concurrent disease, and risks of re-feeding syndrome.

In the natural state, foals will suckle for 35–40 weeks in pregnant multiparous mares, and longer in primiparous mares. Intermittent suckling by the previous year's foal may also occur when a new foal is at foot. Weaning is a gradual process. Enforced weaning prevents comfort responses between the mare and foal and therefore contributes to behavioural and physiological stress responses displayed by both mare and foal (McGreevy, 2012). Foals weaned abruptly show greater signs of stress than foals weaned more gradually (McGreevy, 2012). An abrupt change of diet is thought to contribute to the distress of weaning. Pre-weaning creep feed may reduce weaning stress by provision of mineral content to help meet increased requirements of fatigue, trauma and infection often associated with weaning (Hoffman et al., 1995). However Waters et al. (2002) found foals given concentrate feed after weaning were four times more likely to crib bite than foals on grass-only diets. Appropriate change in gut microbiome composition at weaning is likely critical to the development and function of an appropriate stress response (Mach et al., 2017). Long-lasting effects of weaning method and age are likely to affect learning ability (Nicol 2002).

Free-range horses

Wild and feral horse protection primarily involves keeping population size in balance with water and forage resources. Foaling and population growth rates will only decrease once horses are suffering from starvation. Wild horse mares will continue to foal until body condition reaches 1 or 2 on the Henneke scale or until they are over 20 years old (Kane, 2011). Whilst natural predation keeps numbers in check in a few areas, most require some form of intervention.

The Bureau of Land Management (BLM) and the Forest Service (FS) are responsible for managing the majority of wild horses on public lands in America. Feral horse populations grazed on common land in the UK are owned by individuals. Most common welfare problems relate to limited feed and water availability. Management usually involves reducing competition by either removing domestic livestock or by removing a proportion of the horses. Roundups risk injury and separation or mixing of family groups. Fertility control using porcine zona pellucida antigen and GnRH vaccines have been attempted. Alternative strategies such as supplementary feeding or artificial water supply require consideration to be given to number and location of feeding stations – limited stations causing unnatural congregation of horses will affect spread of infectious disease and parasitism. Horses are less able to learn to adapt to unpredictably failing water stations than to seasonal variation in water supply.

Aside from malnutrition, welfare and health concerns for individual animals usually relate to acquired or congenital physical problems, or to debilitation due to old age. Infectious disease is a threat after capture in a high risk environment of increased numbers of horses in close proximity with higher stress levels, with most being immunologically naïve to infectious respiratory pathogens. BLM and FS land horses are protected by federal laws which prevent sending to slaughter and to wild horse races, and adopters must meet criteria for space and housing. Horses from other geographical areas are offered no such protection. Nevertheless, the BLM has come under criticism for failing to prevent resale of horses to slaughter facilities. Supply greatly exceeds demand, and the BLM has long term holding sites for horses.

Veterinary conditions and drug use

The racing discipline usually requires pre-race examination of horses. However when actions such as shockwave treatment, walking off stiffness, icing swellings on limbs, and administration of non-steroidal anti-inflammatory drugs (NSAIDs) and steroidal anti-inflammatory glucocorticoids are taken in order to pass inspection, then a horse that is unfit to race may be passed as fit to run (David 2009). Relationships between pre-race findings and subsequent racing or training injury (Cohen et al., 1999; Hill et al., 2001) indicate that when pre-existing injuries can be recognised and diagnosed this can allow appropriate action to be taken. Management practices such as shoeing and training intensity have been found to be common factors associated with increased fatality (Kane et al., 1996, 1998; Anthenill et al., 2007). Conservative estimates of horses sustaining a fatal injury during their racing career are 4% in the US (Arthur, 2011). Training fatalities are not usually monitored or reported.

NSAIDs are permitted in US racing, and prohibited under IFHA rules. Repeated intra-articular administration of corticosteroids results in deterioration of articular cartilage (Wernecke et al., 2015). Firing and blistering (application of red hot iron or chemical irritant, usually to the lower limb) is a practice traditionally thought to promote healing by increasing blood flow, and although illegal in many countries, the procedure is still performed. Furosemide (a diuretic) is routinely administered to reduce exercise-induced pulmonary haemorrhage (EIPH – bleeding into the lungs) and is permitted on race days in North America, prohibited on race days under IFHA rules, and commonly used in training. Alternative strategies include preventing horses from drinking prior to running to achieve reduction in vascular volume. The extent of erythropoietin (EPO – a stimulant of red blood cell production) abuse in race horses is difficult to establish since EPO is broken down relatively quickly. Reticulocytes (immature red blood cells) mature within the bone marrow in horses and so are not found in large numbers in peripheral blood and splenic reserve means that a change in measured haematocrit between samples can be normal. Since EPO use is likely to be human-related EPO, there is a risk to the horse of immune mediated anaemia.

Common risks for horses involved in endurance racing include metabolic and musculoskeletal disorders. Contributory factors for these risks include preparation, i.e. under-training (not sufficiently conditioned) or over-training (fatigued), ride frequency, and rest time between competitions, because accumulated stress leads to higher incidence of metabolic failures or lameness. Common emergency conditions at endurance events include exhausted horse syndrome (dehydration, electrolyte imbalances, and glycogen depletion), myositis (a muscle disorder), synchronous diaphragmatic flutter (involuntary contractions of the diaphragm), colic secondary to intestinal atony, and heat stress.

Evaluations are performed by veterinarians before, during and after endurance events. Veterinarians make decisions as to a horse's 'fitness to continue' both in soundness and metabolic capacity. While objective data, e.g. cardiac recovery index (CRI), assist with decision-making, standards and equality of judging can vary across a country and around the world (Loving 2011a). Elevated CRI tends to be associated with fatigue, dehydration, and other indications of exhausted horse syndrome (AERC, 2008; Loving, 2011a). Including 'vet gates' early on during an event minimises the number of later metabolic failures by enabling refuelling and rehydration at a point which accommodates for the lag time for a horse's body to absorb and utilise nutrients (McCutcheon and Geor, 1996). Fewer vet checks during competition (e.g. non-FEI events) present a higher risk for the endurance horse as there are less opportunities to assess whether horses are fit to continue. Australia and France operate an early warning system whereby horses eliminated from competitions for veterinary reasons receive points appropriate to the severity of the disqualifying problem. Horses must only be ridden on shorter rides once a certain number of points are accumulated.

Drug testing is performed at random national and all high-level endurance competitions. There is a drug-free policy. Controlled medication – substances that are regularly used to treat horses – must be cleared from the horse's system by the time of competition. Banned substances are prohibited at all times. There is some debate as to whether NSAID use could improve welfare of horses with mild problems (FEI, 2010a), and there is concern of increased risk of gastric or colonic ulcers or kidney damage when combining dehydration from endurance racing with NSAID use (Loving, 2011a; Tamzali et al., 2011).

Soundness to compete in the eventing disciplines is evaluated at a trot up prior to the start of competition and after the cross country phase. Fitness- and exhaustion-related injuries include EIPH, tendon and ligament injuries, rhabdomyolysis (skeletal muscle cell damage) and trauma from hitting obstacles. Soft tissue injuries typically occur when supporting musculature becomes exhausted and ligaments and tendons are subjected to abnormal forces (Allen, 2011). FEI, USEF and United States Eventing Association (USEA) rules provide for modifying courses in extreme heat to reduce incidence of hyperthermia. Due to a rise in rotational falls frangible pins have become mandatory on certain fence types. Similar to other disciplines – as competing with an injury or unsoundness may increase the likelihood of a catastrophic event – certain medications are prohibited.

Welfare compromise in dressage occurs when attempts are made to force the horse to perform movements for which they are not physically or mentally prepared. Suspensory desmitis (inflammation of a ligament) and tendonitis (inflammation of a tendon) are common ailments within dressage, and more likely if work continues once the horse is fatigued.

There are very few published papers assessing welfare of polo ponies. Injuries are primarily musculoskeletal and traumatic in nature. Commonly, ponies are not fed during the day when working and water may also be withheld. Further issues include poor fitting tack and poor dental care.

Specific welfare issues pertaining to hunter-jumper horses include placing young horses lacking physical maturity under physical stress, with resultant increase in lameness issues such as physitis (swelling around growth plates of long bones), and acute periostitis (inflammation of periosteum, the outer bone surface) (Mitchell, 2011). Despite the prohibition of certain medications, horses may nevertheless be medicated in an effort to mask pain or alter behaviour.

Dexamethasone (a steroidal anti-inflammatory) is permitted as a therapeutic medication in USEF (USEF, 2020) and American Quarter Horse shows (AQHA, 2020). Many hunter and jumper horses receive it ostensibly for allergies, but it is often in reality administered in an effort to control behaviour due to the belief that it makes a horse quiet (Mitchell, 2011). Immune suppression and laminitis are risks from frequent administration. Intravenous thiamine and magnesium sulphate have been used to calm horses for the show ring. Although the substances are not prohibited, the intent is illegal and administration poses a risk for medical complications. Adrenocorticotropic hormone (ACTH) is frequently used to quieten horses. Repeated use could pose a health risk due to stimulation of adrenal glands. Immune stimulants such as bacterial cell wall derivatives have been used as quieting agents as they cause brief febrile response and subsequent quietened behaviour. This form of abuse is not detected in drug testing.

The rodeo industry is self-regulating and the PRCA provides a framework to protect the horse within the rodeo setting, however the PRCA only sanctions a small percentage of all rodeos, and rules require enforcement in order to safeguard the horse.

Cutting horses suffer from exercise induced traumatic diseases such as synovitis (inflammation of a synovial membrane) and capsulitis (inflammation of a joint capsule) of hocks and stifles, tendonitis and desmitis of suspensory apparatus in fore and hind limbs. Conditioning and training to strengthen the musculoskeletal system in young cutting and reining horses can help reduce these injuries. However, the practice of medication and continuation of training means re-injury often occurs due to insufficient recovery time.

Routine injection of stifles and hocks on a scheduled maintenance programme in the absence of any perceptible lameness is commonplace within the cutting, reining, and cow horse industries. Inherited diseases affecting quarter horses include hyperkalaemic periodic paralysis (HYPP – a muscle disorder, elevated blood potassium) and hereditary equine regional dermal asthenia (HERDA – a skin disease). Genetic associations in osteochondrosis (OCD – disturbance of normal cartilage formation) incidence in some breeding lines has also been noted.

Therapeutic medication is allowed in the rules of many North American horse show organisations (USEF, 2020). NSAIDs are permitted for therapeutic purposes up to 12 hours prior to competing under USEF and Equestrian Canada (EC) regulations and within 24 hours of an event under AQHA (2020) regulation. Concurrent use of several NSAIDs is not permitted for 72 hours prior to competition. Welfare issues can present when there is a lameness without diagnosis, or when NSAIDs are given prophylactically, potentially masking a new injury.

City-issued registration identifies Northern American carriage horses used in the tourism industry, and indicates that they have been examined fit by a veterinarian. Periodic veterinary examinations vary in frequency across states and cities. Sick, lame, and inappropriately shod horses are prohibited from working. Time, shift, and area recommendations are designed to reduce risk of traumatic injury by avoiding peak traffic and mandating maximum shift and rest periods. Common health problems of the carriage horse include lameness from continuous concussive work on hard paved ground. Colic is a major cause of death and suspected inadequate deworming and subsequent intestinal vascular damage, and abrupt feeding change together with lack of exercise when trade low, are possible reasons. Other problems include respiratory issues such as asthma, with urban air pollution suggested as a contributing factor (although there is no evidence for this), and pleuropneumonia in new carriage horses who arrive from auctions. Skin problems and harness sores, weather-related problems (most cities have hot weather regulations for carriage horses), trauma due to slipping on uneven pavement, and rub marks from stalls, are also seen. Emergency vet care for horses may not be readily available near major cities as these services are often located in rural areas.

Welfare health concerns relating to the companion horse include obesity with consequent increased laminitis risk, especially if the horse or pony has metabolic or hormonal imbalances; parasite burden; and slow decline due to a combination of dental problems, malabsorption, metabolic problems, chronic pain reducing appetite, and pain induced catabolism. The question of euthanasia and when it should be performed – neither too early nor too late – depend on a person's ability to provide for medical needs.

Working equids include draft animals and those used for transportation. Provision of basic needs such as food, water, safe housing, farriery and veterinary care are likely to be met in higher-income countries. However these needs are met to varying degrees in working animals in low- and middle-income settings. Working equids within the limits of their ability based on their aptitude and fitness, weather conditions, type, and level of task required, is necessary. Tack should be appropriate for the task and not unduly burden or injure the equine. A negative impact on welfare occurs when the horse, donkey, or mule are not provided basic care, and if they are overworked or abused in order to make them work.

The donkey skin trade is almost entirely unregulated or illegal; by 2021, 4.8 million donkeys are killed annually for the Chinese medicine ejiao. Donkeys are often transported in overcrowded trucks for days without food, water, or rest. They may arrive at slaughterhouses with severe injuries and, in some cases, up to 20% of donkeys are dead on arrival at slaughterhouses (The Donkey Sanctuary, 2019). They often can be held for days at slaughterhouses without access to food or water before being slaughtered, often brutally.

PMU ranching farms use harness apparatus to collect urine from mares in foal. PMU contains oestrogen in an orally active form. Federal, provincial, and contractual regulations provide requirements for care of the mares. Welfare concerns include limited access of mares to outside corrals or pasture for six months of the year, and the fate of the foals after forced weaning. Foals and barren mares may be sent to slaughter, although in more recent years North American PMU ranchers have improved marketability of foals for performance activities.

Collection of blood from pregnant mares for extraction of equine chorionic gonadotropin (eCG) for use in farming industries is a poorly regulated practice. Collection takes place between 40 and 120 days of gestation, and in some regions it is common for the pregnancy to be terminated so that the mare can be rebred for a second blood collection period that year. Live foals are routinely weaned prematurely. Horses become weak, emaciated, and die when mares are bled too frequently or if too much blood is collected at any time (Manteca Vilanova et al., 2019). Mares are usually managed extensively on pastures with minimal oversight. Some facilities house mares indoors under intensive conditions. Often horses may be poorly desensitised and habituated to handling, with resultant fear and stress at handling during blood collection.

Unwanted horses

The entire horse industry – including all breeds and disciplines – is responsible for wastage and the unwanted horse problem (Lenz, 2011; Wightman and Mendham, 2021). The large numbers concerned mean they are candidates for neglect. Those slaughtered for meat represent the lowest economic value horses with age, physical disability, and behaviour all factors contributing to their economic worth. Opponents of anti-slaughter legislation argue that unintended consequences could include increased neglect, abuse, and abandonment. However, proponents argue that the availability of slaughter has led to a prevailing culture and attitude that horses are disposable, with horses suffering at and during transportation to slaughterhouses.

Euthanasia is regarded as an acceptable option for unwanted horses by many individuals and organisations involved in welfare, because current rescue and retirement facilities cannot accommodate all unwanted horses. Acceptable methods of euthanasia include injectable barbiturates, gunshot, and penetrating captive bolt (AVMA, 2020); in the UK captive bolts are permitted to stun, and bleeding or pithing is subsequently required.

Alternative options to slaughter include retraining, rescue, adoption, or retirement facilities. Organisations promoting retraining include the Jockey Club in the UK, Thoroughbred Charities of America, and the Thoroughbred Retirement Foundation in the US for racing thoroughbred horses, the AQHA "Greener Pastures" scheme, the Colorado Unwanted Horse Alliance, and the Unwanted Horse Coalition. However, the numbers through these routes form only a small percentage of the horses used in these industries. Owing to the relatively long natural life span of 25–30 years, considerable time, space, and financial commitment is required to care for horses. There is a need for oversight in order to prevent animal hoarders who are unable to provide for their needs, or dealers, from taking horses under false pretences.

A call for change

Studies have shown that most owners believe horses to be sentient and capable of emotions such as pain, fear, or joy (DuBois et al., 2018b; Hötzel et al., 2018). However, belief in sentience does not appear to reflect understanding of welfare issues (DuBois et al., 2018b). A recent survey (Bell et al., 2019) illustrated that horse owners had a lack of knowledge of behaviours associated with negative affective states. McGowan et al. (2010) demonstrated that owners could not identify all clinical signs of ill health in their aged equines and did not seek veterinary advice in all cases when it was warranted. Furthermore, studies have indicated that owner awareness of welfare issues such as keeping horses with social companions or use of training methods identified as causing distress, is frequently at odds with their behavioural practice towards equidae in their care (Hartmann et al., 2015; Bell et al., 2019), suggesting that knowledge was not the only factor involved in poor horse welfare. Other barriers to improving horse welfare may be related to financial constraints, opportunity to change practices, habit, and cultural norms around horse management (Hartmann et al., 2015; Bell et al., 2019).

Horses in our society are still considered personal property and they are frequently kept in a manner that inadequately assures their physical, behavioural, psychological, or emotional welfare needs. Within the horse industry there is too little regard for prioritising the welfare of horses and the horses often lack someone to advocate on their behalf. Given the widespread welfare issues across all disciplines, and that the main beneficiary from the human–horse relationship is the human, it could be argued that the interests of the horse are best served by our leaving them alone – breeding only those who can live sustainably in free-living environments and taking care of remaining domesticated horses as they live out their natural lives. Given this is unlikely to gain popular appeal or to occur in the near future, it is imperative that we make changes to the ways in which we interact with horses in our custodianship if we are to ensure their welfare. A change to current habituation of trainers, owners, and industry to inhumane practices and training methods is necessary, together with engendering empathy for the pain and suffering of the horse. Further to this, regulation and active enforcement of regulation violations with measures for protecting horses in their home environments, is necessary.

Management practices such as medical interventions, shoe type, training practices, or management techniques which contravene any aspect of holistic health of the horse in order to fulfil requirements of a particular industry, must be re-evaluated. Furthermore, requirements of disciplines should be re-evaluated and changed if they encourage such practices. If medication is required in order for horses to perform then the demands of the activity itself must be called into question, and the bench marks for competition should be reviewed. If horses need furosemide to run in the racing industry, they shouldn't be running so often, as far, or at all. Endurance horses ridden in a way that necessitates metabolic treatment reveals an unacceptable consequence of performance demands. While an incentive to ride with emphasis on finishing in good condition is afforded by the Best Condition award, incentives and practice need to go further. Rather than pushing to extremes, would the 'sport' not be more ethical if working within a horse's natural physiological capability? Mandatory log books for endurance horses with sufficient records of sound judgements is surely a minimal requirement to help abolish horses being pushed beyond their capabilities.

Proactive routine investigation is warranted, with trainers, owners, and veterinarians on board, in order to make early diagnoses and prevent situations where injuries are ignored or deemed too low-grade to investigate, which can bring catastrophic results (Arthur, 2011). Race trainers with high fatality rates currently suffer few repercussions. Accurate mandatory reporting of racetrack and training fatalities, with appropriate independent investigation and accountability, is a basic starting point from which to tackle this issue.

The use of overwork, psychotropic medication, and other practices, require change in judging guidance and improvement in ability of judges to evaluate genuine and natural displays of calm and relaxed behaviour, rather than rely on markers which are subject to abuse. Stringent rules and enforcement policies are required. Veterinarians and stewards need better guidelines and support to recognise, monitor and prevent abusive riding and exercise, and to enforce regulations.

Clear identification of horses with microchips and fit-for-purpose electronic systems and databases can assist with compliance of legal requirements such as working hours of carriage horses, and movement and sale of horses with particular reference to illegal travel to slaughter (Wightman and Mendham, 2021).

Across all industries, recognition of pain or discomfort behaviours is necessary, rather than interpreting behaviour as a sign of resistance. Understanding that prevention of expression of such behaviours is likely to further compromise horse welfare, is also necessary. Education along with the introduction and enforcement of robust regulations is essential to promote ethical practices. Methods used to train and work with horses should move towards a true partnership between horse and human, taking account of ethology, and progress to working with consent from the horse.

Horses should be allowed to wean naturally, or at least gradually, in groups, and consideration should be given to pair bonds. Dietary adjustment should be made to at the very least mimic natural foraging. Daily turnout with herd mates should become the norm to allow species-specific behaviour, with a move away from reliance on individual housing systems and abolition of tie stalls.

Emphasis on retraining of horses and increased provision for retirement, with responsibility taken for this by the industries which contribute to the problem, is required. A move towards owners taking lifelong responsibility for the natural life span of their horses would likely necessitate reduction of the problem of unwanted horses at source by responsible breeding, which in some cases may mean no or highly limited breeding. It is simply not acceptable for horses to be malnourished, suffering from physical and psychological stress, or sent to slaughter to uphold human wants. Eventual transition towards horses becoming legal 'wards' with owners considered 'guardians' may in the future place limitations on human use and abuse of horses (Loving 2011b).

Conclusions

We must improve our treatment of equidae. Commonplace forceful training-, management-, and health-associated practices which subordinate equine welfare to human interests require human behaviour change in order to protect the horse from suffering. There needs to be a cultural change, whereby human–horse interaction is dependent on genuine concern for the welfare of the horse, underpinned by understanding of ethology, and prioritisation of physical, psychological, cognitive, social, and emotional needs of the horse. For optimum equid welfare we must communicate empathetically with horses, gain their consent for interaction, and further, allow horses choice and control over their lives.

References

AERC, 2008. Guidelines for judging AERC endurance competitions. *AERC Controlled Judge Handbook 3.0*. Available at: https://aerc.org/static/upload/2009controljudgehb.pdf, accessed on 25 July 2021.

Allen K, 2011. Welfare issues in the event horse. In: CW McIlraith, BE Rollin, eds. *Equine Welfare*. Oxford: Blackwell Publishing Ltd. pp. 463–472.

Anthenill LA, Stover SM, Gardner IA, Hill AE, 2007. Risk factors for proximal sesamoid bone fractures associated with exercise history and horseshoe characteristics in Thoroughbred racehorses. *American Journal of Veterinary Research*, 68(7), pp. 760–771. Available at: doi: 10.2460/ajvr.68.7.760, accessed on 24 July 2021.

AQHA, 2020. AQHA guidelines & rules for drugs and medications American Quarter Horse Association. American Quarter Horse Association. Available at: https://www.aqha.com/documents/82601/129505957/2020+AQHA+Guidelines+&+Rules+for+Drugs+and+Medication.pdf/6179b3b3-26d3-fce3-5abe-0c2159fdec15, accessed on 25 July 2021.

Arthur RM, 2011. Welfare issues in horse racing. In: CW McIlraith, BE Rollin, eds. *Equine Welfare*. Oxford: Blackwell Publishing Ltd. pp. 229–244.

AVMA, 2020. AVMA Guidelines for the euthanasia of animals. American Veterinary Medical Association. Available at: https://www.avma.org/sites/default/files/2020-02/Guidelines-on-Euthanasia-2020.pdf, accessed on 26 July 2021.

Bell C, Rogers S, Taylor J, Busby D, 2019. Improving the recognition of equine affective states. *Animals*, 9(12), p. 1124. Available at: doi: 10.3390/ani9121124, accessed on 10 April 2021.

Cohen ND, Mundy GD, Peloso JG, Carey VJ, Amend NK, 1999. Results of physical inspection before races and race-related characteristics and their association with musculoskeletal injuries in Thoroughbreds during races. *Journal of the American Veterinary Medical Association*, 215(5), pp. 654–661.

David T, 2009. Report of the ARCI regulatory veterinarian committee to the RMTC. Association of Racing Commissioners International. Available at: https://rmtcnet.com/wp-content/uploads/2015/07/NSAID_IA_Cort_Statement.pdf, accessed on 25 July 2021.

DuBois C, Hambly Odame H, Haley DB, Merkies K 2018a. An exploration of industry expert perception of Canadian equine welfare using a modified Delphi technique. *PLoS One*, 13(7). p. e0201363. Available at: doi: 10.1371/journal.pone.0201363, accessed on 10 April 2021.

DuBois C, Nakonechny L, Derisoud E, Merkies K, 2018b. Examining Canadian equine industry participants' perceptions of horses and their welfare. *Animals*, 8(11), p. 201. Available at: doi: 10.3390/ani8110201, accessed on 10 April 2021.

Dyson S, 2021. The ridden horse pain ethogram. *Equine Veterinary Education*, 3. Available at: doi: 10.1111/eve.13468, accessed on 31 July 2021.

FEI, 2010a. The FEI congress on non-steroidal anti-inflammatory drugs (NSAID) usage and medication in the equine athlete. *Fédération Equestre Internationale*. Available at: https://inside.fei.org/media-updates/fei-congress-hailed-invaluable-contribution-debate-use-nsaids-competition, accessed on 24 July 2021.

FEI, 2010b. FEI round table conference. *Fédération Equestre Internationale*. Available at: https://inside.fei.org/media-updates/fei-round-table-conference-resolves-rollkur-controversy, accessed on 25 July 2021.

FEI, 2013. Code of conduct for the welfare of the horse. *Fédération Equestre Internationale*. Available at: https://inside.fei.org/sites/default/files/Code_of_Conduct_Welfare_Horse_1Jan2013.pdf, accessed on 10 April 2021.

Geor RJ, Harris PA, 2013. Obesity. In: RJ Geor, PA Harris, M Coenen, eds. *Equine Applied and Clinical Nutrition: Health Welfare and Performance*. Edinburgh, New York: Saunders Elsevier Ltd. pp. 487–502.

Hartmann E, Bøe KE, Christensen JW, Hyyppä S, Jansson H, Jørgensen GH, Ladewig J, Mejdell CM, Norling Y, Rundgren M, Särkijärvi S, Søndergaard E, Keeling LJ, 2015. A Nordic survey of management practices and owners' attitudes towards keeping horses in groups. *Journal of Animal Science*, 93(9), pp. 4564–4574. Available at: doi: 10.2527/jas.2015-9233, accessed on 23 July 2021.

Heird J, 2011. Abusive treatment and subsequent policy development within various breeds of show horses in the USA. In: CW McIlraith, BE Rollin, eds. *Equine Welfare*. Oxford: Blackwell Publishing Ltd. pp. 245–254.

Henderson AJ, 2007. Don't fence me in: Managing psychological well being for elite performance horses. *Journal of Applied Animal Welfare Science*, 10(4), pp. 309–329. Available at: doi: 10.1080/10888700701555576, accessed on 7 Oct. 2021.

Hill AE, Stover SM, Gardner IA, Kane AJ, Whitcomb MB, Emerson AG, 2001. Risk factors for and outcomes of noncatastrophic suspensory apparatus injury in Thoroughbred racehorses. *Journal of the American Veterinary Medical Association*, 218(7), pp. 1136–1144. Available at: doi: 10.2460/javma.2001.218.1136, accessed on 7 Oct. 2021.

Hoffman RM, Kronfeld DS, Holland JL, Greiwe-Crandell KM, 1995. Preweaning diet and stall weaning method influences on stress response in foals. *Journal of Animal Science*, 73(10), pp. 2922–2930. Available at: doi: 10.2527/1995.73102922x, accessed on 26 July 2021.

Horseman SV, Buller H, Mullan S, Knowles TG, Barr AR, Whay HR, 2017. Equine welfare in England and Wales: Exploration of stakeholders' understanding. *Journal of Applied Animal Welfare Science*, 20(1), pp. 9–23. Available at: doi: 10.1080/10888705.2016.1197776, accessed on 10 April 2021.

Hötzel MJ, Vieira M, Leme D, 2018. Exploring horse owners' and caretakers' perceptions of emotions and associated behaviors in horses. *Journal of Veterinary Behavior*, 29, pp. 18–24. Available at: doi: 10.1016/j.jveb.2018.10.002, accessed on 10 April 2021.

ISES, 2015. Position statement on alterations of the horse's head and neck posture in equitation. International Society for Equitation Science. Available at: https://equitationscience.com/equitation/position-statement-on-alterations-of-the-horses-head-and-neck-posture-in-equitation, accessed on 31 July 2021.

ISES, 2018. Principles of learning theory in equitation. International Society for Equitation Science. Available at: https://equitationscience.com/learning-theory, accessed on 24 July 2021.

ISES, 2019. Position statement on restrictive nosebands. International Society for Equitation Science. Available at: https://equitationscience.com/equitation/position-statement-on-restrictive-nosebands, accessed on 24 July 2021.

Kane AJ, 2011. The welfare of wild horses in the Western USA. In: CW McIlraith, BE Rollin, eds. *Equine Welfare*. Oxford: Blackwell Publishing Ltd. pp. 442–462.

Kane AJ, Stover SM, Gardner IA, Case JT, Johnson BJ, Read DH, Ardans AA, 1996. Horseshoe characteristics as possible risk factors for fatal musculoskeletal injury of thoroughbred racehorses. *American Journal of Veterinary Research*, 57(8), pp. 1147–1152.

Kane AJ, Stover SM, Gardner IA, Bock KB, Case JT, Johnson BJ, Anderson ML, Barr BC, Daft BM, Kinde H, Larochelle D, Moore J, Mysore J, Stoltz J, Woods L, Read DH, Ardans AA, 1998. Hoof size, shape, and balance as possible risk factors for catastrophic musculoskeletal injury of Thoroughbred racehorses. *American Journal of Veterinary Research*. 59(12), pp. 1545–1552.

Kiley-Worthington M, 2011. Equine psychological needs and quality of life. In: CW McIlraith, BE Rollin, eds. *Equine Welfare*. Oxford: Blackwell Publishing Ltd. Pp. 94–112.

King S, Wills L, Randle H, 2019. Early training of foals using the ISES training principles. *Journal of Veterinary Behaviour*, 29, pp. 140–146.

Leitch M, 2011. Welfare in the discipline of dressage. In: CW McIlraith, BE Rollin, eds. *Equine Welfare*. Oxford: Blackwell Publishing Ltd. pp. 332–340.

Lenz TR, 2011. The unwanted horse: A major welfare issue welfare issue. In: CW McIlraith, BE Rollin, eds. *Equine Welfare*. Oxford: Blackwell Publishing Ltd. pp. 425–441.

Longland AC, 2013. Pasture and pasture management. In: RJ Geor, PA Harris, M Coenen, eds. *Equine Applied and Clinical Nutrition: Health Welfare and Performance*. Edinburgh, New York: Saunders Elsevier Ltd. pp. 332–350.

Loving NS, 2011a. Raising welfare standards for endurance riding. In: CW McIlraith, BE Rollin, eds. *Equine Welfare*. Oxford: Blackwell Publishing Ltd. pp. 341–369.

Loving NS, 2011b. The horse as a companion animal. In: CW McIlraith, BE Rollin, eds. *Equine Welfare*. Oxford: Blackwell Publishing Ltd. pp. 255–274.

Mach N, Foury A, Kittelmann S, Reigner F, Moroldo M, Ballester M, Esquerré D, Rivière J, Sallé G, Gérard P, Moisan MP, Lansade L, 2017. The effects of weaning methods on gut microbiota composition and horse physiology. *Frontiers in Physiology*, 8, p. 535. Available at: doi: 10.3389/fphys.2017.00535, accessed on 29 July 2021.

Manteca Vilanova X, De Briyne N, Beaver B, Turner PV, 2019. Horse welfare during equine chorionic gonadotropin (eCG) production. *Animals*, 9(12), p. 1053. Available at: doi: 10.3390/ani9121053, accessed on 28 June 2021.

McCutcheon LJ, Geor RJ, 1996. Sweat fluid and ion losses in horses during training and competition in cool vs. hot ambient conditions: Implications for ion supplementation. *Equine Veterinary Journal. Supplement*, 22, pp. 54–62. Available at: doi: 10.1111/j.2042-3306.1996.tb05032.x, accessed on 7 Oct 2021.

McGowan TW, Pinchbeck G, Phillips CJ, Perkins N, Hodgson DR, McGowan CM, 2010. A survey of aged horses in Queensland, Australia. Part 2: Clinical signs and owners' perceptions of health and welfare. *Australian Veterinary Journal*, 88(12), pp. 465–471. Available at: doi: 10.1111/j.1751-0813.2010.00638.x, accessed on 28 July 2021.

McGreevy P, 2012. *Equine Behaviour: A Guide for Veterinarians and Equine Scientists*. 2nd edn. Edinburgh: Saunders Elsevier.

McGreevy P, Berger J, De Brauwere N, Doherty O, Harrison A, Fiedler J, Jones C, McDonnell S, McLean A, Nakonechny L, Nicol C, Preshaw L, Thomson P, Tzioumis V, Webster J, Wolfensohn S, Yeates J, Jones B, 2018. Using the five domains model to assess the adverse impacts of husbandry, veterinary, and equitation interventions on horse welfare. *Animals*, 8(3), p. 41. Available at: doi: 10.3390/ani8030041, accessed on 10 April 2021.

Mitchell RD, 2011. Welfare concerns in the care, training, and competition of the hunter-jumper. In: CW McIlraith, BE Rollin, eds. *Equine Welfare*. Oxford: Blackwell Publishing Ltd. pp. 370–379.

Nicol C, 2002. Equine learning: Progress and suggestions for future research. *Applied Animal Behaviour Science*, 78, pp. 193–208. Available at: doi: 10.1016/S0168-1591(02)00093-X, accessed on 7 Oct. 2021.

PRCA, 2021. Professional rodeo cowboys association 2021 rule book. Professional Rodeo Cowboys Association. Available at: https://www.prorodeo.org/Portal/Home/PrcaBusiness/2021-PRCA-Rulebook.pdf, accessed on 29 July 2021.

Randle H, Waran N, 2019. Equitation science in practice: How collaboration, communication and change can improve equine welfare. *Journal of Veterinary Behavior*, 29, pp. viii–x. Available at: doi: 10.1016/j.jveb.2018.12.014, accessed on 28 June 2021.

Rioja-Lang FC, Connor M, Bacon H, Dwyer CM, 2020. Determining a welfare prioritization for horses using a Delphi method. *Animals*, 10(4), p. 647. Available at: doi: 10.3390/ani10040647, accessed on 21 July 2021.

Tamzali Y, Marguet C, Priymenko N, Lyazrhi F, 2011. Prevalence of gastric ulcer syndrome in high-level endurance horses. *Equine Veterinary Journal*, 43(2), pp. 141–144. Available at: doi: 10.1111/j.2042-3306.2010.00129x, accessed on 22 July 2021.

The Donkey Sanctuary, 2019. Under the skin. *The Donkey Sanctuary*. Available at: https://www.thedonkeysanctuary.org.uk/sites/uk/files/2019-12/under-the-skin-report-english-revised-2019.pdf, accessed on 29 July 2021.

USEF, 2020. USEF guidelines & rules for drugs and medications. United States Equestrian Federation. Available at: https://www.usef.org/forms-pubs/2Zp2C_YKs4s/2020-equine-drugs-medications, accessed on 25 July 2021.

Waters AJ, Nicol CJ, French NP 2002. Factors influencing the development of stereotypic and redirected behaviours in young horses: Findings of a four year prospective epidemiological study. *Equine Veterinary Journal*, 34(6), pp. 572–579. Available at: doi: 10.2746/042516402776180241, accessed on 7 Oct. 2021.

Wernecke C, Braun H, Dragoo J, 2015. The effect of intra-articular corticosteroids on articular cartilage: A systematic review. *Orthopaedic Journal of Sports Medicine*, 3(5), p. 2325967115581163. Available at: doi: 10.1177/2325967115581163, accessed on 20 July 2021.

Wightman K (Ed), Mendham L (Director), 2021. Panorama: The dark side of horse racing, In: M Lewis, L Kon (Executive Producers), *Panorama, British Broadcasting Cooperation: Nine Lives Media*, accessed on 19 July 2021.

20

NON-DOMESTICATED TERRESTRIAL SPECIES

Miriam A Zemanova

Introduction

Wildlife can be defined as "living things and especially mammals, birds, and fishes that are neither human nor domesticated" (Merriam-Webster 2021). The number of non-domesticated terrestrial species far exceeds the number of domesticated animals. And yet, we still know very little about their well-being. Animal welfare science has traditionally focused on domesticated animals or non-domesticated animals in zoos (Freire and Nicol 2019), whereas wildlife have been the concern of biological conservation, with little attention paid to their welfare until the late 20th century (e.g., Broom 1999). Since the term "wildlife" encompasses a vast range of species, there is a large diversity in behaviour, physiology, and signs of pain. It is, therefore, very difficult to make any generalisations about how poor wildlife welfare manifests. Furthermore, the uncertainty about sentience in some species adds another layer of complexity (Soryl et al., 2021).

With the growth and expansion of human populations over the last centuries, wildlife have been increasingly influenced by human activities. Fraser and MacRae (2011) listed four types of human impact: 1) keeping animals in captivity; 2) causing deliberate harm, e.g., through hunting or pest management; 3) causing direct but unintended harm, e.g., by vehicle collisions, a harvest of agricultural products, or oil spills; and 4) causing indirect harm, e.g., through environmental pollution, loss of habitat, or climate change. Some of these impacts are discussed in the other chapters of this book. This chapter aims at highlighting potential animal welfare issues experienced by non-domesticated terrestrial species in the wild, through wildlife rehabilitation, reintroduction programmes, wildlife research, and the exotic pet trade.

Non-domesticated animals in the wild

In contrast to popular belief, the lives of animals living in the wild can be far from idyllic (Horta 2017). Their well-being may be compromised through starvation, disease, or injury, and the majority of animals die well before reaching their maximum lifespan. Richard Dawkins described the situation in the wild in his book River Out of Eden with the following words (Dawkins 1995, pp. 131–132):

> The total amount of suffering per year in the natural world is beyond all decent contemplation. During the minute that it takes me to compose this sentence, thousands

DOI: 10.4324/9781003182351-24

> of animals are being eaten alive; others are running for their lives, whimpering with fear; others are being slowly devoured from within by rasping parasites; thousands of all kinds are dying of starvation, thirst and disease.

Indeed, it could be considered that most of the wildlife suffering might be a consequence of 'natural' causes, such as resource scarcity limited by the carrying capacity of the ecosystem, droughts, or floods. These factors are, however, exacerbated by human activities. The increasing human population puts immense pressure on ecosystems through the building of infrastructure, pollution, and demand for food resources, and with it associated land use and anthropogenic climate change (Chapter 23). Many practices in forestry, agriculture, or pest control have a significant impact on wildlife welfare.

That humans are responsible for the well-being of animals under their care is a widespread societal view that is often embedded in animal welfare legislation. When it comes to wildlife, there are opposing opinions on whether humans have the duty or even right to manage or assist these animals. Some philosophers have argued that we do not have a moral obligation to intervene in nature because (most) animals would not be considered moral agents (Sapontzis 1984), or because the autonomy and sovereignty of animals should be honoured (Nussbaum 2006). Others have reasoned that we need to expand our circle of moral concern beyond domesticated animals and try to alleviate the suffering of non-domesticated animals living in the wild, i.e., living in natural conditions without human control (Horta 2017). Several forms of assistance to wildlife have been proposed, for instance, the provision of medical care to sick animals, vaccination to prevent diseases, or contraception to control population size and dynamics (Horta 2017; Soryl et al., 2021). One specific approach to help wildlife is wildlife rehabilitation.

Wildlife rehabilitation

The aim of wildlife rehabilitation is to provide care to injured, sick, or orphaned animals so that they could be returned to their natural habitats. Among the reasons for admission are, for example, car strikes, dog and cat attacks, or entanglements (Taylor-Brown et al., 2019). While the primary benefit of wildlife rehabilitation is to help and alleviate the suffering of an individual animal, there are several other advantages, including the improvement of diagnostic and therapeutic practices for wildlife, replenishment of local populations, education, or disease surveillance (Tribe and Orr 2019). Moreover, data from wildlife rehabilitation centres are of great value for species conservation, providing insights into natural and anthropogenic threats for wildlife (Taylor-Brown et al., 2019).

Nevertheless, there are also risks associated with the rehabilitation practice, such as improper medical interventions resulting in poor animal welfare, risk of disease transferred from the rehabilitated and released individual into the population of animals living in the wild, or risks of zoonoses for employees and volunteers of rehabilitation centres (Tribe and Orr 2019). Furthermore, the hazards of the release stage are often underestimated, with the potential for high losses. Hence, post-release monitoring of the released animals is essential to evaluate the success of the rehabilitation program (Mullineaux 2014).

Wildlife rehabilitation requires balancing the well-meaning altruism of people trying to help non-domesticated animals, against the aim of avoiding unnecessary suffering of animals brought into captivity (Mullineaux 2014). As the flight response to humans is a vital survival trait for all wildlife species, every effort must be made to keep human contact to the minimum to prevent the habituation and taming of the animals. Unfortunately, sometimes people bring wild animals to a rescue or rehabilitation facility who do not actually need saving. For instance, Robertson and Harris (1995) radio tracked foxes after release and reported that many 'orphaned' fox cubs were not in fact orphans.

The rehabilitation of non-native, invasive species is a contentious issue, which is handled differently based on local legislation. For instance, the National Wildlife Rehabilitators Association based in the USA defines rehabilitation as "the treatment and temporary care of injured, diseased, and displaced *indigenous* animals, and the subsequent release of healthy animals to appropriate habitat in the wild" (Miller, 2012; emphasis added). Consequently, injured or orphaned individuals that are classified as invasive species might get rejected by most rehabilitation centres. In many countries, it is illegal to release an invasive species into the wild. Therefore, even if the animals receive veterinary care, they have to be either euthanised or stay in captivity. Captivity often leads to substantial animal welfare issues, e.g., due to inappropriate housing conditions or stress caused by the presence of people and other animals (Rivera et al., 2021).

Repatriation to the wild

Animal reintroductions are defined by the IUCN as "the intentional movement and release of an organism inside its indigenous range from which it has disappeared" (IUCN 2013). More generally, they refer to an attempt to restore a population of extirpated species or to increase abundance within a population, in the area that is a part of the species' current or historical range. Species reintroduction programmes are now commonly used to aid conservation efforts across the globe. One example of success is the recovery of the Californian condor (*Gymnogyps californianus*). In 1982, there were only 22 Californian condors left in the wild, as a consequence of habitat loss and pollution. The remaining individuals were brought into captivity and used in a breeding program in 1987. Even though the species remains critically endangered, breeding and reintroduction have been effective and a small population is now thriving in the wild (Wilcove 2000).

However, despite the wide implementation of reintroductions, the success rate of reintroduction programmes varies greatly for different species, environments, and scenarios. Specifically, the success might be influenced by predation risk, habitat quality, number of released individuals, and behavioural traits. Unfortunately, many translocations and reintroductions fail shortly after the release of animals. After release, welfare risks include potential human persecution (especially for large carnivores), injury, hunger, and in social species, the need to re-establish a social structure (Goddard 2020). Harrington et al. (2013) evaluated 199 reintroduction projects and found that two-thirds reported one or more animal welfare issues. Mortality rates of more than half of the released animals were described in 23% of the projects. Furthermore, captive-bred animals often experience more difficulties with coping in the wild after release than wild-caught animals, and have a higher mortality rate (Harrington et al., 2013). For wild-caught animals used in breeding and reintroduction programmes, it is also crucial to consider the welfare implications for the individuals within the source populations. The capture of animals from the wild might have negative consequences in species with complex social structures if key individuals are removed (Goddard 2020). The original populations might be also left genetically depauperate.

Reintroduction of captive-bred predators in particular into their natural ecosystem carries many difficulties, as the ability of the animals to recognise, catch, and kill their prey can be compromised. Many large carnivores come into conflict with people, which sometimes escalates into their persecution (Goddard 2020). On the other hand, there are also issues with the reintroduction of prey species that are naïve to the predator's presence. Associative learning might teach naïve animals about predators and enable them to identify and appropriately respond to them (Clayton et al., 2014). However, whether this training improves the success of reintroduction is not clear. So far, several studies have reported no improvement in the survival rate after release, or have assessed only short-term survival. The welfare of resident wildlife needs to be considered as well. They might be displaced by the reintroduced species, have to compete for resources, or become prey to introduced carnivores (Goddard 2020).

It is important that reintroductions adhere to the internationally accepted standards for animal welfare, and that stress or suffering is minimised. The IUCN Guidelines for Reintroductions and Other Conservation Translocations outline a framework for deciding when a translocation is an acceptable option, on planning a translocation, the feasibility and design, risk assessment, release, monitoring, and dissemination of information (IUCN 2013). Repatriation should be implemented only after a careful consideration of harms and benefits, and conservation goals should not supersede the individual animal's welfare.

Wildlife research

Ecosystems across the globe are experiencing a dramatic extinction process. To fully understand and possibly counter this trend, ecologists, conservation biologists, and wildlife researchers collect data on species distribution, population sizes, or gene flow, all of which are necessary information for effective management. Wildlife research is often glamorised by the media, focusing on charismatic species and on saving their populations. But historically, there has not been enough consideration for the welfare of individual animals. In the seminal work defining the field of conservation biology, Michael Soulé wrote (Soulé, 1985, p. 731):

> It may seem logical to extend the aversion of anthropogenic extinction of populations to the suffering and untimely deaths of individuals because populations are composed of individuals. I do not believe this step is necessary or desirable for conservation biology. Although disease and suffering in animals are unpleasant and, perhaps, regrettable, biologists recognize that conservation is engaged in the protection of the integrity and continuity of natural processes, not the welfare of individuals.

Even today, research in conservation biology is often conducted under the assumption that it has an insignificant impact on the studied wild animals or that any impact is outweighed by the potential benefits to the population or species. Such assumptions however raise concerns for animal welfare (Zemanova 2019, 2020, 2021b). Many research methods traditionally implemented in wildlife research are invasive (i.e., penetrating the skin barrier) or stressful, thus negatively impacting the animal used in research. Our understanding of how research practices affect animal welfare is hindered by the fact that some of the research methods might have delayed consequences, and in many cases, animal welfare implications are simply not known. It is imperative to exercise the precautionary principle, and when in doubt, to always design the study to have the least potential impact (Zemanova 2020). Examples of research activities that have been shown to influence animal welfare are capture, marking, blood and tissue sampling, or attachment of radio transmitters.

Marking of wildlife species is often used to obtain data on behaviour, survival rate, or population size estimation. Practically all marking techniques require capture and some of them include tissue damage through hot- or freeze-branding, or mutilation of limbs with toe-clipping. Toe-clipping is still a commonly used method for marking small species such as amphibians, lizards, and rodents, even though several studies have reported its negative impact on the animal's survival rate and locomotion. Another marking approach is the use of tags or bands. However, these can increase the energetic costs of swimming due to drag in semi-aquatic animals such as seals or penguins (reviewed in Zemanova 2020).

Blood and tissue sampling are commonly used for DNA collection, physiological assessment, or ecotoxicological studies. Although generally considered safe, blood sampling has been linked to a lower survival rate in American cliff swallows (*Petrochelidon pyrrhonota*) (Brown and Brown 2009). Tissue sampling often requires lethal means, causing obvious harm to the individual animal.

To track animal movement, wildlife researchers use GPS collars and harnesses or radio transmitters glued to the skin or implanted into body cavities. There have been several animal welfare issues identified with their use. For example, birds carrying a radio transmitter can get entangled with vegetation and have a lower survival rate. GPS collars on large herbivores have been shown to affect grazing behaviour and decrease the survival rate. If the radio transmitter is attached to the skin, the glue can cause abrasions and lesions. Implanted radio transmitters seem to be particularly problematic, as several cases of mortality have been reported for a range of wildlife species, e.g., brown bear (*Ursus arctos*), European lynx (*Lynx lynx*), or American badger (*Taxidea taxus*) (reviewed in Zemanova 2020).

Even capture alone can be extremely stressful for an animal living in the wild that is not accustomed to being handled by humans. The stress of capture can be reflected in increased cortisol levels, which might skew the results of physiological assessments. In extreme cases, this stress can lead to capture myopathy, which is a metabolic disease often resulting in death. Furthermore, capture can lead to a deteriorated body condition, reduced movement, or a lower survival rate. Traps can also cause injuries, ranging from skin abrasions to broken limbs.

Poor animal welfare is, however, not only an issue for the individual animals affected by wildlife research. It can also result in public outrage and affect the soundness of study results (Zemanova 2021a). Pain leads to behavioural, physiological, and neurobiological changes (Sneddon 2017). Since rigorous science is a prerequisite for good management decisions, it is crucial that the impact of wildlife research on animal welfare is minimised. Unfortunately, there seems to be a lack of education in ethics and animal welfare that could provide basic guidance on how to deal with ethical dilemmas encountered in wildlife research (Zemanova 2017, 2021a). Moreover, the legislation in some countries distinguishes whether a permit is required to conduct the same procedure, e.g., blood sampling, depending on whether its purpose is classified as wildlife research or wildlife management (Lindsjo et al., 2019).

An important milestone in promoting animal welfare in research was achieved in 1959 when Russell and Burch proposed the 3Rs principles (Russell and Burch, 1959). These principles state that scientists should Replace animals with alternative methods whenever possible, Reduce the number of animals in experiments to the minimum, and Refine or limit the pain and distress that animals might be experiencing as a result of the experiment or laboratory housing. The 3Rs principles are nowadays an integral part of legislation in many countries. Even though they were originally proposed and designed for work with laboratory animals – mostly rodents – they are applicable to wildlife research as well – with a few caveats. While laboratory research might use animals as models for human diseases and physiology, the object of wildlife research is the animal itself. Therefore, it would not be possible to use a cell culture – a common replacement approach in laboratory research – to study, for instance, gene flow among kangaroo populations. Furthermore, wildlife is a term encompassing a broad range of species with different ecological and physiological characteristics, making generalisations of guidelines challenging.

Nevertheless, one of the most straightforward strategies to implement the 3Rs is to use non-invasive research methods, but other strategies, such as calculation of the minimum sample size, sharing data and resources, or using anaesthesia and tranquilisation, are also important for good animal welfare (Figure 20.1). Unfortunately, the use of invasive and lethal methods persists. For instance, a recent assessment revealed that on average, only 22% of wildlife genetics studies published between 2017 and 2018 made use of an available non-invasive DNA sampling technique (Zemanova 2019). Even though this review may have not captured all published studies, the

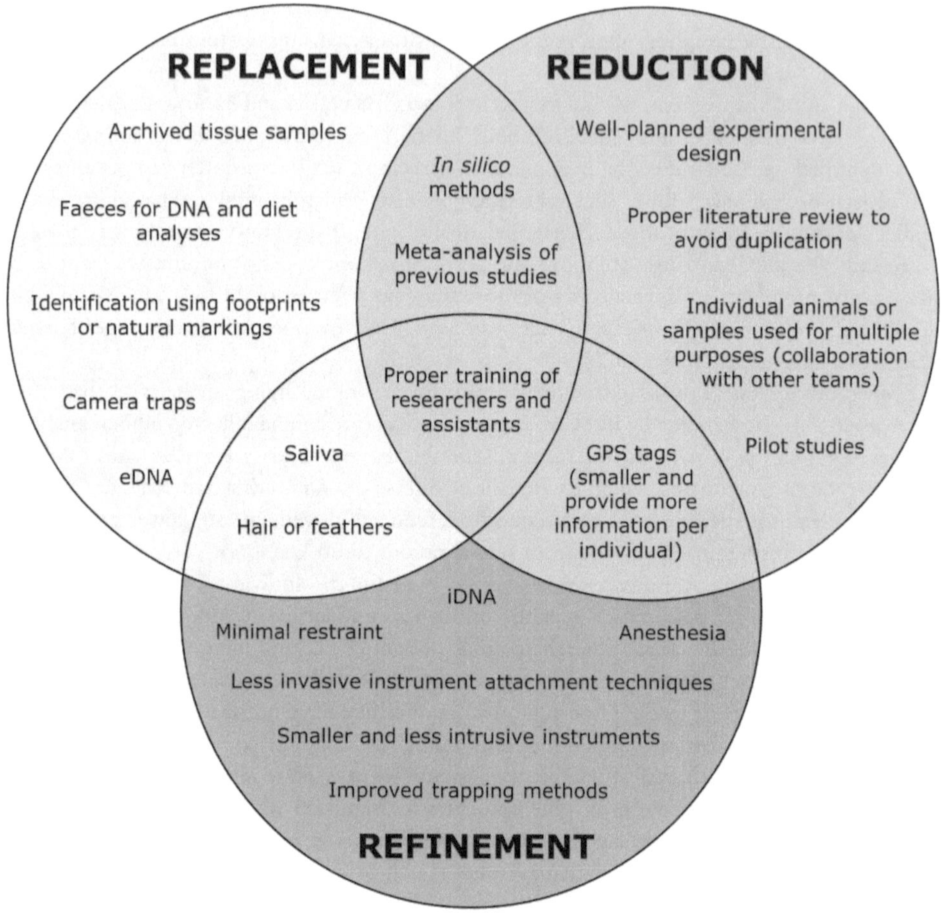

Figure 20.1 Strategies for the implementation of the 3Rs principles (Replacement, Reduction, Refinement) in wildlife research. Overlapping methods for replacement and refinement depend on whether the animal has to be captured or not. Source: Zemanova (2020).

findings indicate that wildlife researchers might struggle to implement non-invasive methods into their work. Sadly, the application of the 3Rs principles to wildlife research has been rather slow. For example, Field et al. (2019) reviewed animal care policies in 206 biodiversity- and wildlife-related journals and found that only 6% required authors to adopt the 3Rs principles in their research. One of the main reasons for the low implementation rate might be a lack of awareness (Zemanova 2021a). The lack of awareness will hopefully be ameliorated with the emergence of the 3Rs guidelines and databases designed specifically for researchers working with wildlife (e.g., https://3RsWildlife.info; Zemanova 2021b).

Exotic pet trade

Most of us are fascinated with wildlife. But for some it is not enough to watch animals in the wild; they want to be able to touch them, cuddle with them, or just flaunt them as status symbols. This desire is reflected in the increasing demand for exotic, i.e., relatively rare or unusual, non-domesticated pets. It was estimated that in the USA, 19.4 million households owned an exotic species in 2013 (Micheli 2014), and in the UK, the exotic pet population amounted to 42 million in 2014 (PFMA 2014). Keeping exotic pets is, however, not limited to the developed countries. For instance, Jepson and Ladle (2005) found that in Indonesia, one is more likely to find an exotic pet in a household than a common domesticated animal, such as a cat or a dog.

Wildlife trade is one of the most prominent contributors to biodiversity loss and a major hindrance to species conservation (Baker et al., 2013). The majority of countries protect threatened and endangered animals through the agreement implemented within the Convention on the International Trade in Endangered Species (CITES). The convention currently lists approximately 5,950 animal species that are prohibited from trading without a license from the authorities (CITES, 2021). Yet, animals traded under the CITES represent only a small proportion of the species bought and sold as exotic pets. Furthermore, the international market in exotic pets is not limited to legal means of supply, and constitutes a significant proportion of the illegal wildlife trade, mostly of birds, amphibians, reptiles, and fish (Schuppli et al., 2014). Illegally traded animals are often smuggled across borders under abhorrent conditions. Small species may be crammed in large numbers into small, airtight containers, resulting in death due to asphyxiation during transport (BBC, 2014). This trend is not easy to combat, because the illegal wildlife trade has become one of the largest sources of income for organised crime. In 2012, it was valued at around 19 billion US dollars a year (WWF and Dalberg, 2012).

Animal welfare issues associated with exotic pet trade are still understudied. Baker et al. (2013) assessed the literature on wildlife trade and found that only 13–25% of the studies reported the impact on animal welfare. In the following, some of the known issues are discussed.

Wild capture versus captive breeding

Any live capture of wildlife poses a risk – not only for the person capturing an animal, who might get scratched or bitten, or acquire a zoonotic disease, but there is a risk of injury also for the captured animal. Natusch and Lyons reported that large numbers of wild-caught reptiles in New Guinea are unsuitable for export due to injury or death (Natusch and Lyons 2012). Since many animals also die in transit, due to dehydration, starvation, crushing, or asphyxiation, many more individuals need to be captured than the number that actually ends up being traded on the market (Baker et al., 2013).

Captive breeding might reduce the pressure on wild populations of favoured exotic species. However, the financial costs of wild capture tend to be much lower than the costs of captive breeding (Burivalova et al., 2017). Additionally, breeding farms can be used to launder illegally caught wildlife. Lyons and Natusch (2011) assessed the trade of the green python (*Morelia viridis*) in Indonesia and were able to trace over 4,000 illegally caught green pythons to a breeding farm. They estimated that a minimum of 80% of the green pythons exported from Indonesia are illegally captured from the wild. Furthermore, some species are difficult to breed. While 90% of freshwater ornamental fish are easily bred in captivity (Andrews 1990), 90% of marine ornamental fish are still wild-caught (Cato and Brown 2008).

Husbandry concerns

The domestication of dogs or cats took thousands of years of artificial selection that resulted in them being well adapted to life as human companions. Conversely, captive-bred, non-domesticated animals have the same needs as their counterparts living in the wild. There is plenty of information about the proper care for domestic animals, such as dogs and cats, and veterinary practices are well versed in treating them. In contrast, there is a lack of specialised veterinary care for exotic animals and such pets may not be easy to care for. For instance, reptiles and amphibians have very specific physiological and behavioural requirements that many owners may not be aware of or do not have the facilities to cater for. Furthermore, these animals may not exhibit stress indicators common in other species (Hernandez-Divers 2001).

Keeping animals in suboptimal settings is a threat to their welfare. Ashley et al. (2014) reported on an investigation at a large international wildlife wholesaler conducted by veterinarians and biologists under the auspices of the Texas state authorities. The investigators found and confiscated over 26,000 animals across 171 species. Sick, injured, or dead animals constituted 80% of the animals on site. The authors identified poor hygiene, inadequate or inappropriate supply of water, food, or heat, and high levels of stress, as the factors contributing to high disease and mortality rates (Ashley et al., 2014). Within households, it has been estimated that the vast majority of pet reptiles are kept under unsuitable conditions, and up to 75% die within a year of purchase (Toland et al., 2021). Cases of incorrect husbandry with dire consequences for animal welfare have been reported also for other types of animals. For example, parrots are one of the most intelligent birds, making them prone to developing stereotypies, i.e., abnormal, repetitive, and seemingly functionless behaviours that are one of the indicators of poor animal welfare (Engebretson 2006). Improper care or lack of attention from cohabiting humans can result in stress – manifested as self-mutilation or feather plucking, injuries from inappropriate housing or poor handling, and lack of vitamins in the diet can lead to metabolic bone disease. Some of the larger species also have a lifespan of up to 80 years (e.g. green-winged macaw), which means that they might need repeated rehoming during their lives (Engebretson 2006).

Zoonoses

Exotic pets can constitute a health risk to other wild animals, domestic animals, and humans. Wildlife is considered to be the source of more than 70% of all zoonotic emerging infectious diseases (Jones et al., 2008), with wildlife trade enabling the spread of initially localised pathogens across the globe. For instance, the vast majority of cases of avian chlamydiosis, a disease that can be transmitted through the air from birds to humans, are the result of exposure to pet birds (Balsamo et al., 2017). The import of wild-caught animals, mixing of species from different regions, often stored together in crowded and stressful conditions, increases the risks of zoonotic outbreaks.

Abandonment or escape

The exotic pet trade facilitates the introduction of non-native species to new regions. Many of the released or escaped exotic species can establish colonies, sometimes with dire consequences for ecosystems. In amphibians and reptiles, the exotic pet trade has contributed the largest number of established non-native species (Lockwood et al., 2019). For example, in Florida, USA, there are at least 140 species of non-native reptiles and amphibians, of which almost 85% originated from the exotic pet trade (Krysko et al., 2011). Exotic bird pets that escape from cages are the main source of avian invasions. It has been estimated that a minimum of 25 exotic parrot species have already established breeding populations in the USA (Uehling et al., 2019).

The reasons owners release their exotic pets into the wild have not been broadly documented, but may include difficulty in taking care of old, large, or aggressive individuals (Lockwood et al.,

2019). Unfortunately, there are limited options for rehoming exotic pets. Specialised sanctuaries have limited capacities, and zoos are often unable to accept animals. This is an important point to make: not all species are suitable to be kept as pets. In summary, according to Schuppli and Fraser (2000), we should always consider: 1) the welfare of the animal, defined as a range of factors captured in the "five freedoms" (Farm Animal Welfare Council 1992); 2) the welfare of others – humans or other animals; and 3) risks to the environment, either in the source region, or in case of introduction of exotic species to new ecosystems.

Conclusions

Consideration of the well-being of wildlife has been a neglected field within animal welfare science. Wildlife welfare has finally been recognised in recent years, driven by the expectations of the general public for humane treatment of wildlife, application of legislation to species living in the wild, and the recognition of wildlife researchers that good welfare of animals used in studies is a prerequisite for robust scientific results. In this chapter, I have highlighted some of the recognised animal welfare issues faced by non-domesticated terrestrial species. While there are still many unknowns when it comes to wildlife welfare, it is clear that we need to be cognisant of the risk that our fascination with wildlife, and our desire to help animals, might be detrimental to their well-being.

References

Andrews C, 1990. The ornamental fish trade and fish conservation. *Journal of Fish Biology*, 37(sA), pp. 53–59.

Ashley S, Brown S, Ledford J, et al., 2014. Morbidity and mortality of invertebrates, amphibians, reptiles, and mammals at a major exotic companion animal wholesaler. *Journal of Applied Animal Welfare Science*, 17(4), pp. 308–321.

Baker SE, Cain R, van Kesteren F, et al., 2013. Rough trade: animal welfare in the global wildlife trade. *Bioscience*, 63(12), pp. 928–938.

Balsamo G, Maxted AM, Midla JW, et al., 2017. Compendium of measures to control *Chlamydia psittaci* infection among humans (psittacosis) and pet birds (avian chlamydiosis), 2017. *Journal of Avian Medicine and Surgery*, 31(3), pp. 262–282.

BBC, 2014. Hundreds of dead animals found at South Africa airport. *BBC News*. Available at: https://www.bbc.com/news/world-africa-25877368 (Accessed on 7 July 2021).

Broom DM, 1999. The welfare of vertebrate pests in relation to their management. In: Cowan, DP and Faere, CJ (eds.) *Advances in Vertebrate Pest Management*. Fürth: Filander Verlag. pp. 309–329.

Brown MB and Brown CR, 2009. Blood sampling reduces annual survival in cliff swallows (*Petrochelidon pyrrhonota*). *Auk*, 126(4), pp. 853–861.

Burivalova Z, Lee TM, Hua F, et al., 2017. Understanding consumer preferences and demography in order to reduce the domestic trade in wild-caught birds. *Biological Conservation*, 209(1), pp. 423–431.

Cato JC and Brown CL, 2008. *Marine Ornamental Species: Collection, Culture and Conservation*. Ames, IA: Iowa State Press.

CITES, 2021. The CITES species. Convention on International Trade in Endangered Species of Wild Fauna and Flora. Available at: https://cites.org/eng/disc/species.php (Accessed on 6 July 2021).

Clayton JA, Pavey CR, Vernes K, et al., 2014. Review and analysis of Australian macropod translocations 1969–2006. *Mammal Review*, 44(2), pp. 109–123.

Dawkins R, 1995. *River Out of Eden: A Darwinian View of Life*. New York: Basic Books.

Engebretson M, 2006. The welfare and suitability of parrots as companion animals: a review. *Animal Welfare*, 15(3), pp. 263–276.

Farm Animal Welfare Council, 1992. FAWC updates the five freedoms. *Veterinary Record*, 131(16), pp. 357.

Field KA, Paquet PC, Artelle K, et al., 2019. Publication reform to safeguard wildlife from researcher harm. *PLoS Biology*, 17(4), p. e3000193.

Fraser D and MacRae AM, 2011. Four types of activities that affect animals: implications for animal welfare science and animal ethics philosophy. *Animal Welfare*, 20(4), pp. 581–590.

Freire R and Nicol CJ, 2019. A bibliometric analysis of past and emergent trends in animal welfare science. *Animal Welfare*, 28(4), pp. 465–485.

Goddard P, 2020. Animal reintroductions: who is safeguarding animal welfare? Available at: https://www.wawcommittee.org/s/Topic_Statement_Reintroductions_Final_Feb_2020.pdf (Accessed on 30 June 2021).

Harrington LA, Moehrenschlager A, Gelling M, et al., 2013. Conflicting and complementary ethics of animal welfare considerations in reintroductions. *Conservation Biology*, 27(3), pp. 486–500.

Hernandez-Divers SJ, 2001. Clinical aspects of reptile behavior. *Veterinary Clinics of North America: Exotic Animal Practice*, 4(3), pp. 599–612.

Horta O, 2017. Animal suffering in nature: the case for intervention. *Environmental Ethics*, 39(3), pp. 261–279.

IUCN, 2013. *Guidelines for Reintroductions and Other conservation Translocations*. Gland, Switzerland. Available at: https://portals.iucn.org/library/node/10386 (Accessed on 20 June 2021).

Jepson P and Ladle RJ, 2005. Bird-keeping in Indonesia: conservation impacts and the potential for substitution-based conservation responses. *Oryx*, 39(4), pp. 442–448.

Jones KE, Patel NG, Levy MA, et al., 2008. Global trends in emerging infectious diseases. *Nature*, 451(7181), pp. 990–993.

Krysko KL, Burgess JP, Rochford MR, et al., 2011. Verified non-indigenous amphibians and reptiles in Florida from 1863 through 2010: outlining the invasion process and identifying invasion pathways and stages. *Zootaxa*, 3028(1), pp. 1–64.

Lindsjo J, Cvek K, Spangenberg EMF, et al., 2019. The dividing line between wildlife research and management - implications for animal welfare. *Frontiers in Veterinary Science*, 6(1), pp. 13.

Lockwood JL, Welbourne DJ, Romagosa CM, et al., 2019. When pets become pests: the role of the exotic pet trade in producing invasive vertebrate animals. *Frontiers in Ecology and the Environment*, 17(6), pp. 323–330.

Lyons JA and Natusch DJD, 2011. Wildlife laundering through breeding farms: illegal harvest, population declines and a means of regulating the trade of green pythons (*Morelia viridis*) from Indonesia. *Biological Conservation*, 144(12), pp. 3073–3081.

Merriam-Webster, 2021. Wildlife. Merriam-Webster Dictionary. Available at: https://www.merriam-webster.com/dictionary/wildlife (Accessed on 30 August 2021).

Micheli R, 2014. Exotic pets: a growing American fad. CNBC. Available at: https://www.cnbc.com/2014/02/10/american-fad.html (Accessed on 25 June 2021).

Miller EA, 2012. *Minimum Standards for Wildlife Rehabilitation*. St. Cloud, MN: National Wildlife Rehabilitators Association.

Mullineaux E, 2014. Veterinary treatment and rehabilitation of indigenous wildlife. *Journal of Small Animal Practice*, 55(6), pp. 293–300.

Natusch DJD and Lyons JA, 2012. Exploited for pets: the harvest and trade of amphibians and reptiles from Indonesian New Guinea. *Biodiversity and Conservation*, 21(11), pp. 2899–2911.

Nussbaum MC, 2006. *Frontiers of Justice: Disability, Nationality, Species Membership*. Cambridge, MA: Harvard University Press.

PFMA, 2014. *Pet Population 2014*. Pet Food Manufacturers' Association. Available at: https://www.pfma.org.uk/pet-population-2014 (Accessed on 24 June 2021).

Rivera SN, Knight A and McCulloch SP, 2021. Surviving the wildlife trade in Southeast Asia: reforming the 'disposal' of confiscated live animals under CITES. *Animals*, 11(2), 439. DOI: 10.3390/ani11020439..

Robertson CPJ and Harris S, 1995. The condition and survival affer release of captive-reared fox cubs. *Animal Welfare*, 4(4), pp. 281–294.

Russell WMS and Burch RL, 1959. *The Principles of Humane Experimental Technique*. London, UK: Methuen.

Sapontzis SF, 1984. Predation. *Ethics and Animals*, 5(2), pp. 27–38.

Schuppli CA and Fraser D, 2000. A framework for assessing the suitability of different species as companion animals. *Animal Welfare*, 9(4), pp. 359–372.

Schuppli CA, Fraser D and Bacon HJ, 2014. Welfare of non-traditional pets. *Revue Scientifique et Technique*, 33(1), pp. 221–231.

Sneddon LU, 2017. Pain in laboratory animals: a possible confounding factor? *ATLA*, 45(3), pp. 161–164.

Soryl AA, Moore AJ, Seddon PJ, et al., 2021. The case for welfare biology. *Journal of Agricultural and Environmental Ethics*, 34(2), p. 7.

Soulé ME, 1985. What is conservation biology? *Bioscience*, 35(11), pp. 727–734.

Taylor-Brown A, Booth R, Gillett A, et al., 2019. The impact of human activities on Australian wildlife. *PLoS ONE*, 14(1), p. e0206958.

Toland E, Warwick C and Arena PC, 2021. Pet hate: exotic pet-keeping is on the rise despite decades of initiatives aimed at reducing the trade of exotic and rare animals. Three experts argue that urgent action is needed to protect both animals and ecosystems. *Biologist*, 59(3), pp. 14–18.

Tribe A and Orr B, 2019. Wildlife rehabilitation practices in Australia. In: Vogelnest, L and Portas, T (eds.) *Current Therapy Medicine of Australian Mammals*. Clayton South, VIC, Australia: CSIRO Publishing. pp. 51–62.

Uehling JJ, Tallant J and Pruett-Jones S, 2019. Status of naturalized parrots in the United States. *Journal of Ornithology*, 160(3), pp. 907–921.

Wilcove DS, 2000. *The Condor's Shadow: The Loss and Recovery of Wildlife in America*. New York: W. H. Freeman and Company.

WWF and Dalberg, 2012. *Fighting Illicit Wildlife Trafficking: A Consultation with Governments*. Gland, Switzerland: WWF International. Available at: https://www.worldwildlife.org/publications/fighting-illicit-wildlife-trafficking-a-consultation-with-governments (Accessed on 25 June 2021).

Zemanova MA, 2017. More training in animal ethics needed for European biologists. *Bioscience*, 67(3), pp. 301–305.

Zemanova MA, 2019. Poor implementation of non-invasive sampling in wildlife genetics studies. *Rethinking Ecology*, 4(1), pp. 119–132.

Zemanova MA, 2020. Towards more compassionate wildlife research through the 3Rs principles: moving from invasive to non-invasive methods. *Wildlife Biology*, 1(1), p. wlb.00623.

Zemanova MA, 2021a. Making room for the 3Rs principles of responsible animal use in ecology: potential issues identified through a pilot survey. *European Journal of Ecology*, 7(2), pp. 18–39.

Zemanova MA, 2021b. New online resource on the 3Rs principles of animal research for wildlife biologists, ecologists, and conservation managers. *Conservation*, 1(2), pp. 106–112.

21
COMPANION FISH

Sabrina Brando

Introduction

Aquatic animals cover many different species and taxa across the world, of greatly varied phylogenetic ages and linkages (Håstein, et al., 2005). Fishes, invertebrates, reptiles, amphibians, and mammals are some of the many aquatic animals living in human care and in the wild, in fresh, brackish, or saltwater, rivers, lakes, estuaries, seas, and oceans among other environments. They have different anatomies, physiology, behaviour, and affective states. Just fishes alone cover 34,000 species according to FishBase (2021). Different chapters in this book cover aquatic animals in a variety of systems such as aquaculture (Chapter 10), hunting, fishing, and whaling (Chapter 16), commercial fisheries (Chapter 17), and those housed in zoos and aquariums (Chapter 15).

This chapter will only briefly cover fishes kept as companion animals, henceforth referred to as pets. It aims to cover a variety of topics and subtopics, to give a sense of various aspects to consider when housing fishes at home, as well as implications for fishes housed elsewhere in human care such as zoos and aquariums. This will consider not only the direct care, monitoring, and assessing of animal welfare, but much of the whole process of how these animals came to a home or facility. The terms 'animal welfare' and 'animal well-being' have both been used over the years to describe the state of the animal (Moberg and Mench, 2000, p. 1), and are used in this chapter interchangeably.

While research on assessing animal welfare for terrestrial animals has been ongoing for many decades, the field of assessing aquatic animals is more recent. For an excellent overview of a science-based assessment of aquatic animal welfare, see Håstein et al. (2005), who identified the following issues in addressing aquatic animal welfare in a consistent manner: the sheer diversity of species of vertebrates and invertebrates, the relative paucity of scientific information, and varying philosophical approaches, policies, guidance and regulations that may influence the provision of optimal welfare and humane practices for aquatic animals. This chapter will cover general fish welfare, fish owners, sourcing, transport, pet stores, care, nutrition, environment, environmental enrichment, and behaviour for fishes as pets.

Companion fish

In 2010 it was estimated that over 350 million fishes were traded annually within the ornamental fish industry (Saxby et al., 2010), however, a more recent estimate is 1.5 billion ornamental fishes handled within trade (Stevens et al., 2017). Ploeg (2012a) indicated that, on a global basis,

 DOI: 10.4324/9781003182351-25

freshwater and marine organisms such as fish, invertebrates, and plants comprise the greatest numbers of animals traded.

Fishes are popular companion animals – both fresh and saltwater species – with many reports and popular publications on why they make such excellent and easy pets (Pets WebMD, 2021). Some aquariums also promote the keeping of certain species of fishes, highlighting that one should first consider if resources, time, and the ability to care for them are available for the duration of their lifetimes. Despite increasing research into the welfare of fish farmed in aquaculture, and the popularity of keeping aquarium fish ('ornamental fish') at home, there is very little research into the optimum living conditions and general welfare of fishes kept as pets.

Warwick (2015) reported that of the 45 million pet fish in the UK, at least 90% will be dead within one year in the home. While worldwide figures could not be located at the time of writing, these could number in the billions, with 150 million in the USA alone (Warwick et al., 2018).

Many different species of fishes and amphibians are kept as companion animals, such as Siamese fighting fish (*Betta splendens*), angelfish (*Pterophylum scalare*), different species of minnows, including the white cloud mountain minnow (*Tanichthys albonubes*), and the zebrafish (*Danio rerio*), neon tetras (*Paracheirodon innesi*), tiger barbs (*Barbus tetrazona*), goldfish (*Carassius auratus*), guppies (*Poecilia reticulata*), and Corydoras catfish (*Corydoras spp.*). Most of the species covered in this chapter are freshwater species, and a few marine species are also mentioned.

General fish welfare

The acquisition process, initially used to source and transport pet fish, has been described as a 'pipeline of death' (Warwick, 2015). He noted that, for pet fish, wild-capture mortality can reach almost 100%, especially at the initial stage.

Once sourced and kept, extrapolating general 'rules' for optimum water quality and many other determinants of welfare is made difficult or impossible due to variability between species, and a lack of research in necessary areas. For example, for the extremely popular ornamental Siamese fighting fishes, it is unknown how much waste is produced per fish (Pleeging and Moons, 2017). This study identified that for these fish, aquaria of limited dimensions, the prevalence of Mycobacterium spp. infections, and the lack of environmental enrichment in the form of sheltering vegetation, are common concerns. The barren environment, and the limited ability to escape, as well as potential for stress due to prolonged visual contact between males in shops and during shows, and aggression to and from conspecifics or other species in the same aquarium, were all listed as factors creating welfare problems. Keeping different species and group sizes together is also a common occurrence. Sloman et al. (2011) concluded that setting scientifically determined guidelines on appropriate species assemblages or stocking densities for ornamental fish is complicated by the plethora of group sizes and combinations found within home aquaria, and noted that many welfare issues could go unnoticed when focusing on single-species studies.

A crucial part of understanding fish welfare and promoting optimal care for fishes at home is understanding characteristics of fish owners and the perceived benefits and challenges of being a fish owner.

Fish owners

Fishes are kept in approximately 10% of Western households and are the most numerous type of pet (Sullivan, 2014). Kidd and Kidd (1999) interviewed home aquarium owners in local stores, comprising 50 men and 50 women, about the benefits and problems of being a fish owner. The expense of the equipment, especially the tanks and equipment to control water temperature, circulation, and chemical balances, tank maintenance, and ensuring tank cleanliness, were all reported as major problems. The calming, relaxation, and stress-reduction and serenity-inducing effects of just watching fish were reported as major benefits by 70% of the respondents. The authors reported that 72% of the respondents considered their fish to be family pets, 32% deemed them to be part of their education, 22% saw them as room decorations, 10% considered them a hobby, and 4% regarded them as companions. This diversity of factors provided insight into the reasons for the contemporary popularity of live inhabitants of home aquaria.

In 2004, Sullivan conducted a survey with fish owners to gain more insights of their perceptions of the main welfare issues affecting their pet fish. Among the 534 owners the most common causes for the death of their fishes were disease and old age. There is likely an underestimation of the role of water quality in fish health. Of all respondents, 27% of fish owners admitted they had limited knowledge of fish care, and rarely sought information or social support for their fish care. Almost all respondents provided structural enrichment such as gravel and shelters for their fish. More knowledgeable owners were more likely to provide real plants, rather than artificial ones.

A study by Jacobson et al. (2000) revealed that participants named their fish within the first week of ownership, and while they were not able to physically interact with them, they were still able to identify personality traits and form a semblance of a relationship with their fish. Recommendations by therapists, counsellors, and other human well-being professionals regularly promote the purchase and care of a pet to improve human well-being. Langfield and James (2009) noted that more interactive pets might not be an option for all people, and reviewed fish ownership as an alternative. Their research used in-depth, semi-structured interviews of nine participants, enquiring about the reasons for and benefits of owning fish as pets, as well as the care and environments of the fish. They discovered that pet fish ownership is a meaningful occupation, providing both a purpose and enjoyment in life, and suggested that fish may be an alternative to a more interactive pet to recommend, if clients wished to own a pet. Taken together with the research mentioned previously conducted in pet stores – highlighting the common lack of up-to-date information and/or staff training – it is clear that considerably more emphasis needs to be placed on providing correct in-store advice and support, to reduce premature death and/or suffering of fishes as pets.

Sourcing

Millions of fishes make their way to people's homes every year. Warwick et al. (2018) discuss how labelling in pet stores could be a possible risk factor – mis-selling exotic animals as "easy" or "beginner" pets could result in new fish owners purchasing pets without appreciating the complexity of their biology and needs. The authors propose that a system should be required that facilitates decision-making at the interface between sale and purchase, and that uses clear, evidence-based labelling. They discuss the development of a pet labelling scheme as a preventative and educational approach, with categories ranging through "easy, moderate, difficult", to "extreme" ("EMODE").

Alongside animal well-being, possible conservation implications and threats to wild populations is another serious concern, and the collection of fishes for the pet trade has been flagged as a major threat in these areas. Only a few studies give a glimpse of the potential impact when considering the many countries fishes are sourced from. Early studies like Chao and Prang (1997) report that nearly 20 million live fishes are exported from the Amazon annually. A first assessment by Raghavan et al. (2013) specifically focused on the trade of endemic and threatened freshwater fishes sourced from two global biodiversity hotspots in India, identifying

30 threatened species comprising 1.5 million individuals which were exported between 2005 and 2012, including endangered species such as *Botia striata*, the vulnerable-listed dwarf pufferfish (*Carinotetraodon travancoricu*), and two species of endangered red lined torpedo barbs (*Puntius denisonii* and *Puntius chalakkudiensis*). The authors reported that several threatened and conservation-concern species were routinely exported from India, with local regulations on fish trade and aquarium collections poorly enforced and managed. They noted that such practices could pose a much more severe threat to freshwater biodiversity than previously recognised, and presented realistic options for better management.

While standards and codes of best practice for handling fishes have been established by the Ornamental Aquatic Trade Association (OATA), little scientific research has been conducted to understand well-being in ornamental fish species. However, lessons may be learned from other sectors involving fish, like aquaculture. Many of the same negative stressors occur in this sector, such as capture, handling, transport, poor water conditions, transport, disease, poor social and physical environment (Chapter 10), and directed research in this sector has resulted in some improvements in welfare standards (Stevens et al. 2017).

An extensive review by King (2019) notes that the ornamental aquatic industry continues to divide opinion in the scientific literature, with practices ranging from unsustainable and over exploitative, such as cyanide fishing – with negative effects on wild populations of reef fish, to more sustainable and responsible practices, that may offer ecological and socioeconomic benefits. King notes that there are differences in freshwater versus marine species, with breeding cycles of freshwater species being less complex. It is estimated that approximately 90% of traded freshwater fish are captive-bred. In contrast, marine species have much more complicated breeding cycles, resulting in 90–95% of traded marine fish being wild caught. The ornamental aquatic industry has a responsibility to ensure that wild caught fish are sourced sustainably, legally, and ethically.

Transport

Fishes can experience high levels of stress and prolonged recovery from transport (Vanderzwalmen et al., 2020). When considering aspects related to transport these can include, for example, the effects of handling, water conditions including water quality, temperature, and social effects. Shaking, packing, hyper-oxygenation, and cold are all stressors associated with live fish transport, presenting a growing concern for fish welfare (Wu et al., 2021). All different types of transport should be considered: longer and shorter, including, e.g. after initial capture from the wild (if relevant), to wholesale and other housing areas, to the pet store, other shops, venues such as hotels, restaurants, or other businesses or homes.

Vanderzwalmen et al. (2020) trialled the addition of "Stress Coat®" – a commercially available water conditioner containing aloe vera, on fish health and behaviour, during both simulated and actual commercial transport. They found that using a water conditioner was associated with a reduction in biting behaviour post-transport in both the simulation and actual scenario and was also associated with a decrease in erratic swimming behaviour. However, no detectable impact of the water conditioner was found on water quality, visible injuries, or the rate of mortality post-transport, and mortality was low throughout. Particularly for reducing the occurrence of erratic swimming and biting behaviours the addition of Stress Coat® to the transport water of ornamental fishes appears to improve behavioural indicators of welfare.

To mitigate the effects of transport stress responses and promote better well-being, the addition of an extract derived from a Chinese herb, I-Tiao-Gung (*Glycine tomentella – GTE*) was tested during live transport of blood parrot cichlid (*Amphilophus citrinellus* × *Cichlasoma synspilum*) and koi (*Cyprinus carpio*) (Wu et al., 2021). During transport, fish ventilation frequency and water quality were stabilised by GTE addition for blood parrot cichlids, and plasma cortisol was downregulated by GTE during the simulated transport in both species.

Chemical stress is only one of the many stressors that fish may experience during transport. Mechanical stress like being moved in a carrier bag is unpredictable. Masud et al. (2019) investigated if carrier bags designed to reduce mechanical disturbance during transport can be used to decrease stress-induced immunosuppression. Guppies are among the most popular tropical ornamental fish traded, and Masud et al. experimentally infected some guppies and transported them. One group were transported in traditional carrier bags, and one group within "Breathing Bags™" designed to reduce mechanical disturbance. The study demonstrated how mechanical disturbance during transport can cause stress-induced immunosuppression and increase the likelihood of contracting infections. However, no significant reduction was found in parasite burdens of fish transported in the Breathing Bags™ compared to standard polythene carrier bags. The authors highlighted the need for specific management procedures which reduce the impact of infectious disease, following routine transport of ornamental fish.

Considering that millions of fishes are carried home from pet stores in plastic carrier bags annually, the negative stress caused by mechanical disturbance during transport warrants further research and attention, including the involvement of pet stores.

Pet stores

Millions of fishes are transported to pet stores across the world, where they are housed in a variety of different environments, handled, further transported, mixed with other species, fed, and are managed for disease in ways which either support or are to the detriment of their well-being. Diseases of aquarium fish are common, but there are very few surveys examining the health of fish exclusively in pet shops. Hongslo and Jansson (2009) conducted a health survey on 720 freshwater fish from a total 30 Swedish pet retailers. It showed that the most common causes of diseases in the fish were infections with parasites and bacteria. The authors highlighted a need for prophylactic treatments while the fish are held in the pet retailers, and the need for improved hygiene routines, both whilst handling fish and when cleaning tanks and equipment, to minimise infections. Regular health monitoring and the timely separation of infected fish into designated special aquariums for treatment, are essential steps in improving overall fish health. A study in pet shops in New South Wales in Australia examined 108 fish that had evidence of morbidity or mortality, in 24 retail outlets. Most fish (70%) had lesions indicating bronchitis – inflammation of the gills, and/or evidence of bacterial infections. Many of the pathogens identified have low host specificity and/or direct life cycles, with potential risks of transmitting disease to native and commercial fish populations. Wickins et al. (2011) advise that those caring for sick ornamental fish should take appropriate precautions to prevent the spread of disease, and should investigate potential pathogens. This survey only examined fish that already presented with morbidity/mortality, so does not indicate the total percentage of fish that were sick, or the incidence of illness.

It continues to be relatively easy to buy fish from retailers. Many will now ask that a tank has already been set up and cycled/filtered for a period (usually one to two weeks) before they will sell fish, and some will offer to do a water test for purchasers. However, in most cases no proof is required to confirm all is in order, before selling any fish. Specialist aquarists likely have the knowledge to give accurate advice to prospective fish keepers, while larger-scale commercial pet retailers that also deal with other species (cats, dogs, etc.) are less likely to have aquarium specialists on staff and may not have the knowledge to adequately care for fish that are live stocked within the stores. This is improving in many stores, for example by offering aquatics training to staff. However, they may still be unable to recognise signs of poor health and welfare in stocked fish, depending on their level of familiarity with the different species kept in store.

In 2020, Alley et al. reported on a survey of pet stores regarding the medical advice provided for pet fish, and the potential impact on welfare, specifically focusing on the quality of advice regarding fish health given by US pet store employees. A random sample of 27 pet store employees in North Carolina were presented with a standardised scenario describing symptoms of a pigment cell tumour in a Siamese fighting fish, via telephone call. 85.2% provided a medical opinion, mostly suggesting an over-the-counter medication, with only 8.7% suggesting a husbandry or environment change, and 13% felt it was an abnormality that did not require medical treatment. Only one employee felt that it was a tumour that needed veterinary attention. Employees not offering any opinion comprised 3.7% of respondents, and 11.1% could not provide an answer and suggested a referral to a veterinarian specialised in exotic species. Even with limited information provided, many stores gave a presumptive diagnosis, including injury, bacterial infection, toxin, or genetic defect. Some considered it not to be a problem. The treatments offered also varied considerably between stores, and diagnoses were being given by untrained personnel. Only two employees requested additional information on water quality, and none of the participants asked about environmental enrichment – it may not have been seen as relevant to this case. Much existing research on aquarium fish focuses on transport, origins, etc., but very little work has been done on their welfare upon arrival at the home aquarium. Fish owners may rely on the point of purchase as a source of information on the care of their animals, so it is important that aquarium retailers provide relevant and up-to-date knowledge, and provide referrals to specialists when warranted.

Care

Freshwater fish are more adaptable and survive better in captivity, e.g. carp such as goldfish and koi, and have become more popular as pet fish. Cecil (1999) provided a broad overview which discusses proper setup and maintenance of equipment such as heaters, coolers, filters, etc., which are essential for long-term fish well-being, along with attention to key environmental variables, such as temperature, lighting, water quality, and pH balance. Cecil reviewed husbandry and husbandry-related diseases in ornamental fishes, including hypo- and hyperthermia from marked temperature changes, hypoxia (low dissolved oxygen), potentially caused by poor surface absorption of oxygen from improperly shaped tanks such as fishbowls, or due to increases in organisms using oxygen in the tank, through overcrowding, algal blooms, or too many plants. Other common concerns include ammonia toxicity, sometimes termed 'new tank syndrome', which can be caused by improper filtration resulting in a build-up of nitrogenous waste in the tank, which can chemically burn fins and gills. Nutritional disease due to improper feeding, and starvation, are especially common in wild-caught fish. Many marine fish have such specialised feeding requirements that it is difficult to adequately provide for them in a captive environment.

Nutrition

Considering the thousands of fish species kept as pets, another key area is nutrition. This includes both the quality and quantity of food provided, as well as presentation and effects on behaviour and general well-being of the fish. Burgess (2018) investigated if increasing the number of feeding locations reduced aggression in two popular species of marine aquarium fish: blue regal tangs (*Paracanthurus hepatus*) and yellow tangs (*Zebrasoma flavescens*). Increasing the number of feeding locations as an enrichment program was correlated with a decrease in aggressive behaviours. Knowledge of diet and feeding behaviour, as well as factors affecting the latter, is fundamental to improving fish welfare (Martins et al., 2012). While this latter study was conducted on farmed fish, it still provides important insights when caring for pet fishes. Brandão et al. (2021) demonstrated that the biology of fish could be used to make detailed husbandry recommendations that are science-based, for individual species kept in captivity, including feeding recommendations. How fishes are fed, their species-specific needs and preferences, and impacts on health, behaviour, and social aspects, should all be considered within an environmental enrichment program.

Environment

While some standards or recommendations exist, such as a general rule to follow for tank size and stocking density of 'one gallon for every one inch of fish', this can be challenging for several reasons, such as varying width and height of fish, tank dimensions, species tolerance to crowding, amount of waste produced, loss of water volume due to addition of ornaments and decorations, etc. Many smaller species use different sorts of shelters, including plants, and/or prefer shade over open areas. Jones et al. (2019) provided two types of enrichment, above-tank shade, and artificial plants, to zebrafish and three-spined sticklebacks (*Gasterosteus aculeatus*). They observed behaviour and shelter preference to see whether shade is a viable method of providing enrichment to aquarium fish. Zebrafish showed no preference for either type of shelter, while the three-spined sticklebacks showed a preference for being in shelter over being in open areas, and preferred shade in preference to the plants. This study is a good example emphasising the importance of species-specific considerations for enrichment, as well as demonstrating that shade is a viable and meaningful enrichment option for certain fish species.

Environmental enrichment

Many home aquaria today feature plants, shelter, rocks, stones, sand, and other items, often referred to as environmental enrichment. While this chapter does not focus on the science and practice of environmental enrichment, contemporary approaches to optimising animal well-being include the provision of appropriate environments, as well as semi-independency from humans (Brando and Buchanan-Smith, 2018). Shelter, plants, and other aspects that create a good environment for animals would be seen as helping to provide good care, and environmental enrichment as helping ensure optimal care. However, many studies on fish welfare approach shelter, plants, and other aspects of the environment, as environmental enrichments, and for consistency they are considered as such, within this chapter.

It can be challenging to predict the effects of environmental enrichments, as species or group size variations may alter the way enrichment is used. For example, Sloman et al. (2011) found that the same item could be used to simply evade competitors, or as a resource worth aggressively defending. Environmental enrichment can mimic natural habitat. For example, Siamese fighting fish live in shallow waters that are densely vegetated, providing places to hide from predators. Enriched environments give the fish greater control over the environment and potentially reduce chronic stress. Interest in this topic is increasing for laboratory fish, but it remains understudied in pet fish (Pleeging and Moons, 2017). What might be the benefits of providing enrichment such as substrates and physical structures for fish? Naslund and Johnsson (2016) conducted a literature review highlighting different enrichment goals for fish kept in different captive environments. For display aquaria, the focus is on promoting natural and active behaviours, as well as encouraging fish to be present in parts of the tank where they are more visible. Making public tanks aesthetically and visually appealing to the public is also a common consideration. Different types of enrichment are outlined, as well as their benefits, for example structural enrichment or visual barriers to reduce aggression and/or wounding of fish, due to ability to hide from aggressors. The authors suggest considering the natural environment of the species, and that more research is necessary to understand the different types of enrichment and these could affect the various species. Research published in 2021 by Jones et al. features a review spanning the last five years, of research into physical enrichment in fish. The authors concluded that enrichment is not adequately described in many studies, and that methodological descriptions are not always complete, e.g. the amount and dimensions of objects are often excluded. Additionally, the ecological relevance (or justification) of enrichment is often excluded, or not made explicit. They proposed a framework for reporting enrichment in captive fish: DETAILS (Dimensions, Ecological rationale, Timing of enrichment, Amount, Input, Lighting and Social environment). Most fish included in this review were salmonids or zebrafish, with a bias towards fish commonly used in aquaculture and research spheres. Ornamental fish were frequently neglected.

Environmental enrichment has been found to also be beneficial for ornamental fishes during transport. Vanderzwalmen et al. (2020) used plastic loops as enrichment for platyfish (*Platypoecilus variatus*) when being transported by road from UK wholesalers to pet stores, to investigate whether providing enrichment during transport impacted behaviour after transport. Behaviour was used as an indicator of welfare immediately following transport, and during a four-week follow-up period. Immediately after transport, the enriched group displayed significantly less erratic swimming behaviours, and in the four weeks after transport the enriched group also displayed less chasing behaviour. Enrichment during transport was found to reduce stress-related behaviours during recovery, and these simple implementations could reduce transport stress for millions of ornamental fishes.

Behaviour

Mixed-species assemblages, different species combinations and group sizes, indicate the diversity of popular approaches to the keeping of pet fishes at home. However, the effects of these practices on fish well-being are poorly understood. This also includes effects on behaviour, with only a few studies in this area. Saxby et al. (2010) investigated the effects on well-being in relation to group size, in four commonly held species, namely neon tetras (*Paracheirodon innesi*), white cloud mountain minnows (*Tanichthys albonubes*), angelfish (*Pterophylum scalare*), and tiger barbs (*Barbus tetrazona*). Species-specific differences and variations between group size were found, including behaviours such as darting, aggression, shoaling, and latency to feed. White cloud mountain minnows and neon tetras demonstrated a decrease in aggression and an increase in shoaling behaviour in large groups, and improved well-being for larger groups for neon tetra, white cloud mountain minnows, and tiger barbs, but no clear link between welfare and group size in angelfish. While the behavioural impact of group size is species-specific, the combination of behavioural parameters may allow for the identification of optimal group sizes for improved welfare for individual species. Sloman et al. (2011) enquired if the size of shoals impacted the behaviour and welfare of fish kept in mixed-species tanks, and manipulated group sizes in mixed-species assemblages, using the same four species as the former study. They observed social behaviours such as shoaling, as well as aggression, in the whole enclosure. The presence of angelfish appeared to have a beneficial effect on the welfare of small shoaling species by reducing aggression, but had little effect on other behaviours. Larger group sizes resulted in an increased tendency to shoal in white cloud mountain minnows and neon tetras. Research as in this study serves to provide insights into welfare issues which may not become apparent within unrepresentative, single-species studies.

The topic of behaviour is closely connected to all other topics mentioned in this chapter, as different behaviours, next to physical health parameters, are key animal-based indicators used in animal welfare assessments.

Conclusions

This chapter has provided an introduction into the complexities of considering and caring for aquatic animals, specifically fishes – both fresh and saltwater species – as popular companion animals. Fishes are kept in approximately 10% of Western households, and are the most numerous type of pet, with 1.5 billion ornamental fishes handled in trade. Difficulties arise in all aspects of fish well-being, from sourcing and transport, to the moment they arrive in the home as a pet, and during ongoing husbandry thereafter. The frequently marginalised status of fish is based on bias rather than science, and seeing fishes as sentient beings changes how they are perceived. More importantly, it highlights the need to reform how fishes are sourced, traded, managed, and kept.

Despite increasing research into the welfare of fish farmed in aquaculture, and the popularity of keeping aquarium fish at home, there is very little research into optimal living conditions, and the general welfare of fish kept as pets. Unlike aquaculture, ornamental fishkeeping involves hundreds of species, with diverse care requirements, and companion animal caregivers may not have the knowledge or access to the required knowledge to provide appropriate care. The mis-selling of fishes as exotic animals labelled as 'easy', or 'beginner' pets can result in new fish owners purchasing pets without appreciating the complexity of their biology or needs. Hobbyist fish owners tend to rely on non-scientific manuals, their own experiences, and information shared with other fish keepers. There are many online forums where information is shared, but such forums often struggle with debates over contrasting opinions, or present conflicting information. Both online and physical pet retailers complicate the issue by presenting fish keepers with a wide variety of aquarium tanks in different shapes and sizes, multiple types of technical equipment (heating, lighting, filtration systems), and aquarium accessories (live plants, plastic ornaments, rocks, etc.). Such products may be available with or without relevant information as to which species they are considered suitable for. Consequently, ornamental fish can be kept many ways. 'Decorative' tanks such as vases and fishbowls were once considered the norm for at-home fishkeeping, despite not providing adequate water circulation for filtration and oxygenation, space for the fish to swim, and being difficult to maintain at appropriate temperatures. This resulted in cities and countries such as Rome, Sweden, and Mexico, banning fishkeeping in bowls.

Available evidence on topics such as transport, environment, enrichment, and behaviour, suggest that fishes require and benefit from the same considerations regarding their well-being as birds and mammals. While not included in this chapter, further evidence for pain and fear in fish, as well as play behaviours and positive interactions with their carers, highlight the need for a holistic approach, which includes physical and psychological aspects, to safeguarding fish well-being. Understanding the challenges posed at each phase, will allow for a constructive approach to conducting more research, as well as implementing changes that can benefit fish well-being everywhere. This does include the necessary cessation of current practices which threaten fish welfare. All efforts to apply scientifically meaningful findings to the improvement of fish well-being and protection should be a high priority. This should include communication aimed at the positive evolution of legislation and standards, and enforcement benefitting fish well-being everywhere, including for fishes kept as companion animals.

Acknowledgements

I would like to thank Max Norman for their help during the preparation of this chapter.

References

Brandão ML, Dorigão-Guimarães F, Bolognesi MC, dos Santos Gauy AC, Pereira AVS, Vian L, Carvalho TB and Gonçalves-de-Freitas E, 2021. Understanding behaviour to improve the welfare of an ornamental fish. *Journal of Fish Biology*, 99(3), pp. 726–739.

Brando S and Buchanan-Smith HM, 2018. The 24/7 approach to promoting optimal welfare for captive wild animals. *Behavioural Processes*, 156, pp. 83–95.

Burgess M, 2018. Welfare implications of an enrichment program for two species of common aquaria fish. *The Plymouth Student Scientist*, 11(2), pp. 309–331.

Cecil TR, 1999. Husbandry and husbandry-related diseases of ornamental fish. *Veterinary Clinics of North America: Exotic Animal Practice*, 2(1), pp. 1–18.

Chao NL and Prang G, 1997. Project Piaba--towards a sustainable ornamental fishery in the Amazon. *Aquarium Sciences and Conservation*, 1(2), pp. 105–111.

FishBase, 2021. https://www.fishbase.se/search.php, accessed 25 March 2021.

Hastein T, Scarfe AD and Lund VL, 2005. Science-based assessment of welfare: aquatic animals. *Revue Scientifique et Technique-Office International des Epizooties*, 24(2), p. 529.

Hongslo T and Jansson E, 2009. Health survey of aquarium fish in Swedish pet-shops. *Bulletin of the European Association of Fish Pathologists*, 29(5), pp. 163–174.

Jacobson G, Sato A and Gilmore B, 2000. Fish aquarium animal assisted therapy and its influence on clients with multiple sclerosis. *Research for Nursing Practice*, 2(1).

Jones NA, Spence R, Jones FA and Spence-Jones HC, 2019. Shade as enrichment: testing preferences for shelter in two model fish species. *Journal of Fish Biology*, 95(4), pp. 1161–1165.

Jones NA, Webster MM and Salvanes AGV, 2021. Physical enrichment research for captive fish: time to focus on the DETAILS. *Journal of Fish Biology*, 99(3), pp. 704–725.

Kidd AH and Kidd RM, 1999. Benefits, problems, and characteristics of home aquarium owners. *Psychological Reports*, 84(3), pp. 998–1004.

King TA, 2019. Wild caught ornamental fish: a perspective from the UK ornamental aquatic industry on the sustainability of aquatic organisms and livelihoods. *Journal of Fish Biology*, 94(6), pp. 925–936.

Langfield J and James C, 2009. Fishy tales: experiences of the occupation of keeping fish as pets. *British Journal of Occupational Therapy*, 72(8), pp. 349–356.

Martins CI, Galhardo L, Noble C, Damsgård B, Spedicato MT, Zupa W, Beauchaud M, Kulczykowska E, Massabuau JC, Carter T and Planellas SR, 2012. Behavioural indicators of welfare in farmed fish. *Fish Physiology and Biochemistry*, 38(1), pp. 17–41.

Masud N, Ellison A and Cable J, 2019. A neglected fish stressor: mechanical disturbance during transportation impacts susceptibility to disease in a globally important ornamental fish. *Diseases of Aquatic Organisms*, 134(1), pp. 25–32.

Moberg GP and Mench JA eds., 2000. *The Biology of Animal Stress: Basic Principles and Implications for Animal Welfare*. Wallingford, Oxfordshire, England: CABI.

Näslund J and Johnsson JI, 2016. Environmental enrichment for fish in captive environments: effects of physical structures and substrates. *Fish and Fisheries*, 17(1), pp. 1–30.

Pets WebMD, 2021. Getting a pet fish. https://pets.webmd.com/getting-pet-fish#1, accessed 2 October 2021.

Pleeging CCF and Moons C, 2017. Potential welfare issues of the Siamese fighting fish (Betta splendens) at the retailer and in the hobbyist aquarium. *Vlaams Diergeneeskundig Tijdschrift*, 86(4), pp. 213–223.

Ploeg A, 2012a. The volume of the ornamental fish trade. In A Ploeg, RR Hensen, & SA Fosså (Eds.), *International Transport of Live Fish in the Ornamental Aquatic industry*. Montfoort, The Netherlands: Ornamental Fish International, pp. 44–57.

Raghavan R, Dahanukar N, Tlusty MF, Rhyne AL, Kumar KK, Molur S and Rosser AM, 2013. Uncovering an obscure trade: threatened freshwater fishes and the aquarium pet markets. *Biological Conservation*, 164, pp. 158–169.

Saxby A, Adams L, Snellgrove D, Wilson RW and Sloman KA, 2010. The effect of group size on the behaviour and welfare of four fish species commonly kept in home aquaria. *Applied Animal Behaviour Science*, 125(3–4), pp. 195–205.

Sloman KA, Baldwin L, McMahon S and Snellgrove D, 2011. The effects of mixed-species assemblage on the behaviour and welfare of fish held in home aquaria. *Applied Animal Behaviour Science*, 135(1–2), pp. 160–168.

Stevens CH, Croft DP, Paull GC and Tyler CR, 2017. Stress and welfare in ornamental fishes: what can be learned from aquaculture? *Journal of Fish Biology*, 91(2), pp. 409–428.

Sullivan ME, 2014. *Fishing for Answers: Improving Welfare for Aquarium Fish* (Doctoral dissertation, University of Western Australia).

Vanderzwalmen M, Carey P, Snellgrove D and Sloman KA, 2020. Benefits of enrichment on the behaviour of ornamental fishes during commercial transport. *Aquaculture*, 526, p. 735360.

Warwick C, n.d. How the exotic pet business has changed from ancient times – and it's not good news! *Veterinary Practice, Exotics*, 8 December, accessed 30 August 2021, https://www.veterinary-practice.com/article/how-the-exotic-pet-business-has-changed-from-ancient-times-and-its-not-good-news.

Warwick C, Steedman C, Jessop M, Arena P, Pilny A and Nicholas E, 2018. Exotic pet suitability: understanding some problems and using a labeling system to aid animal welfare, environment, and consumer protection. *Journal of Veterinary Behavior*, 26, pp. 17–26.

Wickins SC, Dennis MM, Landos M, Šlapeta J and Whittington RJ, 2011. Histopathological survey of lesions and infections affecting sick ornamental fish in pet shops in New South Wales, Australia. *Diseases of Aquatic Organisms*, 94(2), pp. 143–152.

Wu SM, Chen JR, Chang CY, Tseng YJ and Pan BS, 2021. Potential benefit of I-Tiao-Gung (Glycine tomentella) extract to enhance ornamental fish welfare during live transport. *Aquaculture*, 534, p. 736304.

22
MARINE MAMMALS

Sabrina Brando

Introduction

Marine mammals are diverse: they are found in a wide variety of fresh, brackish, and saltwater habitats. Some species have a worldwide distribution, from the Arctic to the Antarctic, while others are more localised. Cetaceans (whales, dolphins, porpoises), pinnipeds (seals, sea lions, walruses), sirenians (manatees, dugongs), sea otters (*Enhydra lutra*), and polar bears (*Ursus maritimus*), rely on aquatic environments to varying degrees throughout their lives. Today, many species of marine mammals are housed in facilities such as aquariums, zoos, and sanctuaries.

This chapter reviews the welfare of marine mammals living in human care, acknowledging the huge diversity of topics and species. It aims to give an indication of the wide variety of aspects to consider when caring for marine mammals in captivity, and the importance of assessing animal well-being. Specifically, this chapter will introduce the types of marine mammals living freely or in captivity, and the types of human impacts they experience. It will then cover regulations, sourcing, and accreditation, research, care, marine mammal well-being, assessments, and behaviour, social life, environmental considerations including enrichment, nutrition, human-animal interactions, training, choice and control, and the use of marine mammals for educational purposes. Finally, it will cover 'swim with dolphins' programmes and ethical considerations, and will provide recommendations for changes in current practices.

Marine mammals

Beginning with the first specialist 'Oceanarium' in 1938, there is a long history of cetaceans being held in captivity. The motivations vary from display, education, military work, and research, to many more. Most are permanent residents; however, other facilities rescue and rehabilitate injured or sick animals before release (Corkeron, 2018). Today, over 2,200 individuals live in zoos worldwide (Species360, 2021); 70% are pinnipeds and 15% are cetaceans (5% *Sirenia*, 9% *Ursus maritimus*, 2% *Enhydra*). The bottlenose dolphin (*Tursiops truncatus*) is the most numerous cetacean, representing 78% of this group, while the most numerous species is the California sea lion (*Zalophus californianus*), which represents 27% of all marine mammals in human care. While the number of facilities keeping cetaceans has declined in Europe and Australasia, they continue to increase in other parts of the world. While it is obvious that there are vast differences between

DOI: 10.4324/9781003182351-26

living in the wild and human care, it is the view of this author that such a life difference does not automatically have to result in adverse welfare outcomes. Throughout this chapter, I will provide a brief overview of some of the current literature, including areas of positive progress, current challenges and concerns, and suggestions for future directions.

Marine mammals in the Anthropocene

Marine mammals face many challenges in the Age of Humans, otherwise termed the Anthropocene. These include boat strikes, ingestion of plastics, trash, or microbeads, the effects of sonar and mass strandings, as well as artificial lighting and other environmental alterations. Whether chemical and noise pollution, marine debris, prey depletion, or ocean acidification, the impact on marine mammals and their environments appear ubiquitous. Davies et al. (2020) investigated the exposure of marine ecosystems to night time light pollution, and whether anthropogenic light is reaching the seafloor in sufficient quantities to have ecological impacts. They found that up to 76% of the three-dimensional seafloor area was exposed to biologically important light pollution. They concluded that light pollution from coastal cities is likely having deleterious impacts on seafloor ecosystems, and that more research into the impacts is urgently needed. Due to mistaken ingestion, as well as potential release of plasticisers and other chemicals, plastics pose significant physiological threats (Panti et al., 2019), as well as risks of entanglement (Fossi et al., 2018). Foraging, mating, nursing, resting, and migrating can all be influenced by underwater noise, by impairing hearing sensitivity, masking acoustic signals, eliciting behavioural responses, or causing physiological stress (Erbe et al., 2018). Brakes and Dall (2016) state that no marine mammal populations remain entirely unaffected by human activities.

Regulations, sourcing, and accreditation

General regulations exist regarding the sourcing and management of marine mammals, but the administration and enforcement of these regulations varies globally. In some countries, such as India, keeping cetaceans for entertainment purposes is banned, while in other regions such as the Caribbean there is a high density of marine entertainment parks. Within the European Union, countries differ in how the EU Zoos Directive has been transposed into national law: in Italy bottlenose dolphins are required to have one day per week free from performances; this is not the case for all EU countries. It's not clear why Italy chose to require this, but many other nations have not. The ability to choose whether to participate or spend time away from people can be beneficial to welfare, and these should be considered as options.

Most marine mammals are listed in the Convention on International Trade in Endangered Species of Wild Fauna and Flora (CITES) Appendix II. The convention includes the requirement that appropriate environments and care are available. It is relevant to discuss that the capture of animals to support the growth of the aquarium industry is ongoing in some parts of the world, such as Southeast Asia – and that capture and holding facilities are largely unregulated in these areas (Corkeron, 2018). As captive breeding programmes are largely successful, capture from the wild should be avoided. Many associations, such as World Association of Zoos and Aquariums (WAZA), have clear positions on where animals may be sourced from: all illegal sourcing which is not cleared through official legislation and conducted under latest best practice, such as drive hunts, is unacceptable.

The policies of associations such as the Zoo and Aquarium Association (ZAA, Australasia), European Association of Zoos and Aquaria (EAZA, Europe) and Association of Zoos and Aquariums (AZA, North America), the European Association for Aquatic Mammals (EAAM,

Europe), and Alliance of Marine Mammal Parks and Aquariums (AMMPA, North America) are key where specific guidelines are not provided by government legislation. However, it is estimated that while there are about 10,000 zoos and aquariums worldwide, only 10% are accredited under the standards of a zoo association. Readers interested in the finer details of animal welfare law covering this area are encouraged to refer to Cao and White (2016), Corkeron (2018), Rose et al. (2017), and Brando et al. (2018).

Research

Marine mammals have long been involved in the study of physiology, cognition, and animal communication, as well as contributing to research in disciplines including physiology, cognition, sensory abilities, ethology, and veterinary medicine. They have also been trained to help us understand problems such as those caused by fisheries bycatch, and anthropogenic oceanic noise.

The field has seen a moderate but growing body of research, with a taxonomic bias towards cetaceans and pinnipeds. Welfare studies have largely focused on bottlenose dolphins, while other species remain poorly studied (Serres et al., 2020b). Peer-reviewed publications are more commonly focused on wild marine mammals than on those in captivity (Hill and Lackups, 2010). This contrasts with the rapid increase in the number of scientific publications on the welfare of other animals, despite opportunities for aquaria to contribute (Romano et al., 2021).

In 2011, Marino and Frohoff proposed a different paradigm for research on cetacean cognition using animals in the wild. They argued that research imposes tremendous stress on captive animals and places burdens on their wild populations. While studying cetacean cognitive function in the wild is an important and interesting endeavour, the types of questions that can be answered are considerably different from the ones applicable to those in human care. Captive animals are extensively trained to voluntarily participate in research projects, often benefitting cognitively and physiologically from the activities as well as providing valuable data. Furthermore, interacting with animals in the wild warrants serious considerations concerning their health and safety, which may be threatened by extensive human contact.

Güntürkün (2014) reviewed the question of whether dolphin cognition is special, concluding that there is not a single achievement that has not also been shown in several other species. The data obtained on communication in dolphins under human care (Herman et al., 1984) constitutes one of the most remarkable examples of animal cognition. Güntürkün noted that keeping an emotional distance from the species we study is key to avoiding unfounded conclusions about the cognitive exceptionality of some species, highlighting that we must not be unduly influenced by public expectations and unsubstantiated arguments. The authors stress the importance of focusing on the animals' needs and perspectives, rather than the advancement of a viewpoint. Jaakkola et al. (2020) discuss the importance of accurate knowledge, and that bias and misrepresentation can undermine productive discourse on animal welfare policy.

Care

As in all animal fields, care provided needs to be considered in relation to well-being: care being what is provided (inputs – which can also refer to resource removal or change), and well-being revolving around the animals' experiences (outputs). Best practice guidelines exist only for a minority of species; at the time of writing, two reference-based husbandry guidelines exist – the AZA *Polar Bear Care Manual*, and the EAZA and EAAM *Best Practice Guidelines for Otaridae and Phocidae*. The EAAM has published bottlenose dolphin standards as a list of provisions without references supporting recommendations. Professional care and well-being manuals should comprehensively cover a wide variety of topics, namely: biology, ecology, diet, anatomy, cognition, social life, and behaviour, as well as habitat requirements, training, and environmental enrichment and human–animal interactions. These recommendations should be based on a 24/7 lifespan approach (Brando and Buchanan-Smith, 2018) that holistically considers factors such as life experiences from birth to death, the roles animals are assigned (for example, presentation, 'swim with' programmes), as well as all the hours that staff are not on site to actively care for the animals. Promoting and facilitating choice and control during the hours staff are absent is key to this approach. The content should be updated on an annual basis, to reflect the latest science, promote evidence-based decision-making, and include case studies, technical drawings, and links to instructive materials.

Marine mammal well-being, assessments, and behaviour

There are many definitions of animal welfare. I would describe animal well-being as comprising the psychological and physical experiences of an individual animal, as perceived by them. Such experiences may vary widely in their 'valence', from very positive to very negative. A variety of inputs may affect well-being. These may act together and may be synergistic rather than simply additive in their consequences. Therefore, evaluation of well-being should consider all such measurable factors, and their interactions. An animal with good welfare has the ability to exert agency over their life to a meaningful extent, and ideally to fulfil all their needs, and to satisfy most of their preferences. To ensure good welfare, the focus should be on promoting agency, choice, and control, and predominantly positive well-being. And the monitoring and assessment of animal welfare should include a focus on their likely experiences.

Many of the high-profile concerns revolve around orcas and bottlenose dolphins – the main concerns being the ethics of confining wide ranging and intelligent species in relatively small tanks, social disruptions (Anderson et al., 2016; Rose et al. 2017), high levels of morbidity and mortality (Jett and Ventre, 2015, Marino et al., 2020), and the green-washing of marine mammal captivity (Visser et al., 2021). In 2019, World Animal Protection (WAP) published a report arguing for the closure of all dolphin venues, which included potential and existing welfare and ethical concerns. Concerns about facilities should be highlighted worldwide, especially in unregulated areas, and it is key that critique is based on evidence and sound reasoning. The WAP report has been criticised as insufficiently science-based, and as containing misleading information (Loro Parque Redaccion, 2019). We cannot compare species without appropriate and relevant evidence, and general and specific concerns, relevance, and opportunities for different taxa need to be specifically researched and discussed. The wide variety of species in human care deserve the time and attention needed, to understand and support their well-being.

There is very little hypothesis-driven research on behavioural indicators of welfare in marine mammals. The mounting pressures from organisations opposed to keeping marine mammals in captivity, and growing concern from the public, have spurred progress in the last decade, particularly in bottlenose dolphin welfare, and predominantly concerning the use of behaviour as a welfare indicator. The most common behavioural welfare indicators are repetitive (possibly stereotypical) swimming, and aggressive or sexual behaviour. However, these reports must be considered with caution as they are based on limited sample sizes (Clegg et al., 2017, Brando et al., 2018). Some repetitive swimming might not be a stereotypy when it is punctuated by the performance of other behaviours. A "C-Well Index" for bottlenose dolphins has been developed through the collaboration of facilities and universities (Clegg et al., 2015). The index contains 36 welfare measures, including animal-based (e.g., body condition scores, stereotypies, swim speed), and resource-based measures (e.g., water quality, diet, shade). Despite some shortcomings, it is an important first step in the development of comprehensive and practical welfare assessment.

Measuring behaviour has proven essential in understanding the experiences of marine mammals. The impacts of such initiatives can vary based on individual and species differences. Across several studies comparing the behaviour of different cetaceans in response to conditions, including disruptions to routine, enrichment, public presence, and other factors, Serres et al. (2019, 2020a) discussed the potential usefulness of behaviour in assessing marine mammal welfare. Other measures of welfare, such as faecal hormones (Serres et al., 2020b), acoustic monitoring (Jones et al., 2021), and personality and subjective well-being scoring (Ubeda et al., 2021), can also be used to evaluate welfare more comprehensively.

Social life

Social life is essential to marine mammal welfare. Social environments provide companionship, play opportunities and social learning, and can also pose challenges such as aggression, monopolisation of resources, and displacement. Maintaining appropriate social groups is essential. To reduce the number of transfers, and associated stress and the changing of social compositions, technologies such as artificial insemination, as well as the development of sperm sexing and assisted reproductive technology, have proven helpful. However, studies of the social dynamics of zoo-housed marine mammals are scarce; instead, studies largely focus on wild conspecifics.

The most common motives of aggression in wild cetaceans are intra-sexual male competition and sexual coercion (Connor et al., 2005). Rake marks, caused by teeth raking along the body, are an important indirect measure of aggression, as not all aggression is observed by staff (Brando et al., 2018). As rake marks are very frequently observed on wild cetaceans and are the expression of normal social behaviour (Lockyer and Morris, 1985), there has been no conclusion about the frequency or severity of rake marks above which welfare is considered to be poor. Nonetheless, rake marks have been proposed as an animal-based measure to monitor dolphin welfare (Clegg et al., 2015).

Social interaction and grouping differ between species; the social lives of dolphins and polar bears, for example, are very different. The line between solitary and social is blurring, as we develop our understanding of what 'social' truly means. Polar bears are typically described as solitary but are often housed socially. Landscape complexity and multiple pathways in exhibits allow animals to make behavioural decisions to spend time together or apart, and can aid in managing aggression. While negative interactions can occur, playing, swimming, and sleeping together are also observed. An important social bond is the one between mothers and infants, aiding cognitive growth and development. Appropriate social structures improve welfare by offering play opportunities, developing problem-solving skills, and facilitating learning.

Environment

Unlike the habitats of captive terrestrial animals, those of marine mammals have not substantially changed over time, with respect to space or environmental complexity. Animals can be housed in open and semi-open sea pen structures, or traditional concrete exhibits (Couquiaud, 2005). While some guidelines and legislation includes information on the design and construction of marine mammal habitats, including pool dimensions, depth, and water volumes for marine mammals based on their adult body lengths and group size, there is little scientific basis to support these. In a review of 44 cetacean facilities from 22 countries, Couquiaud (2005) reported that the average area of traditional pools was 90.5 m^2, while open facilities (i.e., sea pens and lagoons) were much larger, averaging 400 m^2 (range 92–1,633 m^2).

Dolphins have been demonstrated to be more active in open facilities (Ugaz Ruiz et al., 2009; Luna Blasio et al., 2012) and in larger pools in general (Bassos-Hull and Wells, 1997). However, we need to look at the quality of space – not just tank size and volume. The importance of pool dimensions is an intuitive area of research for cetaceans, yet only a few studies have examined pool size preference to date (Myers and Overstrom, 1978; Shyan et al., 2002). Enclosures are less complex compared with modern terrestrial exhibits, although some substrates like bushes, grass and rocks are used, particularly with polar bears and otters. Most habitats are sterile, created in materials such as concrete and acrylic panels, with fake rock formations that do not afford many behavioural opportunities. Modern zoos aim to reconstruct natural-looking habitats that are not only visually appealing, but also provide species-specific behavioural opportunities. Many newly developed marine mammal parks in various regions of the world seem to have used Google Earth to view images from above to use as blueprints for construction, rather than improved and innovative habitat designs. They often seem to have similar, small, and outdated designs.

The expense of water filtration systems and the design of care programmes (e.g., training, enrichment, presentations, etc.), are sometimes provided as explanations for why complexity has been hard to achieve. Increasing complexity with regards to cognitive, behavioural, and physiological opportunities can reap both short-term and long-term welfare benefits. A shift is needed to enhance all marine mammal habitats, with the aim of increasing welfare. For further information on housing and habitat requirements in relation to welfare, see Brando et al. (2018). While increasing complexity has been proposed as a welfare benefit overall, there is little quantitative research identifying which habitat characteristics are most important. For example, using non-invasive tagging systems to measure activity levels, Miller et al. (2021) found that enrichment programmes were most heavily associated with higher distances travelled, and increased energy expenditure, when compared to habitat characteristics.

A variety of sounds common to pools have been proposed as a nuisance, with structures such as pumps potentially creating undesired noise (Couquiad, 2005). Houser et al. (2019) report that fully enclosed pools present the highest noise levels compared to ocean environments, primarily due to the operation of water treatment systems. The authors found that whales themselves were the greatest source of sound, and that the potential for ambient noise to mask communication signals and echolocation clicks appeared to be low. Stevens et al. (2021) presented more nuanced future directions for the evaluation of acoustic welfare in both the wild and captivity, including suggestions for how research in each domain can inform the other.

Environmental enrichment

Any change to a captive environment to enhance welfare could be described as enrichment. There is a bias towards buoyant plastic or rubber objects for marine mammals, as opposed to objects placed under the water surface. Rocks, plants, streaming waters, different depths, and other environmental structures are largely lacking. Enrichment has received little scientific attention: Alligood and Leighty (2015) excluded marine mammals from their meta-analysis of zoo enrichment due to a lack of data. Clark (2013) proposed that cognitive enrichment is a necessary aspect of dolphin management, and should be "relevant, motivating, controllable, and possible to master", which varies based on the species and individual. There is ample evidence to suggest that dolphins can apply behavioural skills to solve problems.

Enriched environments for mammals have been associated with a reduction of stereotypies by as much as 53% of the time (Swaisgood and Shepherdson, 2006), and enrichment was found to be one of the strongest factors determining a reduction in stereotypical behaviour in polar bears (Shepherdson et al., 2013). Stereotypies – invariant, repetitive behaviours – are often attributed to boredom (Fernandez and Timberlake, 2019). As an example, stereotypical behaviours in a study of walruses included repetitive sucking and pattern swimming, and such behaviour was reduced through the introduction of foraging mats and boomer balls containing food. A study of Australian sea lions (*Neophoca cinerea*) found that non-food enrichment objects were interacted with more than those utilising food (Smith and Litchfield, 2010). Scent enrichment further demonstrated a reduction in stereotypical behaviour and an increase in habitat utilisation (Samuelson et al., 2017). Environmental enrichment was reviewed by Makecha and Highfill (2018), including aspects to consider specific to marine mammals, with examples provided. For an example of the importance of considering individual differences, see Eskelinen et al. (2015). For an extensive review, see also Brando et al. (2018).

Nutrition

Nutrition is essential to sustaining healthy and active animals. A diet should meet energetic requirements, not cause nutritional disease, be safe and hygienic, and encourage natural and engaging behaviours. Nutritional requirements are dynamic and vary across lifespans: e.g., pinnipeds require varying changes in fat content during life stages such as breeding and moulting, and dietary regimens should reflect their macronutrient needs (Robbins et al 2022). For more information on, e.g., preventing deficiencies, and on commonly supplied supplements, behavioural and species-specific feeding examples, see Brando et al. (2018).

It is the view of this author that more environmental enrichment activities are often warranted for marine mammals, e.g., through foraging or cognitive tasks. Feeding a variety of fish to captive marine mammals is recommended, to ensure animals do not become conditioned to one diet, in case it becomes less available. Alternative diets may become more commonplace if fish stock declines continue.

Human–animal interactions

Good human–animal relationships are identified as one of the critical requirements for positive animal welfare. Trainers not only focus on building relationships based on positive interactions and reinforcement, but carefully observe animal behaviour and preferences, with attention to the effect of body language, posture, and communication. In addition to husbandry training, animals participate in other forms of formal and informal activities, which are widely accepted as being mentally stimulating, and perceived as rewarding (Clegg et al., 2018). Cetaceans can discriminate between familiar and unfamiliar humans (Hill et al., 2016) and have been observed spontaneously interacting with visitors (Trone et al., 2005; and personal observations of this author).

Clegg et al. (2018) found that an increased frequency of anticipatory behaviours prior to training sessions was significantly associated with pessimistic judgements in cognitive bias tests, which is consistent with previous findings linking higher reward sensitivity with negative affective states. Behavioural patterns are established in response to predictable environmental cues, such as human-controlled periods. These responses should be considered when designing animal welfare assessments (Clegg et al., 2019).

Training

While historically animals were often trained to perform behaviours together with trainers, like rides on the backs of animals, many contemporary facilities focus on education, conservation-minded behaviours, and daily management activities. Animals have been trained to participate in a wide variety of husbandry behaviours to reduce stress. See Brando (2010, 2020) and Brando et al (2016, 2018) for in depth reviews of these. However, although training has had many positive outcomes as cited above, the use of a singular routine training method may be insuffcient for dealing with complex welfare problems, and training does not replace enrichment and habitat complexity needs. Responses to training may also be species specific, e.g., bottlenose dolphins participated better in training than other cetaceans (Serres et al., 2020b). While today most professionals in the field agree that training can be enriching to animals, well-known and practised routines can become monotonous over time, leading to boredom (Melfi 2013). Data is lacking on the motivational challenges encountered by trainers: the possibility of boredom and loss of motivation in the animals warrants further study.

There are many aspects of training requiring further investigation. Brando et al. (2018) propose five complementary areas for future research: exploration of the circumstances under which training is enriching, how facility design can provide choice and control, the welfare effects of human interactions, the effects of no human presence during 'night-time', and what content should be present in welfare, education, and conservation messaging to promote human behaviour change most effectively. Other aspects for research and ethical consideration concern the training methods used, such as negative reinforcement, punishment, or aversive associations through classical conditioning. Further information on marine mammal care and training, with citations of appropriate sources, is available from the International Marine Animal Trainers' Association (IMATA), the Animal Behaviour Management Alliance (ABMA), and Brando et al. (2018).

Choice and control

Studies on choice and control in both humans and animals allude to the positive impact of these factors on welfare, and refer to whether animals are aware of the choices and opportunities in their environments. Animals who can control access to indoor and outdoor areas, activate a shower or control supplementary light, have shown signs of improved welfare. Apart from habitat complexity, enrichment and open access to different areas, choice can also be provided to the animals through different stimuli, like photos, sounds, and objects representing different activities, among which they can choose, as well as through choices about rewards, reinforcers, or fish species (for reviews, see Brando and Buchanan-Smith 2018; Brando 2020). Choices and control can also be given through habitat design and technology. Animals can decide whether they want to participate in an activity by going to a certain point in the pool, or choose what activity they would like to do using abstract concepts such as symbols (Brando, 2009b; Brando et al., 2016). Choice and control should also assist the many activities animals are involved in, including educational presentations. Willingness to participate gives insights into the welfare of the animals, for example by predicting early changes in health status (Clegg et al., 2019).

Education, presentations, and 'swim with' dolphins programmes

I have worked with marine mammals for almost 30 years, from rehabilitation centres to public parks. Colleagues and I have previously advocated changing how animals are portrayed in activities, and what their lives are like outside scheduled events (Brando and Dos Santos, 2007). We argued for a change in presentations within aquaria, highlighting the gap between the goals of contemporary zoos and aquariums, and the content of many of these presentations, such as dolphins jumping through hoops and the use of sea lions in pirate-themed shows. Shows and other activities which are not based on care and respect for the animals are not ethically justifiable. Over the last decade, it seemed we moved to a more hands-off approach for animals housed in human care. Today, however, we see a resurgence of interactive programmes, and while education is part of these programmes, for many the main driver is income (Davis, 1997). Public demonstrations need to reflect the goals of contemporary accredited facilities: conservation, research, and engagement. To connect with audiences and successfully convey these messages, direct contact with the animals is not necessary.

There is increasing evidence that facilities provide learning opportunities for visitors, and contribute towards the development of pro-environmental behaviour (Godinez and Fernandez, 2019), helping to protect biodiversity (Moss et al., 2015). While research into specific outcomes is largely missing, several papers reference increased awareness of marine mammal welfare (Jiang et al., 2007), and emotional connections to wildlife (Luebke et al., 2016). However, these studies do not measure long-term changes in behaviour: one study found that a single visit to the zoo did not produce a measurable effect on adults' conservation knowledge (Balmford et al., 2007). A comparison of dolphin–human interaction practices by Mayes et al. (2004) reported that the quality of the interpretation program was key in impacting attitudes.

EAZA encourages their members to focus on natural, intellectual, problem-solving abilities, and physical attributes of animals, in demonstrations. Practices that provide audiences with a misleading impression of natural behaviours should be avoided, as should the use of props which do not demonstrate natural behaviour. However, activities are perceived as a whole and should be approached holistically: for example, by creating props and environments that are in line with the first criteria. I have previously discussed three criteria that may be used to assess the morality of animal 'entertainment' practices: welfare, flourishing, and dignity. Only if all criteria are fulfilled, should animals be part of such programmes. Most marine mammal programmes today do not fulfil these criteria. For more detailed discussions of the future of public displays using cetaceans, see Brando (2016), Melfi et al. (2020), and Clegg (2021).

Animal welfare guidelines on the use of animals performing in demonstrations have been issued by major zoo organisations. A short description, including that bottlenose dolphins should not be unnaturally provoked for the benefit of the public, was available from the EAAM, but at the time of writing could no longer be accessed publicly. More comprehensive and clear guidelines and aims are available in EAZA and WAZA strategies aimed at ensuring both high welfare and ethical treatment of animals. Guidelines promoting good welfare standards are a good start; however, empirical data is largely lacking on animals' experiences of performing. In 2013, Jensen et al. found that dolphins decreased their activity levels and chose to approach the demonstration area before demonstrations, and increased their vigilance behaviour prior to shows. The 'waiting' behaviour was not considered an abnormal or stereotypical behaviour, and the authors concluded that shows were not perceived negatively by the animals.

While 'swim with' programmes are offered by many facilities globally, there is little research to investigate the impacts of these interactions on welfare, and the effects – good or bad – are poorly understood. One survey found that tourists who complete 'swim with' dolphins programmes largely recalled the physical attributes of the dolphins, and experienced post-purchase dissonance regarding the size of enclosures and unfulfilled expectations (Curtin and Wilkes, 2007). While one study found that there are no detrimental effects on dolphin behaviour (Trone et al., 2005), Kyngdon et al. (2003) reported that dolphins increased the use of a refuge area during 'swim with' programmes. However, the dolphins returned to pre-programme levels of refuge use within 15 minutes. Based on this, the authors concluded that the sessions did not appear to have a detrimental effect. The welfare of dolphins in a seawater facility participating in 'swim with' dolphins programmes was neither found to be compromised nor improved (Brando et al.,

2019). Whether the increased interaction with trainers in 'swim with' sessions is enriching for the animals, needs more research to answer.

Ethics

There are many aspects which warrant ethical reflections, including species kept, types of habitats, and when some forms of captivity may be permissible. Muka and Zarpentine (2021) discussed the Yangtze finless porpoise and vaquita porpoise cases (two highly endangered species), illustrating the ethical issues in conservation. They argued that *some* forms of captivity can be justified – even where they involve risk, as trade-offs between conserving endangered species, and the interests of individual animals, need to be considered.

While bottlenose dolphins are the most studied species in human care, it is important to note that many of the same activities are conducted with other species. We know more about dolphins, and have housed them for longer, than many other animals in human care. When it comes to the ethics surrounding cetaceans in captivity, it might be more of a case of how much we are *willing* to do rather than what *can* be done; especially as the aquatic domain adds a layer of complexity and increases costs considerably. For certain species such as orcas, even with the necessary will, it may not be practically possibly to provide habitats sufficiently large and enriched, to adequately uphold their welfare interests. Perhaps, as for elephants, extensive changes, and innovations in cetacean management could follow a radical evolution in attitudes. There are certainly parallels in the issues driving both elephant and orca exhibit design and management (Maple 2016 p. 156). Organisations housing marine mammals should be investigating and building a different future, not just in habitats, but also in how the animals are exhibited, to convey positive messages (Law and Kitchener 2017).

Anderson et al. (2016) consider confinement in small tanks for commercial purposes ethically indefensible, considering orcas' advanced cognitive capacities and behavioural needs. The authors called for a change in our relationship with orcas "characterized by mutual friendship, understanding, and much greater appreciation of these remarkable creatures" (Anderson et al., 2016). Killer whales who have spent much of their lives in human care are unlikely candidates for successful release programmes. The highly marketed release of Keiko was deemed unsuccessful under the criteria developed for bottlenose dolphins (Wells et al., 1998): he was captured as a calf and kept isolated from conspecifics for much of his life prior to release (Simon et al., 2009), and though physically unrestricted and free to leave when he was released, he continually returned to his human caretakers. Opportunities to improve the management of marine mammals already in human care are possible and should be explored and implemented, and many improvements can be made to habitats. Further capture from the wild should be prevented as much as possible, and should only be considered for true conservation programmes. Marginal non-committed facilities should be phased out.

Multiple zoos have introduced animal-first welfare approaches. Programmes are deemed to be "animal-first" if animals can exert choice and control, remain in the comfort of their habitat, and the program is to the benefit of their welfare in human care (Brando and Herrelko 2021). Animal well-being must be the priority, and the achievement of other goals must be compatible with the maintenance of good welfare. Only if optimal animal well-being can be supported and promoted 24/7 across an animal's lifespan, is the housing of marine mammals morally acceptable.

While there are many ethical concerns associated with captivity, shifting to marine mammal tourism is problematic as well. Whale watching can be one of the most significant economic activities for some coastal communities; however, malpractice can compromise welfare for wild animals, and the sustainability of the practice can be called into question. Bearzi (2017) concludes that ecotourism can be done well if animal welfare remains the main objective – as without the animal, there is no ecotourism.

Recommendations

Marine mammal welfare requires adopting a 24/7 approach, considering all the hours the animals are at the facility, as care staff go home. Whilst training can be enriching, it is not the only enrichment which should be available in animal habitats.

Welfare should be considered throughout the day and night and across the lifespan, and should include aspects like habitat complexity, social needs, nutrition, choice, and control. This approach should be implemented across all pools and habitat areas. Innovations and major improvements in most marine mammal habitats and management should be made. Providing various features such as different pools, possibly vegetation, underwater activities, and haul out areas (depending on the species), as well as having other individuals to interact with in animals' environments, and cognitive challenges, and other dynamics removing stimulus-poor conditions, are all important considerations. Excellence in care and creating semi-autonomous environments combined with additional environmental enrichment, is key. A budget for environmental enrichment, habitat upgrades, and continued staff development, among other relevant aspects, needs to be available.

All marine mammals should be trained using positive reinforcement and should have options to forage outside training sessions. Marine mammals who are participating in presentations and/or interactive programmes should be offered the opportunity to choose to opt in or out. Land or pool sections should be set aside to provide a retreat away from visitors and trainers at any time while participating in a session.

Animals who have been in human care for prolonged periods of time, such as killer whales, should be moved to upgraded habitats or sea pen facilities, including professionally run sanctuaries. It can be difficult if not unethical to release such animals back into the wild, and if so, we are ethically obligated to care for them in sea pen facilities or other suitable habitats. If the necessary changes cannot be made in current facilities, suboptimal and marginal facilities and habitats should be phased out.

More research in all areas of marine mammal welfare care and management is needed from a theoretical as well as a practical perspective. All marine mammal holding facilities should be involved in research to further knowledge and experience regarding marine mammal care and welfare. Research should be a mandatory and fundamental part of each facility, including collaborative research with other facilities, experts, research organisations, and universities. Finally, best practice guidelines that are currently lacking should be developed, and translation into multiple languages would be beneficial.

Conclusions

Marine mammals are diverse, with some species having a worldwide distribution and facing many different challenges in the Anthropocene. In a time where zoos, aquariums, and marine parks are all enquiring into a long past including a colonial legacy, and our relationships with other animals increasingly warrant thorough consideration, we need to scrutinise all our activities and interactions, including considering marine mammals as subjects of their own lives. There is a continued need for extensive conservation and education, as well as animal welfare research.

Evidence-based decisions on how to best provide for optimal animal welfare can have ethical implications, e.g., if data confirm if it is not possible to provide for optimal welfare in human care. At this point an ethical decision on whether to keep each species of marine mammal is necessary. While we find similarities in advanced cognitive and communicative capacities, wide natural ranges, and social networks, what is stated about orcas and other whales isn't necessarily true about dolphins. Minimising the gap in comprehension of marine mammals' perceptions and choices, and minimising uncritical anthropomorphic interpretations, are important. Key areas for future development include the ongoing development of innovative habitat designs, recognition of individual marine mammal needs, and promotion of their autonomy and agency.

Acknowledgements

I would like to thank Elisabeth Hunt from Species360 for the latest information on marine mammals in their database, Dr Heather Hill for her assistance with access to missing literature, Aura Diaz, Melodia Tamayo Moreno, and Grace O'Dea for their help with additional reference materials, and Max Norman for their help with the manuscript preparations.

References

Alligood C and Leighty K, 2015. Putting the "E" in SPIDER: Evolving trends in the evaluation of environmental enrichment efficacy in zoological settings. *Animal Behavior and Cognition*, *2*(3), pp. 200–217.

Anderson R, Waayers R and Knight A, 2016. Orca behavior and subsequent aggression associated with oceanarium confinement. *Animals*, *6*(8), p. 49.

Balmford A, Leader-Williams N, Mace GM, Manica A, Walter O, West C and Zimmermann A, 2007. Message received? Quantifying the impact of informal conservation education on adults visiting UK zoos. In Catalysts for Conservation: A Direction for Zoos in the 21st Century, London, UK, 19-20 February, 2004, pp. 120–136.

Bassos-Hull K and Wells RS, 1997. Effect of pool features on the behavior of captive dolphins. *Marine Mammal Science*, *13*(3), pp. 533–534.

Bearzi M, 2017. Impacts of marine mammal tourism. In DT Blumstein, B Geffroy, DSM Samia and E Bessa (eds.), *Ecotourism's Promise and Peril*, Springer, Cambridge, pp. 73–96.

Brakes P and Dall SR, 2016. Marine mammal behavior: a review of conservation implications. *Frontiers in Marine Science*, *3*, p. 87.

Brando S, 2009. Exploring choice and control opportunities applied in enrichment and training. Oral presentation in International Conference on Environmental Enrichment, Paignton, UK.

Brando S and Dos Santos M, 2007. Mind the gap: the ethical animal show. In EAAM 2007, Antibes, France.

Brando SI, 2010. Advances in husbandry training in marine mammal care programs. *International Journal of Comparative Psychology*, *23*(4), pp. 777–791.

Brando S, 2016. Wild animals in entertainment. In B Bovenkerk and J Keulartz (eds.), *Animal Ethics in the Age of Humans*, Springer, Cambridge, pp. 295–318.

Brando, S., Bowles, A.E., Böye, M., Dudzinski, K.M., van Elk, N., Lucke, K., McBain, J., Singer, R. and Wahlberg, M., 2016. Proceedings of Marine Mammal Welfare Workshops Hosted in the Netherlands and the USA in 2012. *Aquatic Mammals*, *42*(3).

Brando S, 2020. Box B4: marine mammal training. In Melfi VA, Dorey NR and Ward SJ (eds), *Zoo Animal Learning and Training*, Wiley & Sons Ltd, Hoboken, NJ, pp. 197–201.

Brando S and Buchanan-Smith HM, 2018. The 24/7 approach to promoting optimal welfare for captive wild animals. *Behavioural Processes*, *156*, pp. 83–95.

Brando S and Herrelko ES, 2021. Wild animals in the city: considering and connecting with animals in zoos and aquariums. In B Bovenkerk and J Keulartz (eds.), *Animals in Our Midst: The Challenges of Co-existing with Animals in the Anthropocene*, Cham, Switzerland: Springer. pp. 341–360.

Brando S, Broom DM, Acasuso-Rivero C and Clark F, 2018. Optimal marine mammal welfare under human care: current efforts and future directions. *Behavioural Processes*, *156*, pp. 16–36.

Brando S, Kooistra N and Hosey G, 2019. Pre and post session behaviour of captive bottlenose dolphins Tursiops truncatus involved in "Swim-with-Dolphin" events. *Journal of Zoo and Aquarium Research*, 7(4), pp. 195–202.

Cao D and White S 2016. *Animal Law and Welfare: International Perspectives*, Springer, Switzerland.

Clark, F. E. (2013). Marine mammal cognition and captive care: A proposal for cognitive enrichment in zoos and aquariums. *Journal of Zoo and Aquarium Research*, 1(1), pp. 1–6.

Clegg IL, 2021. What does the future hold for the public display of cetaceans? *Journal of Applied Animal Ethics Research*, *1*(aop), pp. 1–39.

Clegg ILK, Borger-Turner JL and Eskelinen HC, 2015. C-Well: the development of a welfare assessment index for captive bottlenose dolphins (Tursiops truncatus). *Animal Welfare*, *24*(3), pp. 267–282.

Clegg ILK, Van Elk CE and Delfour F, 2017. Applying welfare science to bottlenose dolphins (Tursiops truncatus). *Animal Welfare*, *26*, pp. 165–176.

Clegg IL, Rödel HG, Boivin X and Delfour F, 2018. Looking forward to interacting with their caretakers: dolphins' anticipatory behaviour indicates motivation to participate in specific events. *Applied Animal Behaviour Science*, *202*, pp. 85–93.

Clegg IL, Rödel HG, Mercera B, van der Heul S, Schrijvers T, de Laender P, Gojceta R, Zimmitti M, Verhoeven E, Burger J and Bunskoek PE, 2019. Dolphins' willingness to participate (WtP) in positive reinforcement training as a potential welfare indicator, where WtP predicts early changes in health status. *Frontiers in Psychology*, *10*, p. 2112.

Connor RC, Watson-Capps JJ, Scott EM, Sargeant BL, and Mann J, 2005. Aggression in bottlenose dolphins: evidence for sexual coercion, male-male competition, and female tolerance through analysis of tooth-rake marks and behaviour. *Behaviour*, 142 (1), pp. 21–44.

Corkeron P 2018. Captivity. In B Würsig, JGM Thewissen and M Kovacs (eds), *Encyclopedia of Marine Mammals*, Academic Press, London, UK, pp. 183–187.

Couquiaud L, 2005. Survey of international cetacean facilities. *Aquatic Mammals*, *31*, pp. 311–319.

Curtin S and Wilkes K, 2007. Swimming with captive dolphins: current debates and post-experience dissonance. *International Journal of Tourism Research*, *9*(2), pp. 131–146.

Davies TW, McKee D, Fishwick J, Tidau S and Smyth T, 2020. Biologically important artificial light at night on the seafloor. *Scientific reports*, *10*(1), pp. 1–10.

Davis SG, 1997. *Spectacular Nature: Corporate Culture and the Sea World Experience*, University of California Press, San Diego.

Erbe C, Dunlop R and Dolman S, 2018 Effects of noise on marine mammals. In H Slabbekoorn, RJ Dooling, AN Popper and RR Fay (eds.), *Effects of anthropogenic noise on animals*, Springer, New York, pp. 277–309.

Eskelinen HC, Winship KA and Borger-Turner JL, 2015. Sex, age, and individual differences in bottlenose dolphins (Tursiops truncatus) in response to environmental enrichment. *Animal Behavior and Cognition*, *2*(3), pp. 241–253.

Fossi MC, Baini M, Panti C and Baulch S. 2018. Impact of marine litter on cetaceans: a focus on plastic pollution. In Mc Foosi and C Pant (eds), *Marine Mammal Ecotoxicology*, Academis Press, New York, pp. 147–184.

Fernandez, E.J. and Timberlake, W., 2019. Foraging devices as enrichment in captive walruses (Odobenus rosmarus). *Behavioural Processes*, *168*, p.103943.

Godinez AM and Fernandez EJ, 2019. What is the zoo experience? How zoos impact a visitor's behaviors, perceptions, and conservation efforts. *Frontiers in Psychology*, *10*, p. 1746.

Güntürkün O, 2014. Is dolphin cognition special? *Brain, Behavior and Evolution*, *83*(3), pp. 177–180.

Herman, L. M., Richards, D. G., & Wolz, J. P. (1984). Comprehension of sentences by bottlenosed dolphins. *Cognition*, 16(2), pp. 129–219.

Hill, H. and Lackups, M., 2010. Journal publication trends regarding cetaceans found in both wild and captive environments: what do we study and where do we publish? *International Journal of Comparative Psychology*, *23*(3), pp. 414–534.

Hill HM, Yeater D, Gallup S, Guarino S, Lacy S, Dees T and Kuczaj II, SA, 2016. Responses to familiar and unfamiliar humans by belugas (Delphinapterus leucas), bottlenose dolphins (Tursiops truncatus), & Pacific white-sided dolphins (Lagenorhynchus obliquidens): a replication and extension. *International Journal of Comparative Psychology*, *29*(1), pp. 1–21.

Houser DS, Mulsow J, Branstetter B, Moore PW and Xitco M 2019. The characterisation of underwater noise at facilities holding marine mammals. *Animal Welfare*, *28*(2), pp. 143–155.

Jaakkola K, Bruck JN, Connor RC, Montgomery SH and King SL, 2020. Bias and misrepresentation of science undermines productive discourse on animal welfare policy: a case study. *Animals*, *10*(7), p. 1118.

Jensen ALM, Delfour F and Carter T, 2013. Anticipatory behavior in captive bottlenose dolphins (Tursiops truncatus): a preliminary study. *Zoo Biology*, *32*(4), pp. 436–444.

Jett J and Ventre J, 2015. Captive killer whale (Orcinus orca) survival. *Marine Mammal Science, 31*(4), pp. 1362–1377.

Jiang Y, Lück M and Parsons ECM, 2007. Public awareness, education, and marine mammals in captivity. *Tourism Review International, 11*(3), pp. 237–249.

Jones BL, Oswald M, Tufano S, Baird M, Mulsow J and Ridgway SH, 2021. A system for monitoring acoustics to supplement an animal welfare plan for bottlenose dolphins. *Journal of Zoological and Botanical Gardens, 2*(2), pp. 222–233.

Kyngdon DJ, Minot EO and Stafford KJ, 2003. Behavioural responses of captive common dolphins Delphinus delphis to a 'Swim-with-Dolphin' programme. *Applied Animal Behaviour Science, 81*(2), pp. 163–170.

Law G and Kitchener AC, 2017. Environmental enrichment for Killer whales Orcinus orca at zoological institutions: untried and untested. *International Zoo Yearbook, 51*(1), pp. 232–247.

Lockyer C and Morris RJ, 1985. Body scars of a resident, wild bottlenosed dolphin (Tursiops truncatus): information on certain aspects of his behaviour. *Aquatic Mammals, 11*(2), pp. 42–45.

Loro Parque Redaccion, 2019. *Misleading statements of the World Animal Protection's report 'Behind the smile'*, viewed 13 December 2021, https://blog.loroparque.com/loro-parques-response-to-world-animal-protections-report-behind-the-smile/?lang=en.

Luebke JF, Watters JV, Packer J, Miller LJ and Powell DM, 2016. Zoo visitors' affective responses to observing animal behaviors. *Visitor Studies, 19*(1), pp. 60–76.

Luna Blasio A, Valdez Pérez R, Romano Pardo M and Galindo Maldonado FG, 2012. Maintenance behaviour and cortisol levels in bottlenose dolphins (Tursiops truncates) in closed and open facilities. *Veterinaria México, 43*(2), pp. 103–112.

Makecha RN and Highfill LE, 2018. Environmental enrichment, marine mammals, and animal welfare: a brief review. *Aquatic Mammals, 44*(2).

Maple T, 2016. *Professor in the Zoo: Designing the Future for Wildlife in Human Care*. Red Leaf Press, St. Paul MN, p. 156.

Marino L and Frohoff T, 2011. Towards a new paradigm of non-captive research on cetacean cognition. *PLoS One, 6*(9), p.e24121.

Marino L, Rose NA, Visser IN, Rally H, Ferdowsian H and Slootsky V, 2020. The harmful effects of captivity and chronic stress on the well-being of orcas (Orcinus orca). *Journal of Veterinary Behavior, 35*, pp. 69–82.

Mayes G, Dyer P and Richins H, 2004. Dolphin-human interaction: pro-environmental attitudes, beliefs and intended behaviours and actions of participants in interpretation programs: a pilot study. *Annals of Leisure Research*, 7(1), pp. 34–53.

Melfi, V. (2013). Is training zoo animals enriching? *Applied Animal Behaviour Science*, 147(3–4), pp. 299–305.

Melfi, V. A., Dorey, N. R., & Ward, S. J. (2020). *Zoo Animal Learning and Training*. 1st edn. Hoboken, NJ: John Wiley & Sons Ltd.

Miller LJ, Lauderdale LK, Mellen JD, Walsh MT and Granger DA, 2021. Relationships between animal management and habitat characteristics with two potential indicators of welfare for bottlenose dolphins under professional care. *PloS one, 16*(8), p.e0252861.

Moss A, Jensen E and Gusset M, 2015. Evaluating the contribution of zoos and aquariums to Aichi Biodiversity Target 1. *Conservation biology, 29*(2), pp. 537–544.

Muka S and Zarpentine C, 2021. Cetacean conservation and the ethics of captivity. *Biological Conservation, 262*, p. 109303.

Myers WA and Overstrom NA, 1978. The role of daily observation in the husbandry of captive dolphins *(Tursiops truncatus). Cetology*, 29, pp. 1–7.

Panti C, Baini M, Lusher A, Hernandez-Milan G, Bravo Rebolledo EL, Unger B, Syberg K, Simmonds MP and Fossi MC, 2019. Marine litter: one of the major threats for marine mammals. Outcomes from the European Cetacean Society workshop. *Environmental Pollution*, 247, pp. 72–79.

Robbins CT, Tollefson TN, Rode KD, Erlenbach JA and Ardente AJ, 2022. New insights into dietary management of polar bears (Ursus maritimus) and brown bears (U. arctos). *Zoo Biology, 41*(2), 166-175.

Romano TA, Thompson LA, Driscoll MV, Unal E, Tuttle AD, Sirpenski G, Mateleska ME and Wolbrink D, 2021. The role of aquaria in beluga research and conservation. *Polar Research, 40*(S1). https://doi.org/10.33265/polar.v40.5567.

Rose NA, Hancock Snusz G, Brown DM and Parsons ECM 2017. Improving captive marine mammal welfare in the United States: science-based recommendations for improved regulatory requirements for captive marine mammal care, *Journal of International Wildlife Law & Policy*, 20(1), pp. 38–72.

Samuelson MM, Lauderdale LK, Pulis K, Solangi M, Hoffland T and Lyn H, 2017. Olfactory enrichment in California sea lions (Zalophus californianus): an effective tool for captive welfare? *Journal of Applied Animal Welfare Science*, *20*(1), pp. 75–85.

Serres A, Hao Y and Wang D, 2019. Agonistic interactions and dominance relationships in three groups of captive odontocetes: method of assessment and inter-species/group comparison. *Aquatic Mammals*, *45*(5), pp. 478–499.

Serres A, Robeck T, Deng X, Steinman K, Hao Y and Wang D, 2020a. Social, reproductive and contextual influences on fecal glucocorticoid metabolites in captive Yangtze Finless Porpoises (Neophocaena asiaeorientalis asiaeorientalis) and Bottlenose Dolphins (Tursiops truncatus). *Journal of Zoological and Botanical Gardens*, *1*(1), pp. 24–41.

Serres A, Hao Y and Wang D, 2020b. Swimming features in captive odontocetes: Indicative of animals' emotional state? *Behavioural Processes*, *170*, p. 103998.

Shepherdson, D., Lewis, K. D., Carlstead, K., Bauman, J., & Perrin, N. (2013). Individual and environmental factors associated with stereotypic behavior and fecal glucocorticoid metabolite levels in zoo housed polar bears. *Applied Animal Behaviour Science*, 147(3–4), pp. 268–277.

Shyan MR, Merritt D, Kohlmeier NM, Barton K and Tenge J 2002. Effects of pool size on free-choice selections by Atlantic bottlenosed dolphins at one zoo facility. *Journal of Applied Animal Welfare Science*, *5*(3), pp. 215–225.

Simon M, Hanson MB, Murrey L, Tougaard J and Ugarte F, 2009. From captivity to the wild and back: an attempt to release Keiko the killer whale. *Marine Mammal Science*, 25(3), pp. 693–705.

Smith, B.P. and Litchfield, C.A., 2010. An empirical case study examining effectiveness of environmental enrichment in two captive Australian sea lions (Neophoca cinerea). *Journal of Applied Animal Welfare Science*, *13*(2), pp. 103–122.

Stevens PE, Hill HM and Bruck JN, 2021. Cetacean acoustic welfare in wild and managed-care settings: gaps and opportunities. *Animals*, *11*(11), p. 3312.

Swaisgood, R., & Shepherdson, D. (2006). Chapter 9. Environmental enrichment as a strategy for mitigating stereotypies in zoo animals: a literature review and meta-analysis. In G. Mason and J. Rushen (Eds.). *Stereotypic Animal Behaviour: Fundamentals and Applications to Welfare*. Wallingford, UK: CAB International.

Trone M, Kuczaj S and Solangi M, 2005. Does participation in dolphin-human interaction programs affect bottlenose dolphin behaviour? *Applied Animal Behaviour Science*, *93*(3–4), pp. 363–374.

Úbeda Y, Ortín S, Robeck TR, Llorente M and Almunia J, 2021. Personality of killer whales (Orcinus orca) is related to welfare and subjective well-being. *Applied Animal Behaviour Science*, *237*, p. 105297.

Ugaz Ruiz C, Sánchez A and Galindo Maldonado F, 2009. Comportamiento social e individual de un grupo de toninas (Tursiops truncatus) en instalaciones abiertas y cerradas. *Veterinaria México*, *40*(4), pp. 381–387.

Visser IN, Barefoot NN and Spiegl MV 2021. Wildlife conservation and public relations: the greenwashing of marine mammal captivity. In V Carvalho Mocellin (ed.), *Contributions to the Global Management and Conservation of Marine Mammals*. Editora Artemis, Curitiba, Brazil.

PART V

Recent and emerging issues

23

CLIMATE CHANGE, HUMAN–WILDLIFE CONFLICT, AND BIODIVERSITY LOSS

Bob Fischer

Introduction

This chapter introduces some of the ways that human beings compromise the welfare of wild animals. In particular, it focuses on welfare compromises that are due to three broad and overlapping categories of human action: (1) those that affect the environment, such as climate change and habitat fragmentation; (2) human–wildlife conflicts, especially where humans manage wild animals who they view as either threats or nuisances; and (3) species introductions, where humans either intentionally or unintentionally add species to ecosystems.

This task is complicated by two factors. First, the focus here is on *individual* welfare, though much of the available data is on human impacts on populations and species. While there are, of course, reasons to be concerned directly about biodiversity loss generally and extinction events specifically, those phenomena are of indirect interest in this context; they are relevant insofar as they bear on the welfare of individuals. While it is possible to make some general predictions about the welfare implications of population declines – such as increased difficulty finding mates – it is often impossible to assess the welfare implications of particular population declines without more information about the case. Consider, for instance, the introduced rodents on Gough Island, some 2,700 km west of Cape Town in the South Atlantic. The mice eat the young of the critically endangered Tristan albatross (RSPB 2021). The main welfare impact is probably on the chicks who are predated (though there are surely impacts on the adult albatrosses too). A different species introduction case, however, might involve a different pattern of welfare impacts – say, where predation is not the issue, but instead the members of the native species are starving because they are losing in the competition for a scarce food resource. This is the plight of the Coqui, a small tree frog, which is being outcompeted by invasive iguanas on the island of Puerto Rico (Platenberg 2007). So, while impacts on populations and species are generally relevant to individual welfare, they only reveal *that* there are some welfare impacts – not necessarily what the specific welfare consequences are.

The second complicating factor relates to the organisation of the chapter itself. While the division between environmental impacts, human–wildlife conflicts, and species introductions is useful for arranging much of the material that follows, it obscures the fact that these actions have shared drivers. Insofar as the ultimate aim is to consider how humans might reduce negative impacts on wildlife, it is important to consider the underlying factors that lead humans to

DOI: 10.4324/9781003182351-28

act in such harmful ways. So, the chapter begins with a general orientation to human–wildlife relations. After that, the chapter offers a bit more detail about the impacts associated with each category of action. It concludes with a brief discussion of practical recommendations.

Human–wildlife relations: The big picture

The big picture of human–wildlife relations can be broken into two parts. The first is human population growth. At one point, of course, humans were a relatively small group of primates mostly located in East Africa; now, there are roughly 7.8 billion humans on the planet, a number that is twice what it was in 1970 (The World Bank 2019b). The second is humans' historical tendency to advance their interests with an anthropocentric and relatively short-term view of the costs and benefits. In other words, throughout much of human history, humans have largely tried to achieve their goals without much consideration for the impacts of their actions on animals and without giving much weight to the long-term consequences of those actions. So, as the human population grew, human beings (1) expanded their physical reach around the globe; (2) extracted more resources for human use; and (3) created more resources for human use in ways that have largely had negative consequences for wild animals. (What's more, human beings became far more effective resource extractors and creators: per capita gross domestic product has risen dramatically since the Industrial Revolution, climbing by a factor of 13 just in the last 50 years (The World Bank 2019a).

These three activities are interrelated in complex ways, but it is possible to disentangle some of their implications for wild animals. Human expansion involves the development of the infrastructure required to move and house both people and resources, including cities and suburbs, roadways, oil and gas pipelines, submarine communication cables, shipping lanes, and much else. This process has meant that many wild animals have had their habitats destroyed, fragmented, and generally degraded. Moreover, it has meant that many animals have been classified as pests (e.g., elephants who cause crop damage) or threats (e.g., the systematic elimination of big cats in North America and elsewhere). It has also meant that human beings have moved species around the globe, sometimes intentionally (as when, in 1500s, Spanish explorers brought wild pigs to what's now the southeastern US (Mayer & Brisbin 1991) but often unintentionally (as when boats move zebra mussels from one body of water to another (US Department of the Interior Bureau of Reclamation, n.d.)). As a result, many animals face new and particularly fierce competition.

Resource extraction includes everything from the lumber industry to deep-sea drilling; resource creation includes activities like farming, fishing, and energy production. These activities have also meant that many wild animals have had their habitats destroyed, fragmented, and generally degraded as noted. Likewise, they have meant that many animals have been classified as pests or threats. Additionally and crucially, though, these activities have led to climate change, which threatens to radically alter the prospects for many wild animals. Climate change does this directly by making animals vulnerable to conditions for which they are not well adapted (e.g., polar bears who can no longer reach adequate prey because the sea ice forms later and melts earlier each season (NASA Earth Observatory 2020)). Climate change does this indirectly via several mechanisms, including forcing some animals to migrate to new regions where they have to compete with native species, as well as forcing humans to migrate to new regions, which means yet more habitat destruction, fragmentation, and degradation.

Of course, the consequences of human action on wildlife are not uniform. While a certain form of resource extraction, for instance, is likely to harm many wild animals, it is also likely to create opportunities for others. Deer, for instance, sometimes flourish in the wake of forests being clearcut, as clearcutting benefits the shrubs and grasses that deer like to eat (US Department of Agriculture 2003). However, these opportunities tend to be fleeting. When animals benefit from human action, their populations often grow dramatically. And when these increased populations move into regions or use resources that humans value, humans often categorise those animals as pests. In such circumstances, lethal wildlife "management" campaigns regularly follow – a euphemism for killing, exclusion from habitat, and other welfare-compromising actions. So, it is important to evaluate the consequences of human action on wildlife holistically, which generally reveals negative consequences indeed.

There is a simple lesson here: namely, that it would be a mistake to see the issues discussed below as independent events that can be addressed separately. The environmental changes that humans have wrought, the many sources of human–wildlife conflict, and the persistent issue of species introductions, are linked in complex ways by their underlying causes and interventions aimed to mitigate their impacts on wildlife need to be sensitive to these factors. In any case, it is now time to explore these three categories of human action in more detail.

Environment-affecting actions

It is difficult to overstate the extent to which human beings have altered the environment. In the US, for instance, only 6% of virgin forests remain and less than half the wetlands that Europeans first found when they came to America's shores (National Council for Science and the Environment 2008). This is not simply an American phenomenon: more than one-third of the world's land surface is now devoted to agriculture. Moreover, agriculture now claims almost 75% of freshwater resources (UN Report: Nature's Dangerous Decline 2019). The rate of global groundwater stock loss has more than doubled in the last 60 years (American Geophysical Union 2010), and while total global supply remains unknown, it is clear that this rate is unsustainable in many places. Several major aquifers, such as the Ogalalla Aquifer in the High Plains states of the US, are projected to go dry in the next 100 years even if users significantly reduce how much they pump (Scott 2019).

Water is not simply being used at unsustainable rates; it is being polluted as well. In developing countries, it is estimated that some 90% of sewage is discharged directly into local water bodies without prior treatment (UN World Water Development Report 2015), and even in the US, nearly half of rivers and lakes are considered polluted (United States Environmental Protection Agency 2000). Indeed, it may be the case that as much as a third of global biodiversity loss is attributable to pollution in freshwater ecosystems (United Nations Water 2015). There are similar concerns about ocean water, with increasing attention devoted to plastic pollution. Humans create some 300 million tonnes of plastic each year, of which at least 8 million tonnes wind up in the ocean (International Union for Conservation of Nature 2018). These plastics break down slowly into microplastics, which slowly build up in the bodies of marine life and the animals – including humans – who eat marine life.

Environmental change takes many other forms, and human activity is now responsible for significantly altering 75% of terrestrial and 66% of marine environments via the kinds of direct impacts just described (United Nations 2019). However, climate change might be the most dramatic way that humans have affected the planet – and, as a result, animal welfare. On average, the global annual temperature has gone up 0.07°C (0.13°F) per decade since 1880. Since 1981, it has gone up at double that rate. Unsurprisingly, then, new temperature records have been set roughly every 13.5 years from 1900 to 1980; in the last 40 years, those records have been set every 3 years. The five warmest years since 1880 have all occurred after 2015 (Lindsey & Dahlman 2020). (Figure 23.1).

Moreover, there is every reason to think that these trends will continue. Different climate models indicate that Earth's average temperature will be between 2.0–9.7°F (1.1–5.4°C) warmer

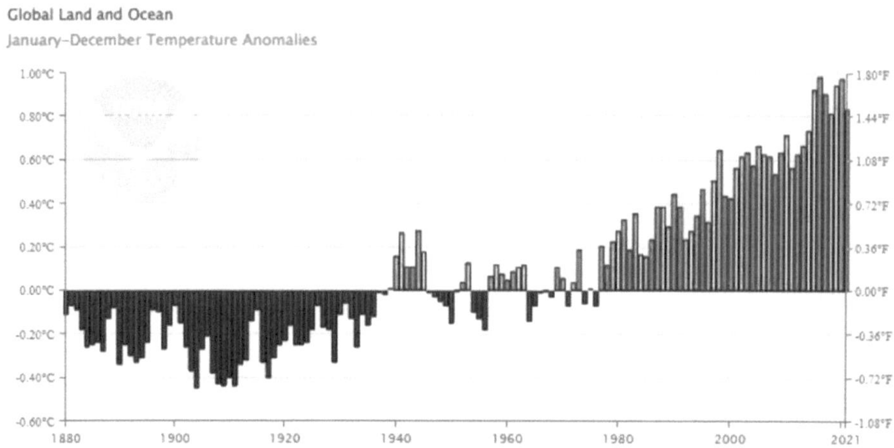

Figure 23.1 Annual Temperature Anomalies, 1880–2021. Credit: NOAA's National Centers for Environmental Information, www.ncdc.noaa.gov/climate-monitoring/.

in 2100 even if significant mitigation efforts are taken, with the precise amount of warming dependent on global energy choices over the next several years (Herring 2012).

Animal agriculture is partially responsible for climate change. Xu et al. (2021) provided explicit estimates of greenhouse gas (GHG) emissions worldwide from plant- and animal-based human food circa 2010. They found that emissions from food systems are responsible for at least 35% of global total anthropogenic GHG emissions and that 57% of food-related emissions were themselves attributable to the production of animal-based food (including livestock feed). Hence, the production of animal-based food appears responsible for at least 20% of anthropogenic GHG emissions. With increased consumption of livestock produce since 2010, a conservative estimate using current production and consumption data could be even higher.

Hayek et al. (2021) argued that a global switch to plant-based diets would itself be sufficient to hit international greenhouse gas reduction targets that are designed to limit global warming to 1.5°C. Much of animal agriculture's impact is due to enteric fermentation – that is, the digestive process that creates methane in ruminants. Feed quality and composition has a significant impact on methane production, and although there is no way to eliminate it, the FAO estimates that it could be reduced by more than 50%. Three other factors account for most of the remaining emissions: the decomposition of manure (which releases both methane and nitrous oxide), feed production (which involves CO_2 emissions at several stages, but most notably the conversion of forest into farmland for feed crops and fertiliser and pesticide production), and the energy consumption associated with the rest of the supply chain (including housing, transporting, slaughtering, and processing animals into retail products).

Demand for such products is expected to increase by roughly 70% by 2050. At that rate of growth, livestock production would still account for a substantial percentage of global greenhouse gas emissions even if the various industries were to pursue aggressive emissions mitigation efforts. If current dietary trends continue, then relative to 1995 levels, methane and nitrous oxide emissions would more than double by 2055 (Popp et al., 2010). As Harwatt (2019) argues, it is difficult to see how current climate change targets can be achieved without shifting towards alternative proteins.

Obviously, animal agriculture is itself a source of extensive animal welfare concerns, though other chapters have those issues as their foci. This chapter attends to the dramatic implications that climate change has had for a wide range of ecosystems.

For instance, arctic sea ice is at its minimum at the end of each summer and that minimum is currently dropping at a rate of 13.1% per decade (NASA 2021a). Likewise, satellite imagery shows that the land ice in Antarctica has been losing mass annually since that study began in 2002. Between melting ice and seawater expanding as it warms, global average sea level has risen roughly 7 inches in the last 100 years (NASA 2021b). This has had devastating effects on coastal ecosystems as they are now experiencing much higher rates of erosion than ever before. Indeed, some island nations, such as the Maldives, are likely to disappear forever – and with them all their native wildlife (Nurse et al., 2014).

Climate change threatens wildlife in several other ways. For instance, climate change can threaten habitat by simply altering a synergy on which a particular species depends. Northern Michigan's Kirtland's Warbler, for instance, nests on the ground under jack pines. Jack pines grow in sandy soil that allows water to drain away rapidly. But climate change is pushing the jack pines further north into less sandy soil that drains more slowly. If, as is predicted, the birds follow the jack pines, they will soon be nesting on ground that will flood periodically before their young can fledge, which means that in 30 to 60 years, the warbler may be extinct (Schneider, Root, & Van Putten 2013).

Because climate change incentivises migration, it can also mean that non-migrating wildlife become vulnerable to new parasites and pathogens that had not previously occupied a particular region. However, it can also impose burdens on migratory species. In some circumstances, climate change can also make existing seasonal migrations more metabolically taxing, as animals are exposed to temperatures that they did not previously encounter. In other cases, climate change can reduce the incentive to migrate, resulting in sedentary populations that are vulnerable to increased rates of infection – a problem that has, for instance, had serious effects on monarch butterfly populations (World Wildlife Fund).

Warren et al. (2013) show that if greenhouse gas emissions continue on their present trajectory, more than half of plant species and over one-third of animal species are likely to lose more than half of their current climatic range in the next 60 years. If emissions were to peak in 2030, then that loss could be reduced by as much as 40%, although that is looking increasingly unlikely given global trends. As a result, it is reasonable to expect a substantial loss of global biodiversity over the course of this century.

Human–wildlife conflict

Climate change harms wild animals in ways that may not always be obvious. Human–wildlife conflicts, however, are far more visible. Perhaps the most obvious examples of human–wildlife conflicts relate to agriculture. Humans actively cull many big cats, wolves, coyotes, foxes, and other predators to prevent them from killing agricultural animals (and pets), effectively extirpating them in many regions. Moreover, species that feed on crops are actively culled as well, including European starlings, sparrows, deer, and mice, through some combination of trapping, hunting, and poisoning (Nyhus 2016). Some of these population control measures have drastically negative impacts on animal welfare, such as using anticoagulant poisons as lethal population management against mice. These chemicals cause internal bleeding that can take hours or days to be fatal (Meerburg, Brom, & Kijlstra 2008), during which the animal suffers considerably.

In all these cases, animals are categorised as nuisances or pests – terms that are typically defined in terms of conflict with human interests. In principle, any species can be so categorised. Tigers, for instance, were long considered pests in China due to the frequency of their attacks on people. In fact, Mao Zedong called for tigers to be slaughtered across China to address this problem, essentially exterminating tigers in the country (Pomfret 2010). Now, of course, many conservationists reject these classifications, insisting that the tiger is a valuable species that ought to be preserved and promoted. Cecil the Lion provides another interesting example of this phenomenon. Cecil was killed by an American trophy hunter in 2015, and when the photos of the kill went viral, there was an immediate and intense public outcry. The hunter became the subject of intense public scrutiny, received death threats, and ultimately had to close his business. By contrast, Goodwell Nzou, a native Zimbabwean, wrote an op-ed in the New York Times where he wrote that: "When I turned on the news and discovered that the messages were about a lion killed by an American dentist, the village boy inside me instinctively cheered: One lion fewer to menace families like mine" (Nzou 2015). As this case illustrates, vantage point clearly matters. Where lions are symbols, they are often beloved; where lions are threats, they are far less popular.

Many conservationists hoped that the creation of protected lands would ease tensions between humans and wildlife. However, neither humans nor wildlife appear to respect the boundaries of protected lands. Humans enter those protected lands to poach wildlife; wildlife leave the protected lands in search of food and mates. Moreover, the status of protected lands is constantly contested, as humans have a persistent interest in converting many of those lands for agricultural use (Extent of Human Encroachment 2019).

There are, of course, many cases where humans are not trying to kill wildlife, yet wildlife die all the same. One familiar example is roadkill. Wildlife habitat is fragmented by roadway construction, which itself has negative impacts on wildlife populations, and vehicle strikes are the main anthropogenic cause of death for many vertebrate species. Indeed, it is estimated that a million or more vertebrates are killed on US roads alone each day (Schwarts, Shilling, & Perkins 2020). There is a similar problem of unintentional killing in the case of energy production. This is not simply a problem with oil and natural gas extraction: wind turbines kill somewhere between 140,000–328,000 birds each year in the contiguous United States (Loss, Will, & Marra 2013) and solar energy production kills between 37,800 and 138,600 birds annually in the same region. (The mechanism is not entirely clear in the case of solar energy production, but one possibility is various parts of solar farms reflect polarised light in a way that makes them appear as waterbodies to birds, who then collide with them (Walston Jr. et al., 2016).)

Species introductions

Humans have also had enormous impacts on wildlife by moving animals to new locations, whether intentionally or unintentionally. Domestication, for instance, led to massive animal translocations. Alfred W Crosby, a professor of history and geography, once wrote that:

> If the Europeans had arrived in the New World and Australasia with twentieth-century technology in hand, but no animals, they would not have made as great a change as they did by arriving with horses, cattle, pigs, goats, sheep, asses, chickens, cats, and so forth. Because these animals are self-replicators, the efficiency and speed with which they can alter environments, even continental environments, are superior to those for any machine we have thus far devised.
>
> *(Crosby 2015)*

Whether or not Crosby is correct about the counterfactual, it is true that Europeans radically altered the landscape of the lands they colonised. Europeans brought hogs to North America that quickly went feral, steadily spreading across what are now the southern US states, outcompeting many native species; Europeans brought hogs to Australia, where they turned into razorbacks after a few generations; and Europeans left hogs on any number of islands in between home and their destinations, just to ensure that they would have ample food when they came back through. Given the rate at which they breed, these animals quickly put pressure on many species that had never had to compete with such omnivores.

Europeans also brought cattle with them. From the 16th to the 19th century, the majority of the cattle in the Americas were likely feral, and Félix de Azara, an 18th-century Spanish military officer, estimated that there were some 48 million feral cattle in the Americas (Azara 1847/2018) – nearly as many the buffalo on the Great Plains, though obviously spread out over a larger region. (They were so common that, during the 1800s in Argentina, people built the walls surrounding plantations out of the skulls of cattle – each property using hundreds of thousands of them.) This obviously placed enormous pressure on other grazers.

In addition to all the species that humans introduced intentionally, there were also the ones that they introduced unintentionally. Rats stowed away on ships and quickly made their way to shore whenever the opportunity arose. They were so plentiful that they actually threatened some colonial aspirations. Jamestown, Virginia was nearly lost to rats in 1609, as thousands of them ate the food stores that the colonists needed to get them through the winter. According to one report from Garcilaso de la Vega, a 16th-century Spanish soldier, rats in Peru "bred in infinite numbers, overran the land, and destroyed the crops and standing plants, such as fruit trees, by gnawing the bark from the ground to the shoots" (Crosby 2015). As with all the other species that humans introduced, rats remade the landscapes they inhabited, provided plenty of food for certain predators, and were devastating for some of their direct competitors.

However, species introductions are not simply a historical phenomenon. The pet trade moves animals around the world, resulting in the presence of "exotic" animals in ecosystems when they escape or their owners release them (e.g., pet fish being released into waterways). In some cases, these animals flourish in their new environs, rapidly challenging the animals who previously occupied that territory (e.g., Burmese pythons in Florida, which have effectively eliminated foxes and two species of rabbits in the Everglades and seriously reduced the numbers of several other mammalian species (McCleery et al., 2015)). Moreover, international shipping continues to move aquatic and terrestrial animals around the world, as aquatic organisms can be trapped in ships' ballast water and insects and other small animals can hide in cargo. In all these cases, animals can find themselves vulnerable to novel pressures for which they are not well adapted, resulting in significant population declines.

Population declines, extirpations, and extinctions

The collective impact of all these actions – those affecting the environment, human–wildlife conflicts, and species introductions – has been dramatic. This is one of the reasons why many scientists now describe the present era as the "Anthropocene", a period beginning with the Neolithic Revolution (some 12,000–15,000 years ago), when humans transitioned from being hunter/gathers to agriculturalists. One of the hallmarks of the Anthropocene is the radical loss of species, often described as the "sixth mass extinction". Each of the earlier mass extinctions involved the loss of 70 – 95% of the species diversity extant at that time, and while current species losses are estimated to be much lower – perhaps around 7% (Régnier et al., 2015) – the rate of species extinctions is thought to be at least 100 times higher than the base rate (Ceballos et al., 2015). Now, roughly 1 million animal and plant species are considered at risk of extinction, many of which are expected to be extinct within the next few decades, including some 40% of amphibian species, over 33% of all marine mammals, and 10% of insect species (UN Report: Nature's Dangerous Decline 2019). To give just a few specific examples, all 11 great shark species have suffered population declines over the last 35 years (Myers et al., 2007), approximately 75% of non-human primate species populations are in decline and 60% are at risk of going extinct

(Estrada et al., 2017), and elephant populations are declining at 8% per year across Africa, mostly due to poaching (Chase et al., 2016).

As discussed earlier, the focus here is on *individual* welfare, not biodiversity. However, species loss implies that all members of the species have died. These deaths, of course, are explained by a wide range of factors, many of which have been identified here: starvation due to lost food sources, predation from introduced species, toxicity from pollution, plastic entanglement or ingestion, and many others. Insofar as these deaths can be attributed to such causes, it seems that humans are responsible for compromising the welfare of an extraordinary number of wild animals.

Conclusions and recommendations

It would be a mistake to adopt too rosy a view of "pristine" ecosystems, untouched by human hands. Healthy, untrammelled ecosystems are not necessarily *high welfare* ecosystems. That is, for any set of metrics used to assess ecosystem health, an ecosystem may score well on those metrics while still containing extensive animal suffering. This is because the welfare of many wild animals is regularly compromised by hunger, thirst, extremes of heat and cold, parasites, predators, stress from the threat of predation, inability to find mates, the loss of young, disease, injury, and any number of other factors. This suffering is not anthropogenic; animals do not experience it because of climate change, habitat fragmentation, or pollution. Instead, this suffering occurs because insofar as natural selection maximises anything, it maximises *fitness*, not the welfare of individuals. If, for instance, animals are more likely to pass on their genes by having large numbers of offspring, then that may well be (and in fact often is) the reproductive strategy that natural selection prefers. This is true even when the number of offspring is far more than can be sustained by the resources in the local environment, which means that many of those offspring will starve; it is true even when the offspring are highly vulnerable to predation, which means that many of those offspring will be consumed by other animals.

At the same time, however, it is crucial to recognise that human beings can *further* compromise the welfare of wild animals, making already-difficult lives even harder. What can be done about this? As suggested earlier, negative impacts on wildlife are ultimately traceable to human population growth combined with humans' tendency to advance their interests with an anthropocentric and relatively short-term view of the costs and benefits of their actions. There is no realistic and all-things-considered desirable scenario in which population growth is radically checked in the next few decades, and it seems doubtful that much can be done to alter central aspects of human decision-making. In such circumstances, and given the pressing nature of the problem, it is difficult to see how any bottom-up, populist solution is feasible.

Instead, top-down, regulatory strategies are the ones to pursue. In particular, it will be important to create regulations that change the system of incentives for various actors such that it is more burdensome to harm wildlife than the alternative. Insofar as climate change is part of the problem, this will probably involve a weighty carbon tax. A carbon tax, of course, is a way of forcing actors to internalise a cost of their activity. Likewise, governments might consider pricing other ways that humans harm wildlife. Imagine an arrangement where harms to *any* species are taxable events, with periodic environmental studies setting costs each year based on the degree to which species are threatened. In principle, this could allow *all* harms to wildlife to have some price – since all wildlife is, in principle, vulnerable to anthropogenic extinction.

There are practical challenges to this proposal. The main hurdle may well be measuring individual causal contributions, not least because such an approach would need to factor in indirect impacts as well (via, e.g., habitat destruction/degradation, air pollution, etc.). In principle, the measurement burden could be pushed onto the organisations being taxed. Underreporting is an obvious concern, but that could be mitigated by periodic audits with steep penalties for failures to disclose wildlife impacts. In any case, it does not matter whether the system catches all bad actors; it only matters that the system disincentivises harm to wildlife long before animals would be eligible for protection under the current model.

Granted, such a strategy may seem far-fetched in current regulatory environments. However, given the nature, scope, and immediacy of the crisis facing wildlife – and the incentives to which humans reliably respond – regulators must find ways to make it far more costly to harm wildlife. It is, of course, an open question whether the carbon tax model is a viable strategy, but it is clear that current strategies are not working. More radical proposals should be explored.

References

American Geophysical Union. 2010. *Groundwater Depletion Rate Accelerating Worldwide* [Press release]. 23 September. Available at: https://news.agu.org/press-release/groundwater-depletion-rate-accelerating-worldwide/.

Azara F, 1847/2018. *Memorias sobre el estado rural del Río de la Plata en 1801.* Alicante: Biblioteca Virtual Miguel de Cervantes.

Ceballos G et al., 2015. Accelerated modern human–induced species losses: Entering the sixth mass extinction, *Science Advances*, 1(5), p. e1400253.

Chase MJ et al., 2016. Continent-wide survey reveals massive decline in African savannah elephants, *PeerJ*, 4, p. e2354.

Crosby AW, 2015. *Ecological Imperialism: The Biological Expansion of Europe, 900–1900.* Cambridge: Cambridge University Press.

Earth Science Communications Team at NASA's Jet Propulsion Laboratory, 2021a. Arctic sea ice minimum. Available at: https://climate.nasa.gov/vital-signs/arctic-sea-ice/.

Earth Science Communications Team at NASA's Jet Propulsion Laboratory, 2021b. *Global Climate Change: Vital Signs of the Planet.* NASA: Climate Change and Global Warming. Available at: https://climate.nasa.gov/.

Estrada A et al., 2017. Impending extinction crisis of the world's primates: Why primates matter. *Science Advances*, 3(1), p. e1600946.

Harwatt H, 2019. Including animal to plant protein shifts in climate change mitigation policy: a proposed three-step strategy, *Climate Policy*, 19(5), pp. 533–541.

Hayek MN et al., 2021. The carbon opportunity cost of animal-sourced food production on land. *Nature Sustainability*, 4, pp. 21–24.

Herring D, 2012. Climate change: Global temperature projections. *Climate.gov.* Available at: https://www.climate.gov/news-features/understanding-climate/climate-change-global-temperature-projections.

International Union for Conservation of Nature, 2018. *Marine Plastics IUCN Issues Brief.* Available at: https://www.iucn.org/sites/dev/files/marine_plastics_issues_brief_final_0.pdf.

Lindsey R. and Dahlman L, 2020. Climate change: Global temperature. *Climate.gov.* Available at: https://www.climate.gov/news-features/understanding-climate/climate-change-global-temperature.

Loss SR, Will T, and Marra PP, 2013. Estimates of bird collision mortality at wind facilities in the contiguous United States. *Biological Conservation*, 168, pp. 201–209.

Mayer JJ and Brisbin IL, 1991. *Wild Pigs in the United States: Their History, Comparative Morphology, and Current Status.* Athens: University of Georgia Press.

McCleery RA et al., 2015. Marsh rabbit mortalities tie pythons to the precipitous decline of mammals in the Everglades. *Proceedings of the Royal Society B*, 282, p. 20150120.

Meerburg BG, Brom FW, and Kijlstra A, 2008. The ethics of rodent control. *Pest Management Science*, 64, pp. 1205–1211.

Myers RA et al., 2007. Cascading effects of the loss of apex predatory sharks from a coastal ocean. *Science*, 315(5820), pp. 1846–1850.

NASA Earth Observatory, 2020. Polar bears struggle as sea ice declines. Available at: https://earthobservatory.nasa.gov/images/146023/polar-bears-struggle-as-sea-ice-declines.

National Council for Science and the Environment, 2008. Beyond old growth: Older forests in a changing world. Available at: https://masswoods.org/sites/masswoods.org/files/pdf-doc-ppt/Beyond_OG.pdf.

Nurse LA et al., 2014. Small islands. In: *Climate change 2014: Impacts, adaptation, and vulnerability. Part B: Regional Aspects, contribution of Working Group II to the Fifth Assessment Report of the Intergovernmental Panel on Climate Change* (Barros, V.R. et al. eds.). New York: Cambridge University Press.

Nyhus P, 2016. Human–wildlife conflict and coexistence, *Annual Review of Environment and Resources*, 41, pp. 143–171.

Nzou G, 2015. In Zimbabwe, we don't cry for lions, *The New York Times*. Available at: https://www.nytimes.com/2015/08/05/opinion/in-zimbabwe-we-dont-cry-for-lions.html.

Platenberg RJ, 2007. Impacts of introduced species on an island ecosystem: non-native reptiles and amphibians in the US Virgin Islands, *Managing Vertebrate Invasive Species*, 39, pp. 168–174.

Pomfret J, 2010. Clock ticks for South China tigers in symbolic year. Reuters. Available at: https://www.reuters.com/article/us-china-tiger/clock-ticks-for-south-china-tigers-in-symbolic-year-idUSTRE61B12R20100212.

Popp A, Lotze-Campen H, and Bodirsky B, 2010. Food consumption, diet shifts and associated non-CO2 greenhouse gases from agricultural production, *Global Environmental Change*, 20, pp. 451–462.

Régnier C et al., 2015. Mass extinction in poorly known taxa. *Proceedings of the National Academy of Sciences of the United States of America*, 112(25), pp. 7761–7766.

RSPB, 2021. Gough Island restoration programme. Available at: https://www.rspb.org.uk/our-work/conservation/projects/gough-island-restoration-programme/.

Schneider SH, Root T, and Van Putten M, 2013. *Wildlife Responses to Climate Change: North American Case Studies*. Washington, DC: Island Press.

Schwartz ALW, Shilling FM, and Perkins SE, 2020. The value of monitoring wildlife roadkill. *European Journal of Wildlife Research*, 66(18), pp. 1–12.

Scott M, 2019. National climate assessment: Great Plains' Ogallala Aquifer drying out. *Climate.gov*. Available at: https://www.climate.gov/news-features/featured-images/national-climate-assessment-great-plains%E2%80%99-ogallala-aquifer-drying-out.

The World Bank, 2019a. GDP per capita (current US$). Available at: https://data.worldbank.org/indicator/NY.GDP.PCAP.CD.

The World Bank, 2019b. Population, total. Available at: https://data.worldbank.org/indicator/SP.POP.TOTL.

United Nations, 2015. The UN world water development report 2015, Water for a sustainable world. Available at: http://www.unesco.org/new/en/natural-sciences/environment/water/wwap/wwdr/2015-water-for-a-sustainable-world.

United Nations Water, 2015. Wastewater management: A UN-Water analytical brief. Available at: https://www.unwater.org/app/uploads/2017/05/UN-Water_Analytical_Brief_Wastewater_Management.pdf.

United States Department of Agriculture, 2003. Managing your woodland for white-tailed deer: Fish and wildlife habitat management guide sheet. Available at: https://www.nrcs.usda.gov/Internet/FSE_DOCUMENTS/nrcs142p2_022286.pdf.

United States Department of the Interior Bureau of Reclamation, n.d. Don't move a mussel! Available at: https://www.usbr.gov/gp/multimedia/publications/zebra_mussel_brochure.pdf.

United States Environmental Protection Agency, 2000. National water quality inventory. Available at: https://www.epa.gov/sites/production/files/2015-09/documents/2000_national_water_quality_inventory_report_to_congress.pdf.

Warren, R., VanDerWal, J., Price, J., Welbergen, J.A., Atkinson, I., Ramirez-Villegas, J., Osborn, T.J., Jarvis, A., Shoo, L.P., Williams, S.E. and Lowe, J., 2013. Quantifying the benefit of early climate change mitigation in avoiding biodiversity loss. *Nature Climate Change*, 3(7), pp.678–682.

Walston LJ et al., 2016. A preliminary assessment of avian mortality at utility-scale solar energy facilities in the United States. *Renewable Energy*, 2, pp. 405–414.

World Wildlife Fund, n.d. Monarch butterflies and climate change. Available at: https://www.worldwildlife.org/pages/monarch-butterflies-and-climate-change.

Xu X et al., 2021. Global greenhouse gas emissions from animal-based foods are twice those of plant-based foods. *Nature Food*, 2(9), pp. 724–732.

24
ANIMAL WELFARE AND HUMAN HEALTH

Cynthia Schuck-Paim, Wladimir J Alonso, and Eric Slywitch

Introduction

Recent decades have seen unprecedented increases in life expectancy for people in all countries, with the average global citizen now expected to live more than 72 years, compared to only 45 years in 1950 (Roser, Ortiz-Ospina and Ritchie, 2013). Just a few generations ago, infectious diseases and a lack of effective medical treatments killed most people at a very young age. Gastrointestinal infections, tuberculosis, influenza, and pneumonia were common causes of death. However, several medical and technological advances that emerged in the 20th century made it possible to control many of the diseases and health hazards that plagued humanity for millennia. Vaccines, antibiotics, and access to sanitation, clean water, and healthcare were among them.

The resulting growth in the human population, along with urbanisation, higher incomes, new technologies, and globalisation were also the catalysts for a massive transformation in the food system and the nature of human–animal interactions in ways completely different than those seen in the past. Industrial processes replaced traditional husbandry practices in the production of animals. The volume and pace of expansion of the population of animals under human custody has been unprecedented, with over 70 billion land animals (FAO, 2018), and an estimated 1–3 trillion fish (Mood and Brooke, 2014), now slaughtered every year. At the same time, the rapid spread of infectious diseases from wild habitats to every corner of the world became possible with the massive movement of animals and people across the globe.

As we discuss in this chapter, these transformations created novel routes for the transmission of pathogens that gained, or re-gained, access to human populations, along with the conditions that are, paradoxically, eroding many of the health advances achieved so far. Livestock health is now considered the weakest link in the global health chain (FAO, 2013), with disease drivers in livestock and wildlife responsible for an increasing share of infectious disease burden (Figure 24.1); human diseases of animal origin (zoonoses) still cause about a billion cases of illness and millions of deaths every year (Karesh et al., 2012). Understanding these drivers, and the inherent link between animals' health, welfare, and human health, is a necessary step to control disease emergence and other global health threats effectively.

In recognising that human and animal health are interdependent and bound to the health of the ecosystems where they exist, the "One Health" approach was created to address these challenges

DOI: 10.4324/9781003182351-29

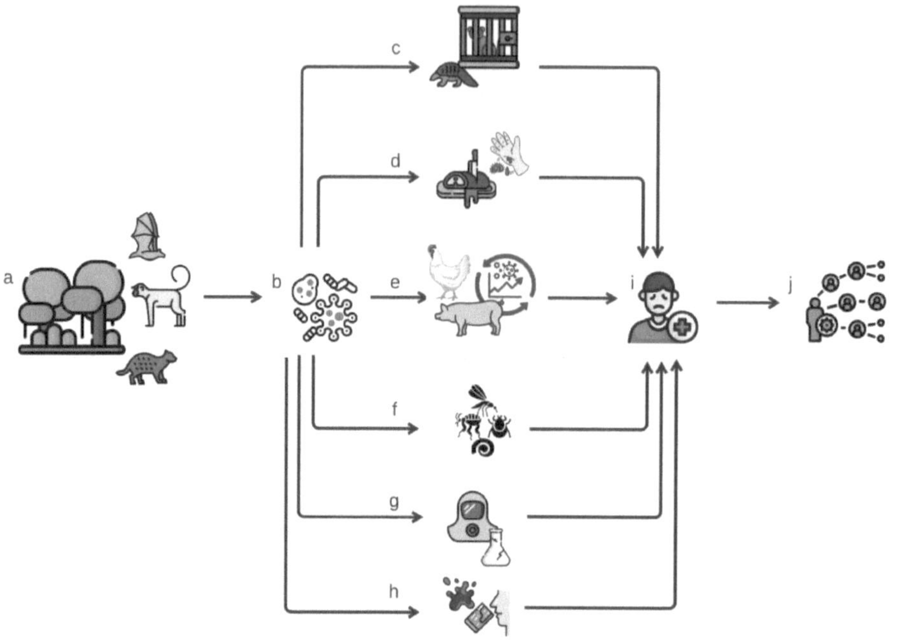

Figure 24.1 Main routes by which pathogens (b) in wildlife (a) can spill over to humans, leading to human disease (i) and epidemics (j); (c) wildlife trade; (d) bushmeat consumption; (e) contamination of livestock (stocking of large numbers of immunocompromised individuals at high densities enables pathogens to acquire higher pathogenicity and transmissibility before reaching humans); (f) vectors (e.g. insects, ticks) are also a major introducer of disease-causing pathogens (e.g. malaria, dengue) into human populations, but these are mostly restricted to the distribution of the vectors; (g) release of laboratory specimens (by accident or bioterrorism); (h) ingestion of infected material. In this chapter we address the routes directly related with animal use (c, d, and e).

(Karesh and Cook, 2009), which are explored in this chapter. We first describe how the impoverishment of animal health and welfare in modern animal production systems has become a major driver for the emergence of communicable diseases at a global level, and then focus on the direct effects of these systems on environmental pollution, the health of workers and of local communities. We also examine the impacts and biosecurity threats associated with wildlife hunting, farming, trafficking, and trade in a globally connected world. The chapter closes with a section on overconsumption of animal-sourced food as a major risk factor for non-communicable diseases.

Intensive animal farming and infectious disease emergence and spread

The emergence of new, or reappearing, infectious diseases is widely recognised as one of the major challenges for global health and socioeconomic development. Despite substantial progress in the reduction of poverty, access to healthcare and sanitation, recent decades have witnessed an increase in the frequency of emerging and re-emerging infectious diseases at a global scale (Jones et al., 2008, 2013). On average, one new infectious disease is recognised every four months, three-quarters of which are zoonotic, originating either through direct contact with wild animal species or with livestock (UNEP, 2020).

Many are the reasons for the acceleration in the emergence of new zoonotic pathogens. Livestock species now constitute more biomass than all wild mammals combined, harbouring a much larger number of zoonotic viruses than their wild relatives (Johnson et al., 2020) – most of which are capable of infecting multiple hosts, including humans (Rohr et al., 2019).

In addition to representing a massive zoonotic reservoir from which newly emerging pathogens can arise and spill over to the human population, or act as a conduit for pathogens from wild animals (Jones et al., 2013), propitious conditions for the emergence and transmission of zoonotic diseases are found in modern animal agriculture. Gains in productivity have stemmed mostly from the selection of fast-growing and highly productive breeds, helped by the development of diets, drugs, and additives designed to maximise the conversion of animal feed into meat, milk, and eggs. Additionally, housing systems involving the confinement of large groups of animals were adopted widely.

This intensification process took a heavy toll on the health of animals and their ability to withstand pathological challenges. Productivity gains that have enabled greater production efficiency have stretched the animals' physiology beyond natural limits, increasing the incidence of metabolic, bone and joint disorders (Grandin, 2014). For example, among chickens raised for their meat (broilers), growth rates have increased by over 400% in less than four decades. Because these birds gain weight so rapidly, the growth of bones and internal organs cannot keep pace with it; musculoskeletal, cardiovascular, and respiratory diseases are highly prevalent (Hartcher and Lum, 2020).

Importantly, when genetic selection is narrowly focused on growth and productivity, critical organ systems and biological functions can be compromised. Crucially, immune function is among them, as energy that would otherwise be used for defence is diverted to growth and reproduction (van der Most et al., 2011). For example, modern broiler breeds have been shown to exhibit damaging over-inflammatory responses to certain disease challenges and insufficient responses to clear other pathogenic challenges (Aylward, 2020). Likewise, strains highly selected for productive traits have shown a decline in humoral immune capacity (Bridle et al., 2006).

The genetic diversity of the immune system is also a determinant of resistance to infectious disease: without genetic variation, the likelihood that some individuals in a population will be resistant to a newly emerging pathogen is greatly diminished. Indeed, epidemiological modelling shows that lower genetic heterogeneity is associated with an increased probability of major (catastrophic) epidemics (Springbett et al., 2003). Yet, modern breeds of animals are extremely homogeneous. For example, poultry breeds have shown diminished polymorphism at loci of the major histocompatibility complex (a group of genes that code for proteins essential for the immune system) associated with pathogen-specific disease resistance (Kaufman et al., 1999; Bridle et al., 2006).

The confinement of large populations of susceptible animals at high densities further increases the likelihood of infectious disease emergence and spread through other mechanisms (Jones et al., 2013). For example, high levels of aerial pollutants such as ammonia and faecal dust, which naturally result from the high volume of animal waste, are found in modern animal operations. Exposure to these pollutants compromises respiratory function, damaging the first barriers of defence against infection by respiratory pathogens (Greger, 2007). Accordingly, in pigs raised intensively, post-mortem findings of lesions in the respiratory tract as a result of pneumonia, pleuropneumonia, pleurisy, and other diseases are pervasive.

It is also well documented that chronic exposure to stress (physical and emotional) has a suppressive effect on the immune system, activating the release of corticotropin-releasing hormone (CRH) by the hypothalamus, a brain region that links the nervous and endocrine systems. Through a cascade of effects, CRH promotes the release of cortisol, a stress hormone known particularly for its anti-inflammatory and immunosuppressive action. Higher levels of cortisol are associated with a lower number of lymphocytes (white blood cells important in the immune response) in blood, also affecting the production of cytokines (signalling molecules of the immune system) and antibodies (Martínez-Miró et al., 2016). Animals exposed to social stressors have also shown alterations in natural killer cell cytotoxicity (cells that kill aberrant cells, such as virally infected cells) and response to vaccination (Proudfoot and Habing, 2015).

Accordingly, there is ample evidence that, as with humans, chronic stress in animals increases disease incidence, the secondary complications of viral infections, and prolongs healing times (Glaser and Kiecolt-Glaser, 2005). For example, physical restraint in mice dramatically enhanced morbidity and mortality following infection with the influenza (H1N1) virus (Luo et al., 2020). In pigs, psychosocial stress is associated with dysregulation of inflammatory processes, neuroendocrine alterations, impaired immune function, and increased susceptibility to disease (Gimsa, Tuchscherer, and Kanitz, 2018). In chickens, chronic exposure to corticosterone downregulates proinflammatory responses and immune function (Kaiser et al., 2009) and in minks (farmed for their fur), lack of housing enrichment induces endocrine and organ changes associated with impaired immunity (Díez-León et al., 2016). Stress-mediated impairment of immunity and increased disease susceptibility is also widely described in fish (Yada and Tort, 2016). Environmental and psychosocial stressors in farmed animals include their stocking at high densities, the deprivation of highly motivated behaviours, limited opportunities for movement, social isolation, maternal deprivation, short sleep periods, limited access to natural light, as well as fear and pain induced by widely employed management practices (e.g. mutilation of body parts, feed restriction, transport between facilities).

Emergence of highly pathogenic viral strains

For decades, one of the greatest concerns of public health officials everywhere has been the possibility of emergence of a highly pathogenic influenza strain achieving sustained transmission in the human population. Influenza viruses that spilled over from animal reservoirs have been responsible for multiple epidemics and pandemics throughout history, including the 1918 "Spanish flu", the 1957 "Asian flu", the 1968 "Hong Kong flu", and more recently the 2009 H1N1 ("swine flu") pandemic (Poovorawan et al., 2013). Avian influenza ("bird flu") is of particular concern, as some subtypes (H5, H7) cause extremely severe illness in humans (Poovorawan et al., 2013). For example, over 700 human infections with H5N1 viruses have been reported so far, and about 60% of the cases have died.

There are many types of influenza viruses circulating in wild animal species, mainly waterbirds. Spillover from these species to humans is not trivial, as many are the adaptations needed for viruses adapted to infect the intestinal tract of aquatic birds, to replicate instead in the respiratory tract of humans. The conditions favouring the selection and spread of these mutations have been achieved in the intermediate hosts between aquatic animals and humans: the pigs and birds we breed for consumption (Poovorawan et al., 2013). Pigs, in particular, having receptors for avian, swine, and human influenza viruses, are regarded as ideal mixing vessels to generate influenza viruses with pandemic potential (Ma et al., 2008). Indeed, the intensification and expansion of pig production at a global level provided multiple opportunities for a strain with mixed genes from avian, human, and swine influenza viruses to become established in this population, leading to the first influenza pandemic of swine origin in 2009 (Trovão and Nelson, 2020).

Similarly, intensive poultry farming has made highly pathogenic avian influenza pervasive. Despite fear that backyard chicken production and other outdoor production systems expose domestic poultry to wild birds (a natural reservoir of influenza viruses), most genetic conversion events from low to highly pathogenic influenza strains were traced back to commercial poultry farms in high-income countries (where intensive poultry farming is ubiquitous) and in countries transitioning to intensive production (Dhingra et al., 2018). Indeed, intensive farming favours the emergence of highly pathogenic strains in multiple ways. To spread in a population, pathogens must multiply within the host, while maintaining the opportunities for transmission by ensuring the host's survival (i.e. maintaining low virulence). This is particularly the case if a pathogen cannot survive for too long in the environment. In such cases, a high level of virulence (causing more severe disease) is costly to the virus, as it may stop spreading when its host dies and contact with other hosts is interrupted (Greger, 2007). Influenza virus survival periods are much shorter outdoors, as it is rapidly inactivated by sunlight and desiccation, making the evolution of high virulence less likely outdoors (Greger, 2007a). Conversely, the longer viral survival period in confined intensive systems increases the likelihood that a highly virulent strain continues spreading in the population. The much higher number of susceptible hosts and contact rates in intensive systems further promote higher virulence, facilitating the spread of viruses causing severe disease.

Hundreds of avian flu outbreaks, involving millions of birds, were detected in commercial poultry flocks in Europe, Asia, Africa, and the Americas in recent years. In some of these outbreaks, humans were also infected (Shi and Gao, 2021). These many new cases provide numerous opportunities for these viruses to mutate or reassort (mix genetic material) with other strains, and at some point acquire the capability of sustained human transmission.

Biosecurity in animal operations

Although large animal operations often rely on biosecurity protocols to reduce zoonotic disease risk, the sheer scale of the outputs of these systems, the common outsourcing of production stages to a variety of independent producers, the transport of live animals, and the many opportunities for contamination during slaughter and processing, make it unlikely that these measures would be sufficient even if they were strictly implemented.

Many are the pathways through which pathogens can spill over from farm animal hosts to the human population. Transmission risk is highest for humans in contact with animals, but under the right conditions pathogens are also capable of surviving for weeks, or months, without a host. During this time they may travel outside farmhouses together with animal waste, water, clothing, equipment, garbage, trucks, bedding, animal vectors (e.g. insects, ticks, rodents), or even through the air, in contaminated aerosol particles expelled by the wind or by ventilation systems.

Besides the inherently challenging nature of mitigating biosecurity risks, failures of compliance with even basic standards of biosecurity are endemic in the industry. Whenever surveyed, biosecurity flaws were found to be widespread, even in developed nations (Racicot et al., 2011). The situation is worse in resource-limited settings: not only is biosecurity expensive, it also requires a clear understanding of strict technical guidelines and behavioural protocols. Seriously risky practices, such as the unsafe disposal of carcasses of dead and sick animals, are common in many places (Negro-Calduch et al., 2013).

Live animal transportation as an epidemic risk

Live animal transportation represents a major epidemic risk. Every year, over two billion animals are loaded onto ships and trucks and sent on national and international journeys lasting from hours to weeks (Levitt, 2020). Among the many welfare challenges to which animals are exposed in these journeys are dehydration, exhaustion, thermal stress, injuries, fear, and even death.

The crowding of animals from multiple origins into poorly ventilated, small, and stressful conditions promotes infectious disease transmission in multiple ways, as well as the opportunity for the mixing of genetically diverse pathogens. In addition to the high contact rate among animals, high pathogen loads are promoted by the immunosuppressive effects of stress, as discussed earlier. Long-distance transport has been also shown to increase the "faecal shedding" of pathogens (their release into the stool). In general, the more pronounced the stress, the higher the levels of pathogens released (Rostagno, 2009).

Predictably, animal trade has long been an effective way of spreading zoonotic diseases. Bovine spongiform encephalopathy ("mad-cow disease") (Hardstaff, Häsler, and Rushton, 2015) and foot-and-mouth disease (Di Nardo, Knowles, and Paton, 2011) are two well-known examples of diseases in which transportation was a primary driver of their spread across borders. Bovine respiratory syncytial virus, infectious bovine rhinotracheitis virus, herpesviruses, bovine parainfluenza, and multiple pathogens associated with gastrointestinal diseases are also known to have their incidences increased during transport (Broom, 2014). Importantly, the rapid expansion in the diversity of influenza A viruses in pigs is attributed to the long-distance live swine trade (Trovão and Nelson, 2020). Naturally, the same risks are present in the trade of wild animal species, as discussed later in this chapter.

Industrial animal farming and antimicrobial resistance

The power of antibiotics to fight infections is rapidly eroding. The emergence of antimicrobial-resistant bacteria is currently one of the biggest threats to global health (Osterholm and Olshaker, 2017). Pathogens associated with serious medical problems, such as tuberculosis, pneumonia, sexually transmitted diseases, urinary tract infections and hospital infections, have now become resistant to several antibiotics. About 700,000 deaths per year already occur due to antibiotic-resistant infections, with an estimated 10 million deaths per year due in 2050 if trends continue unchanged (O'Neill, 2016).

Because bacteria can rapidly adapt, antimicrobial resistance is expected to emerge naturally when bacteria are exposed to antibiotics. Although the misuse and overuse of antibiotics in human medicine accelerated this process, about 70% of the antibiotics sold in the world are not used in humans, but in animals raised in intensive farming systems (Van Boeckel et al., 2019). In these systems, the primary use of these drugs is not the treatment of sick animals, but instead the promotion of growth and/or prevention of infections, to ensure that animals can survive until the slaughter age under the conditions typical of intensive systems (McKenna, 2017). As discussed, intensive farming favours infectious disease emergence and spread. Additionally, animals raised indoors, without contact with the soil, have been shown to lack health promoting gut bacteria that can help maintain mucosal immune homeostasis and limit pathogen colonisation (Mulder et al., 2009). These conditions have created an inherent need for disease prevention – for which antibiotics have been a cheap solution.

As developing nations increasingly intensify animal-food production, antimicrobial resistance is rising rapidly too. Countries such as China, India, Brazil, and Kenya, where meat production increased dramatically, are now hotspots of antimicrobial resistance in animals (Van Boeckel et al., 2019). Antibiotics critical for human medicine are also widely used in intensive fish farming, one of the fastest growing food industries on the planet and now a hotspot for bacterial resistance (Watts et al., 2017). In these farms, infectious diseases are fought by adding large amounts of antibiotics to the water, most of which are also important in human medicine (Done, Venkatesan, and Halden, 2015).

There is ample evidence for a direct causal link between antibiotic consumption in animals and resistance in humans (O'Neill, 2015). The presence of antibiotic-resistant strains of bacteria in animal-sourced foods sold in supermarkets and grocery stores has been reported in nearly every published study that investigated it. For example, in the United States, 75% of the bacteria the Food and Drug Administration found on grocery store meat was antibiotic-resistant (Undurraga, 2018). Often, the same genetic strains present in animal-food samples are those isolated in hospital patients (Wang et al., 2017).

Environmental contamination is another route of infection: bacteria are excreted in the urine and stool of animals still in their active form, making their way to water bodies and the soil, and contaminating other agricultural produce through the use of manure as fertiliser (Founou, Founou, and Essack, 2016). Veterinarians, farmers, slaughterhouse workers, and food handlers can also be contaminated by direct contact with food animals and their products, acting as bridges to spread the resistance in the human population.

In recent years, many countries have regulated the use of antibiotics in livestock, gradually banning their use as growth promoters (McKenna, 2017). However, the line between the use of antibiotics for growth promotion and disease prevention is a blurred one. Additionally, the industry has increased the use of ionophore antibiotics that are not currently categorised as medically important, but have the potential to become effective treatments for serious human infections (ASOA, 2019).

It is important to note, however, that animals raised under typical intensive conditions may see both their health and welfare impoverished in the absence of antibiotics. For example, data on broilers raised in conventional operations, but never given antibiotics, showed a higher prevalence and severity of eye lesions, footpad dermatitis, and airsacculitis (inflammation of the air sacs caused by bacteria) (Karavolias et al., 2018). Therefore, bans on the prophylactic use of antibiotics must be accompanied by simultaneous interventions that ensure both the genetics and living conditions conducive to good health and welfare.

Foodborne infections of animal origin

The impact of foodborne illnesses on global health is far from negligible. Food contamination episodes cause over 600 million cases of illness every year, resulting in 420,000 deaths worldwide (Devleesschauwer et al., 2018). Animal-sourced products are responsible for the greatest share of these cases (Karesh et al., 2012).

For example, chickens are a natural host for Campylobacter species, the most common bacterial cause of human gastroenteritis in the world (WHO, 2018). Salmonella is also responsible for over 50,000 deaths every year (GBD, 2019) following the consumption of contaminated chicken, eggs, and pork. Other important foodborne pathogens coming from animal reservoirs include toxin-producing strains of *Escherichia coli* and *Listeria monocytogenes*.

Nowadays, a major source of contamination of meat is the process of evisceration at the slaughter plants, through which internal organs, especially those in the abdominal cavity, are removed. Needless to say, meat comes from animals who once had a gut, and it is not easy to ensure that faecal matter does not contaminate the animal carcass. Likewise, the use of manure as fertiliser, the contamination of water bodies with animal waste or contact (direct or indirect) with animal products, may also contaminate other products, such as fruits or vegetables.

Poorer animal welfare further increases food safety risks. Stressed animals tend to release more pathogens, such as *E. coli*, Salmonella, or Campylobacter, in their faeces. In pigs, increased feed withdrawal times were associated with increased Enterobacteriaceae and Salmonella in faeces. Likewise, higher stocking densities and stress-inducing conditions (e.g. forced moulting) have been shown to result in increased occurrence, persistence, and spread of Salmonella in poultry (EFSA, 2019).

Water and air pollution

The contamination of water with livestock waste is nowadays a major public health risk. The farming of land animals alone is now among the leading causes of water pollution globally (FAO, 2017).

In addition to leakage from manure lagoons that are poorly constructed or that overflow during precipitation events, the widespread application of animal waste to agricultural crops is another major route of contamination. Livestock excreta contain high quantities of nutrients (e.g. nitrate and phosphorus) that can impact aquatic and marine ecosystems as well as drinking water supplies. Waste from intensive farming systems also carries heavy metals (e.g. zinc, copper, cadmium, lead, mercury), veterinary pharmaceuticals, hormones, and antibiotics (which can promote increased antimicrobial resistance in naturally occurring pathogens in surrounding ecosystems). The rapid expansion of intensive fish farming further adds to the problem, with fish excreta, feed, veterinary drugs (antibiotics, fungicides), and anti-fouling agents similarly polluting downstream ecosystems.

Importantly, animal waste also carries high concentrations of microorganisms harmful to human health, including pathogens such as Campylobacter, Escherichia coli, Salmonella, Clostridium, and parasitic protozoa. Swine waste, for instance, has been found to carry over 100 pathogens associated with human illness (Burkholder et al., 1997).

As a result of activities such as storing and spreading of manure, as well as the fertilisation of crops destined as animal feed (e.g. corn, soybeans), animal-food production is also a major emitter of fine particulate matter in the air, an important risk factor for heart disease, cancer, and stroke (Domingo and Balasubramanian, 2021). High levels of ammonia – an irritant gas emitted from animal waste that reacts with other gases and forms fine dust particles – are particularly conducive to lung function decline and the impairment of the respiratory health of human populations living near intensive animal operations (Borlée et al., 2017). Airborne transmission of zoonotic pathogens that are carried through the air from farms, manure lagoons, and spray fields, is also a concern. For example, antibiotic-resistant bacteria from livestock waste can be dispersed by the wind (McEachran et al., 2015). Respiratory viruses can also travel long distances through the air.

Worker health

The health effects of exposure to animal farming settings are potentiated in farm workers and include a myriad of conditions, such as respiratory disorders, occupational injuries, mental health disorders, and zoonotic infections.

A high prevalence of respiratory disease among farm workers naturally results from the high levels of ammonia, inhalable dust, and endotoxins (an inflammatory substance present in the cell membrane of bacteria) within animal facilities (Dignard and Leibler, 2019). Health effects include impairment of lung function, chronic bronchitis, asthma-like syndrome, among other types of chronic and intermittent respiratory disorders (Von Essen and Auvermann, 2005).

These occupational hazards are also present among slaughterhouse workers. A high level of exposure to bioaerosols, released in the slaughter process, has been associated with a significantly higher incidence of airway disorders in this population (Kasaeinasab et al., 2017). Additionally, an increased prevalence of mental health disorders, such as depression and anxiety, has been identified given the traumatic nature of the work (Slade and Alleyne, 2021). Importantly, exposure to farm animals at any stage of the production chain also translates into a substantially higher risk of infection by zoonotic pathogens. Several studies have reported higher sero-prevalence of pandemic H1N1 influenza, hepatitis E, and highly pathogenic avian influenza H5 and H7 in farm workers (Jones et al., 2013).

Occupational hazards at meat processing plants have also been widely documented, including musculoskeletal disorders and long-term injuries due to the physically intensive and repetitive nature of the work. Processing plants have been also shown to be hotspots of infectious disease transmission, favoured by the low temperatures (to reduce the risk of meat spoilage), high levels of humidity, and high concentration of employees in a closed environment.

Despite the multiple health and safety hazards, reporting of these issues in the animal production chain is rarer than in other work settings – in many countries, workers are migrants, do not speak the local language, or are not legally authorised to work, so fear of reporting system failures is widespread. Poor qualifications and working conditions are also inherently associated with poor stockmanship, which substantially increases the risk of poor animal health and welfare.

Biosecurity threats from wildlife hunting, trafficking, and trade

Extensive capture of wild animals for human use (mainly as a food source) is practised in all countries, for instance in the form of fishing. Although aquatic animals can be reservoirs of pathogens that can infect humans, here we concentrate on those practices that involve species that are evolutionarily closer to us, hence from which pathogen spillover to humans is more likely (Wolfe, Dunavan and Diamond, 2007; Morse et al., 2012).

Bushmeat hunting and consumption

Bushmeat hunting and consumption is sometimes perceived as an extinguishing practice in the fringes of modern civilisation, hence with little significance for global health. This impression is incorrect for two reasons: its actual magnitude, and the exceptionally high biosecurity threat inherent to this activity.

Consider, for example, that in the Congo Basin alone, an estimated 4.9 million tonnes of wild mammals are hunted annually for consumption (Fa, Peres, and Meeuwig, 2002). Hunting is very important for some local populations: cases of childhood anaemia among poor children in Madagascar could triple if bushmeat consumption were eliminated (Kurpiers et al., 2015). Paradoxically, economic development does not necessarily lead to the extinction of this practice, as bushmeat is increasingly consumed as a "gourmet" delicacy by urban populations (Kurpiers et al., 2015).

The disproportionate biosecurity threat posed by bushmeat stems from the privileged pathway that it creates between zoonotic pathogens from wildlife (mainly from ecosystems where pathogen diversity is high (Jones et al., 2008)) and the human population. The risk does not emerge so much from the consumption of the meat (since it is generally cooked), but from the process of hunting, cleaning, and preparing it. During these procedures, the chances of contamination by the body fluids and tissues of the infected animal through small wounds or mucous membranes (e.g. eyes, nose) are not negligible (Kurpiers et al., 2015; Greatorex et al., 2016). In fact, interviewing bushmeat hunters and traders in Sierra Leone, a study found that 38% are cut during prey processing (Subramanian, 2012).

In a globalised world, an infection by a novel pathogen may not be contained to a village in the middle of the jungle. HIV, which most likely emerged as a result of hunting and field-dressing of chimpanzees in west central Africa (Gao et al., 1999), rapidly spread to all corners of the globe, killing approximately 40 million people since 1980. Other epidemics believed to have emerged by this route include the Ebola virus, monkeypox virus, and severe acute respiratory syndrome (SARS) (Wolfe, Dunavan, and Diamond, 2007; Karesh and Cook, 2009; Greatorex et al., 2016).

Wildlife farming and trade

Wildlife trade systems amplify the epidemiological interface between humans and wild pathogens by rapidly dispersing them across borders (Aguirre et al., 2021). The risks associated with the trade of exotic species are not only restricted to the pathogens they carry. Host species themselves can also cause significant damage to the native wildlife of the habitats they are introduced to, which many times in history has led to ecological disasters and food insecurity. These biosecurity threats are aggravated in the illegal trade by the impossibility of sanitary inspections and the particularly brutal conditions in which animals are trafficked (Bezerra-Santos et al., 2021).

When wild animals are instead bred in "wildlife farms", the conditions conducive to pathogen spillover to humans are similar to those discussed for farmed animals, as farmed wild species are similarly confined at high densities in barren facilities that promote immune suppression and infectious disease susceptibility. A heightened risk also emerges from a lack of knowledge of the epidemiological facilitative role of each of the many species in these settings, and the challenging implementation of proper biosecurity protocols. Also, the coexistence of genetically distinct strains may favour the evolution of higher levels of virulence if more virulent strains have a competitive advantage (Nowak and May, 1994).

Recent outbreaks of coronaviruses made supermarket shoppers in the developed world aware of other consumer practices, routine for millions of citizens, and iconically represented by live animal markets. Dead animal tissue decomposes quickly under normal temperature, so despite processes such as salting, smoking, and spicing, bringing live (hunted or farmed) animals to town markets has been one of the most common ways of trading meat before the widespread adoption of refrigeration. The practice of selling live animals for consumption still persists in many nations out of necessity (e.g. lack of electricity) or for cultural reasons (value of freshness and culinary diversity (Greatorex et al., 2016)). However, by mixing a large number of individuals from multiple animal species and diverse origins in a place where thousands of humans join in, these markets offer a privileged epidemiological interface for the spillover of pathogens from wild species to the human population (Greatorex et al., 2016). As an illustrative example, the SARs outbreak of 2002/2003 was traced back to a species of bats in the Yunnan province, but masked palm civets farmed for food and sold in live animal markets acted as the vector of transmission to humans (Shi and Hu, 2008). Likewise, debate on the emergence of SARS-CoV-2 (responsible for COVID-19) at the time of writing has coalesced around two hypotheses, that of a laboratory escape, or a zoonotic emergence – as was the case of all previous coronaviruses that infected humans (Holmes et al., 2021). In the latter case, as with SARS, a live animal market is believed to be the birthplace of the pandemic (Holmes et al., 2021).

Expansion of livestock production near wildlife habitats

Alternated patches of preserved areas and land used for different purposes (including pastures) are a component of many traditional landscapes, increasing biological diversity and socioecological resilience. Nevertheless, they also increase the risk of pathogen spillover directly to humans or intermediated by livestock (Jones et al., 2013). This has been the case, for example, of pig farms in Malaysia, which acted as an intermediate step for the transmission of the Nipah virus from bats to humans. Once pigs are infected, they can transmit the virus to other pigs (particularly in places with a high density of pig farms) and to humans, like any respiratory disease. Nipah killed about 40% of the people who got infected. When livestock is introduced in wild habitats, a "bridge" between wildlife and humans is made (Jones et al., 2013).

Diseases of overconsumption

Besides the global health risks of animal-food production discussed previously, evidence is robust that overconsumption of animal-sourced products is associated with multiple adverse health outcomes, being also a major risk factor for non-communicable diseases.

Meat consumption has increased dramatically over the last five decades, from a global average supply of 26 kg per person in 1960 to over 42 kg (or 63 kg when fish is included) in 2018. In many high- and middle-income countries, the average citizen is now supplied with over 100 kg of animal-sourced products every year. While the intake of meats, eggs, and milk is often much higher than recommended levels, the intake of fruits, vegetables, whole grains, and legumes is much lower (Afshin et al., 2019; Rust et al., 2020), being among the main risk factors for mortality attributable to diet at a global level (Afshin et al., 2019).

Accordingly, higher levels of consumption of meat and dairy products have been associated with a higher incidence of multiple negative health outcomes (Oussalah et al., 2020). For example, well-planned diets with restriction of animal products have been shown to be more effective in the metabolic control of diabetic individuals than well-planned diets containing meat and dairy (Kahleova et al., 2011; Kim, Keogh, and Clifton, 2015). Excluding meat, eggs, and dairy from the diet has been also shown to reduce LDL cholesterol levels by over 35% (Ferdowsian and Barnard, 2009) (equivalent to the effect of using statins in therapeutic doses), as well as reduce the diameter of established coronary stenosis (Ornish et al., 1998) – the plaque buildup in the wall of the arteries that supply blood to the heart and can lead to heart attack or stroke.

Overconsumption of meat and dairy has been also strongly linked with the incidence of cancer. For example, a major umbrella review from 2021 (Huang et al., 2021) on the effect of red meat consumption against cancer outcomes showed an increased risk of overall cancer mortality, non-Hodgkin's lymphoma, bladder, breast, colorectal, endometrial, oesophageal, gastric, lung, and nasopharyngeal cancer. Overall, each 100 g increment in red meat consumption per day was associated with an 11%–51% increased risk of cancer. Consumption of processed meats (those subjected to salting, curing, fermentation, smoking, and other processes to enhance flavour or improve preparation) was also associated with an increased risk of overall mortality, non-Hodgkin's lymphoma, bladder, breast, colorectal, esophageal, gastric, nasopharyngeal, oral cavity, oropharyngeal, and prostate cancer. Specifically, for each increase of 50 g of processed meat per day there was an 8%–72% increase in risk. Indeed, in 2015 the International Agency for Research on Cancer (World Health Organization) had already classified processed meats as having sufficient evidence of carcinogenicity in humans and unprocessed red meat as probably carcinogenic to humans (IARC, 2015). Dairy consumption has also been associated with an increased risk of prostate cancer, with a 9% increased risk for every 50 g of cheese consumed per day and a 3% increase for every 200 g of milk (Aune et al., 2015).

Overconsumption of red meat, eggs, and dairy products has also been shown to increase substantially the intake of carnitine, phosphatidylcholine, and choline, which are ultimately converted into trimethylamine N-oxide (TMAO), a compound associated with an increased risk of virtually all non-communicable chronic diseases, including neurological disorders, intestinal inflammation, chronic kidney disease, Alzheimer's disease, type 2 diabetes, heart failure, stroke, and all-cause mortality (Qi et al., 2018).

The adverse health outcomes associated with the overconsumption of animal foods has also been shown to be a risk factor for communicable diseases. A large study involving healthcare workers from six countries is illustrative, showing that those following dietary patterns low in animal products had significantly lower odds of moderate-to-severe COVID-19-like illness (Kim et al., 2021).

Conclusions and recommendations

We tend to approach each public health problem independently, rather than recognising their common drivers. Animals have served humanity for millennia, but it is necessary to recognise that the way animals are raised and traded nowadays represents a major threat to human health and well-being.

As discussed in this chapter, many of the conditions that translate into poor animal health and welfare are also a threat to public health. Moreover, the extensive human and financial losses associated with infectious disease outbreaks, drug resistance, foodborne illnesses, and the diseases emerging from overconsumption of animal products, make this an enormous economic and social problem too. Enforcing higher animal welfare standards in industry practices, genetic selection, and stockmanship, as well as transparency and independent auditing, will be critical to reduce the risks of emergence and spread of new pathogens, including those with pandemic potential.

Consumption of animal-sourced products is still expected to rise in the next coming decades, further increasing these health risks and their associated costs. While large investments are poured into disease treatment, preparedness efforts, and drug development, we must have this same sense of urgency to accelerate the development of modern methods of food production that can mitigate these risks, and make society more resilient.

References

Afshin A et al., 2019. Health effects of dietary risks in 195 countries, 1990–2017: a systematic analysis for the Global Burden of Disease Study 2017, *The Lancet*, 393(10184), pp. 1958–1972.

Aguirre AA et al., 2021. Opportunities for transdisciplinary science to mitigate biosecurity risks from the intersectionality of illegal wildlife trade with emerging zoonotic pathogens. *Frontiers in Ecology and Evolution*, 9, p. 15.

ASOA, 2019. *Massive use of ionophore antibiotics in poultry production. Alliance to Save Our Antibiotics*. Available at: https://www.saveourantibiotics.org/news/press-release/massive-use-of-ionophore-antibiotics-in-poultry-production/ (Accessed: 28 July 2021).

Aune D et al., 2015. Dairy products, calcium, and prostate cancer risk: a systematic review and meta-analysis of cohort studies. *The American Journal of Clinical Nutrition*, 101(1), pp. 87–117.

Aylward BA, 2020. *A Comparative Evaluation of the Gastrointestinal Immune Response of the Modern and Heritage Broiler Chicken*. PhD thesis. University of Delaware.

Bezerra-Santos MA et al., 2021. Illegal wildlife trade: a gateway to zoonotic infectious diseases, *Trends in Parasitology*, 37(3), pp. 181–184.

Borlée F et al., 2017. Air pollution from livestock farms is associated with airway obstruction in neighboring residents. *American Journal of Respiratory and Critical Care Medicine*, 196(9), pp. 1152–1161.

Bridle BW et al., 2006. T lymphocyte subpopulations diverge in commercially raised chickens. *Canadian Journal of Veterinary Research*, 70(3), pp. 183–190.

Broom DM, 2014. Welfare assessment and welfare problem areas during handling and transport. In Grandin, T (ed.) *Livestock Handling and Transport*. Colorado: CABI, pp. 23–38.

Burkholder JM et al., 1997. Impacts to a coastal river and estuary from rupture of a large swine waste holding lagoon. *Journal of Environmental Quality*, 26(6), pp. 1451–1466.

Devleesschauwer B et al., 2018. The global burden of foodborne disease. In Roberts T (ed.) *Food Safety Economics: Incentives for a Safer Food Supply*. Cham: Springer International Publishing, pp. 107–122.

Dhingra MS et al. 2018. Geographical and Historical Patterns in the Emergences of Novel Highly Pathogenic Avian Influenza (HPAI) H5 and H7 Viruses in Poultry, *Frontiers in Veterinary Science*, 5, p. 84.

Díez-León M et al., 2016. Environmentally enriching American mink (*Neovison vison*) increases lymphoid organ weight and skeletal symmetry, and reveals differences between two sub-types of stereotypic behaviour, *Applied Animal Behaviour Science*, 177, pp. 59–69.

Di Nardo A, Knowles NJ and Paton DJ, 2011. Combining livestock trade patterns with phylogenetics to help understand the spread of foot and mouth disease in sub-Saharan Africa, the Middle East and Southeast Asia. *Revue Scientifique et Technique* , 30(1), pp. 63–85.

Dignard C and Leibler JH, 2019. Recent research on occupational animal exposures and health risks: a narrative review. *Current Environmental Health Reports*, 6(4), pp. 236–246.

Domingo NGG and Balasubramanian S, 2021. Air quality–related health damages of food. *Proceedings of the National Academy of Sciences*, 118 (20), pp. 1–6.

Done HY, Venkatesan AK and Halden RU, 2015. Does the recent growth of aquaculture create antibiotic resistance threats different from those associated with land animal production in agriculture? *The AAPS Journal*, 17(3), pp. 513–524.

EFSA, 2019. Salmonella control in poultry flocks and its public health impact. European Food Safe Authority Panel. *EFSA Journal*. European Food Safety Authority, 17(2), p. e05596.

Fa JE, Peres CA and Meeuwig J, 2002. Bushmeat exploitation in tropical forests: an intercontinental comparison. *Conservation Biology*, 16(1), pp. 232–237.

FAO, 2013. *World Livestock 2013: Changing Disease Landscapes*. Rome. Available at: http://www.fao.org/3/i3440e/i3440e.pdf. (Accessed July 28, 2021).

FAO, 2017. *Water Pollution from Agriculture: A Global Review*. Rome. http://www.fao.org/3/i7754e/i7754e.pdf. (Accessed July 28, 2021).

FAO, 2018. *Production/Livestock Primary*, n.d. Rome: Food and Agriculture Organization of the United Nations.

Ferdowsian HR and Barnard ND, 2009. Effects of plant-based diets on plasma lipids. *American Journal of Cardiology*, 104(7), pp. 947–956.

Founou LL, Founou RC and Essack SY, 2016. Antibiotic resistance in the food chain: a developing country-perspective. *Frontiers in Microbiology*, 7, p. 1881.

Gao F et al., 1999. Origin of HIV-1 in the chimpanzee *Pan troglodytes troglodytes*. *Nature*, 397(6718), pp. 436–441.

GBD (2019). The global burden of non-typhoidal salmonella invasive disease: a systematic analysis for the Global Burden of Disease Study 2017. *Lancet Infectious Diseases*, 19(12), pp. 1312–1324.

Gimsa U, Tuchscherer M and Kanitz E, 2018. Psychosocial stress and immunity-what can we learn from pig studies? *Frontiers in Behavioral Neuroscience*, 12, p. 64.

Glaser R and Kiecolt-Glaser JK, 2005. Stress-induced immune dysfunction: implications for health. *Nature Reviews. Immunology*, 5(3), p. 243–251.

Grandin T, 2014. Animal welfare and society concerns finding the missing link. *Meat Science*, 98, pp. 461–469.

Greatorex ZF et al., 2016. Wildlife trade and human health in Lao PDR: an assessment of the zoonotic disease risk in markets. *PLoS One*, 11(3), p. e0150666.

Greger M, 2007. The human/animal interface: emergence and resurgence of zoonotic infectious diseases. *Critical Reviews in Microbiology*, 33(4), pp. 243–299.

Hardstaff JL, Häsler B and Rushton JR, 2015. Livestock trade networks for guiding animal health surveillance. *BMC Veterinary Research*, 11, p. 82.

Hartcher KM and Lum HK, 2020. Genetic selection of broilers and welfare consequences: a review. *Worlds Poultry Science Journal*, 76(1), pp. 154–167.

Holmes E et al., 2021. The origins of SARS-CoV-2: a critical review. Available at: https://hdl.handle.net/2123/25672. (Accessed: Aug 07, 2021).

Huang Y et al., 2021. Red and processed meat consumption and cancer outcomes: umbrella review. *Food Chemistry*, 356, p. 129697.

IARC, 2015. *IARC Monographs Evaluate Consumption of Red Meat and Processed Meat*. WHO. Available at: https://www.iarc.who.int/wp-content/uploads/2018/07/pr240_E.pdf (Accessed Aug 01, 2021).

Johnson CK et al., 2020. Global shifts in mammalian population trends reveal key predictors of virus spillover risk. *Proceedings of the Royal Society. Biological sciences*, 287(1924), p. 20192736.

Jones BA et al., 2013. Zoonosis emergence linked to agricultural intensification and environmental change. *Proceedings of the National Academy of Sciences of the United States of America*, 110(21), pp. 8399–8404.

Jones KE et al., 2008. Global trends in emerging infectious diseases. *Nature*, 451(7181), pp. 990–993.

Kahleova H et al., 2011. Vegetarian diet improves insulin resistance and oxidative stress markers more than conventional diet in subjects with Type 2 diabetes. *Diabetic Medicine: A Journal of the British Diabetic Association*, 28(5), pp. 549–559.

Kaiser P et al., 2009. Prospects for understanding immune-endocrine interactions in the chicken. *General and Comparative Endocrinology*, 163(1–2), pp. 83–91.

Karavolias J et al., 2018. Raised without antibiotics: impact on animal welfare and implications for food policy. *Translational Animal Science*, 2(4), pp. 337–348.

Karesh WB and Cook RA, 2009. One world: one health. *Clinical Medicine*, 9(3), pp. 259–260.

Karesh WB et al., 2012. Ecology of zoonoses: natural and unnatural histories, *Lancet*, 380(9857), pp. 1936–1945.

Kasaeinasab A et al., 2017. Respiratory disorders among workers in slaughterhouses. *Safety and Health at Work*, 8(1), pp. 84–88.

Kaufman J et al., 1999. The chicken B locus is a minimal essential major histocompatibility complex. *Nature*, 401(6756), pp. 923–925.

Kim H et al., 2021. Plant-based diets, pescatarian diets and COVID-19 severity: a population-based case–control study in six countries'. *BMJ Nutrition, Prevention & Health*, 4, pp. 257–266.

Kim Y, Keogh J and Clifton P, 2015. A review of potential metabolic etiologies of the observed association between red meat consumption and development of type 2 diabetes mellitus. *Metabolism: Clinical and Experimental*, 64(7), pp. 768–779.

Kurpiers LA et al. 2015. *Problematic Wildlife*, Angelici F (ed). Cham: Springer, p. 507.

Levitt T, 2020. Two billion and rising: the global trade in live animals in eight charts. *The Guardian*, 20 January. Available at: http://www.theguardian.com/environment/2020/jan/20/two-billion-and-rising-the-global-trade-in-live-animals-in-eight-charts (Accessed: 27 March 2020).

Luo Z et al., 2020. Novel insights into stress-induced susceptibility to influenza: corticosterone impacts interferon-β responses by Mfn2-mediated ubiquitin degradation of MAVS. *Signal Transduction and Targeted Therapy*, 5(1), p. 202.

Ma W et al. 2008. The pig as a mixing vessel for influenza viruses: Human and veterinary implications. *Journal of Molecular Genetics*, 3, pp. 158–166.

Martínez-Miró S et al., 2016. Causes, consequences and biomarkers of stress in swine: an update. *BMC Veterinary Research*, 12(1), p. 171.

McEachran AD et al., 2015. Antibiotics, bacteria, and antibiotic resistance genes: aerial transport from cattle feed yards via particulate matter. *Environmental Health Perspectives*, 123(4), pp. 337–343.

McKenna M, 2017. *Big Chicken: The Incredible Story of How Antibiotics Created Modern Agriculture and Changed the Way the World Eats.* Washington, DC: National Geographic Books.

Mood A and Brooke P, 2014. Estimating the number of fish caught in global fishing each year. *FishCount.* Available at: http://fishcount.org.uk (Accessed: 06 Aug 2021).

Morse SS et al., 2012. Prediction and prevention of the next pandemic zoonosis. *The Lancet*, 380(9857), pp. 1956–1965.

van der Most PJ et al., 2011. Trade-off between growth and immune function: a meta-analysis of selection experiments. *Functional Ecology*, 25(1), pp. 74–80.

Mulder IE et al., 2009. Environmentally-acquired bacteria influence microbial diversity and natural innate immune responses at gut surfaces. *BMC Biology*, 7, p. 79.

Negro-Calduch E et al., 2013. Assessment of biosecurity practices of small-scale broiler producers in central Egypt. *Preventive Veterinary Medicine*, 110(2), pp. 253–262.

Nowak MA and May RM, 1994. Superinfection and the evolution of parasite virulence. *Proceedings of the Royal Society. Biological Sciences*, 255(1342), pp. 81–89.

ONeill J, 2015. Antimicrobials in agriculture and the environment: reducing unnecessary use and waste. *The Review on Antimicrobial Resistance.* Available at: https://ec.europa.eu/health/amr/sites/amr/files/amr_studies_2015_am-in-agri-and-env.pdf (Accessed: 06 Aug 2021).

ONeill J, 2016. *Tackling drug-resistant infections globally: Final report and recommendations.* Available at: https://amr-review.org/ (Accessed: 29 March 2020).

Ornish D et al., 1998. Intensive lifestyle changes for reversal of coronary heart disease. *JAMA: The Journal of the American Medical Association*, 280(23), pp. 2001–2007.

Osterholm M and Olshaker M, 2017. *Deadliest Enemy: Our War against Killer Germs*. 1st edn. Boston: Little, Brown Spark.

Oussalah A et al., 2020. Health outcomes associated with vegetarian diets: An umbrella review of systematic reviews and meta-analyses. *Clinical Nutrition*, 39(11), pp. 3283–3307.

Poovorawan Y et al., 2013. Global alert to avian influenza virus infection: from H5N1 to H7N9. *Pathogens and Global Health*, 107(5), pp. 217–223.

Proudfoot K and Habing G, 2015. Social stress as a cause of diseases in farm animals: current knowledge and future directions. *Veterinary Journal*, 206(1), pp. 15–21.

Qi J et al., 2018. Circulating trimethylamine N-oxide and the risk of cardiovascular diseases: a systematic review and meta-analysis of 11 prospective cohort studies. *Journal of Cellular and Molecular Medicine*, 22(1), pp. 185–194.

Racicot M et al., 2011. Description of 44 biosecurity errors while entering and exiting poultry barns based on video surveillance in Quebec, Canada. *Preventive Veterinary Medicine*, 100(3–4), pp. 193–199.

Rohr JR et al., 2019. Emerging human infectious diseases and the links to global food production. *Nature Sustainability*, 2(6), pp. 445–456.

Roser M, Ortiz-Ospina E and Ritchie H, 2013. Life expectancy. *Our World in Data*. Available at: https://ourworldindata.org/life-expectancy (Accessed: 23 July 2021).

Rostagno MH, 2009. Can stress in farm animals increase food safety risk? *Foodborne Pathogens and Disease*, 6(7), pp. 767–776.

Rust NA et al., 2020. How to transition to reduced-meat diets that benefit people and the planet. *Science of the Total Environment*, 718, p. 137208.

Shi W and Gao GF, 2021. Emerging H5N8 avian influenza viruses. *Science*, 372(6544), pp. 784–786.

Shi Z and Hu Z, 2008. A review of studies on animal reservoirs of the SARS coronavirus. *Virus Research*, 133(1), pp. 74–87.

Slade J and Alleyne E, 2021. The psychological impact of slaughterhouse employment: a systematic literature review. *Trauma, Violence & Abuse*, 11, pp. 1–12.

Springbett AJ et al., 2003. The contribution of genetic diversity to the spread of infectious diseases in livestock populations. *Genetics*, 165(3), pp. 1465–1474.

Subramanian M, 2012. Zoonotic disease risk and the bushmeat trade: assessing awareness among hunters and traders in Sierra Leone. *EcoHealth*, 9(4), pp. 471–482.

Trovão NS and Nelson MI, 2020. When pigs fly: pandemic influenza enters the 21st century. *PLoS Pathogens*, 16(3), pp. 1–8.

Undurraga D, 2018. *Supermarket Meat Still Superbugged, Federal Data Show*. Environmental Working Group. Available at: https://www.ewg.org/research/superbugs/. (Accessed: 23 July 2021).

UNEP (United Nations Environment Programme), 2020. *Preventing the Next Pandemic: Zoonotic Diseases and How to Break the Chain of Transmission*. Nairobi, Kenya: United Nations Environment Program and International Livestock Research Institute.

Van Boeckel TP et al., 2019. Global trends in antimicrobial resistance in animals in low- and middle-income countries. *Science*, 365(6459), p. eaaw1944.

Von Essen SG and Auvermann BW, 2005. Health effects from breathing air near CAFOs for feeder cattle or hogs. *Journal of Agromedicine*, 10(4), pp. 55–64.

Wang J et al., 2017. Characterization of oqxAB in *Escherichia coli* isolates from animals, retail meat, and human patients in Guangzhou, China. *Frontiers in Microbiology*, 8, p. 1982.

Watts JEM et al., 2017. The rising tide of antimicrobial resistance in aquaculture: sources, sinks and solutions. *Marine Drugs*, 15(6), p. 158.

WHO, 2018. *Campylobacter: Key Facts, Key Facts*. World Health Organisation. Available at: https://www.who.int/news-room/fact-sheets/detail/campylobacter (Accessed: 27 March 2020).

Wolfe ND, Dunavan CP and Diamond J, 2007. Origins of major human infectious diseases. *Nature*, 447(7142), pp. 279–283.

Yada T and Tort L, 2016. 10: Stress and disease resistance: immune system and immunoendocrine interactions. In Schreck CB et al. (eds) *Fish Physiology*. London: Academic Press, pp. 365–403.

25
ANIMAL DISASTER MANAGEMENT

Steve Glassey

Introduction

The Australian Black Summer fires of 2019–2020 that decimated over three billion animals (World Wildlife Fund, 2020) served as a harsh reminder of hazards we humans choose to create. Disasters are not natural, nor are they an event. They are a process manufactured and implemented by people and their choices (Kelman, 2020, p. 15). Definitions of what constitutes a disaster also tend to be anthropomorphic and fail to recognise animals in their terminology, often relegating such sentient beings as environmental impacts or property loss. Humans are increasingly becoming more at risk from natural hazards such as floods, storms, drought, and fires, and this increase is strongly correlated with urbanisation, population growth, and climate change (Haddow et al., 2017). Animals, however, are becoming more vulnerable to these hazards, also through farming intensification, loss of natural habitat, and failing animal-health infrastructure – again all caused by human action. It is only humans – albeit with varying degrees of influence, power, and resources – who can mitigate these risks. This power imbalance places a moral obligation on humans to act to protect animals from the effects of disasters that they have created.

Though sometimes used interchangeably by lay persons, emergencies and disasters are distinctly different. An emergency is an event that threatens life or property, whereas a disaster is an emergency that is beyond existing capacities and requires outside assistance. To avoid confusion with veterinary emergency medicine, animal disaster management is more easily understood when engaging a wide range of audiences from veterinarians to disaster managers. The goal of animal disaster management is to create animal-inclusive, resilient communities.

Why animals matter in disasters

One of the earliest examples of the protection of animals from disaster can be found in the biblical story of Noah's Flood, where Noah and his family were spared by God from a cataclysmic flood after being directed to build an Ark to house themselves and two of every kind of animal (New International Version 2011, Genesis 7). Though science and religion may not agree on the existence of such an Ark, the cultural significance of non-human species being pivotal to the existence of human life within religious texts should not be disregarded.

 DOI: 10.4324/9781003182351-30

It is estimated that more than 40 million animals are affected by disasters annually, with this number increasing in the Anthropocene (Sawyer and Huertas, 2018, p. 2). However, the genesis of animal disaster management in modern times is largely due to the lessons and reforms following Hurricane Katrina. In August 2005, Hurricane Katrina struck the Gulf Coast of the United States of America. In its wake, it left US$110 billion in damage and 1,836 people dead, making it the third-deadliest disaster in US history. This disaster also highlighted the importance of companion animal emergency management, with over 50,000 pets being left behind during the evacuation of New Orleans, and 80–90% of these pets perishing. What was anticipated to be over within a few days turned into a catastrophe and triggered the largest animal rescue operation in US history – an operation that rescued approximately 15,000 pets, supported by some 5,000 volunteers. Prior to 2005, it was Federal Emergency Management Agency (FEMA) policy that pets should be left behind during evacuations. This has now been completely changed with the introduction of the Pets Evacuation & Transportation Standards (PETS) Act. The single most compelling fact for public safety officials to learn from Hurricane Katrina was that approximately 44% of the people who did not evacuate stayed, at least in part, because they did not want to leave their pets behind (Fritz Institute, 2006). Indeed, Heath and Linnabary (2015) reinforce this finding saying that:

> There is no other factor contributing as much to human evacuation failure in disasters that is under the control of emergency management when a threat is imminent as pet ownership. Emergency managers can take advantage of the bond people have with their animals to instill appropriate behavior amongst pet owners in disasters.

The human–animal bond has been the primary focus of animal disaster management, often using the well-documented phenomena of humans placing themselves at risk for animals, as a means to tackle animal welfare concerns through a paradigm of 'saving animal lives, saves human lives'. And this is particularly true of companion and service animals that have benefited the most in terms of regulatory changes to protect them from disaster impacts, despite them being the least vulnerable, given that human guardianship affords them protection. It is the animals that do not have, or have little to no, human–animal bonds, such as wild animals and those exploited for consumption, that are afforded the least levels of protection, making them significantly more vulnerable to the impacts of disaster. Society as a whole generally ranks animals through a sociozoologic system, which classifies animals in a structure of meaning that allows them to define, reinforce, and justify their interactions with other beings (Irvine, 2009, p. 7). This construct of a sociozoological scale gives further weight to the understanding that disasters are not natural; they are manifested by humans, determining which animal species are less important than others, thus making some animals more vulnerable than others. Humans are largely responsible for making animals vulnerable to disaster, but unlike humans, animals often do not have a choice in the construction or exposure of their aggravated vulnerabilities. This vulnerability can be exacerbated by weak animal–health infrastructure which is regarded as a root cause in companion animal disasters (Heath and Linnabary, 2015), along with myriad other complex *wicked problems* within a public policy and planning context (Glassey, 2020a). Even the legal status of animals can contribute to increasing their vulnerability to the effects of disaster. Treated as property, animals are made "legally inferior to people" and therefore "usually afforded low priority in emergency response initiatives" (Best, 2021). The reality of animal disaster laws is that they seldom have little to do with sentience or the welfare of animals; the drivers for such laws are more focused on protecting people through improving human evacuation compliance and preventing humans from returning into hazardous disaster zones to save animals, especially companion animals.

Given the impact on human and environmental well-being arising from animals being affected by disasters and emergencies, the outdated reference to "animal welfare emergency management" by some governments in their emergency planning fails to recognise these relationships and is counter-productive to making animals as a priority in disaster risk reduction, within a One Health or One Welfare environment.

Phases of disaster management

Within the profession of emergency management (also known as disaster management), a life-cycle approach is taken to mitigate hazards, prepare for the impacts of residual risks (the remaining risk after mitigation controls have been applied), respond to disasters to protect life and property, and support affected communities to recover. These are typically known as the four phases of comprehensive disaster management (Haddow, 2011, p. 9), though some countries such as New Zealand refer to these phases as Reduction, Readiness, Response, and Recovery respectively (Glassey and Thompson, 2020).

Prevention phase

Within the context of animal disaster management, the prevention phase includes elimination of the risk or reducing it to an acceptable level, such as banning intensive farming or at least reducing the associated risks, such as not building animal housing facilities on flood plains. Other mitigatory measures include seismic bracing of animal caging systems in regions prone to earthquakes (such as New Zealand), and the installation of fire suppression systems and availability of water for firefighting, to name just a few. However, there is often a residual risk despite these treatments being applied, and therefore preparing for the eventuality of the hazard is required.

Prevention activities can extend to the passage of laws to better afford protection to animals to avoid them being exposed to disaster hazards in the first place. In Texas, under Section 821.077 of the Health and Safety Code, it is illegal to restrain a dog outside and unattended during extreme weather or when such associated weather warnings have been issued (State of Texas, 2007). Though companion animals are less vulnerable than captive production animals, dogs and cats often receive higher levels of legal protection. Again, this illustrates that animals are likely ranked by their attachment with humans, rather than their raw vulnerability alone. Intensively farmed animals such as pigs and chickens are extremely vulnerable to the impacts of disaster. Often these facilities are built on remote and hazard-prone land, which makes the land less expensive and which is therefore perceived to be more profitable to operate a business on. Local ordinances could be used to prevent the building or operation of intensive farms in flood plains, largely eliminating the flood risk to these animals. In 1999, Hurricane Floyd devastated parts of North Carolina. Approximately 2.8 million poultry, 30,500 hogs, 2,000 cattle, and 250 horses drowned during this disaster (Green, 2019, p. 2). In the 2020 Canterbury earthquake, over 20,000 chickens died or were destroyed as their caging systems collapsed (Glassey and Wilson, 2011). The installation of seismic bracing for caging would likely have prevented many of their deaths.

Laboratory animals are seldom considered in disaster management and there is limited research in this area. These animals are always confined to cages, often fully dependent on automated feed, watering, and environmental control for their survival, and when these systems fail, their welfare can be compromised severely. In 2006, a generator failed at the University of Ohio, and when electricity was restored it triggered the heating system and the temperature reached 105°F (40.5°C). Nearly 700 animals died (Irvine, 2009, p. 85). Though some producers may perceive mitigation measures such as automatic fire suppression, backup ventilation systems and seismic bracing to be expensive, disaster risk reduction makes economic sense. According to the United Nations, every dollar invested in risk reduction and prevention can save up to 15 dollars in post-disaster recovery (United Nations Office for Disaster Risk Reduction, 2020a).

Zoos and aquaria also have been impacted by disasters and are often overlooked, with emergency planning requirements generally focused on loss of containment of dangerous animals and protecting the public, rather than the large-scale negative animal welfare impacts on their captive animals that disasters that can have. In 2002, the Prague Zoo was flooded leading to over 150 animals being killed (Irvine, 2009, p. 124), and in the Afghanistan post-war period of 2001, the animals at the Kabul Zoo were left without sufficient care and attention, leaving many to perish from starvation and the following harsh winter conditions (Sawyer and Huertas, 2018, p. 51).

As US and coalition troops withdrew from Afghanistan in August 2021, Kabul, including its municipal zoo, fell under the control of the Taliban. The Asia for Animals coalition (AFA) reported that no animals had been harmed and that the Taliban was ensuring the zoo continued to operate as normal (AFA, 2021). It is unclear if the continued protection of these zoo animals was a conscious decision of the Taliban, whether it be as a lesson from the 2001 post-war period, or even part of their *hearts and minds* campaign to purport a new, changed, and more humane style of governance. The plight of animals during the US withdrawal indeed captured the world's attention and caused outcry when it was alleged American forces had left behind their military service dogs, which was later found to be incorrect. The animals photographed in airline crates at the Hamid Karzai International Airport were in fact dogs from the Kabul Small Animal Rescue who were hoping to have these animals and their staff evacuated (DefenseOne, 2021). Public reaction also successfully pressured the United Kingdom government to allow Pen Farthing – a former British Marine who operated the Nowzad animal sheltering charity in Kabul – to evacuate dozens of dogs and cats to the UK on a privately chartered plane (Washington Post, 2021). Farthing was criticised by government leaders including British Defence Secretary Ben Wallace for supposedly putting the lives of animals ahead of people (Washington Post, 2021).

When the *Aquarium of the Americas* lost backup generator power during Hurricane Katrina, over 10,000 fish suffocated (Irvine, 2009, p. 13). Having resilient infrastructure is key to the survival of captive animals dependent on automated environmental, feeding and watering systems. Similarly, in the 2011 Christchurch earthquake, the Southern Experience Aquarium suffered irreparable damage, and despite rescue efforts an undisclosed number of fish were euthanised due to poor water quality and the generator failing (Potts and Gadenne, 2014, p. 217).

Animals that are at the whim of humans for their survival are most vulnerable to disaster and those that are live-exported by sea are no different. In 2019, the livestock carrier *Queen Hind* capsized with over 14,000 sheep on board bound for slaughter. The conditions on board prior to the capsize were cramped. Despite the efforts of animal rescue specialists from Four Paws and the Animal Rescue and Care Association (ARCA) of Romania, more than 13,820 sheep drowned or died because of the capsizing. It was later found that the vessel had secret floors that would have contributed to overloading, and that affected the vessel's stability (Zee, 2021). The prohibition of live export would have prevented this human-caused disaster.

Preparedness phase

As part of the PPRR framework, disaster planning within the preparedness phase provides an opportunity to improve response effectiveness to protect life and property, as well as reducing the impacts on communities under a pre-agreed approach, which aimed at providing role clarity across organisations. Classic scholars such as Auf der Heide (1989) promote a fundamental principle that emergency plans should be based on *likely*, not *correct* behaviours. From a traditional emergency service perspective, it would be seen as *correct* that, when people are told to evacuate and leave their companion animals behind, they would do so compliantly. However, it is more *likely* that the guardians of these animals when faced with evacuation may refuse to evacuate unless they can take their animals, as experienced in Hurricane Katrina (Irvine, 2009) and disasters such as the Fukushima nuclear incident following the 2011 Japanese earthquake and tsunami (Kajiwara, 2020).

Developing animal-inclusive emergency plans helps to clarify the roles and responsibilities of parties during a disaster. So as not to create dependency and complicate evacuation logistics, it is critical that the guardians of animals take responsibility for their welfare. This responsibility is often enshrined in law, and as disasters are not natural, the obligations on such guardians are not necessarily eroded. In some countries or states, there are additional legal responsibilities for ensuring the safety of animals exposed to foreseen extremes of weather (Glassey, 2018; 2019; 2020b).

Though there are many different models, the Emergency Management Accreditation Program (EMAP) standard is one that is flexible to apply to animal disaster planning at all levels (national, state, local). Using the EMAP standard (2019) as a benchmark, emergency management plans should include the following considerations:

- Program Management, Administration and Finance, and Laws and Authorities;
- Hazard Identification, Risk Assessment, and Consequence Analysis;
- Hazard Mitigation;
- Prevention;
- Operational Planning and Procedures;
- Incident Management;
- Resource Management, Mutual Aid, and Logistics;
- Communications and Warning;
- Facilities;
- Training;
- Exercises, Evaluations, and Corrective Action;
- Emergency Public Education and Information.

In addition to the core standards above, animal-specific considerations should include:

- Lessons from previous emergencies;
- Euthanasia and depopulation;
- Carcass disposal;
- Humane trapping in evacuated areas;
- Feeding in place protocols;
- Veterinary considerations (e.g., zoonotic disease management);
- Disposal of unclaimed displaced animals (such as adoption);
- Animal search, rescue, evacuation, sheltering, body recovery, and decontamination.

Though this chapter does not focus on animal disease management, planning considerations from the Good Emergency Management Practice (GEMP) manual published by the Food and Agriculture Organization of the United Nations (FAO) has useful advice, including advocating that animal-related disaster plans be part of national disaster management arrangements and be able to access related government funding (2011, p. 18). Where countries such as the United States have passed the PETS Act that secures federal funding for companion and service animal emergency management activities, despite reports presented to Parliament, the New Zealand government has continued to exclude animal disaster management from its national disaster response and recovery funding arrangements (Glassey, 2019).

The value in the planning phases is often not the end document, but more so the process that should engage stakeholders to develop a common appreciation of the hazards, and of how a coordinated response should be conducted. Where plans are developed in isolation they typically end up as a *box ticking* exercise, also known as suffering from the "paper plan syndrome" (Auf der Heide, 1989).

Animal disaster management planning approaches are still generally in their infancy, given that in most part until the passage of the US PETS Act in 2006, there were few regulatory drivers for such planning around the world. Much of the planning efforts have focused around adopting human-centric approaches, which makes sense for reasons of compatibility, efficiencies, and giving legitimacy to efforts. However, such adopted planning models were developed and refined for a single species – humans, without due regard to the other species. There are approximately 7,700,000 species of animals on earth (Mora et al., 2011) and this variety of non-human species creates extra challenges for animal disaster planners, who often must develop plans that can accommodate end users (being animals), from a few grams to hundreds of kilograms, that are uncommunicative and likely to hide, escape, or attack. It would appear that helping humans in disasters is easier in comparison.

In 2014, the National Planning Principles for Animals in Disasters (NPPAD) was released by the National Advisory Committee for Animals in Emergencies and endorsed by the Australia-New Zealand Emergency Management Committee (Trigg et al., 2021). The NPPAD provided 8 principles for the planning process and 16 further principles to be included in actual plans. In 2020, it was found that in Australia there was moderate awareness of the principles across stakeholders, and low to moderate implementation of the principles (Trigg et al., 2021). These principles – though developed primarily in Australia – are generally applicable to most other countries and may be of benefit to the planning process.

The preparedness phase could include creating and testing emergency plans for animal housing facilities, public education campaigns around animal disaster preparedness, training animals to be familiar with evacuation processes and transport, carrying out microchipping campaigns, subscription to early warning systems for floods, fires, and the like, and training for animal disaster responders in incident command, wildland fire, and flood safety. This ensures that when the disaster occurs, the response to protect life and property can be at its most effective, which may include pet-friendly evacuation centres, emergency animal fostering, veterinary disaster care, and rescues of animals.

Education, training, and exercising are also critical to the preparedness phase. The range of animal disaster management courses and education programmes is slowly increasing. Information sharing and networking continue to help advance this emerging professional discipline and forums such as the National Alliance for State and Agricultural Emergency Programs (NASAAEP) (Green, 2019, p. 3) and the Global Animal Disaster Management Conference (GADMC) have made significant contributions to promoting animal-inclusive resilient communities.

Complimentary to the range of existing planning approaches, Vieira and Anthony (2021) developed six ethically responsible animal caretaking aims for consideration when developing disaster management plans and policies in the Anthropocene. They include (1) saving lives and mitigating harm; (2) protecting animal welfare and respecting animals' experiences; (3) observing, recognising, and promoting distributive justice; (4) advancing public involvement; (5) empowering care givers, guardians, owners, and community members; (6) bolstering public health and veterinary community professionalism, including engagement in multidisciplinary teams and applied scientific developments. Armed with the Australian NPPAD, the EMAP standard and the six ethically responsible caretaking aims, animal disaster planners now have tools to create effective plans.

Response phase

Although the response phase is often the most publicised, it is often the most short-lived. The window of time to rescue animals before they die of injuries, disease, thirst, or hunger is often small and requires immediate intervention. In agriculture, it is argued that insuring animals may lead to negative animal welfare outcomes, as often the trigger for payment is the death of such animals (Sawyer and Huertas, 2018). It then becomes financially attractive for the guardians of livestock to allow them to perish. However, restocking of herds following disasters has frequently been found to be ineffective, leading to longer-term economic harm to farmers, and there is a driver to encourage early intervention to protect surviving stock as a better alternative (Sawyer and Huertas, 2018).

An example of this ineffective restocking occurred in Myanmar in 2008, following Cyclone Nargis, where areas suffered large losses of working buffalo that were critical to harvesting rice. Without these animals the flood-contaminated lands could not be rendered productive, and so new working buffalo were introduced. However, this restocking programme failed to properly address animal-health considerations and led to the introduction of new diseases and further mortality of such stock (Sawyer and Huertas, 2018). "Poor support for these animals, often worked harder in the aftermath of a disaster, or poorly planned restocking programmes can make a bad situation worse very rapidly" (Sawyer and Huertas, 2018, p. 7). Since the early 2000s humanitarian aid and veterinary professionals started to critically reflect on whether their interventions to protect livestock following disasters were effective. This led the Food Aid Organization of the United Nations (FAO) and other organisations to develop and publish the Livestock Emergency Guideline and Standards (LEGS, 2017). The LEGS manual provides general information and technical standards to improve the quality and livelihoods impact of livestock-related projects in humanitarian situations (LEGS, 2014). However, LEGS focuses on assisting communities in less developed countries and does not provide standards for disaster interventions involving other non-livestock animals such as companion animals.

Where animal rescues are carried out there is often a disconnect between animal interest groups undertaking this function and the human-centric rescue authorities. Often these 'animal rescuers' are spontaneous groups without authority, training or equipment and this *delegitimisation of animal rescue* particularly hinders those specialist animal disaster rescue teams who attempt to seek a legitimate and integrated animal-human disaster response (Glassey, 2021). The delegitimisation of animal rescue is defined as the:

> Sub-optimal response by animal interest groups who respond to assist animals in emergencies or disasters in an unsafe or illegal manner, which consequently makes it more difficult for bona-fide emergency animal rescue groups to be accepted and used by authorities and the community in future interventions.
>
> *(Glassey, 2021)*

Aside from potentially putting human lives at risk, delegitimisation has negative effects for animal welfare through eroding trust between the animal response community and emergency service organisations. Ultimately, this loss of trust and confidence may lead to animal protection in disasters being considered a hindrance rather than an opportunity to improve human and animal safety. Studies have shown that humans do place themselves at risk for the needs of animals, such as breaching cordons to attend to their animals or failing to evacuate if they are unable to take their animals (Heath, 1999; Heath et al., 2001; Irvine, 2009; Glassey, 2010; Potts and Gadenne, 2014; Heath and Linnabary, 2015; Taylor et al., 2015).

During the bushfires in Australia in the summer of 2019 and 2020, the loss of three billion animals gained global attention, as well as responses from domestic and international animal interest groups. Such groups formally or informally identify as 'animal rescue'; however, in the disaster response context, this is confusing and misleading to emergency service organisations. These groups use the term 'animal rescue' whereas it might be more appropriate if 'animal care', 'welfare', or 'rehoming' were used. The use of 'animal rescue' undermines the credibility of emergency services organisations that rescue animals, and some may regard the term 'rescue' as an embellishment of capability.

Unfortunately, the lack of animal-inclusive emergency management planning results in animal interest groups responding to disasters without appropriate authority, training, or equipment, as observed in by Glassey and Anderson (2019) in the Nelson, New Zealand fires of 2019. Even animal interest groups that have a focus on animal disaster response have been found wanting, such as during the summer bushfires where promotional videos showed personnel working with flames and smoke around them, and also without basic protective equipment (Glassey, 2021). The wearing of flame-retardant apparel, safety boots, helmets, goggles, and gloves is a rudimentary requirement for working on firegrounds, as – even days and weeks after the fire has gone through – vegetation and underground fires are common, and create a risk for personnel to step or fall into. The risk of branches and trees falling during and after fires remains substantial and requires helmets to be worn. The use of videos or pictures showing animal interest groups not adhering to basic safety requirements delegitimises animal rescue and reduces the level of confidence and trust of emergency services organisations (Glassey, 2021).

The disconnect is compounded with animal groups setting their own standards for training, often not recognised by public safety agencies. In urban search and rescue operations, internationally accepted search markings placed on collapsed or damaged structures (such as following an earthquake) fail to incorporate animal rescue, leading to confusion when animal rescue groups place their own markings (Glassey and Thompson, 2020).

Another aspect of delegitimsation of animal rescue occurs when animal interest groups respond to an emergency and claim pre-existing animal welfare issues as being caused by, or related to, the event. This could include taking footage of stray animals in a damaged city and suggesting the animal was in need of rescue, when it was, at that time and prior to the disaster, a stray animal; or showing dogs without kennels or being chained up following floods, when the dogs were in these conditions prior to the flood. Such flooding may have exposed these vulnerabilities, but may not have been the cause of such animal welfare concerns. It is argued that prevention is better than post-event response, and animal interest groups wanting to reduce animal vulnerability to disasters could focus efforts on mitigation and strengthening weak animal-health infrastructure to make a sustainable impact on improving animal welfare (Glassey, 2021).

Where animals are rescued from a disaster-affected area, if a guardian is not located, affected animals are often put into temporary accommodation. Disasters by definition exceed local capacity, so often day-to-day facilities such as animal boarding facilities, humane shelters, and pounds may be unavailable due to damage or exceeding capacity, not to mention that often these organisations may also be attending to their own animals and disaster responsibilities. Where possible, existing facilities and service providers should be used as they generally offer higher levels of animal welfare to that of temporary shelters, and their use also stimulates economic recovery. Much has changed in the past decade, with the United States leading many new approaches to emergency companion animal sheltering. Traditional Animal-Only Shelters (AOS) are those where the care of the animals falls to the sheltering team. Animal-Only Shelters can be appropriate in some situations, but they are generally not sustainable when a large number of carers is required, making this approach difficult to scale up for any wide-area disaster. It has also been found that these shelters are 25 times more expensive to operate than Co-Habitation Shelters (CHS) and five times more expensive than Co-Located Shelters (CLS) (Strain, 2018). As animals are separated from their guardians in Animal-Only Shelters, this can increase stress in the animal, which can heighten the risk of disease. Where companion animals are co-located, evacuees are accommodated in a building nearby to where the animals are housed, allowing guardians to maintain care and responsibility for their pets. This provides routine and sense of purpose and increases the guardian–animal interaction time. The other option – which is just gaining traction in the US – is co-habitation, where humans and their companion animals are housed as a single-family unit. This often leads to reduced stress in both the animal and the human, as pets often provide a familiar psychosocial coping mechanism and animals are typically more settled and quieter. The lack of providing suitable, pet-friendly sheltering leads not only to poor animal welfare outcomes, but also can compromise human safety – especially for those with strong attachments to their animals. This was the case following the 2011 Japanese earthquake, tsunami, and nuclear disaster, where lonely elderly people were left with no option but to sleep in their cars near evacuation centres that did not permit animals, only to be socially isolated, suffer hypothermia in the winter, and, on one occasion, Deep Vein Thrombosis (DVT) from cramped sleeping and sitting conditions (Kajiwara, 2020, p. 66). Accepting that 'Feeding in Place' can also be an alternative to emergency animal sheltering in some circumstances, the bottom line is that Co-Habitated Sheltering is the gold standard (Green, 2019, p. 147).

The lack of pet carriers has been linked as a causal factor in evacuation failure (Heath, 1999, p. 209), particularly for those with multiple small animals. It is now common practice for specialist animal disaster response charities like Animal Evac New Zealand to go into areas likely requiring evacuation or under evacuation notice and distribute pet carriers to improve evacuation compliance. This leads to better human and animal safety outcomes (Glassey and Anderson, 2019).

When confronted with the need to evacuate, some households may even intentionally partially evacuate to leave someone behind to attend to their animals, whilst the remainder leave for safety (Taylor et al., 2015). Where animals have been left behind in an evacuated disaster zone, many often return to rescue or attend to their animals, which may put themselves or public safety responders at risk, as in the 2010 Haiti earthquake (Sawyer and Huertas, 2018, p. 10), Canterbury earthquakes (Potts and Gadenne, 2014), and Edgecumbe flood (Glassey et al., 2020).

It is common for humans to put themselves at risk to protect their animals or act protectively, such as in the case of the Weyauwega train derailment in 1996. Following the derailment of a train carrying large quantities of hazardous materials, the entire Wisconsin township consisting of 1,022 households was hastily evacuated. Within a couple of days, pet owners attempted to breach the cordon to rescue their animals. Frustrated owners on 'behalf of the animals' then phoned through a bomb threat to the emergency operations centre. This led to significant negative media attention which prompted the state Governor to order the National Guard to enter with armoured vehicles to assist with the rescue of hundreds of pets left behind (Irvine, 2009, p. 38).

The loss of companion animals in particular can have devastating mental health impacts. Hunt et al. (2008) found that survivors of Hurricane Katrina were just as likely to suffer post-traumatic impacts from losing their companion animal as they were from losing their home. Disasters can also draw out the worst in humanity and create opportunities to exploit those vulnerable in the community by individuals, such as *disaster paedophiles* who use the state of chaos to traffic unaccompanied minors (Montgomery, 2011). Animals too can be vulnerable from similar abuse as observed in Hurricane Harvey with reports of *disaster rustling* and *disaster hoarding*, the latter involving animal hoarders who used the disaster as an opportunity to restock their hoard (Glassey, 2018).

Recovery phase

Even as the response phase commences, so should the initial planning for the recovery phase. Recovery can be also described as the regeneration of the community, and this phase also needs to include considerations for animals and their welfare. This often can include the supply of animal-friendly rental accommodation, reunification of displaced animals, and restoration of veterinary and animal welfare services. Recovery should *build back better*, and the United Nation's definition, which is human-centric, is defined as:

> The use of the recovery, rehabilitation and reconstruction phases after a disaster to increase the resilience of nations and communities through integrating disaster risk reduction measures into the restoration of physical infrastructure and societal systems, and into the revitalization of livelihoods, economies and the environment.
>
> *(United Nations Office for Disaster Risk Reduction, 2020b)*

The lack of post-disaster, pet-friendly accommodation has constantly been identified as an issue, from Haiti where, following the 2010 earthquake, internally displaced persons in tented camps were unable to have their companion animals (Sawyer and Huertas, 2018, p. 10), to those who returned to radioactive exclusion zones near Fukushima to secretly attend to their animals, or were sleeping in their vehicles in freezing winter conditions with their animals, as animals were not allowed in temporary mass shelters (Kajiwara, 2020). Similarly, in Christchurch following the 2011 Canterbury earthquake, pet-friendly accommodation became very scarce, forcing owners to relinquish their animals, causing much distress for both humans and animals (Potts and Gadenne, 2014).

The stressful impacts on people and animals during and following a disaster can be suffered for months. Those people who respond to help disaster-affected animals, from volunteer rescuers to professional veterinarians, are not immune from the impacts of being exposed to the distressing experiences often found in a disaster. In a global study of veterinary disaster responders, it was found that 51% exhibited behavioural health issues during their response and up to 6 months afterwards (Vroegindewey and Kertis, 2021). It is important for anyone considering becoming involved in animal disaster response to have access to psychological first aid training and resources.

The recovery phase should also include a process to reflect upon the response, and even on the recovery. Commonly following a response, an After Action Report (AAR) is written following a debrief of organisations involved in the response. The AAR is an important first step in the lessons management process, which aims to improve not only subsequent responses, but enhancements to the wider phases of comprehensive emergency management. Largely, AARs are not mandatory, nor is the format, content, and dissemination. Though AARs are critical to improving subsequent responses, which should lead to better public safety and animal welfare outcomes, they are seldom shared, often due to fear of deficiencies bringing political embarrassment or reputational harm.

The lessons identified in AARs are unfortunately seldom learned. A study by Glassey et al. (2020) found that only 7% of applicable lessons were learned in the context of animal disaster response arising from the 2017 Edgecumbe Flood, to the 2019 Nelson Fires. The comparative analysis of AARs for both these events found that common problems related to training, capability, law, policy, planning, information management, and incident management, were repeated, and lessons seemingly not learned. The assumption that lessons are learned from previous disasters requires closer examination.

Recommendations

To improve animal welfare in disasters, much work is needed. Firstly, reducing animals' vulnerability to hazards must be made a priority. As part of a comprehensive emergency management approach, frameworks to create animal-inclusive community resilience must include evidence-based laws and policies. Such frameworks need to ensure guardians take primary responsibility for animal welfare in disasters, but must also provide for the monitoring and performance of government and partner organisations who facilitate and coordinate animal disaster management. There is currently no system to compare the effectiveness of animal disaster management frameworks across countries. It is recommended that the Animal Protection Index (World Animal Protection, 2020) be revised to include an animal disaster management indicator, or that a global animal disaster management index is developed similarly to the National Capabilities for Animal Response in Emergencies (NCARE) as developed by the American Society for the Prevention of Cruelty to Animals (Spain et al., 2017). Model laws for animal disaster management should also be developed and considered as part of the revised or new indices. Other frameworks such as the Five Domains (Mellor, 2017) could benefit from further research with respect to their application to animal disaster management.

There also needs to be more of a concerted effort to mainstream animal disaster management, away from being an "animal issue". The One Health – One Welfare approaches offer opportunities to connect animal and human welfare, and environmental sustainability, all in the context of disaster management and in line with international disaster risk reduction frameworks such as the Sendai Framework (Dalla Villa et al., 2020). Travers et al. (2021) also give recommendations to enhance the linkage between One Health and animal disaster management, including:

> five overlapping spheres of action: (i) integrate pets into disaster management practice and policy; (ii) create pet-friendly environments and related policies; (iii) engage community action in disaster management planning; (iv) develop personal skills by engaging owners in capacity building and (v) reorient health and emergency services toward a more-than-human approach.

Maybe the answer is developing a 'One Rescue' paradigm that recognises the benefits and opportunities for public safety when animals are integrated into disaster planning by human-centric authorities, such as having fire and rescue services coordinate animal disaster response to ensure an integrated approach, avoiding duplication of effort, and levering capacity from trained and equipped animal disaster responders, effectively acting as force multipliers. This approach positions the protection of animals not as an after-thought in disasters, but a core function that will lead to better human and animal safety outcomes. This shift also would require those from the 'animal' side to step up and gain more credibility within the disaster management profession, through completion of emergency management training, qualifications, and credentials such as the Certified Emergency Manager (CEM®) to supplement animal welfare or veterinary backgrounds. Likewise, those in the human-focused 'disaster management side' need to better understand the importance and benefits of including animals in disaster arrangements, through professional development such as World Animal Protection's PrepVet course and FEMA Independent Study courses on companion animal and livestock emergency planning.

Conclusions

Millions of animals are disaster-impacted every year and this will continue to grow as humans make choices that increase the vulnerability of such animals to an expanding range of hazards, exacerbated through climate change, intensification of animal farming, urbanisation, weak animal-health infrastructure, and poor animal disaster management arrangements. As long as society fails to improve the status quo of animal disaster management, not only is animal welfare compromised, but the safety, well-being, and livelihoods of humans are too. To mitigate these impacts, a coordinated effort to better integrate animal and human disaster management systems, along with improved mechanisms for accountability at all levels, is required. Around eight million species globally are depending on humans to have the moral compass to step up and address these vulnerabilities, and such action cannot come soon enough.

References

Asia for Animals, 2021. Kabul zoo updates. https://www.asiaforanimals.com/kabul-zoo [accessed on 4 September 2021].

Auf der Heide E, 1989. *Disaster Response: Principles of Preparation and Coordination*. St Louis: C.V. Mosby Company. Available from: https://erikaufderheide.academia.edu/research#papers [accessed 12 September 2021].

Best A, 2021. The legal status of animals: A source of their disaster vulnerability. *Australian Journal of Emergency Management*, 36(3), pp. 63–68. DOI: 10.47389/36.3.63.

Dalla Villa P, Watson C, Prasarnphanich O, Huertas G and Dacre I, 2020. Integrating animal welfare into disaster management using an 'all-hazards' approach. *Revue Scientifique et Technique* (International Office of Epizootics), 39(2), pp. 599–613.

DefenseOne, 2021. No US military dogs were left behind in Afghanistan, DOD says. Available from: https://www.defenseone.com/threats/2021/08/no-us-military-dogs-were-left-behind-afghanistan-dod-says/184984/ [accessed on 4 September 2021].

Emergency Management Accreditation Program, 2019. The EMAP standard. Available from: https://emap.org/index.php/what-is-emap/the-emergency-management-standard [accessed on 8 August 2021].

Food and Agriculture Organization of the United Nations (FAO), 2011. *Good Emergency Management Practice: The Essentials*. 2nd edn. (Honhold N, Douglas I, Geering W, Shimshoni A & Lubroth J, eds). FAO Animal Production and Health Manual No. 11. Rome, Italy: FAO, 131 pp. Available from: http://www.fao.org/3/a-ba0137e.pdf [accessed on 14 August 2021].

Fritz Institute, 2006. Hurricane Katrina: perceptions of the affected. Available from: http://www.fritzinstitute.org/PDFs/findings/HurricaneKatrina_Perceptions.pdf [accessed 12 September 2021].

Glassey S, 2010. Recommendations to enhance companion animal emergency management in New Zealand. Wellington: Mercalli. Available from: https://animaldisastermanagement.blog/resources/ [accessed 12 September 2021].

Glassey S, 2018. Did Harvey learn from Katrina? Initial observations of the response to companion animals during Hurricane Harvey. *Animals*, 8(47), pp. 1–9. DOI: 10.3390/ani8040047.

Glassey S, 2019. *No Animal Left Behind: A Report on Animal Inclusive Emergency Management Law Reform*. Wellington: Animal Evac New Zealand.

Glassey S, 2020a. Animal welfare and disasters. *Oxford Encyclopedia of Crisis Analysis*, Oxford: Oxford University press. pp. 1–26. DOI: 10.1093/acrefore/9780190228637.013.1528

Glassey S, 2020b. Legal complexities of entry, rescue, seizure and disposal of disaster-affected companion animals in New Zealand. *Animals*, 10(9), pp. 1–12. DOI: 10.3390/ani10091583.

Glassey S, 2021. Do no harm: Challenging conversation about how we prepare and respond to animal disasters. *Australian Journal of Emergency Management*, 36(3), pp. 44–48. Available from: https://knowledge.aidr.org.au/resources/ajem-july-2021-do-no-harm-a-challenging-conversation-about-how-we-prepare-and-respond-to-animal-disasters/ [accessed 31 July 2021].

Glassey S and Anderson M, 2019. *Operation Nelson Fires: After Action Report*. Wellington, NZ. Available from: http://www.animalevac.nz/wp-content/uploads/2019/08/Animal-Evac-NZ-AAR-Nelson-Fires-2019-isbn-ready.pdf. [accessed 31 July 2021].

Glassey S and Thompson E, 2020. Disaster search markings need to include animals. *Australian Journal of Emergency Management*, 35(1), pp. 69–74.

Glassey S and Wilson T, 2011. Animal welfare impact following the 4 September 2010 Canterbury (Darfield) earthquake. *Australasian Journal of Disaster and Trauma Studies*, 2011(2), pp. 1–16. Available from: https://www.massey.ac.nz/~trauma/issues/previous.shtml [accessed 12 September 2021].

Glassey S, Rodrigues Ferrere M, and King M, 2020. Lessons lost: A comparative analysis of animal disaster response in New Zealand. *International Journal of Emergency Management*, 16(3), pp. 231–248. DOI: 10.1504/IJEM.2020.113943.

Green D, 2019. *Animals in Disasters*. 1st edn. Oxford: Butterworth-Heinemann.

Haddow GD, Bullock JA and Coppola DP, 2017. *Introduction to Emergency Management*. 6th edn. Oxford: Butterworth-Heinemann.

Heath SE, 1999. *Animal Management in Disasters*. St. Louis, Missouri: Mosby.

Heath SE, Kass PH, Beck AM and Glickman LT, 2001. Human and Pet-related risk factors for household evacuation failure during a natural disaster, *American Journal of Epidemiology*, 153(7), pp. 659–665.

Heath SE and Linnabary RD, 2015. Challenges of managing animals in disasters in the U.S. *Animals*, 5(2), pp. 173–192. DOI: 10.3390/ani5020173.

Hunt M, Al-Awadi H and Johnson M, 2008. Psychological sequelae of pet loss following Hurricane Katrina. *Anthrozoos*, 21(2), pp. 109–121.

Irvine L, 2009. *Filling the Ark: Animal Welfare in Disasters*. Philadelphia, PA: Temple University Press.

Kajiwara H, 2020. *Surviving with Companion Animals in Japan: Life after a Tsunami and Nuclear Disaster*. Cham, Switzerland: Springer Nature.

Kelman I, 2020. *Disaster by Choice: How Our Actions Turn Natural Hazards into Catastrophes*. Oxon, UK: Oxford University Press.

LEGS, 2014. *Livestock Emergency Guidelines and Standards*. 2nd edn. Rugby, UK: Practical Action Publishing.

LEGS, 2017. About LEGS. Available from: https://www.livestock-emergency.net/about-legs/ [accessed 4 September 2021].

Mellor DJ, 2017. Operational details of the five domains model and its key applications to the assessment and management of animal welfare. *Animals*, 7(8). p. 60. DOI: 10.3390/ani7080060.

Montgomery H, 2011. Rumours of child trafficking after natural disasters. *Journal of Children and Media*, 5(4), pp. 395–410.

Mora C, Tittensor DP, Adl S, Simpson AGB and Worm B, 2011. How many species are there on earth and in the ocean? *PLoS Biology*, 9(8), pp. 1–8.

New International Version, 2011. *Biblegateway.com*. Available from: https://www.biblegateway.com/passage/?search=Genesis 7&version=NIV. [accessed 5 August 2021].

Potts A and Gadenne D, 2014. *Animals in Emergencies: Learning from the Christchurch Earthquakes*. Christchurch: Canterbury University Press.

Sawyer J and Huertas G, 2018. *Animal Management and Welfare in Natural Disasters*. 1st edn. New York: Routledge.

Spain CV, Green RC, Davis L, Miller GS and Britt S, 2017. The national capabilities for animal response in emergencies (NCARE) study: An assessment of US States and Counties. *Journal of Homeland Security and Emergency Management*, 14(3). p. 20170014. DOI: 10.1515/jhsem-2017-0014.

State of Texas, 2007. Texas health & safety code. Available from: https://statutes.capitol.texas.gov/docs/hs/htm/hs.821.htm [accessed 1 September 2021].

Strain M, 2018. Co-habitated human/pet shelter toolkit, 2018. Available from: https://animaldisastermanagement.files.wordpress.com/2021/09/strain-2018-co-habitated-humanpet-shelter-tookit.pdf [accessed 4 September 2021].

Taylor M, Burns P, Eustace G and Lynch E, 2015. The preparedness and evacuation behaviour of pet owners in emergencies and natural disasters. *Australian Journal of Emergency Management*, 30(2), pp. 18–23.

Travers C, Rock M and Degeling C, 2021. Responsibility-sharing for pets in disasters: lessons for one health promotion arising from disaster management challenges. *Health Promotion International*, 2021, pp. 1–12. DOI: 10.1093/heapro/daab078.

Trigg J, Taylor M, Mills J and Pearson B, 2021. Examining national planning principles for animals in Australian disaster response. *Australian Journal of Emergency Management*, 36(3), pp. 49–56. DOI: 10.47389.36.3.49

United Nations Office for Disaster Risk Reduction, 2020a. Funding. Available from: https://www.undrr.org/about-undrr/funding [accessed on 3 February 2021].

United Nations Office for Disaster Risk Reduction, 2020b. Terminology: Build back better. Available from: https://www.undrr.org/terminology/build-back-better [accessed on 3 April 2021].

Vieira ADP and Anthony R, 2021. Reimagining human responsibility towards animals for disaster management in the Anthropocene. In Bovenkerk B and Keulartz J, eds. *Animals in Our Midst The Challenges of Co-existing with Animals in the Anthropocene*. Cham, Switzerland: Springer Nature, pp. 223–254. Available from: https://link.springer.com/book/10.1007%2F978-3-030-63523-7 [accessed 12 September 2021].

Vroegindewey G and Kertis K, 2021. Veterinary behavioural health issues associated with disaster response. *Australian Journal of Emergency Management*, 36(3), pp. 78–84. DOI: 10.47389.36.3.78.

Washington Post, 2021. A Royal Marine rescued animals from Afghanistan in a mission dubbed 'Operation Ark.' Available from: https://www.washingtonpost.com/nation/2021/08/30/pen-farthing-afghanistan-animal-rescue/ [accessed on 4 September 2021].

World Animal Protection, 2020. Methodology: Animal protection index. Available from: https://api.worldanimalprotection.org/methodology [accessed on 4 April 2021].

World Wildlife Fund, 2020. Australia's 2019–2020 Bushfires: The wildlife toll (interim report). Available from: https://www.wwf.org.au/news/news/2020/3-billion-animals-impacted-by-australia-bushfire-crisis#gs.wz3va5 [accessed 15 August 2021].

Zee J, 2021. Animal transport disasters: Queen hind sheep rescue in Romania. In Global Animal Disaster Management Conference. Available from: https://gadmc.org/speakers/profile/?smid=410 [accessed on 15 August 2021].

PART VI

Animal ethics and law

26
ANIMAL ETHICS

Cheryl Abbate

Introduction

If one thing is true, it's that humans frequently exploit animals. We hurt and kill billions of animals for food each year. Each year, millions of animals are harmed in research facilities around the world. Vast numbers of animals are exploited for human entertainment, including those used in aquariums, circuses, rodeos, zoos, dog fighting, horse racing, dog racing, bullfighting, and dog sledding. Millions of animals are killed by hunters each year, and millions more are bred and slaughtered in fur farms. Thousands of majestic animals, including lions, leopards, rhinos, elephants, and Cape buffalos, are killed each year in the name of trophy hunting (Humane Society of the United States, 2016). Millions of "pets" are confined to human homes. Many animals are forced to "work" for humans, including police dogs, guide dogs, draft horses, and herding, guarding, and hunting animals. And, in many cases, our exploitation of animals causes them tremendous suffering. In many cases, our exploitation of an animal takes away from them their one and only valuable life. And, in many cases, our exploitation of animals yields only trivial benefits, such as entertainment or taste. Where did we go so wrong as a species? How is it possible that the majority of humans, most of whom don't seem to be callous or heartless, regularly participate in such wholly *unnecessary* practices that cause frequent and widespread suffering to the most innocent, defenceless, and vulnerable of creatures?

As with any egregious and normalised injustice, our species-level moral failing can be attributed to our faulty belief systems. For instance, sometimes we deny obvious facts about animal experiences and animal suffering, such as the fact that animals are harmed on intensive farms. Often, these faulty beliefs are attributed to what Anthony Appiah (2010) refers to as *strategic ignorance* (also known as *cognitive dissonance*), which is the intentional avoidance of truths that would require us to acknowledge the evils in which we are complicit.

But often our mistreatment of animals stems from a different kind of rational failure: fallacious reasoning that involves a jump from a descriptive claim about what humans have historically done and currently do to animals to a normative conclusion about what humans are morally permitted to do to animals. As Appiah (2010) points out, when it comes to morally horrendous human practices that were once normalised in societies, defenders of the practices often invoked tradition. For instance, those who defended slavery tried justifying the practice on the grounds that "We've always had slaves". And we commonly see a similar failure of reasoning when it

DOI: 10.4324/9781003182351-32

comes to widespread human practices that cause frequent and widespread suffering to animals, such as when people attempt to defend animal farming by saying, "We've always eaten meat".

For the most part, we understand that treating *humans* unjustly in the past doesn't justify treating *humans* unjustly in the present. Yet, when it comes to practices that involve the mistreatment of animals, there is widespread failure to acknowledge that past practices don't justify present practices. It's thus especially important that, in our discussions about the ethical treatment of animals, we acknowledge the distinction between *descriptive "morality"* and *normative morality*, as humans often mistakenly assume that descriptive claims about (common) human practices that involve animals entail normative claims about the moral treatment of animals.

Descriptive claims are attempts (which can be successful or unsuccessful) to describe how the world is. In other words, descriptive claims are assertions (which can be true or false) about how the world *is*. We make plenty of descriptive claims about animals. We point out that cougars are fast. We say that bears have an incredible sense of smell. Moreover, we make descriptive claims about (common) human practices that involve animals, such as:

- Humans have always eaten animals;
- Torturing cats is illegal;
- Most people don't abuse puppies.

Normative claims (also referred to as prescriptive or evaluative claims) are assertions about how the world *ought to be*. Normative *morality* is thus concerned about (1) what we are *morally* obligated to do, (2) what we are *morally* forbidden from doing, and (3) what we are *morally* permitted to do. Examples of normative moral claims about animals include:

- Eating animals is permissible;
- It's morally wrong to torture cats;
- Abusing puppies is immoral.

In discussions about moral philosophy, ethicists are primarily concerned with arriving at *true normative claims about morality*. Often, this involves what Bernard Rollin (2006) refers to as a "rational criticism" of descriptive morality. And we do this by providing sound moral arguments, which contain at least one descriptive claim and one moral claim. After all, a fundamental tenet of morality is that you cannot derive an "ought" from an "is", which means you cannot derive a normative claim from a descriptive claim.

Often, our false moral beliefs about animals stem from the "is-ought fallacy", which assumes that because things are a certain way, it's permissible for them to remain that way. For instance, consider the following arguments, which are frequently used in attempts to defend meat-eating:

P1) Humans have always eaten animals.
C) Therefore, it's permissible for humans to eat animals.

P1) It's legal for humans to eat animals.
C) Therefore, it's permissible for humans to eat animals.

To see the flaws in these arguments, we just need to identify and evaluate the moral principles that are implicitly assumed. In the first argument, the underlying moral assumption is that *it's always permissible for humans to do what humans have done in the past*. In the second argument, the underlying moral assumption is that *it's always permissible for humans to do what is legal*. Both principles are clearly false, as evident by the many obvious counter-examples. Consider, for instance, that humans have always oppressed other humans. It doesn't follow from this that oppressing humans is justified. And consider, for instance, that it was once legal to own slaves. Even so, it was *never* morally permissible to own slaves.

So, how do we arrive at *true* normative judgments about the *moral* treatment of non-human animals? The first thing we need to do is set aside descriptive claims about what it is legal to do to animals, what humans think about animals, and what humans have done to animals in the past. Then, we need to critically analyse questions related to moral status, personal morality, and professional ethics.

Moral status

Those of you reading this chapter are full moral agents, and you thus have moral obligations. This means that there are certain things that you, *in virtue of being a moral agent*, are morally obligated to do and there are other things that you, *in virtue of being a moral agent*, are morally forbidden to do. One fundamental question in ethics is this: *to whom do our moral obligations apply*? This is to ask: who are the subjects of our *moral* obligations?

We certainly have moral obligations to other moral agents. For instance, I'm certainly morally obligated not to murder my fellow moral agents. But moral agents aren't the *only* subjects of our moral obligations. Consider, for instance, that insofar as there is nothing young infants are obligated to do, infants aren't moral agents. We wouldn't, for instance, say that an infant violates her "moral obligation" or acts "wrongly" if she were to bite her mother's hand, even if this hurts the mother. But even though infants aren't moral agents, we still have moral obligations to infants. For instance, I'm certainly morally obligated not to go around killing infants for fun.

Note that *one* reason why it would be morally impermissible to murder an infant is that this would surely harm the infant's parents. But, in addition, and more importantly, it would be morally impermissible to murder an infant because this would harm, and moreover wrong, the infant *herself*. This is to say that there are obligations I owe *directly* to infants themselves. If you agree that I have moral obligations that are owed directly to infants, this means that you agree that there are creatures other than moral agents who have *moral status*. As Marry Warren puts it:

> To have moral status is to be morally considerable, or to have moral standing. It is to be an entity toward which moral agents have, or can have, moral obligations. If an entity has moral status, then we may not treat it in just any way we please.
>
> *(Warren, 1997, p. 3)*

One of the most important goals of moral discourse is determining what kind of creatures have moral status. And to determine which creatures have moral status, we must ask: what is the criterion of moral status?

The most popular approaches to moral status assume that only *individuals*, as opposed to "wholes" (e.g., species and ecosystems), have moral status. Yet, there are competing accounts of moral status that disagree with one another about which kinds of individuals have moral status. One popular approach to moral status is Strong (Egalitarian) Anthropocentrism, which is the view that all and only humans have *equal* moral status. A virtue of this approach is that it acknowledges that *all humans are equal*. In doing so, it condemns sexism, racism, and ableism, insofar as these "isms" violate the moral principle of human equality.

While Strong (Egalitarian) Anthropocentrism has its virtues, it's fundamentally flawed insofar as it fails to condemn the "ism" that applies to our treatment of other animals: *speciesism*, a term coined by Richard Ryder (2011). Like sexists and racists, speciesists violate the basic moral principle of equality. Just like sexists believe that men have greater moral worth than women because of their sex, and just like white people who are racist against black people believe that white people have greater moral worth than black people because of their race, speciesists believe that humans have greater moral worth than other animals because of their species membership. But, as many animal ethicists, such as Tom Regan (1983) and Peter Singer (1975) compellingly argue, species membership, like race and sex, is an arbitrary characteristic to appeal to in ethics. Why, for instance, would your pain matter more than the pain of my cat simply because you are, but she is not, a member of the species *Homo sapiens*? Pain is pain, and, special considerations aside, it is equally bad if it occurs in a man, woman, white person, black person, abled human, disabled human, cat, and so forth.

In an attempt to avoid the charge of speciesism, those who endorse Strong Anthropocentrism attempt to identify a morally relevant difference between humans and non-human animals that would explain why humans have, but animals lack, moral status. For instance, Immanuel Kant (1785) endorses a Rationality or Personhood Approach, which is the view that all and only rational creatures have moral status. Those who endorse this kind of approach, including Kant, often assume that all and only humans have moral status because they (wrongly) assume that all and only humans are rational agents. Yet, an obvious objection is that not all humans are rational agents, such as very young, severely mentally disabled, and very elderly humans. But surely these humans have moral status, insofar as it would be wrong to torture them for fun.

Because the Rationality Approach has such counterintuitive implications for human infants, some strong anthropocentrists endorse the Potential for Rationality Approach, which is the view that all and only creatures who have the *potential* for rationality or, as Don Marquis (1989) puts it, a "future like ours", have moral status. This approach can accommodate the intuition that young humans have moral status, insofar as they have the potential for rationality or a "future like ours". Yet, one serious problem for this approach is that it can't accommodate the intuition that severely mentally disabled humans have moral status. After all, most severely mentally disabled don't have the potential for rationality or a "future like ours", insofar as their mental disabilities are usually *irreversible*. But surely all severely mentally disabled humans have moral status. After all, it certainly would be wrong to torture them just for fun.

In recognising that very young humans have moral status, despite that they lack rationality, and that severely mentally disabled humans have moral status, despite that they lack the potential for rationality, most ethicists endorse Sentio-centrism, which is the view that all and only sentient creatures, that is, creatures who have the capacity to suffer, have moral status. This view accommodates the intuition that human infants and severely disabled humans have moral status, insofar as they are sentient. But it implies that many other animals – sentient non-humans – have moral status, too.

Some might argue that Sentio-centrism is absurd because it allegedly entails that humans and mice have equal moral worth. Yet, not all approaches to moral status are egalitarian. That is, one might endorse Sentio-centrism, while still maintaining that humans have greater moral worth than mice. David DeGrazia (2008), for instance, argues that moral status isn't an all-or-nothing notion and that it rather comes in degrees. *Weak* Anthropocentrism, for instance, admits that creatures other than humans have moral status, while insisting that humans are more morally considerable than non-humans with moral status.

Another line of response is to insist that moral status is an egalitarian notion, which means that a *creature* either has moral status or it doesn't, while pointing out that the *lives* of humans usually have more value than the lives of most non-human animals, insofar as human lives usually have more opportunities for satisfaction than do the lives of other animals. This was essentially the position of Regan (1983), who argues that while (most) non-human animals and humans *themselves* have equal moral value, their *lives* aren't always equally valuable. On his view, the most valuable lives are those with the greatest opportunities for satisfaction. This is why, for instance, if we find ourselves faced with the choice to save the life of either a child or the life of a 90-year-old person, we ought to save the life of the child. Although children and the elderly have equal moral status, special considerations aside, a child's life is more valuable than the life of a 90-year-old person simply because a child's life has more opportunities for satisfaction. Regan then goes on to argue that the lives of rational humans usually have more opportunities for pleasure than do the lives of animals, insofar as rational humans are, while non-humans aren't, able to enjoy the satisfaction of bringing impartial reasons to bear on their moral decision-making. This then would explain why, even if humans and mice have equal moral status, it still would be, special considerations aside, obligatory to save a (rational) human over a mouse if we find ourselves faced with the choice to save only one.

Personal morality and ethical theory

Once we determine who has moral status, we must then ask: how should we treat those with moral status? What do we owe to them? Should we treat them as creatures who can never be harmed to benefit others? Or should we try to maximise the overall good of all creatures with moral status, allowing some to be harmed in the name of social utility?

We'll now consider what morality demands of us by virtue of us being moral agents. We might call this *personal morality*. This is not to say that morality is subjective or dependent on the personal beliefs of individual moral agents. Rather, it's to say that there are certain things that morality demands of us *by virtue of our being persons or moral agents*. And to answer the question of what morality demands of us in virtue of our being persons or moral agents, we must turn to a discussion about ethical theory. After all, ethical theories provide us with fundamental moral principles that we can use to consistently address a wide array of moral problems we encounter as moral agents.

Utilitarianism

Utilitarianism is a popular ethical theory that is a kind of *teleological* or *consequentialist* theory. *Teleological* (or consequentialist) moral theories evaluate the rightness of an act solely on the basis of the act's consequences or the act's "ends". In its most basic form, utilitarianism endorses the principle of equality, insofar as it is concerned *equally* about *everyone* who can be benefited or harmed by an act, as opposed to ethical egoism, which is concerned only about how an act impacts *the agent performing the act*.

The most basic version of utilitarianism is referred to as *classical utilitarianism*. What we morally ought to do, according to classical utilitarianism, is maximise net *happiness*, which refers to the amount of pleasure minus the amount of pain (and, on the flipside, we ought to minimise net suffering). On this view, the ultimate principle of morality is the Principle of Utility: *Perform those acts that maximise net happiness for all those affected.*

Because animals are sentient, our actions have the potential to harm or benefit them. Classical utilitarianism thus has much to say about our interactions with animals. In particular, it calls for a radical transformation, and, in some cases, a complete eradication, of common human practices that involve the exploitation of animals. Consider, for instance, what classical utilitarianism has to say about intensive farming. According to a classical utilitarian, if we want to determine whether or not it is morally justified to raise animals on intensive farms for food, we would have to determine (1) how many sentient creatures (humans and animals) will experience pain from this (and how much pain); (2) how many sentient creatures (humans and animals will experience pleasure from this (and how much pleasure); and then we must ask (3) do the benefits (total pleasures produced) outweigh the costs (total pain caused)? And even if the total pleasures produced do outweigh the total pain caused, we must ask: is there an alternative way to get comparable pleasures without causing so much pain? So in determining if raising animals on intensive farms is justified, we must consider:

(1) Who is harmed by intensive farming?
    - Billions of animals on these farms, who experience suffering of various forms throughout their lives (Welfare Quality®, 2019);
    - Humans and animals who are harmed indirectly by the ecological harm that is a side effect of intensive farming (think climate refugees and future generations) (Letcher, 2021);
    - Humans and animals who live in neighbouring communities who suffer from the pollution from these farms (Carter, 2016; Kresge and Strochlic, 2007; Wallinga, 2004);
    - Workers on these farms and slaughterhouses, who may be physically injured on the job and may suffer psychological trauma (Dillard, 2008; Woorall, 2004);
    - Humans whose health suffers from the over-consumption of animal-based foods (Garrett, 2007).
(2) Who is benefited by intensive farming?
    - Humans who get pleasure and sustenance from eating intensively farmed products;
    - The corporations that own intensive farms and slaughterhouses and their shareholders;
    - The workers who receive a source of income from working on these farms and slaughterhouses.

And, as mentioned, a utilitarian would also consider:

(3) Is there a way to produce some of the benefits of intensive farming without causing the listed harms?

Many argue that, when it comes to the pleasures of eating animal meat, humans can derive comparable gustatory pleasures by eating plant-based meats and that these plant-based meat alternatives may be even better for consumers, because they are often healthier. Jeremy Garrett (2007), for instance, argues that even if humans enjoy great gustatory pleasure from eating animals, the long-term benefits of a healthier, longer life, which Garett claims are consequences of plant-based lifestyles, surely outweigh this pleasure.

It should also be noted that even though intensive farms employ many people, if intensive farming becomes obsolete, new jobs will be created elsewhere, and these jobs may be even better than slaughterhouse work, since they will be safer and won't require the psychological trauma that can be a side effect of being an active participant in causing trauma and the death of others (Dillard, 2008; MacNair, 2002). Relatedly, major meat corporations, such as Tyson Foods, are planning to soon replace line-workers with robots (Wallace, 2020), so if intensive farms become obsolete, not many jobs will be "lost", given that robots will inevitably replace human workers in meat-packing facilities.

A classical utilitarian would use this kind of ethical assessment for all other moral issues involving animals, including the use of animals in biomedical research, the use of animals in entertainment, the hunting of animals, and so forth. While classical utilitarians usually condemn the harming of animals for trivial purposes (such as the use of animals for entertainment, food, or fashion simply because the pain caused to animals isn't outweighed by the pleasures humans receive), they may also say that we are morally required to harm animals in certain circumstances, such as when harming animals is necessary (or the most efficient way) to prevent significant harm to others. For instance, if, in some situation, we can save many human lives only by performing research on some smaller number of animals (in the least painful way possible), this research will be justified – although whether animal research is effective and efficient at achieving such public health benefits is very debatable (Chapter 13). For classical utilitarianism, the ends justify the means, which entails that if we can prevent significant harm from occurring only by doing something less harmful, we are justified in doing something less harmful (even though it's harmful).

Rights theory

One important objection to classical utilitarianism is that it can't account for the "rights" of the individual. After all, there are many cases in which killing or seriously harming someone is the most effective way to maximise net happiness, but surely morality rarely requires us to kill a person just to promote the social good. Consider, for instance, the infamous organ harvesting objection. As the objection goes, because classical utilitarianism is committed to the view that we always ought to maximise net happiness, it entails that if we could kill one person and harvest their organs to save five people, we ought to do so. This strikes many people as counter-intuitive, given the widely shared intuition that people have moral rights over their bodies, lives, and freedom, which can't be violated in the name of social utility. And this is what explains the wrongness of killing-for-organs.

Underlying the intuition about inviolable human rights is the fundamental assumption that humans have *inherent value*, which means that they are valuable in-and-of-themselves, regardless of their usefulness to others and regardless of whether others recognise their value. And because humans have inherent value, they ought to be treated as if they have this value. So, on the rights view, the fundamental moral principle is the Respect Principle, which says that we ought to treat creatures with inherent value in ways that respect their inherent value. This essentially means that we ought not to treat creatures with inherent value as if they lacked inherent value, that is, as if they are mere resources or mere means to maximising the social good.

Rights theory is a *deontological* approach to morality. According to *deontological* theories, which are often referred to as *duty-based* approaches to ethics, when we evaluate the morality of an act, we should focus on the *motives* behind the act instead of focusing exclusively on the act's consequences. For instance, in the rights view, if we treat creatures with inherent value with respect, we've fulfilled our moral obligation, and this is true even if we fail to maximise net utility. Moreover, when we treat creatures with inherent value disrespectfully, we violate morality's demands, and this is true even if we maximise net utility.

Some rights theorists think that rationality grounds inherent value. Yet, it's commonly noted that not all humans are rational, but, still, they have inherent value. The severely mentally disabled, for instance, aren't rational, but, still, we shouldn't reduce them to the status of mere resources, such as by killing them to harvest their organs. According to Regan, humans, including the severely mentally disabled, have inherent value, and they have this value not because they are rational, but because they are subjects of a life (SOL). As he explains:

> [I]ndividuals are subjects-of-a-life if they have beliefs and desires; perception, memory, and a sense of the future, including their own future; an emotional life together with feelings of pleasure and pain; preference-and welfare-interests; the ability to initiate

> action in pursuit of their desires and goals; a psychophysical identity over time; and an individual welfare in the sense that their experiential life fares well or ill for them.
> *(Regan, 1983, p. 243)*

And since humans have inherent value *because* they are SOLs, we can conclude that being a SOL is what grounds inherent value. After all, it would be arbitrary to say that humans, such as the severely mentally disabled, have inherent value *because* they are SOLs, while denying that other creatures who are SOLs have inherent value. And since many non-human animals are SOLs, many non-humans, too, have inherent value.[1] Animal rights theory thus extends the logic of human rights to our treatment of animals, insisting that any plausible human rights framework logically entails that non-human animals have basic inviolable rights, too.

Most animal rights theorists, such as Corey Wrenn (2012, 2016), take the rights view to entail an *abolitionist approach*, which is the view that justice demands that we abolish any and all practices that reduce animals to the status of mere resources. Such an approach is at odds with the traditional (utilitarian) welfarist approach to animal protection, which, according to Robert Garner (1993, p. 336), "holds that human exploitation of animals is justified provided that any suffering inflicted is necessary". On this view, it's permissible to use animals for human gain, so long as we don't cause the animals to suffer unnecessarily. Someone who endorses this approach to animal welfarism, then, would advise that while we need not abolish animal-abusing industries, we should still clean them up or reform them by making the lives of the exploited animals as painless as possible.

"Animal welfarism" and "animal rights" are often presented as opposing positions, but this isn't theoretically correct. After all, animal welfarism is the general view that we ought to improve the welfare of animals, and it's perfectly consistent to believe both that (1) animals should have good welfare; and (2) the rights of animals ought to be respected. So, we should consider that there are at least two different subset of animal welfarism: (1) *utilitarian welfarism* and (1) *abolitionist welfarism*.

The foundational belief of abolitionist welfarism is that exploiting creatures with inherent value is always wrong, regardless of how pain-free this exploitation is and regardless of how much others might benefit from the exploitation. As Regan (1985, p. 13) insisted, "you don't change unjust institutions by tidying them up". Animal rights theorists thus emphasise that the fundamental problem with animal exploitation is not *how* we use animals (human and non-human), but rather *that* we use them as our resources to begin with. And this approach is quite intuitive when we consider the implications for human slavery. After all, the morally appropriate response to human slavery isn't just to improve the lives of slaves on plantations. Rather, the only ethically appropriate response to slavery is the complete eradication of it, and this is true even if humans would enjoy great, irreplaceable benefits from slavery. Using this logic, animal rights theorists like Regan and Wrenn argue that the morally right response to animal farming isn't just to improve the lives of farmed animals by giving them more space, better food, or "nicer" deaths. Rather, the only ethically appropriate response to animal farming is the complete eradication of it. As Wrenn (2016) puts it, we need *vegan abolition*, not "kinder" oppression.

The rights approach is said to have radical implications for *all* industries using animals – not just the animal farming industry (Regan, 1983; Wrenn, 2016). For instance, both Regan and Wrenn agree that the animal entertainment industry, the animal clothing industry, the hunting industry, and even the animal research industry should be abolished, and not merely reformed. This means that even if we could produce life-saving therapies by performing research on animals, doing so is still immoral. As Regan (1983, p. 347) insists, "justice *must* be done, though the heavens fall".

Contractualism

Contractualism is the view that justice, in some sense, is based upon a social agreement or contract (actual or hypothetical). Arguably, the most compelling and influential account of contractualism is the one advanced by John Rawls (1971), who argues that the principles of justice we should follow and accept are those principles that free, equal, and rational people would agree to under circumstances that are fair. In describing what such fair circumstances would look like, Rawls asks us to imagine a *hypothetical* "original position", in which a group of *hypothetical* people are behind what he calls a "veil of ignorance". Behind this veil, the hypothetical people don't know anything about themselves; they don't know their race, sex, sexual orientation, intellectual abilities, strength, place in society, social status, income, profession, fortune, religion, their conception of the good, and so forth. According to Rawls, the choices that rational people would make in this hypothetical situation determine the principles of justice. He goes on to argue that people behind the veil would agree to two principles of justice: (1) the Equal Liberty Principle, which says that every person should have an "equal right to the most extensive basic liberty compatible with a similar liberty for others", and (2) the Difference Principle, which says that "social and economic inequalities are to be arranged so that they are ... reasonably expected to be to everyone's advantage". And, as Rawls emphasises, the Difference Principle entails that social inequalities are justified *only if* they advantage the least advantaged.

To say that the hypothetical people in the original position are rational is to say that "they are concerned to put themselves in as advantageous a position as possible after the lifting of the veil of ignorance" (Rowlands, 1997, p. 237). Keeping this in mind, it seems obvious that the hypothetical people in the original position wouldn't agree to principles that permit or perpetuate racial injustice, deny various rights to women (such as the right to vote), and so forth. After all, the hypothetical people in the original position don't know their race or their sex, so it would be irrational for them to agree to a principle that promotes racism or sexism when they may very well turn out to suffer the effects of such principles. But while the hypothetical people in the original position don't know their race or sex, there seems to be one thing about themselves they do know: they are fully rational. After all, they wouldn't be able to reflect upon and debate about principles of justice if they were, for instance, a baby, a severely mentally disabled human, or a non-human animal.

It seems, then, that a rational person under a veil of ignorance wouldn't necessarily agree to principles of justice that afford serious moral protection or rights to non-human animals. Consider, for instance, the issue of animal exploitation. Hypothetical persons in the original position (1) know that they are not non-human animals; and (2) think that they very well may end up being someone who stands to benefit from animal use. Given this, it seems that a rational person under a veil of ignorance wouldn't agree to a principle that grants animals the right not to be used as resources for human gain.

So, in Rawls' account of justice, "contractors" seem to be the only subjects of rights, duties, and justice, and thus his account of justice seems to exclude non-human animals from the sphere of justice. Rawls himself agrees that non-human animals won't enjoy entitlements of justice in his view, but he doesn't see this as a problem, given that humans will still owe duties of compassion to other animals. But perhaps, contrary to Rawls' own interpretation of contractualism, Rawls' theory of justice actually does extend protections of justice to other animals. This is the position of Mark Rowlands (1997), who begins his animal-friendly interpretation of Rawls' theory of justice by highlighting and appealing to Rawls' *intuitive equality argument*, which Rawls uses in defence of his two principles of justice.

A fundamental assumption of the intuitive equality argument is that *it's unfair for one's fate to be determined by undeserved inequalities*, and this entails that it's unjust that some people are benefited by undeserved inequalities. So, if someone has some property P that they haven't done anything to merit (i.e., P is undeserved), then P is morally arbitrary and possessors of P are neither entitled to P nor entitled to the benefits of P. For instance, because most people are born into a particular social and economic group and because people are born with athletic skills, good looks, and high intelligence, people haven't done anything to merit their socioeconomic class group and natural talents and capacities, and thus they aren't entitled to the benefits of these properties. So, this is why the hypothetical people in the original position must choose the principles of justice behind a veil of ignorance. After all, if I know that I have property P, I, being self-interested, will agree to principles of justice that benefit people with P, even if they harm people who lack P, and this would be unfair.

How does any of this entail an animal-friendly approach to contractualism? Rowlands (1997, p. 242) compellingly argues that rationality, just like socioeconomic class and natural talents and gifts, is an undeserved, and thus arbitrary, property, given that "[a] person plays no role in deciding whether or not she is going to be rational; she either is or she is not. The decision is not hers, but nature's". Rowlands (1997, p. 243) thus argues that, according to Rawls' intuitive equality argument, "knowledge that one is a rational agent should be bracketed off in the original position". And if hypothetical persons in the original position don't know if they will be rational agents, then they will agree to principles that take this into account. Rowlands thus concludes that Rawls' theory of justice doesn't entail that non-rational creatures, including non-human animals, will be excluded from the sphere of justice. Indeed, if the hypothetical people in the original position behave rationally, they'll grant sentient non-rational creatures the same justice-based protections they grant to rational creatures.

If Rowlands is right, what does this imply for animals? Well, it seems that extending the Equal Liberty Principle to non-human animals would entail that human liberty should be limited by the liberty of other animals. Our freedom to eat, for instance, will end where the basic rights of animals begin. Moreover, our freedom to perform biomedical research will end where the basic rights of animals begin. Presumably, an animal-friendly interpretation of Rawls would have abolitionist commitments. And, perhaps more interestingly, an animal-friendly interpretation of Rawls would often, if not always, require that we prioritise animals over humans in moral decision-making about distributive justice. After all, given the special vulnerabilities of animals, they are arguably the least advantaged. This means that, according to the Difference Principle, it would (usually) be justified to devote extra resources to protecting other animals and promoting their basic interests. So much for Anthropocentrism!

Virtue ethics

Utilitarianism, rights theory, and contractualism attempt to provide us with abstract moral principles that we can use to consistently navigate any conceivable moral issue. A compelling alternative to what we might call "rule morality" is virtue ethics, which advocates a case-by-case, or "contextual", approach to morality that is fundamentally concerned with these kinds of questions: what kind of person should I be? What are good character traits and how do I cultivate them? What does it mean to act viciously, and how can I avoid being vicious?

According to virtue ethicists, our moral goal is to be virtuous. Virtuous people possess a wide array of virtues, which are character traits that are good to have that are manifested in habitual action. Common examples include traits like compassion, justice, empathy, sympathy, temperance, integrity, courage, and practical wisdom. So, if we want to know what to do when we find ourselves confronted with a moral dilemma, virtue ethicists would simply advise us to do what the virtuous person would do. Moreover, as Rowlands (2012) notes, according to virtue ethics, a morally good person is one who not only *acts* (stably) as a virtuous person would, but a morally good person is one who *feels* (stably) in the way a virtuous agent would.

Virtue ethics thus "directs us to think about the rights and wrongs of our treatment of non-human animals in terms of virtues and vices rather than in terms of consequences, or rights and duties" (Hursthouse, 2011, 119). So, consider how a virtue ethicist would approach the issue of meat-eating. According to Hurtshouse, the virtue ethicist will first ask: is vegetarianism a virtuous practice? Is it something a virtuous person would embrace? In answering this question, we must first acknowledge that when we eat meat, we are usually a party to a large amount of avoidable animal suffering. And, according to Hursthouse, compassion requires us to feel sorry for the animals on the farms and to regret the harms they endure, honesty requires us to admit that we (at least those of us with access to plant-based alternatives) don't need to eat meat, and temperance requires that we not pursue pleasure when doing so requires that we act viciously. Relatedly, as Hursthouse points out, it would be callous to shrug off the frequent and widespread suffering on intensive farms. And because the practice of intensive farming is arguably cruel (Anomaly, 2015; Vining, 2008); when we eat intensively farmed animals, we are party to cruelty. Hursthouse thus concludes that it would normally be greedy and self-indulgent to pursue meat-eating pleasures when we know that the animals we eat endure frequent and widespread suffering on animal farms.

On the other hand, virtue ethicists acknowledge that there are unusual circumstances in which humans have no other choice but to eat other animals if they are to survive. Consider, for instance, people who live in the Horn of Africa. Many, if not most, cannot grow sufficient amounts of plant-based nutrients. These people, then, need to kill and eat animals to survive. And, as Hursthouse reasonably claims, people who kill and eat animals under these conditions aren't callous or deficient in compassion.

When it comes to animal research, a virtue ethicist will first point out that there are different kinds of animal research, so there isn't a straightforward answer to the question of "is animal research virtuous?" Consider, first, cosmetics testing. As Hursthouse compellingly argues, testing cosmetics on animals is usually cruel, for the same reasons intensive farming is: suffering is inflicted on a being for an unjustifiable reason. When it comes to cosmetic testing and the purchasing of cosmetics that were tested on animals, vices of vanity and self-indulgence are also relevant. But on the other hand, animal research, theoretically, could be morally acceptable on a virtue ethics framework, when the following conditions obtain: (1) there is good reason to think that the research will save many human lives; (2) performing this research *on animals* is necessary (or the most efficient way) to save human lives; (3) the suffering of the animals involved is minimised as much as possible and their welfare is positively promoted as much as possible (for instance, they are well-fed, given opportunities to play and exercise, afforded opportunities to bond with conspecifics, and so forth); and (4) the animals used are given some form of restitution after the research ends (if the experiments don't *require* their deaths).[2] As in the case of eating animals in situations involving real necessity, *if*, theoretically speaking, we could save thousands of human lives only by performing research on a handful of animals, it doesn't seem "callous" or "cruel" to perform this research, especially if we do all that we can to give the animals involved the best possible lives. Yet, *practically speaking*, few, if any, instances of animal research will be justified on a virtue ethics framework, given that animal research usually is a highly inefficient, unreliable, and arguably counterproductive methodology for attempting to advance human health care (Engel, 2012; Knight, 2022).

Conclusions

Plausible approaches to morality accept that sentience is sufficient for moral status and that we, at the very least, ought not to seriously harm animals, unless we have a weighty moral reason for doing so. Yet, as discussed, we often seriously harm animals just for pleasure or entertainment, forgoing ways of pleasuring and entertaining ourselves that don't involve the harming of other creatures. Moreover, when we seriously harm animals, we often end up harming *humans*. It thus is unsurprising that the four most compelling approaches to morality (contractualism, rights theory, utilitarianism, and virtue ethics), call upon us, as a species, to radically transform the way we treat other animals.

Notes

1 While it's unclear whether some animals, such as insects, are SOLs, we can be confident that birds and mammals over the age of one are SOLs, according to Regan.

2 Although animals are usually killed after experiments end, it's not clear whether they usually *need* to be killed or whether they are killed because disposing of animal carcasses is easier than rehoming them.

References

Anomaly J, 2015. What's wrong with factory farming? *Public Health Ethics*, 8(3), pp. 246–254.

Appiah A, 2010. What will future generations condemn us for? *Washington Post*. Available at: https://www.washingtonpost.com/wp-dyn/content/article/2010/09/24/AR2010092404113.html (Accessed 29 June 2021).

Carter C, 2016. Vegan soul: Moving beyond (animal) meat in black communities. In Donaldson B and Carter C, eds. *The Future of Meat Without Animals*. London: Rowman & Littlefield, pp. 217–228.

DeGrazia D, 2008. Moral status as a matter of degree? *The Southern Journal of Philosophy*, XLVI, pp. 181–198.

Dillard J, 2008. A slaughterhouse nightmare: Psychological harm suffered by slaughterhouse employees and the possibility of redress through legal reform. *Georgetown Journal on Poverty Law & Policy*, 15(2), pp. 391–408.

Engel M, 2012. The commonsense case against animal experimentation. In Garret J, ed. *The Ethics of Animal Research*. Cambridge: MIT Press, pp. 215–236.

Garner R, 1993. Political animals: A survey of the animal protection movement in Britain. *Parliamentary Affairs*, 46(3), pp. 333–352.

Garrett J, 2007. Utilitarianism, vegetarianism, and human health: A response to the causal impotence objection. *Journal of Applied Philosophy*, 24(3), pp. 223–237.

Humane Society of the United States, 2016. Trophy hunting by the numbers. Available at: https://www.hsi.org/wp-content/uploads/assets/pdfs/report_trophy_hunting_by_the.pdf (Accessed 4 August 2021).

Hursthouse R, 2011. Virtue ethics and the treatment of animals. In Beauchamp T. and Frey R, eds. *The Oxford Handbook of Animal Ethics*. Oxford: Oxford University Press, pp. 119–143.

Kant I, 1785 (1998). *Groundwork of the Metaphysics of Morals*, Gregor, M. (trans.), Cambridge: Cambridge University Press.

Knight A, 2022. Scientific and educational animal use. In Knight A, Phillips C and Sparks P, eds. *Routledge Handbook of Animal Welfare*. Abingdon, Oxon: Routledge.

Kresge L and Strochlic R, 2007. *Clearing the Air: Mitigating the Impact of Dairies on Fresno County's Air Quality and Public Health*. Calif: California Institute for Rural Studies.

Letcher T, 2021. *The Impacts of Climate Change*. Amsterdam: Elsevier.

MacNair R, 2002. *Perpetration-Induced Traumatic Stress: The Psychological Consequences of Killing*. Westport: Greenwood Publishing.

Marquis D, 1989. Why abortion is immoral. *The Journal of Philosophy*, 86(4), pp. 183–202.

Rawls J, 1971. *A Theory of Justice*. Harvard University Press.

Regan R, 1983. *The Case for Animal Rights*. Berkeley: University of California.

Regan R, 1985. The case for animal rights. In Singer P, ed. *Defense of Animals*. New York: Basil Blackwell, pp. 13–26.

Rollin B, 2006. *Science and Ethics*. Cambridge: Cambridge University Press.

Rowlands M, 1997. Contractarianism and animal rights. *Journal of Applied Philosophy*, 14(3), pp. 235–247.

Rowlands M, 2012. Virtue ethics and animals. In Protopapadakis E, ed. *Animal Ethics: Past and Present Perspectives*. Berlin: Deutsche Nationalbibliothek, pp. 29–38.

Ryder R, 2011. *Speciesism, Painism and Happiness: A Morality for the Twenty-First Century*. Exeter: Academic Imprint.

Singer P, 1975. *Animal Liberation*. New York: Avon Books.

Vining J, 2008. Animal cruelty laws and factory farming. *Michigan Law Review First Impressions*, 106(5), pp. 123–127.

Wallace A, 2020. Tyson and other meat processors are reportedly speeding up plans for robot butchers. *CNN Business*. Available at: https://www.cnn.com/2020/07/10/business/tyson-meatpacking-plants-automation/index.html (Accessed 9 October 2021).

Wallinga D, 2004. Concentrated animal feeding operations: Health risks from air pollution. Institute for Agriculture and Trade Policy. Available at: www.iatp.org/documents/concentrated-animal-feeding-operations-health-risksfrom-air-pollution (Accessed 9 October 2021).

Warren M, 1997. *Moral Status: Obligations to Persons and Other Living Things*. Oxford: Oxford University Press.

Welfare Quality®, 2019. *Practical Strategies for Improving Farm Animal Welfare: An Information Resource*. Lelystad: Welfare Quality® Consortium.

Woorall M, 2004. Meatpacking safety: Is OSHA enforcement adequate? *Drake Journal of Agricultural Law*, 9, pp. 299–321.

Wrenn C, 2012. Abolitionist animal rights: Critical comparisons and challenges within the animal rights movement, *Interface*, 4(2), pp. 438–458.

Wrenn C, 2016. *A Rational Approach to Animal Rights Extensions in Abolitionist Theory*. London: Palgrave.

27

ANIMAL LAW – HISTORICAL, CONTEMPORARY, AND INTERNATIONAL DEVELOPMENTS

Ian Robertson and Paula Sparks

Ancient philosophy, Roman law classification, key points in history

The development of law in the Western hemisphere has its roots in Roman law, which in turn was shaped and influenced by early Greek philosophers such as Plato, Socrates, and Aristotle, for whom all of nature was ordered in a hierarchical manner. This would provide the basis for the concept of the Great Chain of Being, which supposed a hierarchical ladder of beings with God at the top of the pyramid, progressing downwards towards the kings, nobles, commoners, animals, and other life forms. The same sense of organisation was applied to civil society which was organised on a hierarchal basis with Greek men at the top of the social pyramid, extending downwards to women and children. Slaves and animals were considered unable to reason and regarded as "living tools" for the benefit of society, rather than forming part of it (Wise, 2014).

This early world view of animals as a usable commodity found its way into Roman law which categorised animals as ownable property (Kelch, 2012). The Romans were renowned for their sense of order and written codification. Roman law divided the world into categories of persons, things, and actions where the classification of "persons" applied only to certain classes of humans. According to the Romans, legal persons held rights and therefore, on the world view persisting at the time, personhood was reserved to only those beings considered to be capable of exercising free will. On this basis women, children, slaves, mentally incompetent humans, and animals were all excluded from being persons but fell into the category of legal "things", meaning that, if owned, they would be classified as "property" (Wise, 2014).

A legal "thing" belonged to the rights holder and could not exercise any legal rights because as property, they did not possess legal rights. All inanimate objects were things, but so too were animate objects like women, children, and animals.

The world view propounded by these early philosophers persisted over the generations and became embedded in early Christian doctrine, which also advocated for a hierarchical social order with the authority of the monarch being derived from God, who had dominion over all living things (Wise, 2014). As with the early philosophers, the early Christians also believed that animals were created for the purpose of humans (Kelch, 2012).

 DOI: 10.4324/9781003182351-33

While there were voices advocating for an approach of stewardship in respect of animals, the predominant and stronger view advanced the concept that mankind held a position of dominion which entitled people to use animals as they pleased. For example, Saint Thomas Aquinas (1225–1274) is widely recognised for supporting the view that man being made in the image of God was above other animals; the natural order was that animals were created for the benefit of humans (Robertson, 2015).

This world view of animals being distinct and different from humans, continued largely over the centuries, lending credence from philosophers such as Rene Descartes (1596–1650) who promoted the concept of animals as autonomous beings, lacking language, thought, and self-consciousness (Robertson, 2015).

Britain's 18th and 19th centuries marked a time of considerable attention to social justice issues including the treatment, protection, and voice of groups including women, children, and animals. As part of this overall social justice movement the moral status of animals became the subject of discourse amongst the jurists and free thinkers of the time. The philosopher and founder of utilitarian theory, Jeremy Bentham (1748–1832) challenged the prevailing orthodoxy, questioning the underlying reasons for excluding animals from moral consideration, including the neglect of the interests of animals in legislation. He is widely attributed for questioning the morally relevant distinction between animals and humans and suggesting that animals should be assessed on the basis of their capacity for suffering rather than the animal's ability to reason or communicate (Bentham, 1789).

In Britain, the writer, John Lawrence (1753–1839) was another early proponent of the ethical treatment of animals. His book, "A Philosophical and Practical Treatise on Horses, and on the Moral Duties of Man Towards the Brute Creation" (Lawrence, 1796) argued that "the rights of beasts [animals] be formally acknowledged by the state, and that a law be framed upon that principle, to guard and protect them from acts of flagrant and wanton cruelty, whether committed by their owners or others" (Lawrence, 1796, p. 123). This was a radical position for the day, not only because it challenged traditional thinking, but also in calling for protection from cruelty to be extended to animals abused by their owners, whose property rights over those "owned" animals were at the time unqualified and unassailable on the basis of God given rights (Robertson, 2015).

Despite this philosophical awakening of interest in the moral consideration of animals, law's inherently conservative and retrospective style of development meant that legal protection for animals was still absent from formalised laws around the world. To the extent that any legal protections existed, these were for the benefit of the owner's interest in animals as their property.

The very first legislation (The Cruel Treatment of Cattle Act 1822 (3 Geo. IV c. 71), often referred to as "Martin's Act", after Richard Martin MP, who introduced the bill, made it an offence to "wantonly and cruelly beat or ill-treat [] [any] horse, mare, gelding, mule, ass, ox, cow, heifer, steer, sheep or other cattle". Other domesticated animals and all wild animals were excluded from scope. Despite the fact that the very first piece of law applied a very narrow focus and protected only certain species from beating and "ill treatment", it represented a momentous legislative animal law reform because it fettered the traditional unlimited entitlements of property owners. Furthermore, the law reform protecting sentient animals ("animals that have the ability to feel or experience") from experiencing unnecessary suffering set the stage for progressively wider anti-cruelty protections throughout the world.

As a consequence of the breadth of the British Empire, English law spread throughout the world via its colonies. For example, there was a parallel animal law development in the United States, which saw the first anti-cruelty legislation passed in 1867 (an Act for the more effectual prevention of cruelty to animals N.Y. Rev. Stat. ch. 375, §§ 1-10 (1867).

Similar to Martin's Act, the US animal law reform was the result of tireless campaigning by its champions. In the United States the law reform was largely led by Henry Bergh, who, like Richard Martin, also played an important role in its enforcement after enactment of the first anti-cruelty legislation. Similar to the position in Britain, the purposes behind this law focused on the common cruelty that was inflicted on animals typically seen being driven through streets or pulling carts. Unlike its earlier British counterpart though, it also imposed some limited positive duties of care to prevent an animal from suffering, such as providing water to impounded animals and imposed penalties for abandoning infirm and disabled animals in a public place. It also criminalised practices such as cock fighting and bull and bear baiting on the basis that they were specifically identifiable activities which caused animals to unnecessarily suffer (Kelch, 2013; Freeberg, 2020).

This was also an era of scientific discovery, and the contribution of Charles Darwin (1809–1882) has been widely credited for introducing a scientific element to the previous largely, philosophical, and theological debate (Robertson, 2015; Kelch, 2013), his theory of "natural selection" (Darwin 1871) undermining the Great Chain of Being theory that had propped up the notion of animals being created for human use (Wise, 2014).

National, domestic, and municipal animal welfare and anti-cruelty law

These early legislative reforms constituted the first in a series of laws that would characterise the incremental development of animal protection measures, targeting unnecessary acts of cruelty towards animals.

In Britain, the Protection of Animals Act 1911 consolidated early legislation and broadened the scope of the 1822 Act. The duty of care reflected a continued focus on prohibiting specific identified acts of animal cruelty however that benchmark was set to shift with the advent of the Five Freedoms (Five Freedoms, Farm Animal Welfare Council).

In the modern landscape of animal welfare legislation, there is usually protection through primary legislation (enacted by the legislature) and secondary legislation (using power derived from the legislature) which may be supported by codes of practice or guidance, reflecting scientific advice about animal husbandry or care. Whilst such codes generally do not form part of the law, breach may be used as evidence to support prosecutions under animal welfare legislation, and conversely, compliance with codes may be a defence.

Most jurisdictions throughout the world which have any kind of animal protection law (and some still have none) currently function on a legal model of "anti-cruelty" law which prohibits owners or persons in charge of animals causing them to feel or experience unnecessary suffering. It's helpful to recognise that this is a two-part test.

From a compliance and enforcement perspective, the first test asks, "did the animal suffer?" If the animal did not experience pain or distress, then the conclusion must be that no breach of a legal obligation has occurred. On the other hand, if the animal is assessed as having suffered, then the second limb of the two-part test is activated and asks, "was the suffering necessary?"

While societal world views regarding necessity vary enormously, the law broadly considers the nature, duration, and severity of any suffering experienced by the animal in context of the relevant species, use, and circumstances of the suffering. In England and Wales, the Animal Welfare Act 2006 provides the court with specific guidance on how to interpret "necessity", and a plethora of secondary legislation and supporting authoritative evidence provides further guidance as to what constitutes good practice and scientific knowledge to further assist the courts in applying contemporary standards of law.

The Animal Welfare Act 2006 also requires owners and persons in charge of animals to "take such steps as are reasonable in all the circumstances to ensure that the needs of an animal for which he is responsible are met to the extent required by good practice" (Animal Welfare Act, 2006, Section 9). This duty of care obligates owners and persons in charge of animals to meet the physical and behavioural needs of animals. The "needs" are defined as an animals' need for a suitable environment, suitable diet, ability to exhibit normal behaviour patterns, to be housed with or apart from, other animals, and to be protected from pain, suffering, injury, and disease. The imposition of this positive duty of care enables intervention by compliance personnel in the event of "likely" suffering. This authorises, for example, appointed officers to remove an animal from a situation where its welfare needs are not being met or are "likely" to be unmet.

Science has always been a key informant to the development of animal law and of particular importance in Britain was the Brambell Report (Brambell, 1965), (the product of a parliamentary committee chaired by Professor Roger Brambell and tasked with examining the welfare of farmed animals) which established the concept of the Five Freedoms (Five Freedoms, Farm Animal Welfare Council) as the basis of good animal welfare, specifically freedom from thirst, hunger, and malnutrition; freedom from discomfort; freedom from pain, injury and disease; freedom to express normal behaviour; and freedom from fear and distress (Robertson, 2015).

The principles of the "Five Freedoms" underpinned law's extension from prohibiting blatant acts of cruelty by adding obligations to prevent suffering with "positive duties of care" and empowering enforcement to intervene where an animal was deemed "likely" to suffer.

Robertson and Goldsworthy (2021) argue for law reform that further extends law's duty of care by applying the contemporary scientific authority of the Five Domains (Mellor, 2019) to include within animal welfare legislation, an added responsibility for animals' positive states (i.e., comfort, interest, and pleasure). This is consistent with the objectives and messaging of organisations and science recognising that in terms of the animal's welfare and quality of life experience, less pain is not the same as more pleasure.

One of the assumptions made about animal welfare and anti-cruelty legislation is that the word "animal" applies to all animals. That assumption often results in confusion, misunderstanding and frustration for those who are less than familiar with the methodology applied by law in respect of "animals".

Although biology and other disciplines may collectively refer to all non-humans as animals or even identify humankind as an animal, there are two critical points to understanding the "legal animal". First, the law universally distinguishes humankind from the rest of the animals, so through the legal lens, a human is not an animal. Secondly, it is not the case that all other creatures that are not humankind fall into the legal category of animal. Indeed, the definition of an animal may differ across legislation, reflecting the uses, purposes and circumstances of the specific animal being referred to.

Additionally, animal welfare legislation primarily protects species scientifically proven to be capable of suffering. Historically, this has meant that invertebrate species have usually been excluded from animal protections. This is shifting, however, with the development of scientific knowledge about the sentience capacities of invertebrate species and in some countries specific measures have been adopted for the protection of crustaceans and cephalopods (Norway, Switzerland, Austria, New Zealand, and some other states and regions across the world have some form of legislative protection for crustaceans (Crustacean Compassion, 2018). This change may be accelerated by a recent independent report from the London School of Economic and Political Science in England, whose findings about the sentience of decapod crustaceans and cephalopods concluded there was "strong scientific evidence" of the sentience of cephalopods and crustaceans (Birch et al., 2021). In consequence, the UK Government has included these species within the definition of 'animals' in the Animal Welfare (Sentience) Act 2022 (Defra, 2021). It is uncertain when or how this acknowledgement will translate into adequate or any practical legal protections.

Consistent with the concept that legal duties could only be applied to animals that were reliant on their human caregiver, early legislative protections were afforded only to domesticated and captive animals and did not extend to their "wild" counterparts. Similarly, today, the degree, nature, and extent of the legal responsibility predominantly reflects the degree, nature, and extent of the animal's reliance for its well-being on the human caregiver. For example, the responsibilities of a person (where a "person" is either an individual or an organisation) for an animal that totally relies on that person for its food, water, shelter, and medical care, are traditionally different to the responsibilities of persons for wildlife that largely fend for themselves.

Legal responsibilities prescribed under animal protection/welfare legislation largely continue to reflect the principle that protections and responsibilities reflect the use and purpose of the animal concerned rather than the species. The very same rabbit, for example, will be the subject of significantly contrasting legislative protections, policies and procedures dependent upon whether it is kept as a pet rabbit, a rabbit used in research, or classified as a "pest" that exists in the wild.

Reflecting this approach, animal welfare legislation may explicitly make exceptions for certain categories of animal use. For example, in England anything done in the course of recreational fishing is also excluded from the anti-cruelty provisions of the Animal Welfare Act 2006. This is not because fish are not recognised as capable of experiencing pain (as vertebrate species they would ordinarily fall within scope), but the exemption reflects a consensus of support for fishing as a recreational activity. Countries also vary in the extent to which exceptions are made to animal welfare legislation for certain commercial practices and methods of husbandry.

Properly resourced and authorised enforcement is critical to giving practical effect to animal protection laws. Animal "welfare" legislation typically authorises appointed officers to intervene both when an animal has actually suffered and when an animal was deemed "likely" to suffer. England's legislation authorises judicial transfer of ownership of an animal in certain cases of anti-cruelty or poor welfare, even before conclusion of proceedings. Scotland has gone one step further and provides administrative powers to rehome animals without the need for judicial approval, subject to compliance with prescribed procedural safeguards (Animals and Wildlife (Penalties, Protections, and Powers) (Scotland) Act, 2020).

Compliance standards vary between jurisdictions, and it is common that animal protection advocates campaign for higher penalties to deter offending. For example, commentators in India criticised the level of fines for breaches of the Prevention of Cruelty to Animals Act 1960 (see Chapter 31). In England the maximum penalty for the most serious offences against animals (including torture and killing) was increased in 2001 (The Animal Welfare (Sentencing) Act 2021) from six months' to five years' imprisonment.

Not everyone is entitled to provide legal representation in respect of the animal's legal interests. In public law, that entitlement of standing, or *locus standi* (the right to appear in court), is traditionally a prerequisite condition whereby the party seeking a legal remedy must demonstrate to the court that they have sufficient connection to the legal issue or that they have suffered harm. In Britain, the courts have taken a comparatively liberal approach to the standing of parties to bring representative actions on behalf of animals (*R v Secretary of State for Foreign and Commonwealth Affairs, ex parte World Development Movement Ltd* [1995] 1 WLR 386)). In contrast, for example, America's higher bar has thwarted some NGOs seeking to challenge decisions of state and public authorities (Chapter 33; Sunstein, 1999).

Law reform

Animal law standards and compliance vary significantly between countries. There are still countries that have little to no codified law and operate primarily simply on natural law principles where, if the animal is inadequately looked after, it will fail to thrive and may even die with consequences to the animal's owners.

Many countries around the world today operate on principles of anti-cruelty law prohibiting acts or omissions resulting in an animal suffering unnecessarily. The anti-cruelty law in some countries operates on a basic standard of animal "protection" law that simply criminalises blatant acts of cruelty.

Countries with more advanced law have moved beyond animal protection law by extending legal responsibilities ("duty of care") to protect animals not just from blatant acts of cruelty but also from "likely" suffering thereby creating animal "welfare" law.

Schaffner (2010) identifies problems with existing law that warrant updating related to animal's lack of standing and barriers to legal representation; human-centric classification of animals, and the broad discretion built into interpretation and enforcement of animal laws where animals are one, but not the only, stakeholder. Limitations of the existing legal paradigm, exemptions for exploitative categories of animal use, and giving greater priority to enforcement have also been issues identified as warranting updating (Sunstein and Nussbaum, 2004).

Schaffner (2010, pp. 123–129) also addresses the efficacy of and evidence base behind breed-specific provisions in dog control legislation. Although there is well-established science validating that temperament is a heritable trait and that certain breeding lines are more aggressive than others, breed-specific legislation ("BSL") has attracted criticism for unfairly resulting in dogs being euthanised or subject to other controls on the basis of breed, rather than individual characteristics. An independent report commissioned by the UK Government suggests there is "a broad consensus within the literature that breed does not, by itself, provide an evidence base for addressing "dog dangerousness". (Nurse et al., 2021, 8, p. 65). The report makes recommendations for a wider range of strategies for dog bite prevention, reflecting the multi-causality of dog attacks.

Law reform is an established process that has the objective of ensuring law is updated and fit for the needs of a modern society and a number of proposals have been advanced seeking to utilise the law to elevate standards of animal welfare. Thus, attempts have been made to draft model animal welfare legislation as a beacon of good practice. For example, the Model Animal Welfare Act (World Animal Net: http://worldanimal.net/images/stories/documents/Model_AWA/WAN-Model-Animal-Welfare-Act.pdf).

Others propose more radical change. Francione (2006) advocates abolishing the property status of animals completely, ending human exploitation and allowing animals greater autonomy. Francione's view is that "the property status of animals means that their interests will virtually always be ignored whenever it will benefit humans and despite the many laws that supposedly protect animals" (Francione, 2006, p. 77).

Favre (2009) does not propose abolishing the property status of animals but has alternatively proposed creating a new category of "living property"' which limits the unqualified rights of the owner. A similar approach has recently been adopted in Spain, where the criminal and civil codes now explicitly recognise animals as "sentient beings" as a special form of property, whose interests can be considered by the courts when dealing with issues concerning allocation or disposal of property (Proposición de Ley de modificación del Código Civil, 2021).

There are attempts being made through the courts to secure personhood for certain animals where there is strong scientific evidence that they can exercise "practical autonomy". For example, attorney, Steven Wise is arguing before the courts that certain species who can exercise practical autonomy should be recognised as legal persons and entitled to certain basic liberty rights such as freedom from torture and enslavement (Wise, 2010; Knight, 2022).

Robertson and Goldsworthy (2021) demonstrate that extending law's duty of care by adding an obligation to provide the animal with the opportunity to experience positive states to existing anti-cruelty responsibilities, provides immediate practical benefits to animals.

International law and bodies

International law refers to the collective body of rules accepted by nations that are derived from instruments including such as international conventions (for example, international treaties or agreements), customs, general legal principles, and case law. Treaties, which are perhaps the most commonly understood sources of international law, vary in nature, and can represent either bilateral or multilateral agreements between nation states, who enter into such arrangements voluntarily.

However, as a consequence of state sovereignty there is no singular universal equivalent of legislation that applies legal responsibilities, accountabilities, and liabilities to sovereign states. Parallel mechanisms of responsibility exist only when a sovereign state has become a ratified member to an international organisation.

As a point of note and consideration, membership does not infer blanket acceptance of each organisation's rules. In fact, nations may enter a reservation against certain provisions, enabling them to sign up to some, but not all provisions in a treaty where there is a lack of political will to sign up to the whole.

As a consequence, it is helpful for the reader to view the subject of international "law" as a body of responsibilities that is significantly different to the concept of national legislation.

Foundational to the operation of rules in the international sector, is the concept that "There is no supra-national legislature empowered to create laws binding on the global community, nor any international police force to ensure compliance with such rules as have been established" (Bowman et al., 2010, 25).

International laws are typically concerned with issues that transcend national boundaries, such as environment protection and climate change. In the field of animal law, similarly, international law has traditionally focused on issues such as conservation and biodiversity loss or importantly regulation of trade.

Amongst the diverse world views regarding the role of animals in the national and international human–animal relationship, there are those who take the view that the interests of animals are under-represented on the global stage. However, as the inseparable relationship between animals, people, and their shared planet is increasingly recognised in programmes such as the One Health initiative of the United Nations (an international organisation established by charter in 1945), the issue of standards of animal welfare, and impacts on human society, has resulted in greater attention being given to the issue of animal welfare. However, to date, the welfare of animals as a distinctly separate policy objective does not feature in any of the UN programmes and animal interests in and of themselves are not represented.

United Nations

The UN's 2030 Agenda for Sustainable Development was adopted in 2015 by all 193 Member States of the United Nations and is described as a "plan of action for people, planet and prosperity" (Sustainable Development Goals, 2015, Preamble). Animals are notably absent as a separate specific stakeholder action in the UN's Agenda which aspires to "a world in which humanity lives in harmony with nature and in which wildlife and other living species are protected" (Sustainable Development Goals, 2015).

The Agenda identifies 17 Sustainable Development Goals (SDGs) as "a universal call to action to end poverty, protect the planet, and ensure that by 2030 all people enjoy peace and prosperity" (Sustainable Development Goals, 2015).

The goals include such matters as ending poverty, hunger, and gender inequality, and creating sustainable cities and communities. Those that implicitly involve animals as part of the integrated approach are goals 13 ("Climate Action"), 14 ("Life Below Water", to "Conserve and sustainably use the oceans, seas and marine resources for sustainable development"), and 15 ("Life on Land", to "Protect, restore, and promote sustainable use of terrestrial ecosystems, sustainably manage forests, combat desertification, and halt and reverse land degradation and halt biodiversity loss").

Although the Sustainable Development Goals do not recognise the intrinsic value of animals, this is not to say that they lack the potential to indirectly improve the lives of animals through recognition of the role of animals in the inseparable human–animal–planetary relationship. However, Verniers and Brels (2021) raise an implicit concern that, since the goals are intended to be transformative, the absence of animal welfare from the goals will likely result in a slower pace of positive change at a global level, since whilst the goals are themselves "non-binding and aspirational in nature", they point to the fact that "these aspirations are the impetus to expedite understanding in hard law" (Vernier and Brels, 2021, p. 49). Further, absent a specific goal to improve animal welfare, there is a risk, given the inseparable relationship between humans and other animals, that benefits to humankind incidental to improving animal welfare will be missed.

As Verniers (2021) points out, the absence of animal welfare as a singular subject from the goals was commented upon in the Global Sustainable Development Report 2019 (Messerli et al., 2019), authored by an independent group of scientists appointed by the Secretary-General of the United Nations. In a section identifying key issues missing when the 2030 Agenda was drawn up, the report states:

> Animal welfare: The clear links between human health and well-being and animal welfare is increasingly being recognised in ethics and rights-based frameworks. Strong governance should safeguard the well-being of both wildlife and domesticated animals with rules on animal welfare embedded in transnational trade.
>
> *(Messerli et al., 2019, p. 117)*

An example of how the link between human and animal health is being recognised is the One Health program (Centers for Disease Control and Prevention (CDC, nd), One Health), which explicitly recognises the interconnectedness of human and animal health. As such, its focus is on issues such as food safety, zoonotic risk, and antibiotic resistance, and recognises a shared susceptibility to disease and environmental contamination. While this program focuses on health issues, it reflects similar social, economic and environmental initiatives that recognise the inseparable relationship between people and animals on their shared planet.

This lacuna in consideration of the intrinsic value of animals and mechanisms to reflect those interests in policy making at the UN has prompted Ambassador Muhamed Sacirbey, Former Bosnian Foreign Minister and Ambassador to the United Nations to call for the appointment of a Special Representative (colloquially termed, an Animal Ambassador) to "seek to represent human perspectives on animal welfare" (Sacirbey, 2014).

There have also been calls, variously, for an 18th Sustainable Development Goal that addresses the welfare of animals (Visseren-Hamaker, 2020) or alternatively modification of the Sustainable

Development Goals to encompass animal health and welfare (Verniers and Brels, 2021), and for a One Welfare approach extending to animal welfare, which would extend and complement the One Health approach (Pinillos, 2018).

World Organisation for Animal Health

The World Organisation for Animal Health retains the acronym OIE, from its predecessor the Office International des Epizooties, which was established by International Agreement in 1924. It is an inter-governmental organisation concerned with improving global animal health and welfare, initially being established to combat animal diseases. It has 182 members and maintains relations with others.

The OIE is under the authority of a World Assembly of Delegates designated by the Governments of Member States. Unlike other international bodies, the OIE has formally recognised the importance of animal welfare and is responsible for setting international standards of animal welfare.

In 2017 all Member States adopted the OIE Global Animal Welfare Strategy (OIE, 2017), the four pillars of which are the development of international animal welfare standards, communication and awareness raising about animal welfare, capacity building and training of veterinary services, and implementation of OIE animal welfare standards and policies.

For the purposes of the strategy, the OIE defines animal welfare as "the physical and mental state of an animal in relation to the conditions in which it lives and dies" (OIE Terrestrial Code, 7.1.1) and its guiding principles include the "Five Freedoms". The Aquatic Code similarly suggests international standards for the welfare of farmed fish, excluding ornamental species (OIE Aquatic Code).

As White (2013) points out, however, the OIE's focus remains animal disease with only 11 standards for animal welfare incorporated into Terrestrial and Aquatic Codes. Furthermore, the standards are broad and aspirational, rather than precise and they lack mandatory language. Indeed, the OIE relies upon voluntary commitments by Member States to develop and adhere to its principles and there is no mechanism for monitoring compliance, or for enforcement (its dispute resolution mechanism is similarly voluntary).

World Trade Organization

The World Trade Organization (WTO) is an inter-governmental organisation that regulates international trade with its goal "to ensure that trade flows as smoothly, predictably and freely as possible" (World Trade Organization, 2021). It was established by the "1994 Marrakesh Agreement Establishing the World Trade Organization" and re-enacted the General Agreement on Tariffs and Trade (GATT) signed in 1947 and containing measures to minimise trade barriers such as quotas, subsidies, and tariffs.

The WTO rules are binding on all Member States, and it has a Dispute Settlement Body that deals with complaints; recommendations are binding.

Trade liberalisation is a fundamental principle, and trade restrictions based upon the manner in which a product is produced (known as "process and production methods" (PPMs) are not permitted, unless they change the character of the product. If the product remains unchanged, a restriction in trade will thereby need to be justified under one of the exceptions to the rules. This has historically inhibited nations from restricting imports on the basis that they have been produced to lower standards of animal welfare standards.

Recent jurisprudence from the WTO is however encouraging. The leading case in this area is EC-Seal Products case 2014 (European Communities, 2014), concerning a decision by the

European Union to prohibit the import and marketing of seal products in the EU with certain exceptions for Inuit communities, reflecting the concerns of EU citizens about inhumane suffering inherent in seal hunts. The Appellate Body of the WTO upheld a finding of the Panel that the trade restriction was justified in order to protect public morals, as defined under GATT Article XX(a).

CITES and wildlife trade

While the 19th century saw anti-cruelty legislation starting to emerge for the protection of some domesticated species, wildlife was left largely unprotected.

The early attempts at regulation of human activities impacting upon wildlife were motivated by conservation, not for the intrinsic benefit of those species but predominately to preserve the species for the protection of the animals as a resource (for example, see the Convention for the Protection of Birds Useful to Agriculture 1902; Bowman et al., 2010).

The Second World War was an impetus for international cooperation. The United Nations was established to maintain international peace and security and a number of specialist agencies were created including UNESCO (UN, Educational, Scientific, and Cultural Organisation) which created the IUPN (International Union for the Protection of Nature) re-named IUCN (International Union for Conservation of Nature and Natural Reserves), which publishes a Red List of species (as a guide to conservation status of plants and animals).

UN wildlife treaties include an International Convention for Regulation of Whaling (1946), an International Convention for Protection of Birds (1950), and the Atlantic Treaty Systems (1959). In 1979 the Bonn Convention was signed by Parties, agreeing on measures to protect the habitats and migration routes of migratory species.

The 1973 Washington Convention on the International Trade in Endangered Species (CITES, 1973) is a multilateral agreement to protect and conserve endangered species. CITES regulates international trade in animals and plants threatened with extinction through the operation of three Appendices onto which at risk species are placed. CITES does not completely prohibit international trade but rather controls trade through the system of import and export restrictions; furthermore, those restrictions only apply to international trade and not domestic trade in protected species.

CITES requires nations to establish national authorities to administer provisions of the Convention. Each state is required to designate a Scientific Authority to provide scientific evidence underpinning decisions about whether trade is sustainable and a Management Authority to issue permits/certificates. The Management Authority is responsible for determining that species subject to Appendix 1 restrictions, will not be used primarily for commercial purposes.

Nations who are Parties to CITES are responsible for implementation and enforcement. However, some still do not have the necessary national laws in place for implementation (a legislative status table is maintained by CITES: National Legislation Project; see CITES Legislative Status Table, 2021). Even in countries with the full implementation measures in place, enforcement remains a problem and the illegal wildlife trade is a significant issue (Nurse and Wyatt, 2020; Wyatt et al., 2021).

European Union

The European Union is an economic and political union of 27 Member States, forming a single trading block. Britain was a member until its withdrawal in 2020 after its citizens voted in a referendum in 2016 to leave the EU. The EU legislation through a series of regulations and directives in key areas concerning animal protection, including wildlife (Birds and Habitats Directive, EU Invasive Alien Species Regulation, and other wildlife laws), research (animal testing regulations), agriculture (regulations around live exports, slaughter regulations, farm subsidies, and on-farm welfare) and the movement of domesticated animals across borders (such as pet passports and ban on import/export of cat and dog fur).

The EU is established by a series of treaties, and it is given legal competence to take make animal welfare measures by Article 13 of the Treaty of the Functioning of the European Union (TFEU) which states that:

> In formulating and implementing the Union's agriculture, fisheries, transport, internal market, research and technological development and space policies, the Union and the Member States shall, since animals are sentient beings, pay full regard to the welfare requirements of animals, while respecting the legislative or administrative provisions and customs of the Member States relating in particular to religious rites, cultural traditions and regional heritage.

Historically, the status of animals was as goods or products only, so animals farmed were classed as "agricultural products" and there was no legal competence to take measures reflecting the fact that they are sentient beings. Animal advocacy groups campaigned for express recognition that animals have an enhanced status as "sentient" so that policy objectives (such as free trade) could be balanced against the objective of having regard to animal welfare.

However, Article 13 did not contain any procedure or mechanism obligating members to show that they had taken animal welfare into consideration in determining domestic policy or place any prominence on animal welfare when weighing it up as a policy consideration against other policy objectives. The instrument's recognition of animal sentience was significant in promoting discussion about animal "sentience"; however, in order to realise the full potential of the legislative recognition it's been argued that a clear directive applying responsibilities for the negative and positive states of the animal is necessary to evolve policies and practices beyond continued anti-cruelty responsibilities (Goldsworthy and Robertson, 2021).

Britain did not carry across Article 13 in the European Union (Withdrawal Act) 2018 but the government has enacted the Animal Welfare (Sentience) Act 2022. This legislation explicitly recognises that animals are sentient and seeks to have regard to their welfare needs when formulating and implementing policy by establishing an Animal Welfare Committee of animal welfare experts, to advise Ministers about the animal welfare impacts of prospective policy. This is a similar approach to that of New Zealand which has recognised animal sentience in national legislation that also established a National Animal Welfare Advisory Committee (NAWAC) and the Netherlands, which has established the Dutch Council on Animal Affairs (Council on Animal Affairs) to advise Ministers.

Council of Europe

The Council of Europe (CE) is an inter-governmental body established by the Statute of the Council of Europe and signed in London, 5 May 1949. It is not a part of the European Union, but a separate institution with an overlapping, but different membership. The CE has passed a number of important Conventions concerning animals (Council of Europe).

The CE has no legislative power and cooperation is entirely voluntary. There is no sanction available for nations who sign or ratify a Convention and then fail to implement or comply with its provisions. However, the EU and Member States may formally approve Conventions as a body or adopt a Convention as a new Community law and this is often a route by which a CE Convention becomes embedded in the laws of the EU or its Member States, and in this way, it has played an important role in raising standards of animal protection throughout Europe.

Conclusions

National law and international agreements governing the human–animal relationship all revolve around the principle of responsibility. The inseparable relationship between animals, people, and our shared environment means that the level of responsibility applied by the law for the life experience of animals also affects us, our children, and our world – today and for generations to come.

References

Bentham J, 1789. An introduction to the principles of morals and legislation,. 1st ed.. London: Payne and Son

Bowman M, Davies P and Redgwell C, 2010. *Lyster's International Wildlife Law*. Cambridge University Press.

Birch J, Burn C, Schnell A, Browning H and Crump A, 2021. *Review of the Evidence of Sentience in Cephalopod Molluscs and Decapod Crustaceans*. London School of Economics and Political Science.

Brambell FWR and Technical Committee to Enquire into the Welfare of Animals kept under Intensive Livestock Husbandry Systems, 1965. *Report of the Technical Committee to Enquire into the Welfare of Animals Kept Under Intensive Livestock Husbandry Systems*. HM Stationery Office.

Centers for Disease Control and Prevention (CDC), One Health CDC, n.d. Available at https://www.cdc.gov/onehealth/index.html [Accessed 12.12.2021].

CITES, 1973. Available at https://cites.org/eng/legislation [Accessed 12.12.2021].

CITES Legislative Status Table, 2021. Available at https://cites.org/sites/default/files/documents/legislation-status/legislation-status.pdf [Accessed 15.12.2021].

Council of Europe, n.d. Available at https://www.coe.int/en/web/conventions/home [Accessed 21.12.2021].

Council on Animal Affairs. Available at https://english.rda.nl/ [Accessed 30.12.2021].

Crustacean Compassion. 2018. Our open letter, 31 January 2018. Available at https://www.crustaceancompassion.org.uk/open-letter [Accessed 16.12.2021].

Darwin C., 1871. *The Descent of Man and Selection in Relation to Ser.* London: Murray.

Defra, 2021. Available at https://www.gov.uk/government/news/lobsters-octopus-and-crabs-recognised-as-sentient-beings [Accessed 12.12.2021].

European Communities: Measures Prohibiting the Importation and Marketing of Seal Products, 2014. *World Trade Organization*. Available at https://www.wto.org/english/tratop_e/dispu_e/cases_e/ds401_e.htm [Accessed 30.12.2021].

Farm Animal Welfare Council/Farm Animal Welfare Committee. Five freedoms. Accessible at https://webarchive.nationalarchives.gov.uk/ukgwa/20121010012427/http://www.fawc.org.uk/freedoms.htm [Accessed 12.12.2021].

Favre D, 2009. Living property: A new status for animals within the legal system. *Marq L. Rev.*, 93, p. 1021.

Francione GL, 2006. Animals, property, and personhood, In *People, Property, or Pets?* Purdue University Press, p. 77.

Freeberg E, 2020. *A Traitor to His Species: Henry Bergh and the Birth of the Animal Rights Movement*. Hachette UK.

Goldsworthy D and Robertson I, 2021. *Recognising and Defining Animal Sentience in Legislation: A Framework for Importing Positive Animal Welfare Through the Five Domains Model.*

Kelch TG, 2012. A short history of (mostly) Western animal law: Part I. *Animal Law Review.*, 19, p. 23.

Kelch TG, 2013. A short history of (mostly) Western animal law: Part II. *Animal L.aw Review*, 19, p. 347.

Knight A, 2022. Scientific and educational animal use. In Knight A, Phillips C and Sparks P, eds. *Routledge Handbook of Animal Welfare*. Abingdon, Oxon: Routledge.

Lawrence J, 1796. *A Philosophical and Practical Treatise on Horses, and on the Moral Duties of Man towards the Brute Creation: Comprehending the Choice, Management, Purchase, and Sale of Every Description of the Horse;*

the Improved Method of Shoeing: Medical Prescriptions and Surgical Treatment in all Known Diseases (Vol. 1). Re-published 1810. London: Printed for Sherwood, Neely, and Jones, p.132. Available at https://wellcomecollection.org/works/sc32cjds [Accessed 16.04.2022].
Mellor DJ, 2019. Welfare-aligned sentience: Enhanced capacities to experience, interact, anticipate, choose and survive. *Animals*, 9(7), p. 440.
Messerli P, Murniningtyas E, Eloundou-Enyegue P, Foli EG, Furman E, Glassman A, Hernández Licona G, Kim EM, Lutz W, Moatti JP and Richardson K, 2019. *Global Sustainable Development Report 2019: The Future Is Now–Science for Achieving Sustainable Development*. United Nations, New York: Department of Economic and Social Affairs, United Nations Publications.
National Animal Welfare Advisory Committee (NAWAC). Available at: https://www.mpi.govt.nz/animals/animal-welfare/national-animal-welfare-advisory-committee/ [Accessed 30.12.2021].
Nurse A and Wyatt T, 2020. *Wildlife Criminology*. Bristol University Press.
Nurse A et al, 2021. Investigation of measures to reduce dog attacks and promote responsible ownership amongst dog owners with dog control issues in the UK. Middlesex University. Available at http://randd.defra.gov.uk/Default.aspx?Module=More&Location=None&ProjectID=19861 [Accessed 17.01.2021].
OIE, 2017. https://www.oie.int/en/adoption-of-the-first-oie-global-strategy-on-animal-welfare/ [Accessed 09.01.2022].
OIE Terrestrial Code Online Access, n.d. Available at https://www.oie.int/en/what-we-do/standards/codes-and-manuals/terrestrial-code-online-access/ [Accessed 15.12.2021].
OIE Aquatic Code Online Access, n.d. Available at https://www.oie.int/en/what-we-do/standards/codes-and-manuals/aquatic-code-online-access/ [Accessed 15.12.2021].
Pinillos RG ed., 2018. *One Welfare: A Framework to Improve Animal Welfare and Human Well-being*. Cabi.
Proposición de Ley de modificación del Código Civil, accessible at https://www.congreso.es/public_oficiales/L14/CONG/BOCG/B/BOCG-14-B-157-1.PDF [Accessed 15.12.21].
Robertson IA, 2015. *Animals, Welfare and the Law: Fundamental Principles for Critical Assessment*. Routledge.
Robertson I and Goldsworthy D, 2021. Will Victoria Miss the sentience doorway and have a 2021 animal law reform that is outdated even before it is enacted? Available at SSRN 3861352.
Sacirbey, 2014. Do Animals Need a UN Ambassador? Ambassador Muhamed Sacirbey, Former Bosnian Foreign Minister and Ambassador to the United Nations, Contributor. *Huffington Post* (07.04.2014). Available at https://www .huffpost.com /entry /do -animals -need -un -ambass b 5558663 [Accessed 09.07.2021].
Schaffner JE, 2010. *An Introduction to Animals and the Law*. Springer.
Sunstein CR, 1999. Standing for animals (with notes on animal rights). *UCLA Law Review.*, 47, p. 1333.
Sunstein CR, and Nussbaum MC, eds. 2004. *Animal Rights: Current Debates and New Directions*. Oxford University Press.
Sustainable Development Goals, 2015. United Nations General Assembly. 2015. *SDGs Transform Our World, 2030*. Available at https://sdgs.un.org [Accessed 15.12.2021].
UN, Educational, Scientific and Cultural Organisation.
Verniers E., 2021. Bringing animal welfare under the umbrella of sustainable development: A legal analysis. *Review of European, Comparative & International Environmental Law*, 30(3), pp. 349–362.
Verniers E and Brels S, 2021. UNCAHP, one health, and the sustainable development goals. *Journal of International Wildlife Law & Policy*, 24(1), pp. 38–56. DOI: 10.1080/13880292.2021.1923731
Visseren-Hamakers IJ, 2020. The 18th sustainable development goal. *Earth System Governance*, 3, p. 100047.
White S., 2013. Into the void: International law and the protection of animal welfare. *Global Policy*, 4(4), pp. 391–398.
Wise SM, 2010. Legal personhood and the nonhuman rights project. *Animal Law Review*, 17, p. 1.
Wise SM, 2014. *Rattling the Cage: Toward Legal Rights for Animals*. Hachette+ ORM.
World Trade Organisation, 2021. About WTO. Available at https://www.wto.org/english/thewto_e/thewto_e.htm [Accessed 15.12.2021].
Wyatt T., Maher J., Allen D., Clarke N. and Rook D., 2021. The welfare of wildlife: an interdisciplinary analysis of harm in the legal and illegal wildlife trades and possible ways forward. *Crime, Law and Social Change* 77: 69–89 –1

28

KEY ANIMAL LAW IN AUSTRALIA

Meg Good and Jed Goodfellow

Introduction

This chapter provides a brief overview and critique of the legal and regulatory framework governing animal protection in Australia. Given its strong reliance on animal agriculture and high rate of companion animal ownership, unsurprisingly in Australia the regulation of the human–animal relationship has become a significant and political issue. Increasingly, the Australian public are questioning the treatment of animals in animal use industries and the adequacy of animal protection under the law. This chapter explains how animal welfare is regulated within Australia's federal government structure, and puts forward recommendations for necessary structural and institutional regulatory reforms.

What is the legal status of animals in Australia?

Under the law in Australia, most animals are the property of human or corporate owners; a legal status that has significant implications for how they are valued, protected, and managed. However, this does not necessarily apply in the context of animals living in the wild, whose legal status may differ according to context and jurisdiction (Cao 2015, pp. 69–101).

Animals deemed property are the subjects of human property rights and therefore not rights-holders themselves, rendering them reliant on animal welfare legislation to protect their most essential interests. Yet unlike close neighbour New Zealand, under most of Australia's animal welfare laws, the sentience of animals is not expressly recognised. Only one jurisdiction, the Australian Capital Territory (ACT), has formally recognised the sentience of animals under its primary animal welfare legislation (*Animal Welfare Act 1992* (ACT)). The Act's objectives state that "animals are sentient beings that are able to subjectively feel and perceive the world around them" (s 4A (1) (a)). Further, it recognises that they have "intrinsic value" and "deserve to be treated with compassion and have a quality of life that reflects their intrinsic value" (s 4 A (1) (b)).

More jurisdictions look set to follow the ACT's lead, which may be viewed as a "shift away from categorising [animals] as property" (Kotzmann, 2019). While not altering their property status, legislative acknowledgment that animals have intrinsic value – that is, value independent of their value to humans – certainly renders them unique as a class of property (Kotzmann,

DOI: 10.4324/9781003182351-34

2020). However, if this does signify the beginning of a shift away from a pure property paradigm for animals it is unclear what alternative legal status may be substituted. Although there has been significant debate within the Australian animal protection community regarding the possibility of recognising animals as "legal persons", there have been no attempts in Australian parliaments or courts to challenge the fundamental legal status of animals under the law. Accordingly, their status as property seems unlikely to change in the near future.

How is animal welfare regulated in Australia?

Reflecting the legal status of animals generally, the Australian regulatory framework for animal welfare rests on the dual presumption that using animals for most human purposes is acceptable, and that causing pain and suffering to animals can be justified in various circumstances.

Federal division of power

Australia has a federal system of government in which legislative power is divided between the national Commonwealth Government and the eight state and territory governments. Although there is potential to use indirect heads of legislative power, there is no direct power to pass laws with respect to animal welfare and protection at the national level under the *Australian Constitution* (Bruce, 2018, 75). For this reason, the majority of Australian animal welfare law exists at the state and territory level. However, the Commonwealth has responsibility over animal welfare as it relates to international trade, such as the live export of animals, the regulation of export abattoirs, and the international trade in wildlife.

Regardless of its limited direct power, it is possible for the Commonwealth to take a greater role in animal welfare and protection, such as by leading and coordinating the states and territories to develop nationally consistent animal welfare standards. Historically, however, the Commonwealth has opted to limit the scope of its involvement, which has led to a lack of national consistency in Australia's animal welfare laws and policies. In fact, the absence of federal leadership and oversight was a significant factor contributing to the "D" ranking Australia received in World Animal Protection's latest 2020 revision of the global Animal Protection Index ("API") (World Animal Protection, 2020).

Engagement with international animal welfare standards and initiatives

Australia has provided its support to most of the few international agreements and initiatives that exist in the international animal welfare space. For instance, Australia has a consistent track record of engagement with the OIE's Assembly, Specialist Commissions, and Animal Welfare Standards. The API concluded that although at the state and territory level, "the OIE's animal welfare standards are broadly covered in legislation or policy", the fact that Australia no longer has a national animal welfare strategy "may act as a barrier for the OIE standards to be fully implemented" (World Animal Protection, 2020). The Australian Government has also pledged its in-principle support for the proposed Universal Declaration on Animal Welfare, and signed relevant international treaties such as the Convention on Biological Diversity (1992), Convention on International Trade in Endangered Species of Wild Fauna and Flora (1973), Convention on Wetlands of International Importance Especially as Waterfowl Habitat (1971), and the International Convention for the Regulation of Whaling (1946).

Governance framework

With the exception of the ACT and South Australia, animal welfare law and policy in Australia is governed by state, territory, and Commonwealth agriculture ministers and departments. Table 28.1 provides an overview of the responsible departments in each jurisdiction. The inherent issues with delegating animal welfare regulatory responsibilities to ministers and departments of agriculture are discussed in further detail below under "How can the legal and regulatory system be reformed?"

With the exception of Western Australia, all states and territories have established animal welfare advisory committees. However, only half of these are formally recognised in legislation. The Australian Capital Territory, Northern Territory, South Australia, and Tasmania establish animal welfare advisory committees under their animal welfare legislation, while New South Wales, Victoria, and Queensland have non-statutory committees. Unlike animal welfare advisory committees in other nations like New Zealand, none of the committees in Australia are empowered to make animal welfare standards or codes. The committees are merely advisory bodies and their advice and reports are not published. There is also minimal to no interaction between the various state and territory committees and Australia has no national animal welfare advisory body to promote national consistency.

There are limited political or government mechanisms specifically designed to provide oversight and accountability for the administration of animal welfare law. At the federal level, the Commonwealth Government established the Inspector-General of Live Animal Exports to oversee the Department of Agriculture, Water, and Environment's performance in regulating the live animal export trade following an animal welfare disaster onboard a live sheep voyage to the Middle East in 2017. General oversight and accountability mechanisms which can be employed to scrutinise animal welfare regulatory actions include Ombudsman bodies, Freedom of Information laws, federal and state parliamentary scrutiny committees, and of course the relevant tribunals and courts within the Australian judiciary.

Table 28.1 Principal animal welfare legislation and responsible government departments

Jurisdiction	*Principal legislation*	*Responsible department*
Australian Capital Territory (ACT)	*Animal Welfare Act 1992*	Transport Canberra and City Services
New South Wales (NSW)	*Prevention of Cruelty to Animals Act 1979*	Department of Primary Industries
Northern Territory (NT)	*Animal Welfare Act 1999*	Department of Industry, Tourism and Trade
Queensland (Qld)	*Animal Care and Protection Act 2001*	Department of Agriculture and Fisheries
South Australia (SA)	*Animal Welfare Act 1985*	Department for Environment and Water
Tasmania (Tas)	*Animal Welfare Act 1993*	Department of Natural Resources and Environment Tasmania
Victoria (Vic)	*Prevention of Cruelty to Animals Act 1986*	Agriculture Victoria
Western Australia (WA)	*Animal Welfare Act 2002*	Department of Primary Industries and Regional Development

Animal welfare legislation

As Table 28.1 provides, all states and territories have developed their own animal welfare legislation (hereafter referred to collectively as the "Animal Welfare Acts"). The Animal Welfare Acts are broadly consistent in terms of key elements and practical outcomes, however, there are quite significant differences in the framing and drafting of offences, penalties, and enforcement powers. Like other Commonwealth nations, Australia inherited its animal welfare laws from the United Kingdom and therefore similar animal welfare principles apply.

Cruelty offences

As a starting point, all of the Animal Welfare Acts prohibit animal cruelty. While different drafting is employed, the general meaning of animal cruelty under the legislation is the infliction of unnecessary harm upon an animal. In determining when harm caused to an animal is unnecessary in the circumstances, Australian courts apply a proportionality test derived from the English common law, which considers both the object of the harm and the means used to achieve that object, with particular focus on the amount of pain or suffering caused and whether it was proportionate to the object sought to be achieved (*Ford v Wiley* (1889) *23 QB 203*; *Department of Local Government & Regional Development (WA) v Emanuel Exports Pty Ltd, Graham Richard Dawes and Michael Anthony Stanton, Magistrates Court of Western Australia, 8 February 2008*).

If a disproportionate level of pain and suffering is found to have been inflicted, the impugned act or omission will amount to animal cruelty. By implication, this means the infliction of pain and suffering upon an animal is not in itself an offence. It is only when that pain and suffering is considered unnecessary that it becomes an offence. All of the Animal Welfare Acts then go on to expressly list a range of other conduct that falls within the meaning of animal cruelty, such as beating, terrifying, overriding, overworking, fighting, abandoning, and many others.

Duties of care

In addition to prohibiting animal cruelty, the Animal Welfare Acts impose duties of care upon persons responsible for the welfare of animals under their control. These duties may be framed in a positive sense (i.e. "a person in charge must provide … ") or in a negative sense (i.e. "a person in charge must not fail to provide … "). Currently, most Animal Welfare Acts in Australia frame the duties in the negative, as elements of the general cruelty offence, but there is now a trend towards separating out the duty requirements from the cruelty offences and framing them as a positive "duty of care". This reframing is important as it enables inspectors to act proactively to prevent animal neglect before the animal has suffered, improving animal welfare outcomes and better serving the intent of the legislation. The duties generally relate to the provision of appropriate food and water, shelter and living conditions, treatment of disease or injury, exercise and opportunities to express normal behaviour, and handling.

Defences and exemptions

The vast majority of animals used for commercial or other instrumental purposes are covered by industry codes or standards that operate as defences or exemptions to the application of the cruelty or duty provisions contained in the Acts. This effectively sets up a two-tiered system of animal welfare, where farmed animals are subject to lower standards of protection compared to those within a domestic companion context who enjoy the full protections set out in the legislation. Unlike in New Zealand's *Animal Welfare Act 1999*, there is no requirement in Australia for codes and standards to be consistent with the purpose and duties of the principal Animal Welfare Acts. This creates an incoherent legislative framework in which subordinate legislation authorises conduct that is inconsistent with the purpose and intent of the legislation under which it is made.

How is animal welfare regulated with regard to specific areas of animal use?

Farmed animals

Requirements relating to the welfare of farmed animals in Australia are prescribed in industry-specific "Model Codes of Practice", first developed in the 1980s and 1990s. The Model Codes are voluntary (except in South Australia), although compliance with their provisions affords a defence to prosecution for general animal cruelty or breach of duty of care offences in most jurisdictions. These defences effectively render practices, which may otherwise be considered cruel and prosecutable offences, immune from criminal liability.

However, the Model Codes are now being progressively replaced by the "Australian Animal Welfare Standards and Guidelines", which are designed to be enforceable, with the Standards prescribed as mandatory regulations under state and territory law (Animal Health Australia, 2020). Progress on converting the Model Codes into the enforceable Standards and Guidelines has been very slow. Despite the conversion process commencing in 2005 (Australian Government, 2009), only the *Australian Standards and Guidelines for the Land Transport of Livestock* have been fully implemented in all state and territory jurisdictions. Standards and Guidelines for sheep, cattle, and saleyards were developed between 2016 to 2018, but at the time of writing, only three out of eight states and territories have implemented them.

The Model Codes and Standards and Guidelines have been criticised by animal welfare groups for failing to lift the bar on animal welfare standards beyond current industry practice. The current Model Codes and Standards and Guidelines continue to permit practices that harm animal welfare, such as extreme confinement systems like battery cages for layer hens and sow stalls for pigs. They also continue to permit painful husbandry practices without pain relief, including tail docking, castration, spaying, beak trimming, hot iron branding, de-horning, and mulesing.

Legislative requirements relating to Australia's live animal export trade are set out in the *Australian Standards for the Export of Livestock* ("ASEL") made under the *Australian Meat and Livestock Industry Act 1997* (Cth) and the *Export Control (Animals) Rules 2021* ("ECAR") made under the *Export Control Act 2020* (Cth). The ASEL outline prescriptive requirements relating to the transport process from farm, to port, to the conditions onboard the vessels including stocking densities, ventilation requirements, veterinary treatment, and reporting obligations. The ECAR attempt to regulate the treatment of exported animals in the importing countries through the Exporter Supply Chain Assurance System ("ESCAS"). The ESCAS places obligations on exporters to ensure exported animals only go through pre-approved supply chains that meet OIE standards. In practice, many animals are "leaked" outside of approved supply chains and breaches of OIE standards occur frequently (Australian Government, 2021). Even if compliance with the standards was fully achieved, they permit slaughter without stunning which means that Australian animals can be slaughtered overseas whilst fully conscious.

Companion animals

The welfare of companion animals is regulated at the state and territory level primarily by the general cruelty and duty of care provisions of the principal Animal Welfare Acts (see above).

More prescriptive Codes of Practice apply for the commercial breeding and sale of companion animals in most state and territory jurisdictions. Unlike the Codes and Standards for farmed animals, the Codes for the breeding and sale of companion animals are developed on a state-by-state basis and therefore vary considerably in scope, detail, and the level of protections afforded. These Codes generally regulate the conditions in which dogs can be bred and sold commercially, including licensing of the breeding and sale facility and business, kennel specifications, maximum litters, veterinary requirements, and record keeping.

Further regulations have been introduced in Victoria to effectively prohibit large-scale commercial puppy farming by capping the number of breeding dogs for any facility to ten, requiring any facility seeking to breed over ten dogs to seek Ministerial approval, and banning the sale of dogs (other than rescue dogs) through pet stores (*Domestic Animals Act 1994*, Pt 4. Div. 3AA). While other jurisdictions have introduced prescriptive requirements for dog breeding, they still permit large-scale intensive dog breeding operations.

Animals used in research

The use of animals in scientific research is governed by the *Australian Code for the Care and Use of Animals for Scientific Purposes* (the "Scientific Code"). As with the Codes and Standards for farmed animals, the Scientific Code is developed at the national level (by the National Health and Medical Research Council) and is implemented and enforced at a state and territory level under the Animal Welfare Acts, with the exception of NSW, which has established a separate legislative regime for animal research under the *Animal Research Act 1985* (NSW). The Scientific Code sets out the principles of the "3Rs" – replacement, reduction, and refinement – and the requirements for establishing Animal Ethics Committees. However, the effectiveness of the Code for protecting animal welfare has been critiqued on numerous grounds, including the nature of some of the experiments it permits. Moreover, the adequacy of its implementation is brought into question when considering the very large numbers of animals used in Australia for scientific research every year. While national statistics are not collated in a uniform way, estimates suggest Australia has one of the highest per capita rates of animal use for research in the world at approximately 3.2 million animals in 2015 (Taylor and Rego Alvarez 2019, 204).

Wild animal welfare protection

Animals living in the wild

Although the Commonwealth has adopted some responsibility for wild animal welfare, it is managed primarily by state and territory governments (Cao, 2015, pp. 245–255). Most animal welfare legislation technically applies to animals living in the wild, however its application is often significantly limited or in some cases, completely excluded (White, 2009, pp. 238–242). Nature conservation legislation may also address wild animal welfare, however its primary focus is the preservation of species and ecosystems, rather than protection of individual animal welfare (Thiriet, 2009, p. 270).

Commentators argue that the overall effect of this legislative regime is the creation of a "hierarchy of protection", preferencing native animals with higher conservation status first, common native animals second, and common introduced species last (Thiriet, 2009, 270; White, 2009, 251). As a consequence, wild animal welfare regulation fails to acknowledge the equal capacity for suffering of all sentient animals (Thiriet, 2009, 270; White, 2009, p. 256).

The treatment of kangaroos is a prime example of the hierarchy in action. Although kangaroos are native animals protected under nature conservation legislation, both commercial and non-commercial kangaroo shooting are legally permitted. This is due to government perception that certain kangaroo species are "common" native animals, whose populations need to be reduced – a view that some scientists question (Boom and Ben-Ami, 2010). Nationally each year on average two million kangaroos are commercially shot across five state jurisdictions – a figure which includes an estimate of the number of dependent young killed by shooters or left to die in the wild. Government support for the killing of these protected native species is premised on a belief that it is acceptable to manage native animals which are perceived to exist in abundance via lethal means. This is despite the fact that strong opposition has mounted both in Australia and globally against all forms of kangaroo shooting, largely due to the unavoidable and significant welfare risks associated with the shooting of free-ranging animals in the wild.

Wild animals in captivity

Wild animals are held in captivity in Australia for a variety of purposes, including for conservation, education, and entertainment. Exhibiting captive wild animals is regulated at the state/territory level, with different regimes applying across the jurisdictions (Bruce 2018, pp. 169–178). A vast range of species are held captive and exhibited across the country, although the New South Wales Government recently introduced a ban on the breeding and importation of cetaceans (*Biodiversity Conservation Regulation 2017*, s 2.8A).

In 2019, national standards and guidelines for the exhibition of animals kept in facilities (such as zoos) were endorsed by the states and territories (*The Australian Animal Welfare Standards and Guidelines for Exhibited Animals* (2019)), aiming to create more consistency in the standard of care across Australia. However, these standards do not cover circuses, which are governed by separate circus animal codes and standards regulated at a state and territory level. Local councils also possess the power to determine whether circuses may be held on their land, with over 40 opting to ban circuses that use wild animals (RSPCA South Australia, 2021). Despite this, keeping and exhibiting wild animals for entertainment purposes in mobile exhibitions remains legal (except in the ACT for certain species), although presently there are no circuses using wild animals in Australia.

Animals in entertainment

In addition to exhibition in zoos, circuses, and aquaria, animals are used for entertainment in Australia in a range of other contexts, including rodeos, horse racing, greyhound racing, and film/theatre. These industries are regulated at the state/territory level and different legislative standards apply across the country (Bruce, 2018, pp. 179–188). Certain problematic practices are banned in specific jurisdictions. For example, the Australian Capital Territory has prohibited rodeos and NSW has banned jumps racing. Some practices have been banned in all jurisdictions, such as animal fighting events or using animals as bait and lures for animal racing.

How can the legal and regulatory system be reformed?

The legal and regulatory system for animal protection in Australia arguably does not adequately represent and safeguard the interests of animals. Australia lacks a consistent national approach, with the Commonwealth government adopting minimal responsibility and leadership.

Various practices which cause animal pain and suffering are legalised, and those protections that do exist are often applied on a differential basis, failing to acknowledge the equal capacity for pain and suffering shared by all sentient animals. Substantive law issues are compounded by the fact that enforcement of animal welfare legislation is under-resourced and inadequate for achieving the overall legislative objects. Arguably there is significant need for reform, yet achieving law reform in this area has proven to be difficult and slow-moving.

Reform challenges and successes

Reform attempts face a broad range of challenges, including problematic law reform processes, close relationships between government and industry, lack of public awareness, lack of political support, and the existence of powerful lobby groups campaigning on behalf of animal use industries. For this reason, generally only small incremental reforms are able to pass through the law reform process, frustrating the achievement of broader systemic change. As a result, the reform dialogue is often reduced to discussions about how to *regulate* inherently harmful activities, rather than *eliminating* them.

Complete bans have mainly been achieved where the practice in question was very minor or non-existent within the relevant jurisdiction. For example, legislation restricting the use of animals for cosmetics testing was introduced, yet this type of testing was not actually being used in Australia. Similarly, the Australian Capital Territory (ACT) banned certain intensive farming practices that were almost completely absent in the jurisdiction. The ACT ban prohibited two significant intensive farming practices – specifically, the use of battery cages for egg production and the use of sow stalls for pig production. However, at the time the ban was implemented there was only battery cage operator and no intensive piggeries in the ACT (Brennan, 2014). A further example from NSW is the introduction of a ban on the importation and breeding of cetaceans for captivity at a time where the only remaining dolphinarium in NSW had already ceased the practice.

Despite their limited practical impact, these achievements represent an important shift in how animals are valued. They also create models for reform for other jurisdictions and make it much harder for governments to roll back on these hard-won protections in the future. However, in practice they all had minimal impact on the status quo, whilst in contrast proposed reforms with significant implications both for vested interests and animal welfare continue to fail despite public support. For instance, although a majority of the Australian public are opposed to live animal export (Sullivan, 2019), and it is clear that regulation cannot address the trade's inherent animal welfare challenges, the industry continues to receive government support largely due to its perceived economic benefit for the country.

It has proven extremely difficult to achieve meaningful reform in relation to the practices of well-established animal use industries, such as horse racing, greyhound racing, and intensive animal agriculture. Due to these challenges, some important reforms are largely excluded from the reform agenda, deemed to be politically unfeasible. These include banning all intensive animal agriculture practices across the country and introducing animal rights recognition into law.

Despite the challenging law reform context, some impactful reforms have been achieved, largely due to the efforts of animal protection advocates and organisations. These include Tasmania's ban on establishing new battery cage operations and qualified ban on sow stalls, the ACT's ban on greyhound racing, Victoria's effective prohibition on puppy farming and requirement for pain relief in mulesing, and bans on recreational duck hunting across various jurisdictions. As is often the case, a number of reforms were achieved as reactionary measures by government in response to animal cruelty exposés (McEwan, 2019, p. 8). These reactionary law reform measures, although important for improving animal welfare, are reflective of a broader systemic failure to adequately *prevent* animal suffering.

Need for institutional and structural reform

Animal protection campaigns in Australia have traditionally taken a single-issue approach, pursuing reforms in regard to specific practices, such as tightening the regulations for greyhound racing or banning live export. However, the animal protection community is now starting to place increasing focus on more foundational institutional and structural challenges impeding the progress of animal welfare policy and regulation. In particular, the impacts of close relationships between government and animal use industries on animal welfare policy, and the inadequate frameworks governing the development of animal welfare standards. Proposed reforms in these areas aim to reduce the high degree of influence and control of agricultural institutions over Australia's animal welfare policy framework, and to introduce more transparent and accountable processes for policy development.

Independent offices of animal welfare

One particularly prominent national campaign in this regard is the movement to secure the introduction of an independent statutory entity dedicated to animal welfare at the national level. A key aim of this reform is to address the issues raised by the influence of animal use industries, and to reduce conflicts of interest in the animal welfare policy process. Numerous versions of the reform have been debated, however they are all variations of an independent federal statutory entity dedicated to animal welfare policy and standards development.

Although such a body could not be granted power to enforce animal welfare law (due to constitutional limitations), it could coordinate the development of national animal welfare policy and standards in consultation with the states and territories. This is precisely what the Australian Productivity Commission proposed in its landmark report on the regulation of Australian agriculture in 2017 (Australian Government Productivity Commission, 2017). After identifying several deficiencies with the current governance and regulatory arrangements, including the perception of conflicting interests within agriculture departments, the Productivity Commission recommended the establishment of an Australian Commission for Animal Welfare (ACAW).

It was envisaged that the Commission would oversee the development of national animal welfare standards and monitor the performance of state and territory governments in implementing the standards. However, the Australian Government rejected the need for such a body, raising uncertainty about its proposed constitutional basis and claiming the Productivity Commission had not established a viable role for the proposed ACAW (Australian Government, 2019).

Separately from the Productivity Commission's report, the Australian Government did establish an Inspector-General of Live Animal Exports Act 2019 (Cth) to oversee the Department of Agriculture's performance in regulating the trade with a clear focus on animal welfare (*Inspector-General of Live Animal Exports Act 2019*). While this entity does not perform the role proposed for the ACAW, it is nevertheless an example of an independent statutory entity with a focus on animal welfare that could be expanded and built upon in the future with further resourcing and simple amendments to the enabling legislation. To address equivalent issues at a state and territory level, similar statutorily independent bodies have been proposed to administer state and territory animal welfare law in the form of state Animal Welfare Authorities. These Authorities could replace the role of departments of agriculture in administering animal welfare legislation at the state level.

Such institutional reforms are not a panacea for the lack of government commitment to addressing current animal welfare challenges requiring urgent attention. However, properly constituted, they could significantly reduce the influence of conflicting political and economic interests currently dominating the policy domain, and facilitate a renewed and balanced focus on upholding current standards and introducing meaningful reforms.

The degree of control exerted by agricultural institutions over animal welfare policy and law has been shown to constitute a form of "regulatory capture", in which the responsible agencies consistently act in the interests of the livestock and other animal use industries in a way that deviates from the public interest the regulation is designed to serve (Goodfellow, 2016). This is largely due to the fact that the dominant purpose of these institutions is to promote productive and profitable primary industries and otherwise serve the interests of the agricultural sector.

Recent examples of regulatory capture in the animal welfare policy process demonstrate the need for a more independent regulator in this policy domain. One particularly significant example is in relation to the development of Australia's national Standards and Guidelines for the Welfare of Poultry, led by the New South Wales Department of Primary Industries (DPI). In 2017, Freedom of Information documents revealed (despite denial by the government) that, during the process of developing the standards, DPI regulators held secret meetings with poultry industry executives. The purpose of these meetings was to develop strategies for navigating the draft standards through stakeholder consultation meetings and to arrange for industry executives to vet the proposed independent chairperson for the meetings prior to their appointment (Thomas and Branley, 2017; Ellis, 2018). DPI had also removed reference to the importance of allowing poultry to perform normal patterns of behaviour from the draft standards because this was not compatible with the continued use of barren battery cages for egg-laying hens, a practice the egg industry was seeking to protect (Thomas and Branley, 2018).

Another recent example of regulatory capture involved the federal Agriculture Minister's last-minute intervention on behalf of the Australian cattle industry to override plans to reduce stocking densities onboard live export ships. Following a two-year review of the *Australian Standards for the Export of Livestock* from 2018–2019, modest reductions in stocking densities to allow cattle to lie down and better access food and water troughs were recommended for implementation by 1 November 2020. However, after vigorous lobbying from the cattle industry, federal Minister for Agriculture, the Hon. David Littleproud MP, intervened three days before the commencement date to shelve the density reductions and re-insert the old density levels in the revised standards, saving the industry an estimated $40 million per year (Australian Government, 2020).

These are just two specific examples of a systemic approach by government whereby private economic interests are prioritised over the public interest in achieving clear, science-based animal welfare improvements (for a full account of the effects of regulatory capture, see Goodfellow 2015). By delegating responsibility for animal welfare policy exclusively to agricultural departments, the system is subject to capture by institutional design. Accordingly, institutional reform, such as introducing independent offices of animal welfare, is required to remove conflicting political and economic interests and introduce greater independence and evidence-based objectivity into the process.

Transparent and accountable animal welfare standard-setting process

A second related campaign is the call for more transparent and accountable animal welfare standard-setting systems. The aim of this reform is to ensure that animal welfare standards adequately reflect public opinion, and are consistent with best available science. One way to help achieve this outcome is by creating a more formalised standard-setting system that better protects the interests of animals. Unlike countries such as New Zealand, where the process for developing national standards and regulations is prescribed in law (see, *Animal Welfare Act 1999*, Pt. 5 and s 183A), Australia's process is undertaken in an ad hoc fashion where one jurisdiction voluntarily nominates to take on the responsibility of managing the development process for a given set of standards. There is no legislation governing how the process takes place, what factors need to be considered, nor what criteria need to be met. This is problematic, as in the absence of appropriate formalised procedures, agricultural institutions are able to exert significant influence over the process.

These institutions include state and federal ministers for agriculture, departments of agriculture, and peak livestock industry representative bodies; together forming an "exclusive policy community" controlling the development of animal welfare standards (Goodfellow, 2015, pp. 171–173). This is especially the case regarding welfare standards for farmed animals. The key incentives and priorities for these institutions are often misaligned with those of protecting and improving animal welfare, particularly in circumstances where welfare proposals conflict with industry productivity goals (Goodfellow, 2016, pp. 215–216).

When it comes to setting standards of animal care, the Australian agriculture policy community typically adopts a narrow conception of animal welfare which focuses primarily on basic health and biological functioning outcomes. The affective states domain of animal welfare (that is, how the animal is subjectively feeling) has traditionally been downplayed or dismissed by this policy community as being subjective or hypothetical, or at best, indeterminate (Goodfellow 2015, pp. 237–240). This is not consistent with contemporary animal welfare science which now takes a more holistic approach to welfare assessment in which both the physical and mental states of the animal are considered (Mellor, 2017). Introducing formal legislative criteria for standards development processes may help to prevent this limited conception of animal welfare from frustrating the achievement of science-based welfare standards that more closely align with the interests of animals.

While there may arguably be constitutional barriers to enacting a national animal welfare law to govern the standard-setting process, no such barriers exist for state and territory governments to introduce standards development processes and criteria within their legislation. Amendments could be made to the regulation-making powers within state animal welfare legislation requiring the responsible authorities to be satisfied that any proposed standards are based on relevant scientific evidence and have taken into account community expectations.

The amendments could also include a requirement that all standards proposed are consistent with the substantive duty of care provisions contained in the principal animal welfare legislation. This could improve the coherence of the legislative framework by reducing the level of inconsistency between that which is permitted in the principal legislation relative to the practices permitted in the subordinate industry standards. To further enhance accountability, the legislation could also require all proposed standards to be tabled in the relevant state parliament to encourage high level political debate and greater awareness of animal welfare issues among politicians and the general public.

An example of how such provisions improve accountability in the standards development process can be seen in the recent New Zealand case of *The New Zealand Animal Law Association v The Attorney-General* [2020] NZHC 3009. Section 183A of the NZ *Animal Welfare Act 1999* requires any practices prescribed in regulations that do not meet the substantive animal care obligations of the Act, including the obligation to ensure that the behavioural needs of animals are met, to be brought into line with those obligations within a ten-year period. This provision allowed the New Zealand Animal Law Association (NZALA) and SAFE (NZ) to challenge the legality of regulations that permitted the use of farrowing crates and mating stalls which prevent pigs from expressing their behavioural needs.

The High Court of New Zealand agreed with NZALA and SAFE, finding that the Minister of Agriculture and the National Animal Welfare Advisory Committee had acted illegally when they failed to phase out farrowing crates and mating stalls within the requisite time period. The Court declared the regulations to be unlawful and invalid, and in response one month later the New Zealand Government announced its commitment to phasing out the use of these crates and stalls by 2025. Prescribing similar criteria and improved processes in Australian state and territory animal welfare legislation could significantly enhance the transparency and accountability of the current standards development framework and lead to more robust and consistent animal welfare standards.

Conclusions

Although Australia is one of the world's most economically advanced nations, it has failed to implement an adequate framework for animal protection. Due to the country's federal structure and a lack of national leadership, the Australian legal and regulatory regime for animal welfare and protection is fragmented across the states and territories. Although this fragmentation has created an inconsistent approach to animal welfare, there are some commonalities between the jurisdictions. Almost all fail to recognise the sentience of animals, and they each permit a wide range of practices that cause animal pain and suffering. They also share broader structural and institutional challenges necessitating regulatory reform.

Institutional reform is required to remove conflicting political and economic interests and introduce greater independence and evidence-based objectivity into the animal welfare policy process. This could be achieved through the establishment of federal and state/territory statutory entities dedicated to animal welfare policy and standards development. A related necessary reform is the introduction of a transparent, consistent, and accountable animal welfare standard-setting process. Although achieving these reforms would not address all of the major deficiencies in the Australian animal welfare regime, it would represent a fundamental step in the right direction.

References

Animal Health Australia, 2020. Australian animal welfare standards and guidelines. http://www.animalwelfarestandards.net.au. [Accessed on 5 November 2021].

Australian Government, 2009. Australian animal welfare strategy: Development of Australian standards and guidelines for the welfare of Livestock: Business plan. http://www.animalwelfarestandards.net.au/files/2011/01/Animal-Welfare-Standards-and-Guidelines-Development-Business-Plan.pdf. [Accessed on 5 November 2021].

Australian Government, 2019. Productivity commission inquiry into the regulation of Australian agriculture: Australian government response. https://www.pc.gov.au/inquiries/completed/agriculture/agriculture-government-response.pdf. [Accessed on 5 November 2021].

Australian Government, 2020. Export advisory notice 2020/23: Notification of changes to ASEL 3.0: Cattle by sea and all livestock by air. https://www.agriculture.gov.au/sites/default/files/documents/ean-2020-23.pdf. [Accessed on 5 November 2021].

Australian Government, 2021. Regulatory compliance investigations. https://www.agriculture.gov.au/export/controlled-goods/live-animals/livestock/regulatory-framework/compliance-investigations/investigations-regulatory-compliance. [Accessed on 5 November 2021].

Australian Government Productivity Commission, 2017. *Regulation of Agriculture: Inquiry Report*. https://www.pc.gov.au/inquiries/completed/agriculture/report. [Accessed on 5 November 2021].

Boom K and Ben-Ami D. 2010. Shooting our wildlife: An analysis of the law and policy governing the killing of Kangaroos. THINKK The Kangaroo Think Tank, University of Technology Sydney. http://thinkkangaroos.uts.edu.au/publications.html. [Accessed on 5 November 2021].

Brennan M. 2014. ACT bans battery cages and sow stalls. *FlagPost*, Australian Parliamentary Library. https://www.aph.gov.au/About_Parliament/Parliamentary_Departments/Parliamentary_Library/FlagPost/2014/February/ACT_bans_battery_cages_and_sow_stalls. [Accessed on 5 November 2021].

Bruce A, 2018. *Animal Law in Australia: An Integrated Approach*. 2nd edn. Sydney: LexisNexis Butterworths.

Cao D, 2015. *Animal Law in Australia*. Sydney: Thomson Reuters.

Convention on Biological Diversity, June 5, 1992, 1760 U.N.T.S. 79.

Convention on International Trade in Endangered Species of Wild Fauna and Flora, March 3, 1973, 993 U.N.T.S. 243.

Convention on Wetlands of International Importance Especially as Waterfowl Habitat ('Ramsar Convention'), February 2, 1971, 996 U.N.T.S. 245.

Department of Local Government & Regional Development (WA) v Emanuel Exports Pty Ltd, Graham Richard Dawes and Michael Anthony Stanton, 2008. *Magistrates Court of WA, Magistrate CP Crawford*, 8 February 2008.

Ellis E, 2018. Governments can't be trusted to deliver welfare standards for chickens. *The Conversation*, 12 February. https://theconversation.com/governments-cant-be-trusted-to-deliver-welfare-standards-for-chickens-90091. [Accessed on 5 November 2021].

Ford v Wiley, 1889. 23 QBD 203.

Goodfellow J, 2015. *Animal welfare regulation in the Australian agricultural sector: A legitimacy maximising analysis*. PhD diss., Macquarie University.

Goodfellow J, 2016. Regulatory capture and the welfare of farm animals in Australia. In D Cao and S White, eds. *Animal Law and Welfare: International Perspectives*. Switzerland: Springer International Publishing.

International Convention for the Regulation of Whaling, December 2, 1946, 161 U.N.T.S. 72.

Kotzmann J, 2019. ACT's new animal sentience law recognises an animal's psychological pain and pleasure, and may lead to better protections. *The Conversation*, 3 October. https://theconversation.com/acts-new-animal-sentience-law-recognises-an-animals-psychological-pain-and-pleasure-and-may-lead-to-better-protections-124577. [Accessed on 5 November 2021].

Kotzmann J, 2020. Recognising the sentience of animals in law: A justification and framework for Australian states and territories. *Sydney Law Review*, 42(3), pp. 281–310.

McEwan A, 2019. Animal law and policy reform in Australia. Voiceless, the animal protection institute. https://voiceless.org.au/animal-law/animal-law-and-policy-reform/. [Accessed on 5 November 2021].

Mellor D, 2017. Operational details of the five domains model and its key applications to the assessment and management of animal welfare. *Animals*, 7(8), p. 60.

RSPCA South Australia, 2021. Good news: Australia no longer has circuses using wild animals. https://www.rspcasa.org.au/australia-circus-animal-free/. [Accessed on 5 November 2021].

Sullivan K, 2019. Vote compass finds almost two-thirds of Australian voters support a ban on live animal exports. *ABC News*, 26 April. https://www.abc.net.au/news/2019-04-26/vote-compass-live-exports-almost-two-thirds-support-ban/11046230. [Accessed on 5 November 2021].

Taylor K and Alvarez LR. 2019. An estimate of the number of animals used for scientific purposes worldwide in 2015. *Alternatives to Laboratory Animals*, 47(5–6), pp. 196–213.

The New Zealand Animal Law Association v The Attorney-General, 2020. NZHC 3009.

Thiriet D, 2009. Recreational hunting: Regulation and animal welfare concerns. In P Sankoff and S White, eds. *Animal Law in Australasia: A New Dialogue*. Sydney: Federation Press.

Thomas J and Branley A, 2017. Egg farmers accused of colluding with government department to sabotage moves to outlaw battery hens. *ABC News*, 21 December. https://www.abc.net.au/news/2017-12-21/egg-farmers-accused-of-colluding-with-nsw-government/9229242. [Accessed on 5 November 2021].

White S, 2009. Animals in the wild: Animal welfare and the law. In P Sankoff and S White, eds. *Animal Law in Australasia: A New Dialogue*. Sydney: Federation Press.

World Animal Protection, 2020. *Australia. Animal Protection Index*. https://api.worldanimalprotection.org/country/australia. [Accessed on 5 November 2021].

29
KEY ANIMAL LAW IN CHINA

Deborah Cao

Introduction

This chapter attempts to outline the key areas and signposts or pointers of the current laws and regulations regarding animals and animal welfare in the People's Republic of China (PRC). Given the vast size of China, its huge population, and the large number of animals living, working, and dying in China, legal protection or the lack thereof for these animals is a serious matter, not only in China but for the international community as well.

Animal protection in contemporary China

As is widely known and acknowledged both in China and elsewhere, animals in general do not fare well in the PRC, and there is little effective legal protection for most of them against human abuse and mistreatment (for detailed discussions of legal and cultural issues related to animals in China, both past and present, see Cao, 2015). This is so despite some laws, regulations, and other measures in place for protecting various types of animals, and despite the fact that some of the penalties for wildlife crimes are severe. A basic legal regime relating to animals exists in China, consisting of a constitutional provision, national laws, subordinate administrative regulations, provincial and local implementing measures, and various binding interpretations and opinions by the highest court and relevant legal and administrative authorities.

Constitutional and legal status of animals

Since the founding of the PRC in 1949, a number of laws and regulations concerning animals and animal protection have been adopted in the last few decades. To start with, the Chinese Constitution (1982) mentions animals once. It states that "the State ensures the rational use of natural resources and protects rare and valuable animals and plants" (Art 9). This means that, fundamentally, in the Chinese legal order, rare and valuable animals or protected wildlife are classified as natural resources. As such, they are the object of utilisation for human benefit and, at the same time, the animals also warrant human protection. This is the basic thinking running through Chinese laws related to animals, that is, utilisation and protection, and often utilisation takes priority over protection. As objects of property under Chinese law, animals including

 DOI: 10.4324/9781003182351-35

protected wildlife do not have legal rights (similar to the situation in most countries). Different from many Western countries, animals in China are not yet recognised as sentient beings or beings with intrinsic values of their own (For a discussion of traditional Chinese philosophy and culture which recognises animal sentience, see Cao, 2018b and Shih and Singer, 2018). The laws and regulations related to animals consequently are primarily designed to protect humans and human interests or human utilisation of animals as resources, not focusing on animal welfare. Paradoxically, for some animals as private or state property, this is the only legal protection that Chinese law affords them. For instance, for companion animals such as dogs, this is the only way by which owners of such animals can exert a claim for their non-human family members as there is no other legal recourse to protect them.

Anti-cruelty laws

There is no anti-cruelty law for any type of animals, whether domestic or wild, under Chinese law. There is no animal welfare law either, although some provisions in animal-related laws are for the improvement and protection of animal welfare (for a recent survey of the attitudes of the Chinese population towards animal welfare, see Carnovale et al., 2021). Although non-governmental or non-official efforts have been made continuously to introduce such a law over the last decade, China still does not have any legal provision to prevent cruelty and abuse to animals at the national or local levels, except one jurisdiction, Hong Kong – which has retained the anti-cruelty law inherited from the previous English administration. China is one of the few countries in the world today that does not have anti-cruelty laws, which is a major defect in China's legal system (For studies on the animal laws and cruelty related offences in various countries, including Australia, Israel, South Africa, Brazil and others, see Cao and White, 2016). This means that most animals including companion animals, farm animals, and wild animals used for entertainment and other purposes have little or no legal protection in China (For discussions of animal protection in China and various challenges, see Li and Davey, 2013).

Wildlife

Wild animals, in particular, state-protected wildlife, have more legal protection compared with all other types of animals in China. Such laws consist of a constitutional provision, various national laws, subordinate administrative regulations, provincial and local wildlife implementing regulations, various measures of the relevant legal and administrative authorities, and binding interpretations and opinions by the highest court and other legal and administrative authorities (see Cao, 2015). Among these are the Wildlife Protection Law (WPL, 1988) and, to a lesser extent, Environmental Protection Law (1989), Fishery Law (1986), Customs Law (1987), and Forestry Law (2000). There are also subordinate regulations and measures involving wildlife protection, health and quarantine authorised under the major national laws. Additionally, the Criminal Code (1997) has provisions on wildlife protection and on the criminal liabilities including penalties for illegal hunting, killing and unlawful dealing (i.e., sale, transport, and purchase) of protected wild animals. China is also a party to a number of international treaties and bilateral treaties pertinent to wildlife, including the Convention on International Trade in Endangered Species of Wild Fauna and Flora (CITES), Biodiversity Convention and Wetland Convention, among others.

The most important basic national law for wildlife is the WPL (1988), enacted in pursuance of the Constitution (1982) (For further discussions of the WPL (1988), see Sharma, 2005, Cao, 2015, Li, 2021). This law is to both protect wildlife and legalise the domestication and utilisation of wildlife and the development of the wildlife industry in China. It sets out the guiding principles and framework for wildlife management, utilisation, and protection, that is, for the conservation of rare and valuable animals and endangered species as well as the development and utilisation of wildlife resources (Art 1). These objectives embody the basic Chinese official position that wildlife species are natural resources owned by the state and their protection is necessary. The WPL (1988) also aims to promote ecological balance through wildlife management and protection as stated in its most recent amended version. The WPL (1988) has been amended a number of times and is currently under review for further amendment. It has been suggested that despite the various amendments and changes, the wildlife laws and policies in China today including the WPL are basically an updated version of the government policies in the 1960s, that is, the essence of the law remains unchanged despite the many changes to its provisions (Liang, 2016).

In pursuance of the WPL (1988), there are various subordinate regulations and administrative measures relating to wildlife protection, including the Regulations for the Implementation of Terrestrial Wildlife Protection (1992), setting out procedures and rules for domestication and breeding operations of protected wildlife, the Regulations on the Implementation of Aquatic Wildlife Protection (1993), Regulations on Management of the Import and Export of Endangered Wild Fauna and Flora (2006), Circular on the Ban of Trade of Rhino Horns and Tiger Bones by the State Council (1993), Urban Zoos Management Regulations (1994) and Natural Reserves Management Regulations (1994). There are also interpretations and opinions by the highest court in China, the Supreme People's Court (SPC), and the highest prosecution authority, the Supreme People's Procuratorate (SPP), on the specific provisions of the above-mentioned national laws and on the application of the laws in wildlife-related court trials, in particular in determining crimes and appropriate punishment (for recent studies of wildlife crimes in China, see Cao (2016, 2019), and van Uhm (2019, 2020); for discussion of Chinese wildlife law and politics, see Li, 2021). There are also regulations at the provincial level for the implementation of the national laws, authorising local authorities and specifying local conditions.

In short, the WPL (1988) provides the legal basis for both protection and utilisation of wildlife and endangered species. It also authorises and legalises the utilisation and trade of protected wildlife for various purposes through a licensing system. As a result, the law has been used to give a green light to various forestry authorities to license farmers and traders for wildlife exploitation. This licensing system also helped traffickers and restaurants to pass off illegally acquired wild animals, and enabled individuals and commercial enterprises to breed Siberian tigers, moon bears, and other species even though these animals are special state-protected animals.

A recent important development in the aftermath of the COVID-19 pandemic is that since early 2020, a comprehensive national ban has been imposed on the illegal trade of "terrestrial wildlife", prohibiting the sale and eating of wild animals, in an effort to stem the traditional practices of eating wild animals in China – a positive and necessary move (see www.npc.gov.cn/englishnpc/lawsoftheprc/202003/e31e4fac9a9b4df693d0e2340d016dcd.shtml. For further discussions of COVID-19 and wildlife protection in China, see Huang, Wang, Yang et al., 2021).

Animals used for research and testing

According to the Chinese authorities, some 12 million animals are used for scientific purposes nationwide each year, including mice, rabbits, guinea pigs, dogs, and primates. China is a leading producer and exporter of animals used for research. One of the earlier national legislation for animals in China is the Regulations for the Administration of Laboratory Animals that came into effect in 1988, establishing the basic rules for animals used in education, research, and testing. In 1992, the Detailed Implementation Rules for the Administration of Laboratory Animals Used in Medical Experiments and the Standards for Medical Experiments Using Animals (1992) was published by the Ministry of Health. In 2016, China released its first national standards governing the treatment of animals in research in an effort to improve both conditions for animals in research and China's prospects for international research collaborations (see Bayne, Wang, and Pang, 2018; Cao, 2015, 2018a; McLaughlin, 2016).

As mentioned earlier, Chinese laws and regulations related to animals have been enacted with the utilisation of the animal resources as their primary objectives. It is also true for the regulation of animals used in research, or perhaps more so because of the international research culture and ethical standard practices for the recognition and publication of scientific research experiments. China enacted the laws very much with this in mind. Accordingly, China also has various guidelines and standards for quality control of animals used in experiments at the national and local levels. The national Laboratory Animal Quality Management Measures and the Laboratory Animal Licensing Management Measures were issued in 1997 and 2001 respectively. Unrelated to animal welfare, they concern only the quality control of animals used in research. The quarantine of imported and exported animals for research purposes is covered by the Law on the Quarantine of Animals and Plants in Imports and Exports (1991).

Of the regulatory measures, the most important are the Regulations for the Administration of Laboratory Animals (1988) and the Guidelines for the Humane Treatment of Laboratory Animals (2006) (see Cao, 2015; Kong and Qin, 2010). The stated purpose of the 1988 Regulations is to ensure the quality of animals used in labs (Article 1). Nevertheless, they provide some protection to animals used in education, research, and testing. The most recent development is that in August 2021, the Chinese Ministry of Science and Technology released a draft for proposed amendments to the Regulations for public consultation. A significant proposed change is that the amended Regulations, if the proposed amendments are adopted, will include "ensuring animal welfare" as one of the legislative objectives. The draft states, "The regulations are made for the purposes of standardising the management of laboratory animals, guaranteeing the quality of laboratory animals, ensuring animal welfare, maintaining public safety, and meeting the needs of scientific research, economic construction and social development" (Art 1). This, if adopted, will be the first time that "ensuring animal welfare" is written as a legislative objective in Chinese laws. The draft changes to the Regulations also include expanded rules on public safety and infectious disease control in research facilities and specifications on genetic modifications.

Relevantly, the proposed draft has a new section on the management of animal welfare ethics, including the provisions that the use of laboratory animals shall follow the 3Rs principles (replacement, reduction, and refinement of laboratory animal use), and research facilities and individuals engaged in laboratory animal work shall take effective measures to meet the requirements of laboratory animal welfare and animal research ethics. It also requires that laboratory animal welfare ethics committees must be established to conduct "independent, fair and just and scientific ethics review and supervision", although it does not specify how this would be carried out. Another new proposed provision is that the transport of laboratory animals must ensure the welfare requirements of the animals.

Some of these proposed provisions are found in the earlier policy document – Guidelines for the Humane Treatment of Laboratory Animals (2006). In the Guidelines, there are more specific provisions for the "humane treatment of laboratory animals", for instance, taking effective measures in the course of laboratory animal care and use so that such animals will not suffer from unnecessary harm, hunger, thirst, discomfort, fear, torture, disease, and pain and the animals can achieve their natural behaviour; the animals will receive good management and care; they will be provided with a clean and comfortable living environment, and sufficient and healthy food and water, and pain and suffering will be lessened or avoided. The Guidelines also state that the humane treatment of the animals includes the promotion of the 3Rs principles and that such animals should be used scientifically, rationally, and humanely.

As there are no criminal offences under Chinese law for mistreating animals, a violation of the Guidelines which is a policy directive may result in professional disciplinary action, but how this may happen is unclear. There is little public information related to disciplinary measures for violation or breach of the relevant ethical rules in this area. There is no provision for penalties in the proposed amended Regulations either for mistreatment of animals or violation of animal ethics rules. Thus, a major issue regarding animals used in research is the lack of transparency and information to the public. This is true in China as well as other countries where animals are still used for research and testing purposes (For detailed discussions, see Cao, 2015).

Companion animals

In recent years, there has been a growing interest in animal legal protection in China, both at the official and community levels propelled by the grassroots animal rescue efforts (For discussions of the Chinese animal protection movement as an emerging social movement, see Cao, 2015). In 2009–2011, a non-official legislative proposal in the form of anti-cruelty law was drafted and submitted to the Chinese national legislature and central government (The full text in English translation of the Prevention of Cruelty to Animals Law of the PRC (Experts' Draft Proposal) is found at http://aldf.org/downloads/ChinaCrueltytoAnimalsProposal3-10.pdf. See also Chang, Michaels, Littlefair, and Li, 2010). Separately, numerous delegates to the Chinese National People's Congress (the equivalent to the Parliament) have submitted proposals for legislating against animal cruelty each year since 2011. However, there is no official indication, either positive or negative, from the Chinese legislative body as to its plan or intention for legislation in this area.

In China, companion animals are generally regarded as the personal property of owners, although there is no law explicitly saying so. As such, registered dogs may have certain limited indirect legal protection as the property of their owners – the only legal protection as far as dogs are concerned, as mentioned earlier.

In most Chinese cities, there are dog management regulations issued by the local city governments. They are not designed for dog protection or welfare purposes but for dog management. Most of these regulations provide for the responsibilities by the owners such as registration and vaccination of dogs and leashing dogs outdoors, various dog bans, including banned breeds or types of dogs, designated zones where dogs are not to be raised or taken for walks and restrictions on the number of dogs a household can have. These restrictions vary from city to city. Some city regulations also have provisions on dog adoption.

A related issue is the eating of cats and dogs as food. Despite their emergence and growing popularity as family companions, cats and dogs are still being eaten in China. However, in the past decade, more Chinese have come to accept that animal abuse, and eating cats, dogs, and wildlife is unacceptable. Many people are starting to oppose the eating of dogs despite the availability of dog dishes in restaurants. Their objections usually include animal cruelty, food safety concerns, cat and dog theft and related crimes. Some Chinese, especially the young, believe dog and cat eating is barbaric. During the last decade, some legislators at the city, provincial, and national levels have proposed banning cat and dog eating, often generating wide interest and heated debates, but none have succeeded until 2020 in the cities of Shenzhen and Zhuhai in southern China when Shenzhen became the first Chinese city to ban the eating of cats and dogs followed by Zhuhai. The ban is a part of a wider clampdown on the wildlife trade following the outbreak of COVID-19. What stands out in the Shenzhen law is the additional prohibition of eating cats and dogs within the law banning the sale and eating of wild animals (see Cao, 2020; Cao, 2021).

Farm animals

China has been a member of the OIE since 2007, but China's laws and regulations have not incorporated all the OIE animal welfare standards, in particular for farm animals, such as transport, slaughter, and production systems (For discussions of the Chinese animal situation and international treaties, see Chang 2020. For a comparative study of EU animal regulation and China, see Chang, 2000). In China, for farm animals, there are some laws and regulations for animal husbandry, transport, and slaughter, with the primary objective of protecting humans and human interests, not animals or their welfare. For instance, the Animal Husbandry Law (2005) was enacted

> for the purposes of regulating the production and business operations of stockbreeding, ensuring the quality and safety of livestock and poultry products, protecting and reasonably utilising the genetic resources of livestock and poultry, protecting the legitimate rights and interests of the stockbreeding producers and business operators, and promoting the sustainable and sound development of stockbreeding.
>
> *(Art. 1, Animal Husbandry Law (2005))*

Other related laws include Animal Epidemic Prevention Law (1997) which was last amended in 2021, and Food Safety Law (2009) that have some provisions related to animals. Other relevant laws include the Regulations on Feed and Feed Additives Management (2001), Regulations on Administration of Veterinary Drugs (2004), Regulations on the Prevention and Control of Pollution from Large-scale Livestock and Poultry Breeding (2014), and Regulations on the Management of Pig Slaughter (2008), which are currently being amended. These regulations do not mention or regulate farm animal welfare or their protection because the overall concern and focus of such Chinese laws are human interests and utilisation of animal resources. There are no specific laws for the rearing of pigs, chickens and hens, or cattle and calves, or for protecting their welfare, although humane pig slaughter including stunning has been supported by the Chinese authorities in recent years, for instance, the Technical Standards for Pig Humane Slaughter released by the Ministry of Agriculture in 2008 encouraging humane slaughter practices. A new development is the publication of a series of farm animal welfare standards for pigs, cattle, chickens and hens, and other farm animals in recent years, but these are not government directives or regulations, and have no official or legal implementation status (the farm animal welfare requirements are developed under the auspices of the International Cooperation Committee of Animal Welfare (ICCAW) in China, and the full texts (in Chinese and English) are found at www.iccaw.org.cn/plus/list.php?tid=69).

Other captive animals used for human purposes

China has a regulatory regime for other types of wildlife in captivity, including regulating the use of protected wildlife. One regulation is the Implementation Regulations on Terrestrial Wildlife

Protection (1992). The Regulations provide the operational framework for extensive utilisation of wildlife including protected wildlife, legalising and promoting a number of wildlife industries in China: wildlife as laboratory animals, for zoos and circuses, and for agriculture. For the regulatory regimes for wildlife in zoos and circuses and animals used in the fur industry, there are no specific laws in China that cover the breeding or treatment of wildlife used in the fur industry. China is one of the major producers of furs today, but there is little transparency for such commercial activities, and there have been reports of serious abuse of such animals over the years (see Cao, 2015). The relevant government policy giving consideration to animals bred and raised in the fur industry is the Guiding Opinions on the Promotion of Sustainable Development of Wild Fauna and Flora (2004). This document prohibits and restricts the hunting of wildlife resources in the wild for direct use for commercial purposes. Furthermore, the Interim Provision on Technical Management Related to the Domestication, Breeding and Utilization of Wild Fur Animals (2005) is another specific, relevant policy directive for all forestry departments in different cities and provinces to implement. This provides the technical specifications and animal welfare standards for the fur animal breeding industry. The Interim Provision has a number of guidelines on animal welfare though most sections pertain to fur quality control.

As for animals used in zoos and for entertainment, there are the City Zoo Management Regulations (2004) which regulate the management of zoos but also include some sanitary, environmental, and enrichment requirements for animals kept in zoos. There have also been directives from the national authorities since 2010 that animals should not be used for performing purposes in zoos, but these are not largely unimplemented (for further discussions, see Cao, 2015).

Conclusions

Despite the widespread animal welfare problems for many animals in China, recent years have seen some positive developments, including outlawing the ivory industry since 2018 (see Xiao et al., 2021), the ban on the trade of wild animals for eating throughout the country since 2020 and the ban on eating cats and dogs in Shenzhen and Zhuhai. These efforts suggest that the Chinese society and Chinese people's attitudes towards animals are gradually changing. They should give us hope that Chinese practices towards animals can change for the betterment of animal welfare in response to contemporary ethical standards and laws.

References

Bayne K, Wang J and Pang W, 2018. Oversight of animal research in China. In J Guillén, ed. *Laboratory Animals: Regulations and Recommendations for the Care and Use of Animals in Research*. Cambridge, MA: Academic Press, pp. 263–291.

Cao D, 2015. *Animals in China: Law and Society*. London: Macmillan.

Cao D, 2016. Wildlife crimes and legal protection of wildlife in China. In D Cao and S White, eds. *Animal Law and Welfare: International Perspectives*. London: Springer, pp. 263–278.

Cao D, 2018a. Ethical questions for research ethics: Animal research in China. *Journal of Animal Ethics*, 8(2), pp. 138–149.

Cao D, 2018b. Wild game changer: Regarding animals in Chinese culture. *Harvard Review of Philosophy*, XXV, pp. 147–168.

Cao D, 2019. Crimes against animality in international law. *L'Observateur des Nations Unies*, 45(2), pp. 179–191.

Cao D, 2020. Global risks of intensive animal farming and the wildlife trade. *Animal Sentience*, 5(30), p. 2.

Cao D, 2021 . A positive small step in the treatment of animals in China. *Journal of Animal Ethics*, 11(2), pp. 1–3.

Cao D and White S eds., 2016. *Animal Law and Welfare: International Perspectives*. London: Springer.

Carnovale F, Jin X, Arney D, Descovich K, Guo W, Shi B, and Phillips CJ, 2021. Chinese public attitudes towards, and knowledge of, animal welfare. *Animals*, 11(3), p. 855.

Chang J, 2000. *A Comparative Study of Animal Welfare Law in China and the European Union* (in Chinese). Beijing: Kexue huanjing chubanshe.

Chang J, 2020. China's legal response to trafficking in wild animals: The relationship between international treaties and Chinese law. In A Peters, ed. *Studies in Global Animal Law*. Berlin: Springer, pp. 71–79.

Chang J, Michaels G, Littlefair P and Li H, eds. 2010. *Animal Protection Law of the PRC and Prevention of Cruelty to Animals Law of the PRC: Experts' Proposal and the Public Response*. Beijing: China Environmental Science Press.

Huang Q, Wang F, Yang H, Valitutto M and Songer M, 2021. Will the COVID-19 outbreak be a turning point for China's wildlife protection: New developments and challenges of wildlife conservation in China. *Biological Conservation*, 254, 1–5.

Kong Q, and Qin C, 2010. Analysis of current laboratory animal science policies and administration in China. *ILAR e-Journal*, 51, pp. e1–e10.

Li PJ, 2021. *Animal Welfare in China: Culture, Politics and Crisis*. Sydney: Sydney University Press.

Li PJ and Davey G, 2013. Culture, reform politics, and future directions: A review of China's animal protection challenge. *Society & Animals*, 21, pp. 34–53.

Liang Z, 2016. Keywords of the wildlife protection law: An textual analysis (in Chinese). *Dongfang zaobao*, December 11. Retrieved from https://www.huanbao-world.com/a/zhengce/2019/0128/80723.html

McLaughlin K, 2016. China finally setting guidelines for treating lab animals. *Science*, March 21.

Sharma C, 2005. Chinese endangered species at the brink of extinction: A critical look at the current law and policy in China. *Animal Law*, 11, pp. 215–254.

Shih C and Singer P, 2018. Animal welfare: A Buddhist-utilitarian dialogue. *Harvard Review of Philosophy*, XXV, pp. 169–181.

van Uhm DP 2019. Chinese wildlife trafficking networks along the silk road. In T Wing Lo, Dina Siegel, Sharon I Kwok, eds. *Organized Crime and Corruption Across Borders*. London: Routledge, pp. 114–133.

van Uhm DP 2020. Wildlife trafficking and criminogenic asymmetries in a globalised world. In A Brisman and N South, eds. *Routledge International Handbook of Green Criminology*. London: Routledge, pp. 529–542.

Xiao L, Lu Z, Li X, Zhao X and Li BV, 2021. Why do we need a wildlife consumption ban in China? *Current Biology*, 31(4), pp. R168–R172.

30
KEY ANIMAL LAW ACROSS EUROPE

Debbie Legge

Introduction

Any discussion of animal welfare legislation in Europe[1] must consider the relevant European Union (EU) law (European Commission, n.d.b). This is set out in two framework treaties: the Treaty on European Union (TEU) and the Treaty on the Functioning of the European Union (TFEU). Europe is a continent of those countries inside and outside the EU (Europa, n.d.) (World Atlas, 2021). Those within will have a degree of conformity if the law is set out in treaties and regulations that have direct effect and affect, which means that they become part of the Member State's legal systems with no additional legislation required. Directives form the rest of legislation in the EU and leave to the Member State discretion in meeting the directive's aims (European Commission, n.d.d). So, whilst there may be some conformity in some areas of animal welfare legislation, in others there is less so. Implementation may also vary within the states due to the constitutional make up of each country. For those European countries outside of the EU the situation becomes more fragmented. Some non-EU countries such as Belarus have no animal welfare legislation, although it does have legislation on animal health. It is not clear what the situation is in Moldova. Bosnia and Herzegovina and Russia do have some animal welfare legislation, and the Ukraine is aligning its animal welfare law with the EU. The UK's law is in the process of being reformed (DEFRA, 2021a).

The basis for the legal systems in European countries can be from Roman law, the Germanic tradition, Scandinavian, or the common law (Carozza, 2015). This leads to different constitutional and legal bases for each European country's approach to animal welfare legislation. Many European countries have a constitution and a civil and penal code. Many have a federal system of government, or law may be devolved into nations such as in the UK, leading to differing legislative coverage.

The chapter begins by considering the legislative or constitutional recognition of sentience; the adherence to international conventions and/or compliance with international animal protection laws, animal welfare, and protection against cruelty. It will then consider specific areas, such as animals used in farming, in zoos and aquaria, companion animals, animals used for sport and recreation, scientific research, the protection of wild animals, and will conclude on the overall protection of animals in Europe.

DOI: 10.4324/9781003182351-36

Legislative or constitutional recognition of animal sentience

Article 13 of the TFEU covers animal sentience which EU countries have applied in varying degrees, for example, the Netherlands has gone further. Outside of the EU Switzerland has not specifically mentioned sentience but considers the physical and mental state of animals and focuses on their dignity (Animal Welfare Act 2005).

Animals have traditionally been seen as "things" rather than "beings", and the issue of legal personality differs across Europe. The Dutch civil code (Article 3:2a) states that animals are not "things" whilst others such as Austria, the Netherlands, the Czech Republic, and Germany have taken legal steps to recognise legal personality. Of non-EU countries, Switzerland and Moldova have also introduced similar provisions. Switzerland amended its constitution so that animals were acknowledged as "beings" rather than "things" (Federal Constitution of the Swiss Federation (1999)). Some European countries, such as Austria, Germany, and Slovenia, have included animal protection within their constitution (WAP, n.d.). The UK has recognised sentience (DEFRA, 2021b).

Whilst it is important to have animal sentience and welfare recognised, there also needs to be accountability for decision making. The legal structure of the EU is split between the Commission, the Council, the Parliament, and the European Court of Justice (European Commission, n.d.e). Law on animal welfare has three legal bases: the first relates to a level playing field in agriculture and fisheries, which means that provisions for ensuring animal welfare standards are consistent across the EU. The second is trade and ensuring that consumer products are safe. Finally, human health and veterinary public health (Eurogroup, 2021).

There are varying legal structures used by European countries for ensuring political or government accountability for animal welfare and bodies providing scientific/ethical advice. Some countries such as Belarus have limited political or governmental accountability for animal welfare issues, with no ministry responsible for animal welfare or for specific issues such as experimentation. Animal health is covered by the Department ofVeterinary and Food Control within the Ministry of Agriculture and Health. Others such as Austria have an integrated approach, with a Federal Ministry of Health which cooperates with the Federal Ministry of Agriculture, Forestry, Environmental, and Water Management on animal welfare. Each of the nine states has an ombudsperson for animal welfare with independent non-governmental members. The Animal Protection Commission advises the Ministry of Health on animal welfare issues. The scientific Animal Protection Council and the Animal Protection Enforcement Advisory Board monitor compliance with animal welfare legislation (WAP, n.d.). In Denmark the Ministry of Environment and Food is responsible for animal welfare legislation, a national committee of agricultural organisations, research institutes, animal welfare societies, and relevant ministries. Denmark has an Animal Welfare Council, the Ethical Council for Animals, a Danish Centre for Animal Welfare, and the Council concerning the Keeping of Certain Animals Ministry of Food, Agriculture, and Fisheries of Denmark (n.d.).

Reform

Whilst the move to recognise animals as sentient by the EU Member countries and others, and the inclusion of animal welfare measures or anti-cruelty measures for animals in some constitutions, is to be welcomed and encouraged, there needs to be a wider recognition of animals as "beings" rather than "things" enshrined in law throughout Europe. Animal welfare in the EU and in many countries is split between different bodies and ministries. The Austrian and Danish models provide good examples of an integrated approach. Having a minister or better a ministry for animal welfare would seem a sensible approach. It is important for countries to have a national advisory board/commission to provide independent advice to governments, which includes a wide range of bodies.

Adherence to international conventions and/or compliance with international animal protection laws

In relation to the OIE animal welfare standards many EU states meet and often go beyond these, for example in relation to broiler chickens. Many EU countries have transposed the OIE animal welfare principles and standards into legislation. Sweden, Denmark, and Austria have effective bodies for enforcement: in Sweden the Swedish Centre for Animal Welfare and the 3Rs and in Denmark the Animal Welfare Commission and the 3Rs bodies whilst Austria has the Animal Protection Commission. France, Spain, and Germany have enforcement mechanisms in place in relation to the OIE standards that have been transposed into legislation. Germany is the European observer on the steering committee. Other EU countries such as Poland have a lack of commitment to implementing and enforcing EU legislation related to these standards. Italy has incorporated these standards, but the EU Commission has had to take enforcement action to uphold some of the law relating to these areas. Romania has covered EU requirements only and penalties are weak. In some non-EU countries such as Switzerland and Russia not all provisions have been transposed into national legislation. Some countries such as Ukraine are members and participate in the OIE but have not yet implemented all the requirements (WAP, n.d.).

In relation to the Universal Declaration of Animal Welfare (WAP, n.d.), many EU states and the EU Commission have supported this initiative. Within the EU, Austria, Poland, and the Netherlands have provided government support for this. Other European states such as Switzerland have also given full government support to it. Some European countries such as Ukraine, Russia, and Belarus have not made a pledge of support.

The EU and some European states outside of the EU have signed up, and or ratified and implemented the European Conventions on animal welfare. These include: the European Convention for the Protection of Animals kept for Farming Purposes No 087, during International Transport No 193, for Slaughter No 102, of Pet Animals No 125, and for the Protection of Vertebrate Animals used for Experimental and other Scientific Purposes No 123.

Reform

It is to be hoped that more European countries will support these soft legal instruments that cover animal welfare in Europe.

Animal welfare and protection against cruelty

Article 13 of the TFEU has a very general welfare requirement which leaves much discretion to the individual Member State in implementation. Many European countries within the EU such as Austria and the Netherlands have legislation covering animal welfare. Austria's Animal Welfare Act (2004) covers all vertebrates, cephalopod, and decapod crustaceans and is seen as progressive in terms of its approach to animals. The Netherlands Animals Act (2011) establishes a duty of care for animals based on the five freedoms. There is a duty to act and there is a duty of care including protections from anxiety and stress to helpless animals, which applies to all animals including animals in the wild. They are also protected through the Nature Conservation Act (2017), but this does not cover hunting. Some such as Belarus have no overall animal welfare legislation. Some countries outside of the EU but inside the Schengen Zone such as Switzerland, Lichtenstein, Norway, and Iceland also have Animal Welfare Acts.

Many European countries have laws protecting animals from cruelty. Denmark has legislation prohibiting specific forms of animal cruelty. It creates a duty of care and covers failures to act as well as deliberate acts of abuse. It covers physical and psychological well-being, as does the law in Sweden which has similar provisions that cover both animals in captivity and wild animals (WAP, n.d.).

Some EU countries such as Romania have undefined laws. Some European countries such as Belarus have no laws preventing animal cruelty. Austria's Animal Welfare Act (2004) goes beyond EU requirements and covers all animals in relation to cruelty but there are exceptions in relation to wild animals that are hunted or fished and for non-stunning in religious slaughter. Outside of the EU, Switzerland protects against a wide variety of animal cruelty and not only prohibits inflicting pain, suffering, and harm but also anxiety and protecting an animal's dignity and a failure to act (Animal Welfare Act (2005) and Animal Protection Ordinance (2008)).

Some countries such as the Netherlands and Finland have police resources dedicated to the enforcement of animal cruelty law. In England and Northern Ireland, punishment for the most serious cases of cruelty is five years, as is the case in Ireland (Animal Health and Welfare Act, 2013). Greece has recently increased its punishment to ten years (Kokkinidis, 2021).

Reform

Animal cruelty provisions should cover failure to act as well as deliberate acts of abuse and deal with both physical and psychological well-being. It should cover all sentient animals and their late-foetal developmental stages with effective punishment and enforcement of the law. A duty of care should apply to all sentient animals including wild animals.

Animals used in farming

For EU Member States the law on the welfare of animals used in farming has been based on the 'five freedoms' set out in the European Convention for the Protection of Animals kept for Farming Purposes (1976). These are freedom (i) from hunger and thirst; (ii) discomfort; (iii) pain, injury, and disease; (iv) to express normal behaviour; and (v) from fear and distress. These are the basis for the EU directive on the protection of animals kept for farming purposes (Council directive 1998/58/EC). It includes as animals, "fish, reptiles or amphibians bred or kept for the production of food, wool, skin or fur or for other farming purposes" (Article 2). They should not suffer any unnecessary pain and suffering (Article 3). It also covers the conditions

> under which animals (other than fish, reptiles or amphibians) are bred or kept, having regard to their species and to their degree of development, adaptation and domestication, and to their physiological and ethological needs in accordance with established experience and scientific knowledge
>
> *(Article 4)*

The Commission has announced that it will propose legislation to phase out cages for many farmed species.

There are also directives covering specific animals, including minimum standards for the protection of calves (Council directive 2008/119/EC), pigs (Council directive 2008/120/EC), laying hens (Council directive 1999/74/EC), and broiler hens (Council directive 2007/43/

EC). Veal crates have been banned, and sow stalls are to be phased out although they can still be used as farrowing crates before birth and for 28 days after (European Commission, 2021b). In the EU piglets cannot be castrated without anaesthetic, but this only applies to piglets being castrated who are at least seven days old. Beak trimming can take place without anaesthetic for broiler chickens. Calves can be isolated under eight weeks old, and animals can be transported for over eight hours· In relation to the EU, the transport of animals is set out in a regulation (Council regulation (EC) No 1/2005) as are the rules for slaughter (Council regulation (EC) No 1099/2009). The UK is proposing to ban the export of live animals for slaughter and fattening (DEFRA, 2021d).

Some European countries go further than EU standards, Austria is phasing out farrowing crates by 2033, limiting the stocking density of broiler chickens and banning the tethering of cattle and calves. It also limits transport within Austria to 4.5 hours, and long-distance travel over 8 hours is forbidden. Sweden imposes increased space requirements, a maximum of eight hours transport for slaughter and stipulates that cows must be kept at pasture in the summer. Beak trimming and tail docking are not allowed. The Netherlands has banned enriched cages for egg laying hens and set a four-day limit for the use of sow gestation stalls. There is more space for pigs than the minimum EU requirements and it forbids concrete fully slatted floors. The UK bans sow stalls throughout a sow's pregnancy. Denmark has outlawed the slaughter of non-stunned animals (WAP, n.d.).

Whilst there is EU legislation in this area there are still concerns over the welfare of farmed animals, and there are other specific concerns over some forms of production. Ducks and geese are not protected by species-specific provisions and the law is contradictory. Ducks and geese need to reach minimum weights (Commission regulation (EC) No 543/2008) which would seem to contradict the Farm Animal Welfare Directive (Council directive 98/58/EC). Most Member States have implemented a ban on the production of foie gras. But some countries such as France still allow it. Some devolved regions in Belgium have banned it for example, Flanders will ban the production of foie gras from 2023 (Eurogroup, n.d.a).

Under EU law, the welfare of farmed fish is covered by EU legislation only during rearing, transport, and slaughter. The welfare of wild-caught fish is not covered at all. The EU has recently published updated guidelines for fish farms which include provisions for the welfare of fish (European Commission, 2021c).

The use of animals such as horses and donkeys for draught raises welfare concerns (The Donkey Sanctuary, n.d.). In Europe the welfare of horses used as meat and other products is an issue. The biggest exporters of horses for slaughter are the Netherlands, Poland, and France (Eurogroup, n.d.m). The countries where horse meat is eaten include France, Italy, and Belgium. EU legislation in this area relates to consumer protection (Eurogroup, n.d.d). A regulation covers the transport of horses for meat (regulation (EC) No 853/2004), but this raises welfare concerns within and outside of the EU.

The EU has published a Farm to Fork strategy, with new EU animal welfare rules and an aim to consider the whole food supply chain. It has an action plan of 27 legislative and non-legislative measures to be enacted from 2020 to 2024. It is revising animal welfare legislation, including on animal transport and the slaughter of animals, working in conjunction with new plans and guidance on aquaculture and considering options on animal welfare labelling "to better transmit value through the food chain" (European Commission, 2020). The European Commission has stated that legislation is proposed in 2023 to ban caged animal farming by 2027 (Abnett, 2021).

It is not clear whether these reforms will cover other forms of farming such as for fur. Cat and dog fur are banned in the EU (regulation (EC) No 1523/2007) but not fur from rabbits, minks, or foxes (Eurogroup, n.d.e). There are no EU guidelines on humane slaughter for fur. 11

Member States such as Austria, Croatia, and Slovenia have totally or partially banned or strictly regulated fur farming, sometimes with phasing-out periods, as in the Netherlands and Ireland, in relation to mink farming (Eurogroup, n.d.f). Denmark is phasing out fur farming in relation to foxes but not mink although it is currently banned until 2022 (Murray, 2020). Outside the EU, Norway has a phased approach regarding mink farming. Other countries such as Spain, Belarus, Poland, and Russia allow it. Switzerland and Italy allow fur farming with some minimum standards of care. Sweden allows fox, mink, and chinchilla farming with regulations on breeding and keeping these animals for fur although the last chinchilla farm closed in 2014 and the controls on fox fur farming set very strict conditions which have effectively phased it out (Humane Decisions, n.d.).

Reform

It is to be hoped that the new Farm to Fork strategy will consider some of the concerns raised by the Eurogroup, who call in their No Animal Left Behind campaign for a strategy to improve the welfare of farmed animals. The end to cages is a welcome step forward. There should also be a ban on fur farming in Europe (Fur Free Alliance, 2021) and on the import and export of fur products. The welfare of farmed fish also needs to be improved. The use of horses as meat and in other products should be banned.

Animals in zoos and aquaria

Zoos are regulated in the EU by Council directive 1999/22/EC and a non-binding good practice document (European Commission, 2015). The directive sets out measures for the licensing and inspection of zoos to respect conservation and protection measures including appropriate accommodation (Council directive, 1999/22/EC).

Council regulation (EC) No 1/2005 allows the transportation of wild animals for rehabilitation or reintroduction for example in captivity. The transportation should not cause injury or undue suffering and if the animals are wild, timid, or dangerous then written instructions should be given on care and feeding, etc., and acclimatisation to the mode of transport prior to the journey.

Outside the EU many countries have laws covering zoos, but these need better enforcement as in the case of Poland and Romania. In some such as Belarus it is limited to veterinary services to protect animals against extreme natural and other factors. In the UK there is a proposal to amend the Zoo Licensing Act to improve the regulations and contribute to conservation (DEFRA, 2021d).

In terms of other animals in captivity, 14 EU countries keep cetaceans in captivity and 14 do not. Of these three, Croatia, Cyprus, and Slovenia have banned dolphinaria outright (Eurogroup, n.d.p). Outside of the EU, Switzerland imposed an import ban on dolphins which led to the closure of its last dolphinarium. The UK also has no dolphinarium but lacks legislation imposing this. At a regional level Brussels has banned the keeping of sea mammals in tanks for human entertainment and is the first in Europe to include seals, sea lions, and walruses. Orcas are found in Tenerife and Russia (WDC, n.d.a).

Reform

There needs to be a more comprehensive licensing system for zoos, with frequent and consistent licensing review periods that also cover sanctuaries and rescue centres, and effective enforcement mechanisms through a minimum number of inspections (Eurogroup, n.d.p). There should be a focus on the animals' mental well-being and the provision of enclosures large enough for social interaction and to express normal behaviour. In some countries such as Austria and the Netherlands the anti-cruelty measures as well as the Animal Welfare Acts apply to animals in zoos, and this should be encouraged across all European countries (OIE). There should be a ban on the keeping, display, and breeding of marine mammals in captivity and where this has stopped legislation should ensure that no more can open.

Companion animals

Regulation (EC) No 998/2003 sets out the law on the transportation of companion animals. It covers kept and wild animals including animals and products within the Union and from the Union, and non–commercial movements of pet animals into a Member State from another Member State, or from a third country or territory (European Commission, n.d.c). These new rules take over the rules from the existing regulation (EU) No 576/2013 on the non-commercial movements of pet animals. However, there is a transitional period until 21 April 2026, during which regulation (EU) No 576/2013 will continue to apply. Regulation (EU) 2016/429 covers companion animals in relation to the control or eradication of strays (Eurogroup, n.d.j). There are provisions for stray animals laid down by the World Organisation for Animal Health (OIE) and the International Companion Management Coalition (ICAM).

In terms of the welfare of companion animals, legislation is left to the Member States. In France it is an offence to abandon companion animals. In the Netherlands, there is a free spay and neuter service for dogs (Eurogroup, n.d.j) and it has police resources dedicated to animal cruelty, as does Finland. In relation to stray companion animals the identification and registration of dogs is required in 22 of the 27 Member States but only Belgium, France, and Greece require it for cats (Eurogroup, n.d.b). In the countries outside of the EU, Belarus allows the culling of stray animals and there has been concern about the treatment of strays in other countries such as Romania. Slovenia has some joined-up provisions to stop the abandonment of dogs including anti cruelty measures, laws against abandonment, guidance on euthanasia, requirements for owners to care for their animals and to keep them on a lead, controls on the breeding and selling of animals and limits on the numbers of animals that can be owned. There is a compulsory microchipping programme for dogs whilst the animal rescue centres also take part in education. Switzerland and Sweden have similar provisions that cover stray dogs, but the situation for cats allows for them to be killed (Tasker, n.d.). The UK has a licensing system for dog breeding and pet sales, has banned the commercial third party selling of puppies and kittens, and protects service animals. It is introducing laws on puppy smuggling and compulsory cat microchipping as well as other proposals (DEFRA, 2021d) and has introduced a pet abduction offence (DEFRA, 2021e).

Reform

It would seem sensible that there is a coordinated approach to the welfare of companion animals, as in Slovenia for dogs, that includes provisions that discourage and regulate breeding including a ban on certain features such as ear cropping, put controls on the way animals are sold and transported, provide for microchipping and a national owner database with licencing and prohibit abandoning any companion animal. Stray animals should be covered by anti-cruelty and welfare legislation. There have been calls for an EU-wide mandatory system for the identification and registration of cats and dogs, more control, and tougher sanctions against those supplying false pet passports. There are also calls for a common EU definition of puppy and kitten farms and EU breeding rules for pets are needed while EU countries should be encouraged to put in place registers of authorised breeders and sellers. People should be encouraged to adopt, rather than buy, companion animals (European Parliament, 2020). There are also calls for humane stray animal population management to be embedded in all EU Member States (Eurogroup, n.d.j).

Animals used for sport and recreation

There are welfare concerns on the use of animals such as horses and donkeys for recreation (The Donkey Sanctuary, n.d.), including the breeding and transportation of equines and responsible ownership (Eurogroup, n.d.i). The EU has a regulation on identification and ownership (regulation (EU) 2016/429) and a lifetime passport within the EU. The single lifetime identification document will only be required under Commission regulation (EU) 2015/262 until 28 January 2022 when regulation (EU) 2019/6 will be applicable., but "the tracing of equines outside the EU remains a continuous challenge" (Eurogroup, n.d.l). The UK is considering further protections for equines and horse racing (DEFRA, 2021d).

In terms of other "sporting" use of animals, France and Spain allow bullfighting and France allows cockfighting. Greyhound racing takes place in Ireland and the UK (Welfare of Racing Greyhounds Regulations 2010) but the UK is considering further protections (DEFRA, 2021d).

Directive 2009/147/EC and Council directive 92/43/EEC apply to the hunting of wild animals and birds and there is a European Charter on Hunting and Biodiversity (European Commission, n.d.g). Many European countries allow hunting, and whilst some European countries have banned hunting with dogs, some such as Austria and the Netherlands allow this. In Poland hunting with dogs is allowed with the permission of the landowner. Outside of the EU, Switzerland allows hunting, subject to welfare provisions in some provinces, but the Canton of Geneva has banned all hunting. The UK allows trail hunting, which needs to be further regulated (RSPCA, 2021) or hunting banned outright. Many countries have a licensing system, some with examinations. The breeding of birds and other animals, such as deer for shoots, is also a welfare issue as is the intensive breeding and release of game birds. There are concerns over the shooting and hunting of hares (RSPCA, 2021). The UK is considering legislation for a hare close season (DEFRA, 2021d).

There is no EU law on circuses, so it is left to individual European countries to develop the law in this area. Some countries have banned the use of wild animals in circuses (Eurogroup, 2020) whilst others restrict the use either of all or exclusively of wild animals (Eurogroup, n.d.n). In some European countries, such as Belarus, there are no restrictions on the use of animals for fun fairs or circuses. The problem with individual countries developing their own law is that circuses regularly move between Member States. Therefore, collective action through EU-wide regulation is required to address the problem (Eurogroup, n.d.n) and a wider European solution is needed.

Reform

It would seem clear that the use of animals for "sport" or "recreation" should be banned throughout Europe particularly in the case of cock fighting, bullfighting, and greyhound racing. At the very least the cruellest forms of hunting such as with dogs, live baiting, poisoning, trapping, falconry, and hunting with bows should be banned. Horse and greyhound racing have welfare issues and at the very least there should be a registration system for equine premises and regular inspections by outside bodies. The use of animals in circuses should be banned throughout Europe.

Animals used in scientific research

Animal experimentation in the EU is governed by directive 2010/63/EU based around the 3Rs: Replacement (fostering the use of alternative methods), Reduction (trying to use fewer animals for the same objective), and Refinement (efforts to minimise pain and suffering) (Eurogroup, n.d.h). There has been a delay in Member States passing legislation to implement the directive, and there is little monitoring of this or the effectiveness of using animals (Eurogroup, n.d.k).

The European Commission has been working on databases on alternative approaches and the European Parliament has called for the use of animals to be phased out. Many countries are also looking into alternatives such as the Danish Consensus Platform for 3R Alternatives to animal experimentation, a collaboration between the industry and animal welfare organisations. Many countries have animal ethics committees which evaluate the usefulness of animal experiments; however, their compositions and effectiveness vary substantially. Sweden's ethics committee has several different stakeholders, but it does not include animal welfare organisations and wild animals can be used in experiments. Switzerland and Austria have similar systems (WAP, n.d.). Denmark has a wide variety of stakeholders on its ethics committees.

There is a ban in the EU on cosmetics testing on animals (regulation (EC) No 1223/2009). There is no cosmetic testing in the EFTA countries. Turkey is cruelty-free. The Ukraine and Russia are phasing out animal testing (Grum, 2019). However, some testing on ingredients only used for cosmetics may happen under chemicals regulations after a decision made by the European Chemicals Agency (Eurogroup, n.d.g).

Reform

The 3Rs is a welcome framework for regulating animal experiments and it is to be hoped that more European countries follow the lead in banning the use of great apes and other primates. It is also to be hoped that they look to the Danish example in terms of multi-stakeholder inclusion on ethics and other animal experimentation regulatory bodies. The EU needs to close the loophole that allows for procedures likely to cause pain or distress, that are severe, or prolonged which are currently permitted under the directive.

Protection of wild animals

The protection but not welfare of wild animals and birds in Europe is set out by the Bern Convention (European Council 1979). It has been signed by the EU and 50 European countries. The EU's Natura 2000 (European Commission, n.d.f) and the Emerald Network have been set up under the Bern Convention. There are two wider initiatives that also impact on conservation: the European Green deal (European Commission, n.d.a) which includes the 2030 Biodiversity strategy and a forest strategy. The protection of wild animals is also covered by EU directives on birds (directive 2009/147/EC) and seals (Council directive 83/129/EEC). Other provisions protect their habitat (Council directive 92/43/EEC) or ban methods of trapping such as leg holds (Council regulation (EEC) No 3254/91) or glue traps for birds (*Judgment of the Court (First Chamber) of 17 March 2021 (request for a preliminary ruling from the Conseil d'État – France) – Association One Voice, Ligue pour la protection des oiseaux v Ministre de la Transition écologique et solidaire* (2021) Case no. C-900/19 OJ C 54, 17.2.2020). However, glue traps for mice are still widely used although the UK has introduced a ban (DEFRA, 2021c).

The EU supports the moratorium on whaling by the International Whaling Commission (IWC), n.d. a) and whaling is banned in EU waters. However, whales, dolphins, and porpoises are suffering from death in nets (bycatch) in European waters as current EU measures are inadequate, poorly implemented, and enforced (Groves, 2020). Greenland and the Faroe Islands are not covered by the law in Denmark, although it represents these areas on the IWC, but there is no commercial whaling in these areas although dolphins are killed in the Faroe Islands (Berry 2021). Outside the EU, Iceland, Monaco, Norway, Switzerland, and the Russian Federation are members of the IWC (IWC, n.d.b). Norway still hunts the minke whale which takes place under an "objection" to the global ban on commercial whaling (Groves, 2021). It is reported that Iceland is not hunting whales anymore (WDC, n.d.b). There is one region in Russia where an aboriginal subsistence hunt is allowed (IWC, n.d.c). Commercial seal hunting takes place in Greenland under its own legislation (OIE). The UK has proposed to ban the shark fin trade (DEFRA, 2021d).

The EU implements the Convention on the Trade of Endangered Fauna and Flora (CITES) through Council regulation 338/97/EC, which provides controls on the sale and possession of wild animals, birds and plants found within the territory of the EU, as well as CITES-listed species. The Euro group for Animals argues that the regulation needs stronger controls "including the designation of species, import controls, transport and housing, as well as internal EU trade" (Eurogroup, n.d.o). In 2016, the EU released its wildlife trafficking action plan (European Commission, n.d.h). The issue of invasive alien species had been covered by regulation (EU) No 1143/2014 but there is concern that it is vague on animal welfare (Eurogroup, n.d.f).

Some countries have implemented stricter regulations on the sale or import of some wildlife products. France has implemented a ban on the import of lion trophies. In terms of ivory the EU only permits the sale of antique or "pre-Convention" ivory that was acquired before elephants were included in the CITES appendices (Council regulation (EC) No 338/97) but the law on this area is being strengthened (European Commission 2021a). Luxembourg, France, the Netherlands, Belgium, and the UK have all adopted, or are set to adopt, stricter measures on the trade in ivory (Eurogroup, n.d.o). The UK proposed to ban the import of hunting trophies from endangered animals and ensure that import and export does not threaten conservation (DEFRA, 2021d).

Whilst the EU has set up the Committee on Trade in Wild Fauna and Flora, the Scientific Review Group, and the Enforcement Group (Eurogroup, n.d.o)), it seems that enforcement is needed as EU Member States are being used as transit countries to smuggle illegal ivory from elephants poached in Africa with the destination of Asia, and this will be covered by new legislation (European Commission 2021a).

The other issue in terms of the protection of wild animals is the keeping of wild animals as pets. Wild animals kept as pets are covered by CITES. Some EU States such as Belgium, the Netherlands, Luxembourg, Malta, and Croatia have adopted legislation including Positive Lists: lists of species that can be kept and traded as pets. Other EU States such as Spain, Austria, and Germany have negative lists of banned species, but these lack clarity, because of the emergence of new species, which importers can then exploit. Some European countries such as Belarus have no legislation (WAP, n.d.). Outside of the EU, Ukraine's law needs better enforcement whilst in Romania there is law but the keeping of animals is seen as a status symbol. Switzerland has a list of species that need permission and some that need specialist certificates (Swiss Animal Protection Ordinance (2008) Articles 39 and 40). The UK proposes to introduce a ban on keeping primates as pets, with interim zoo standards in the meantime (DEFRA, 2021d).

Hunting is allowed in many countries in Europe (European Commission, n.d.g). The hunting of animals for "sport" has been covered above but there are also welfare concerns over how wild animals are killed when they are culled for other reasons.

Reform

Both international law and EU legislation in this area focuses on conservation rather than the protection of individual animals. There is a need for an offence of cruelty to all animals. There should also be a European ban on all hunting that is not for subsistence; a ban on the import and export of hunting trophies, the import and sale of animal fur, the trade in wild animals and the keeping of primates as pets (RSPCA, 2021).

Conclusions

Overall, when compared to other regions, animal welfare in Europe would seem to be well supported by the law but not consistently so. However, for those countries outside of the EU this coverage can be variable and limited. Within the EU there can be a gap between Member States in terms of implementation and enforcement. Some areas, such as the use of animals in circuses, are poorly protected by EU law making the law in the Member States more significant. Europe-wide there is a huge divergence in the way animal welfare is regulated and protected, and the Council of Europe's initiatives are one way that a Europe-wide solution to some of the issues raised above could be met.

Acknowledgments

This chapter has relied heavily on, and is indebted to, the World Animal Protection Index (WAP n.d.), Eurogroup for Animals (Eurogroup, n.d. a–q), and the Global Animal Law database.

Note

1 For EU law, see EU law – EUR-Lex (europa.eu). The European Commission has an EU Platform on animal welfare see EU Platform on Animal Welfare (europa.eu).

References

Abnett K, 2021. Caged animal farming must end in EU, European Commission says. *Reuters*. Available at: Caged animal farming must end in EU, European Commission says | Reuters (Accessed 21 September 2021).

Berry G, 2021. Even locals outraged as 1400 dolphins die in Faroese hunt. Available at: Even locals outraged as 1400 dolphins die in Faroese hunt - Whale and Dolphin Conservation (whales.org) (Accessed 21 September 2021).

Carozza P, 2015. European law. Available at: European law | Britannica https://www.britannica.com/topic/European-law (Accessed 13 September 2021).

DEFRA, 2021a. Action plan for animal welfare. Available at: Action Plan for Animal Welfare: GOV.UK (www.gov.uk) (Accessed 15 September 2021).

DEFRA, 2021b. Animals to be formally recognised as sentient beings in domestic law. Available at: Animals to be formally recognised as sentient beings in domestic law: GOV.UK (www.gov.uk) (Accessed 21 September 2021).

DEFRA, 2021c. Government backs Bill banning the use of glue traps for pest control. Available at: Government backs Bill banning the use of glue traps for pest control - GOV.UK (www.gov.uk) (Accessed 21 September 2021).

DEFRA, 2021d. Government launches second Animal Welfare Bill to protect pets, livestock and wild animals. Available at: Government launches second Animal Welfare Bill to protect pets, livestock and wild animals - GOV.UK (www.gov.uk) (Accessed 21 September 2021).

DEFRA, 2021e. Pet abduction to be made new criminal offence in crackdown on pet theft. Available at: Pet abduction to be made new criminal offence in crackdown on pet theft - GOV.UK (www.gov.uk) (Accessed 21 September 2021).

Eurogroup for Animals (Eurogroup), 2020. The show can't go on: The fight for an EU-wide ban on wild animals in circuses. Available at: The show can't go on: The fight for an EU-wide ban on wild animals in circuses | Eurogroup for Animals (Accessed 15 September 2021).

Eurogroup for Animals, 2021. The new animal health law: A toolbox for powerful solutions. Available at: The new Animal Health Law: A Toolbox for Powerful Solutions | Eurogroup for Animals (Accessed 13 September 2021).

Eurogroup for Animals (Eurogroup), n.d.a. Banning force-feeding in foie gras production. Available at: Banning force-feeding in foie gras production | Eurogroup for Animals (Accessed 13 September 2021).

Eurogroup for Animals, n.d.b. Cats and dogs are neglected by the law. Available at: Cats and dogs are neglected by the law | Eurogroup for Animals (Accessed 13 September 2021).

Eurogroup for Animals, n.d.c. Ensuring that all farmed species are covered by EU animal welfare legislation. Available at: Ensuring that all farmed species are covered by EU animal welfare legislation | Eurogroup for Animals (Accessed 13 September 2021).

Eurogroup for Animals, n.d.d. Equine meat imports and labeling | Eurogroup for Animals (Accessed 13 September 2021).

Eurogroup for Animals, n.d.e. Fur farming. Available at: Fur farming | Eurogroup for Animals (Accessed 13 September 2021) also Fur Free Alliance (n.d.) Fur Bans. Available at: Fur bans (furfreealliance.com) | (Accessed 13 September 2021).

Eurogroup for Animals, n.d.f. Invasive alien species. Available at: Invasive Alien Species | Eurogroup for Animals (Accessed 16 September 2021).

Eurogroup for Animals, n.d.g. It's time to take action for animals in laboratories. Available at: It's time to take action for animals in laboratories | Eurogroup for Animals (Accessed 15 September 2021).

Eurogroup for Animals, n.d.h. Promotion of the 3Rs and non-animal research methods. Available at: Promotion of the 3Rs and non-animal research methods | Eurogroup for Animals (Accessed 13 September 2021).

Eurogroup for Animals, n.d.i. Responsible ownership of equines. Available At: Responsible ownership of equines | Eurogroup for Animals (Accessed 13 September 2021).

Eurogroup for Animals, n.d.j. Strays. Available at: Strays | Eurogroup for Animals (Accessed 13 September 2021).

Eurogroup for Animals, n.d.k. The use of animals in science is still too high. Available at: The use of animals in science is still too high | Eurogroup for Animals (Accessed 13 September 2021).

Eurogroup for Animals, n.d.l. Traceability of equines | Eurogroup for animals (Accessed 13 September 2021).

Eurogroup for Animals, n.d.m. Transport of equines intended for slaughter. Available at: Transport of equines intended for slaughter | Eurogroup for Animals (Accessed 13 September 2021).

Eurogroup for Animals, n.d.n. Wild animals in circuses. Available at: Wild animals in circuses | Eurogroup for Animals (Accessed 13 September 2021).

Eurogroup for Animals, n.d.o. Wildlife trade and trafficking. Available at: Wildlife Trade and Trafficking | Eurogroup for Animals (Accessed 16 September 2021).

Eurogroup for Animals, n.d.p. Zoos and Dolphinaria. Available at: Zoos and dolphinaria | Eurogroup for Animals (Accessed 13 September 2021).

Europa, n.d. Countries. Available at: Countries | European Union (europa.eu) (Accessed 13 September 2021).

European Commission, 2015. EU zoos directive good practice document. Available at: https://ec.europa.eu/environment/nature/pdf/EU_Zoos_Directive_Good_Practices.pdf (Accessed 13 September 2021).

European Commission, 2020. A farm to fork strategy for a fair, healthy and environmentally-friendly food system. Communication from the EU Commission, COM (2020) 381.

European Commission, 2021a. Commission proposes new measures to ban trade in ivory. Available at: Commission proposes new measures to ban trade in ivory (europa.eu) (Accessed 21 September 2021).

European Commission, 2021b. European Citizen's initiative: Commission to propose phasing out of cages for farmed animals. Available at: Commission to propose phasing out of cages for farm animals (europa .eu) (Accessed 21 September 2021).

European Commission, 2021c. Final communication from the commission to the european parliament, the council, the european economic and social committee and the committee of the regions strategic guidelines for a more sustainable and competitive eu aquaculture for the period 2021 to 2030 {swd (2021) 102 final} 1.

European Commission, n.d.a. A European green deal. Available at: A European Green Deal | European Commission (europa.eu) (Accessed 15 September 2021).

European Commission, n.d.b. Animal welfare. Available at: Animal welfare (europa.eu) (Accessed: 13 September 2021).

European Commission, n.d.c. Available at: Animal Health Law (europa.eu) (Accessed 13 September 2021).

European Commission, n.d.d. Law. Available at: Law | European Commission (europa.eu) (Accessed 13 September 2021).

European Commission, n.d.e. Law-making process. Available at Law-making process | European Commission (europa.eu) (Accessed 13 September 2021).

European Commission, n.d.f. Natura 2000. Available at: Natura 2000 - Environment - European Commission (europa.eu) (Accessed 15 September 2021).

European Commission, n.d.g. Sustainable hunting initiatives. Available at: Sustainable Hunting - Conservation - Environment - European Commission (europa.eu) (Accessed 15 September 2021).

European Commission, n.d.h. The EU approach to combat wildlife trafficking. Available at: The EU Approach to Combat Wildlife Trafficking - Environment - European Commission (europa.eu) (Accessed 16 September 2021).

European Parliament, 2020. Pet trafficking: Measures against the illegal puppy business. Available at: Pet trafficking: measures against the illegal puppy business | News | European Parliament (europa.eu) (Accessed 15 September 2021).

Fur Free Alliance, 2021. Animal protection organisations welcome Member States' call for an end of fur farming in the EU: Fur free alliance. Available at: Animal protection organisations welcome Member States' call for an end of fur farming in the EU - Fur Free Alliance (Accessed 21 September 2021).

Global Animal Law Association, n.d.. Universal declaration of animal welfare (UDAW). Available at Universal declaration of animal welfare (UDAW) (GlobalAnimalLaw.org) (Accessed 13 September 2021).

Groves D, 2020. WDC report critical of inadequate EU regulations as over 1,000 dolphins a month face death in nets this winter. Available at: WDC report critical of inadequate EU regulations as over 1,000 dolphins a month face death in nets this winter - Whale and Dolphin Conservation (whales.org) (Accessed 16 September 2021).

Groves D, 2021. Over 100 whales killed in opening days of Norway hunt season. Available at: Over 100 whales killed in opening days of Norway hunt season - Whale and Dolphin Conservation (Accessed 16 September 2021).

Grum T, 2019. Global ban on animal testing: Where are we in 2019? Available at: Global ban on animal testing: where are we in 2019? (cosmeticsdesign-europe.com) (Accessed 15September 2021).

Humane Decisions, n.d. Lists of countries that have banned fur farming. Available at Cities and Countries That Have Banned Fur Farming and Fur Sales (humanedecisions.com) (accessed 20 September 2021).

International Whaling Commission, n.d.a. Description of the aboriginal subsistence hunt in Chukotka, Russian Federation. Available at: Russian Federation (iwc.int) (Accessed 16 September 2021).

International Whaling Commission, n.d.b. IWC. Available at: IWC | International Whaling Commission (Accessed 16 September 2021).

International Whaling Commission, n.d.c. Member map. Available at: Member Map (iwc.int) (Accessed 16 September 2021).

Kokkinidis T, 2021. Greece introduces new regulations for pets; Stricter penalties for abuse. Available at: Greece Introduces New Regulations for Pets; Stricter Penalties for Abuse (greekreporter.com) (Accessed 16 September 2021).

Ministry of Food, Agriculture and Fisheries of Denmark, n.d. Animal welfare legislation. Available at: Animal Welfare Legislation (foedevarestyrelsen.dk) (Accessed 20 September 2021).

Murray A, 2020. Coronavirus: Denmark shaken by cull of millions of mink. *BBC News*. Available at: Coronavirus: Denmark shaken by cull of millions of mink - BBC News (Accessed 13 September 2021).

Royal Society for the Protection of Animals (RSPCA), 2021. No animal left behind- proposal for an Animal Welfare Strategy. Available at: 2fd41066-48e4-5608-c304-d460c8d03d32 (rspca.org.uk) (Accessed 13 September 2021).

Tasker L, n.d. WSPA-RSPCA-International-stray-control-practices-in-Europe-2006-2007. Available at: WSPA-RSPCA-International-stray-control-practices-in-Europe-2006-2007.pdf (stray-afp.org) (Accessed 13 September 2021).

The Donkey Sanctuary, n.d.. Issues. Available at: The issues donkeys face today | The Donkey Sanctuary (Accessed 13 September 2021).

Whale and Dolphin Conservation (WDC), n.d.a. Home. Available at: Home: Whale and Dolphin Conservation (whales.org) (Accessed 16 September 2021).

Whale and Dolphin Conservation, n.d.b. Whaling in Iceland. Available at: Whaling in Iceland - Whale and Dolphin Conservation (whales.org) (Accessed 16 September 2021).

World Animal Protection, (WAP), n.d. Animal protection index. Available at: World Animal Protection | Animal Protection Index (Accessed 11 June 2021).

World Atlas, 2021. European Countries that are not members of the European Union. Available at: European Countries That Are Not Members Of The European Union - WorldAtlas (Accessed 13 September 2021).

31
KEY ANIMAL LAW IN INDIA

Sonia Shad and Yashprada Joglekar

Introduction

India has amongst the richest traditions of respecting animals and treating them with dignity and reverence. Its culture bears an ancient history deeply rooted in spirituality which believes that animals have souls. Indian culture is strongly influenced by Ahimsa (non-violence) towards all living beings. The great king Ashoka (304–232 BC) is the first known king to officially make the welfare of animals a central tenet of his administration, and his rock edicts are the first to articulate basic rights for animals (Rich, 2008).

In India, both the state and the central governments can frame laws. Schedule VII of the Constitution of India lays down a division of subject matters in three lists, on which legislation may be passed by the state and central government. Prevention of Cruelty to Animals is on the Concurrent List, i.e., both the state and central governments can legislate on this subject (Entry 17, Concurrent List, Schedule VII, The Constitution of India, 1950). Animal Husbandry falls under the State list (Entry 14, 15, 16, 21, State List, Schedule VII, Constitution of India, 1950). In the event of a repugnancy between the laws made by the Central and State Legislature, the Central Law will override the State Law. A State Law passed subsequent to the Central Law will prevail, however, if it has received Presidential assent under Article 254, Constitution of India.

India has a central anti-cruelty-specific legislation that applies to all animals, and rules framed thereunder that address specific issues with animal welfare. Due to the interconnected nature of animal welfare and other human rights and public health concerns, including child welfare, labour welfare, food safety, and environmental protection, in order to appropriately address animal welfare, it is often important to look at allied laws. Animal law in India is, therefore, an amalgamation of constitutional, criminal, and civil law that impacts not only animal welfare but also human welfare and environmental conservation.

Due to the expansive nature of the legal framework related to animal protection, this chapter provides only a cursory overview of the central animal protection law in India. Laws relating to issues that require an analysis of allied laws, or state laws like animal sacrifice, or the prohibition of cow slaughter have therefore not been covered.

 DOI: 10.4324/9781003182351-37

Constitutional and legal status of animals in India

Animals in India are technically considered to be legal property and do not have legal personhood. Animals have been provided some protection under statutory law and through the judicial precedents set by the Supreme Court and High Courts. The judiciary often utilises an eco-centric lens while addressing issues relating to animal welfare in accordance with Article 51A(g) (Article 51A(g), The Constitution of India, 1950. "—It shall be the duty of every citizen of India— to protect and improve the natural environment including forests, lakes, rivers and wild life, and to have compassion for living creatures;") and Article 48A (Article 48A, The Constitution of India, 1950. "The State shall endeavour to protect and improve the environment and to safeguard the forests and wild life of the country") of the Constitution of India. The Supreme Court of India, in *Sachidananda Pandey v. State of West Bengal and Ors* 1987 AIR 1109 held that while the Directive Principles of State Policy and Fundamental Duties in the Constitution are not justiciable, the burden to ensure that these principles are upheld to the extent possible does not rest solely with the Legislature. The Supreme Court further elaborated that the judiciary is obligated to factor the guiding principles enshrined in the Constitution while making decisions relating to animal welfare, to the extent possible and as appropriate given the circumstances of the case.

The Supreme Court in *AWBI v. Nagaraja & Ors*. (2014)7 SCC 547 (Para 72) recognised that animals have intrinsic worth, honour, and dignity beyond their usefulness to humans. The Apex court in *People for Animals v. Md. Mohazzim* (2015) SCC On Line Del 9508. (Para 5), in accordance with *inter alia* Article 51A(g), recognised birds' right to fly.

India is a member of the World Organisation for Animal Health (OIE) and the Supreme Court of India has read the Five Freedoms[1] in OIE's Terrestrial Animal Health Code, into S. 3 and S. 11 of the Prevention of Cruelty to Animals Act, 1960 (PCA Act). These sections read with Article 21 and Art. 51A(g) of the Constitution of India are considered to be the Magna Carta of animal rights in India. The Supreme Court in *AWBI vs Nagaraja* (2014)7 SCC 547 also placed responsibility on the State Governments, the Ministry of Environment and Forests, and the Animal Welfare Board of India to ensure that these statutory rights provided to animals are protected and safeguarded.

Any citizen in India has the right to report a violation of the statutory protection afforded to animals. Further, all animal welfare organisations and citizens have *locus standi* to seek judicial enforcement of the statutory rights of animals in public interest (Kansal, 2016). Public Interest Litigation ("PIL") is an anomaly of the Indian legal system. It allows any citizen or group that is not directly affected to raise matters of public concern before the High Court or Supreme Court. It is often used to protect the rights of minorities, the environment, or disadvantaged groups whose needs have not been addressed appropriately. PILs have been an effective tool to address significant and grievous harm to animals and the environment and have led to significant legal protection.

Anti-cruelty legislation

The Prevention of Cruelty to Animals Act, 1960

The Prevention of Cruelty to Animals Act, 1960 ("PCA Act") is the primary anti-cruelty legislation in India. This law applies to all animals, including but not limited to farmed animals, wild animals, companion animals, and animals used in research. The primary objective of this legislation as mentioned in the preamble of this statute is to prevent the infliction of unnecessary pain or suffering on animals.

S. 2(f) of the PCA Act defines an "owner" for the purpose of this Act, as any person that owns the animals, has custody, or charge of an animal with or without the original owner's consent.

S. 3 of the PCA Act imposes a duty of care upon every person who has charge of an animal, to ensure their well-being and prevent unnecessary pain and suffering. It is a protective and preventive provision that confers no right on the "owner" but confers duties and obligations on them.

S. 11(1) of the PCA Act, 1960 makes it a criminal offence to treat an animal with cruelty. The section specifically lists acts of violence, confinement or restriction of movement to an area that is insufficient for the well-being of the animal or for an unreasonable amount of time, deprivation of basic needs, failure to ensure proper medical care, mutilation or killing of an animal by any means that is unnecessarily cruel, animal fighting, and subjecting an animal to unnecessary pain or suffering as offences under the Act. S. 11(1)(a) of the PCA Act, provides a broad scope for what acts are covered under this law and provides protection from a plethora of acts that are not specifically addressed under the Act. Anyone that treats or causes or, being an "owner", permits any act that causes unnecessary pain or suffering to an animal would be liable for criminal penalties under the Act.

As noted, the OIE Five Freedoms have also been read into these sections and violation of any of these freedoms would therefore be considered as an act of animal cruelty in accordance with Indian law.

S. 11(3) lists exceptions to S. 11(1) including the "destruction" of an animal in accordance with the law, or for human consumption as long as the method of killing the animal was carried out without the infliction of unnecessary pain and suffering and in the manner prescribed by law. The Supreme Court has also highlighted that the PCA Act is a welfare legislation for sentient beings over whom humans have significant power, and therefore when it comes to the application of this welfare legislation, the "species' best interest" has to be kept in mind, subject to "just exceptions out of human necessity" (*AWBI vs Nagaraja* (2014) 7 SCC 547 (Para 12)). This is a good stepping stone towards a more equitable legal system where the basic needs of animals are prioritised over avoidable human wants, however the term "necessity" is subjective.

Any police officer above the rank of a sub-inspector has the power to enter into any premises without a prior warrant and seize any animal against which an offence has taken place or suspected to have taken place (Ss. 32, 34, PCA Act, 1960).

The penalties under the act are minimal and rarely act as a sufficient deterrent. The fines range from 50 INR to 1,000 INR (0.7–13.3 USD) and although there are provisions for imprisonment, for a period of 3 months to 2 years in certain circumstances, these have rarely been used. Only the offences listed in S. 11(1) (l)(n)(o) which deal with mutilating or killing an animal in an unnecessarily cruel manner, committing an act in furtherance of animal fighting, or participating or promoting an event where animals are released for shooting, and S. 12 which deals with using any means that are harmful to the animal, including injection of substances, with the objective of increasing lactation or permitting such actions on an animal in their care, are cognisable. Offences that are considered to be serious in India are classified as cognisable offences. They can be investigated by the police upon being reported and do not require an arrest warrant to be issued by a magistrate.

The Animal Welfare Board of India ("the Board") has been constituted in accordance with the PCA Act. The Board primarily acts as an advisory body, but it also has additional functions in furtherance of the objectives of the PCA Act, including setting up animal shelters, education, coordination of associations and bodies working in furtherance of animal welfare (S. 9 PCA Act, 1960). Through consequent legislation and judicial pronouncements, the AWBI's role has evolved to include *inter alia*, licencing, registration, monitoring, and creation of Standard Operating Procedures in furtherance of the objectives listed under the act and rules framed thereunder.

The PCA Act (S. 38 PCA Act, 1960) also confers powers on the Central Government to make rules under this act in furtherance of the objectives of the PCA Act.

The Indian Penal Code, 1860 (IPC)

The IPC defines an animal as any living being that is not a human (S 47. IPC, 1860). S. 428 of the IPC deals with the offence of killing or maiming an animal valued above INR. 10 (0.13 USD) and S. 429 deals with mischief by killing or maiming any cattle or poison or otherwise rendering useless an animal of any value above INR 50 (0.66 USD) The penalties under this section are stronger and include imprisonment and/or a fine. This does not apply to animals that are slaughtered or killed in accordance with and in the manner prescribed by the law.

As per S. 503 IPC 1860, read with Art. 51A (g) and Art. 21 of the Constitution of India 1950, individuals who attempt to prohibit someone from legally carrying out their fundamental duty of compassion towards animals, can be held liable for the offence of criminal intimidation.

Prevention of Cruelty to Animals (Care and Maintenance of Case Property) Rules, 2017

The Prevention of Cruelty to Animals, (Care and Maintenance of Case Property) Rules, 2017 apply when an offence under the PCA Act or the Rules framed thereunder is committed against an animal.

The Magistrate decides where the animal is to be housed. It can only be housed in an infirmary, pinjrapole (animal shelter), a Society for Prevention of Cruelty to Animals (SPCA) (Prevention of Cruelty to Animals (Establishment and Regulation of Societies) Rule 2001, framed under subsection (1) of Section 38 of the Prevention of Cruelty to Animals Act, requires the establishment of an SPCA in every District), an animal welfare organisation recognised by the Animal Welfare Board of India or gaushala (cow shelter) during the pendency of the litigation (R.3 (a), The Prevention of Cruelty to Animals, (Care and Maintenance of Case Property) Rules, 2017). The accused must also provide a bond for the care of the animal during the pendency of the suit, if the accused fails to do so, the animal will be forfeited (R5(1), The Prevention of Cruelty to Animals, (Care and Maintenance of Case Property) Rules, 2017). The owner, in lieu of a bond, can also voluntarily relinquish an animal but the same is permanent and does not affect the criminal charges against him (R.7, The Prevention of Cruelty to Animals, (Care and Maintenance of Case Property) Rules, 2017). If the accused is found guilty or pleads guilty the magistrate shall deprive him of ownership and forfeit the animal to the organisation/body already possessing custody of the animal (R. 8, The Prevention of Cruelty to Animals, (Care and Maintenance of Case Property) Rules, 2017).

Animals who are forfeited and animals whose owners have been deprived of custody in accordance with the law, are to be disposed of or adopted in the manner prescribed under the Rules (R9, The Prevention of Cruelty to Animals, (Care and Maintenance of Case Property) Rules, 2017). Adoption here does not mean the transference of rights of ownership. The adoptee is only the lawful guardian of the animal and will not have the rights generally possessed by the owner of the animal but will have the duty to take all responsible measures to ensure the well-being of such animal and to prevent infliction upon such animal of unnecessary pain or suffering (R9 (8), The Prevention of Cruelty to Animals, (Care and Maintenance of Case Property) Rules, 2017).

Laws relating to transportation of animals

The transportation of animals is significantly regulated. The Transport of Animals Rules, 1978 covers the transportation of a number of animal species including companion animals like dogs and cats, monkeys, cattle, equines, poultry, etc. by different modes of transportation including, road, rail, and air. Apart from general requirements like a fitness certificate, and restrictions on transportation of ill, injured, or animals in the later stages of pregnancy, these rules also list specifications for select species listed under the Act.

The Transport of Animals on Foot Rules, 2001 lays down the conditions under which animals can be transported on foot and the welfare requirements for the same.

The Food Safety and Standards (Licensing and Registration of Food Business) Regulations, 2011 (FSSR, 2011) prescribe the manner in which the transport of animals for the purpose of slaughter must be carried out.

The Central Motor Vehicles (Eleventh Amendment) Rules, 2015 require vehicles transporting livestock animals to obtain a licence for the same from the Regional Transport Officer for vehicles modified to suit the requirements for transportation of animals in accordance with R.125E (2). No motor vehicle meant for the transportation of animals can carry other goods (R.125E The central Motor Vehicles Rules, 1989). Under the Motor Vehicles Act, the police have the power to seize the vehicles that fail to comply with the requirements prescribed under law (S. 207 Motor Vehicles Act, 1998).

Although transportation comprises only a portion of the life of animals, particularly domesticated animals, these laws have a significant role to play in protecting animals. Most often these violations are evident and provide individuals with the ability to identify animals subject to cruelty and report the same without trespassing on any private property.

Laws relating to farmed animals

India's per capita meat consumption is the second-lowest in the world (Mittal, 2018). In spite of this low figure, India is still one of the largest producers of meat, dairy, and eggs in the world, (Shahbandeh, 2021; TNN, 2017) and contrary to common opinion, over 70% of the population in India is non-vegetarian (Census India, 2014). Farmed animal welfare is a pressing issue in India because of the sizeable number of animals in the animal husbandry industry. In India there are currently 1.39 billion farmed animals of which 852 million are poultry (DADH Annual Report 2020–2021).

The only animals that can be slaughtered for consumption within the territory of India are Ovines (sheep), Caprines (goats), Suillines (pigs), Bovines (cows), Poultry and Fish (FSSAI Notification, Dated 6.08.14). Rabbits were added to the list of animals that are permissible for consumption and slaughter in 2017 (FSSAI Notification, Dated 12.08.17). Many states have additional restrictions on the slaughter of cows and consumption of beef.

The laws relating to slaughter are fairly comprehensive. Every stage including transportation for slaughter, pre-slaughter handling, the act of slaughter, processing, packaging, and sale is covered by the Food Safety and Standards (Licensing and Registration of Food Businesses) Regulations, 2011 and the Prevention of Cruelty to Animals (Slaughterhouse) Rules, 2001 has additional welfare requirements. The penalties under the Food Safety and Standards Act, 2006 and rules framed thereunder are higher than those under the PCA Act and the 2011 Regulation includes requirements for animal welfare during the process of slaughter.

Only healthy animals can be slaughtered, and each animal must be examined by a veterinarian and provided with a fit for slaughter certificate (R.3 Prevention of Cruelty to Animals (Slaughter House) Rules, 2001). All animals have to be stunned prior to slaughter and no animal can be slaughtered in sight of another (A (3.6) Part IV, Schedule 4, Food Safety and Standards (Licensing and Registration of Food Businesses) Regulations, 2011; R.6, Prevention of Cruelty to Animals (Slaughter House) Rules, 2001)). Slaughterhouses must be separate from meat shops and no meat should be sold directly to consumers from the slaughterhouse (A (2) Part IV, Schedule 4, Food Safety and Standards (Licensing and Registration of Food Businesses) Regulations, 2011).

Slaughter at live animal markets is not legal in accordance with the Food Safety and Standards Act, 2006 and rules framed thereunder. Despite this, India is leading in the wet market share compared to other Asian markets. As per a 2017 report by the Department of Animal Husbandry and Dairying, around 95% of slaughter for chicken meat in India was carried out in wet markets (Ministry of Agriculture and Farmers' Welfare, 2017).

Another significant welfare concern with respect to farmed animals is the conditions in which farmed animals are raised. Although the provisions of the PCA Act apply to all animals, the lack of holistic regulation regarding minimum space and welfare requirements for each species has led to rampant disregard of animal welfare in intensive animal agricultural units.

The Law Commission of India in Report No. 269 titled "Transportation and House-keeping of Egg-Laying Hens (Layers) and Broiler Chickens" (Law Commission, 2017) recognised the insufficiency of the current legal framework in effectively regulating the poultry industry, and recommended rules in accordance with the PCA Act, to bridge the lacunae in the law. The issue of lack of regulation in the poultry industry is currently sub-judice in the Delhi High Court. The Delhi High Court has placed a moratorium on the use of battery cages in poultry facilities set up after the order passed on September 5, 2018, recognising that the space requirements fail to meet the minimum welfare standards prescribed by law (*Federation of Indian Animal Protection Organisations (FIAPO) & Anr. v. State of Uttar Pradesh & Ors.* W.P.(C) 9056/2016. Order dated 5 September 2018).

There is insufficient regulation for the welfare of fish, and there are no specific stunning requirements or welfare standards prescribed for fish used for human consumption under law.

Laws relating to companion animals

Laws relating to stray animals

India has approximately 62 million stray dogs and 1 million stray cats in India (PTI, 2021). The management of stray animals is primarily governed by the Animal Birth Control (Dogs) Rules, 2001 ("ABC Rules") and the Prevention of Cruelty to Animals Act, 1960 (PCA).

The ABC Rules, 2001 prescribe the manner in which stray dogs have to be dealt with for humane spay and neuter. Management of stray dog populations can only be carried out in a manner that not only complies with the law but is also prescribed by the law (*Animal Welfare Board of India v. People for Elimination of Stray Troubles. & Ors*, SLP(C) 691/09, 18 November 2015). Thus, these rules prescribe the only legal way for any person or body, government or otherwise, to handle stray dogs.

The ABC Rules prescribe two methods to deal with the stray dog population: immunisation and sterilisation (Rule 3(3), Animal Birth Control Rules, 2001). Individuals and Government

Authorities cannot cull dogs as a means for population control. Rule 9 of the ABC Rules allows for euthanasia of stray dogs in the prescribed manner in only two circumstances: (1) if the animal is incurably ill, or (2) if the animal is mortally wounded (Rule 9, Animal Birth Control Rules, 2001 and Section 35, Prevention of Cruelty to Animals Act, 1960).

It is also important to note that Rule 13 of the ABC Rules clearly states that any act, rule, regulation, or by-law made under any law by any state or local authority will not apply if the provisions are more irksome, inconvenient, or harmful to the animal. Only those which result in greater animal welfare, i.e., which ensure a higher degree of well-being and care, are legally valid.

The Board has issued guidelines (AWBI, Circular No.3-3i202 r-2022tPCA dated 03 June 2021) for the sterilisation and immunisation of stray cats. These guidelines are to be read harmoniously with the ABC Rules, 2001. This means that similar animal welfare standards are prescribed for both, stray dogs and cats, although the medical or surgical procedures and requirements differ.

Feeders and Caretakers of stray animals are also protected by the law as they play a vital role in assisting authorities in vaccinating and immunising animals and in reducing human animal conflict (Animal Welfare Board of India, 2015). The High Court of Delhi issued guidelines for the maintenance of community animals and stated that it is the responsibility of every Resident Welfare Association or Municipal Corporation if there is no RWA, to ensure that every community dog in every area has access to food and water when there are no caregivers or feeders in the area (Para 129, *Dr. Maya D Chablani v. Smt Radha Mittal & Ors.* 2021 SCC Online Del 3599).

Laws relating to breeding and sale of companion animals

The Prevention of Cruelty to Animals (Dog Breeding and Marketing) Rules, 2017 requires all breeders, regardless of the size of the venture, to be registered by their respective state governments in accordance with these rules (R3. Prevention of Cruelty to Animals (Dog Breeding and Marketing) Rules, 2017). Any person who has been convicted of an offence relating to animals cannot be issued a registration certificate under these rules (R. 7(b) Prevention of Cruelty to Animals (Dog Breeding and Marketing) Rules, 2017).

Animals for sale have to be healthy and inoculated and must be microchipped. These animals cannot be displayed in public for immediate sale. The animals cannot be sold to a pet shop that does not possess a licence. If the breeder is selling to an individual, the individual must be screened for their ability to care for the animals, financially and otherwise, and the breeder must check on the status of all dogs sold by him on a yearly basis (R8 Prevention of Cruelty to Animals (Dog Breeding and Marketing) Rules, 2017).

All pet shops have to be registered and individuals convicted of an offence under any law relating to animal welfare are not eligible to obtain registrations under the Prevention of Cruelty to Animals Act (Pet Shop) Rules, 2018. Animals in pet shops must have access to food, water, clean and sufficient space, and veterinary care, and cannot be displayed in the window or outside the shop (R. 7(1) Prevention of Cruelty to Animals Act (Pet Shop) Rules, 2018). If animals are left in the pet shop at night, there must be sufficient staff to attend to them (R(7)(2)(l) Prevention of Cruelty to Animals Act (Pet Shop) Rules, 2018).

Laws relating to wild animals

The Wild Life (Protection) Act, 1972 ("WLPA") is the predominant legislation that protects wildlife in India. In addition to the WLPA and the PCA Act, the Customs Act, 1962, Indian

Forest Act, 1927, Forest Conservation Act, 1980, and Biological Diversity Act, 2002, *inter alia* impact the protection of wildlife in India.

The WLPA places wildlife into six schedules. Wild animals are classified based on the degree of protection offered to them from Schedule I–Schedule V in decreasing order of protection. Animals under Schedule V are deemed to be vermin and have little to no protection under the provisions of the Act (S. 62, The Wildlife Protection Act, 1972).

Chapter III of the WLPA deals with hunting. Hunting of any wild animal listed in Schedule I–IV (S. 9, WLPA) is prohibited except in the following circumstances:

Special purposes like education, scientific research, scientific management, collection of specimens, and in the case of snakes for the production of anti-venom. A prior permit must be issued for the same by the Central Government in the case of Schedule I animals and the respective State Government, in the case of animals listed in Schedule II–IV (S. 12, WLPA).

Animals listed in Schedule I can only be hunted if the animal is a threat to human life or has become so diseased/disabled that its recovery is not possible. The animal should not be killed, unless the Chief Wild Life Warden is satisfied that capture, translocation, and tranquilisation is not possible. Further, captured animals should not be kept in captivity unless release is not possible and the reasons for the same are recorded in writing (S. 11(1)(a), WLPA).

Animals listed in Schedule II–IV can only be hunted with permission from the Chief Wild Life Warden or an officer authorised on their behalf if the animal is a threat to human life or property or has become so diseased/disabled that its recovery is not possible (S. 11(1)(b), WLPA).

It is important to note that capture, tranquilisation and translocation or placing the animal in custody if translocation is not possible is the first course of action, only if the above are not possible can the Chief Wild Life Warden determine that in that particular instance killing can be permitted and the reasons for the same must be recorded in writing. There are also alternative human wildlife conflict management practices being developed and used in India and hunting should not be a preferred solution to address this issue.

Any wild animal killed or wounded in self-defence is the property of the government. Killing in self-defence is not considered an offence as long as the person was not committing an offence under the act at the time where such defence was necessary (S. 11(2), (3), The Wildlife Protection Act, 1972).

Chapter IV of the WLPA deals with the designation and protection of Sanctuaries and National Parks.

Chapter V deals with the trade of wild animals, articles and trophies. Any wild animal, killed, bred, hunted, or in captivity in contravention of the WLPA and all instruments and tools used in furtherance of the same, including vehicles traps, etc. are deemed to be the property of the Central Government (S. 39(1), The Wildlife Protection Act, 1972). Any individual in possession of such an item must declare the same within 48 hours to the nearest police officer or authorised official and hand over the article/animal to the appropriate officer or official (S. 39(2), The Wildlife Protection Act, 1972).

Possession or custody of any captive animal or product related to the same, in Schedule I or Part II of Schedule 2 without an ownership certificate is prohibited (S. 40(1), The Wildlife Protection Act, 1972). Individuals are not allowed acquire, receive, keep in custody possess, sell, or offer for sale or otherwise transfer or transport any animal specified in Schedule I or Part II of Schedule without permission from the Chief Wildlife of Animals.

It is illegal to capture a wild animal, under the WLPA, Section 48(b)(i). Acquisition and receipt of animals listed in these schedules after the commencement of the WLPA Amendment Act, 2002 is prohibited except by way of inheritance (S. 40(2A), The Wildlife Protection Act, 1972). Upon inheritance of such an animal after the notification of this Act, a declaration must be made to the Chief Wildlife Officer or an authorised official within 90 days of receiving the same (S. 40(2B), The Wildlife Protection Act, 1972). Unfortunately, even though elephants are listed under Schedule I and a certificate of ownership is mandatory, the proviso under S. 40(2) exempts live elephants from the purview of S. 40(2A) and (2B). This exemption has made the enforcement of the provisions against illegal capture, sale, and trade of elephants in India significantly harder.

India has been a party to the Convention on International Trade in Endangered Species of Wild Fauna and Flora ("CITES") since 1976. However, there is no domestic law that regulates possession or trade of exotic species listed in the CITES appendices within the territory of India (Vijairaghavan et al., 2021). In 2020, the Ministry of Environment Forest and Climate Change issued an advisory which in furtherance of developing an inventory of exotic live species in India, allowed individuals to voluntarily declare animals listed in Appendix I, II, and III of CITES within six months of the advisory without any supporting documentation (I (a)–(c), Advisory for Dealing with Import of Exotic Live Species in India and Declaration of Stock). The Chief Wildlife Warden after appropriate verification is obligated to provide an online certificate of possession within six months of the declaration (I (G), Advisory for Dealing With Import of Exotic Live Species In India and Declaration of Stock). As of May 2021, 43,693 applications for amnesty have been made under this advisory from 30 States and Union Territories (Vijairaghavan et al., 2021).

The penalties under the WLPA are significantly more when compared to the PCA. The penalties for habitual offenders are also higher (S. 51, The Wildlife Protection Act, 1972). The WLPA is a strong legislation for the protection of wildlife in India, but it is imperative for additional regulation to address the lacunae in the law with special emphasis on compliance with international conventions ratified by India.

Laws relating to use of animals for scientific or educational purposes

This section covers only the specific animal welfare legislation of the regulatory framework, and it is important to note that the international standards and ancillary laws, rules, guidelines, and standards play a significant role in the regulation of use of animals for scientific and educational purposes.

S. 14 of PCA, specifies that the PCA Act, does not render unlawful any experiments on animals that are for the purpose of advancement of discovery or knowledge that will be useful for saving or prolonging life or combating a disease or reducing suffering of humans, animals, or plants. The term experimentation on animals, as used in the Prevention of Cruelty to Animals Act, 1960, includes animals used in research, testing, and education. The Breeding of and Experiments on Animals (Control and Supervision) Rules, 1998 and the PCA Act provide basic welfare requirements for animals being used for scientific or educational purposes.

The Committee for the Purpose of Control and Supervision of Experimentation on Animals (CPCSEA) has been constituted under the provisions of the PCA Act and the rules framed thereunder (S 15, 16, Prevention of Cruelty to Animals Act, 1960). It is mandatory that all institutions that breed or use animals for scientific or educational purposes must register themselves with the CPCSEA (Rule 3, 4, 5, Breeding of and Experiments on Animals (Control and Supervision) Rules, 1998).

Each institution that uses animals for scientific or educational purposes has to form an Institutional Animal Ethics Committee. This body is responsible for inspecting the concerned institution's animal facilities and reviewing and approving research protocols (Rule 13. Breeding of and Experiments on Animals (Control and Supervision) Rules, 1998; CPCSEA Standard

Operating Procedures (SOP) for IAEC). The institution must only use the lowest number of animals, on the lowest phylogenetic scale to achieve an accuracy of 95%. If there are non-animal alternatives available, there must be sound justification as to why the alternatives were not used (Rule 9, Breeding of and Experiments on Animals (Control and Supervision) Rules, 1998).

Animal testing for cosmetic products is prohibited in India (Rule 148-B, Drugs and Cosmetics (2nd Amendment) Rules, 2014; Rule 39(7), The Cosmetics Rules, 2020). No cosmetics that have been tested on animals after 2014 can be imported (Section 135-B, Drugs and Cosmetic Act, 1945; Rule 18(4), The Cosmetics Rules, 2020).

The use of animals in medical and pharmacy educational institutions for the purpose of teaching in undergraduate courses is prohibited and computer assisted modules are the alternative prescribed (Establishment of Medical College Regulations, 2013 (Amendment); Pharm.D (Amendment) Regulations, 2014).

Laws relating to performing animals

R.2(h), of the Performing Animals (Registration) Rules, 2001, defines a "Performing Animal" as an animal which is used for the purpose of any entertainment. Entertainment includes films, circuses, and any animal events into which the public is admitted.

At every event with performing animals the exhibitor or trainer must have a Performing Animals Registration Certificate from the Animal Welfare Board of India which contains all the details about how the animals are to be used in the event (Rule 3, Performing Animals (Registration) Rules, 2001). Dog shows, animal rides, animal races, and animal sports require a registration certificate as detailed in the aforementioned laws, as they all fall under the ambit of performing animals as defined in rule 2(h) of the Performing Animals (Registration) Rules, 2001.

When s. 2(7A), 2(39), 38H and 42 of the WPLA are read together, wild animals that perform, fall under the definition of circus, which comes under the term Zoo in the act. Therefore, wild animals, cannot perform in any circus without recognition by the Central Zoo Authority, and compliance with the housing and other needs detailed in the Recognition of Zoo Rules. This means that all local wildlife animal performances like snake charming and dancing bears, etc. are illegal. The use of bears, monkeys tigers, panthers, lions in performances is prohibited in accordance with the Ministry of Social Justice and Empowerment Notification G.S.R 619(E) dated 14.10.1998. Bulls were added to the prohibited list of animals used for performance by the Ministry of Environment and Forest Notification G.S.R. 528(E) dated 11 July 2011.

The ban on using certain animals as performance animals was upheld in *Nair N.R. and Ors. Vs. Union of India and Ors.* AIR 2000 Ker 340. Bull races are also in violation of Sections 3, 11(1), and 22, PCA. The Supreme Court in *AWBI vs A Nagaraja & Ors* specifically stated that bulls are not performing animals.

Animal fighting is strictly prohibited and is in violation of Ss. 3, 11(1)(m) & 11(1)(n) and other provisions under Section 11(1) depending on the nature of the cruelty experienced and in the case of use of prohibited animals, S. 22, PCA.

Conclusions

Animal law in India is fairly expansive but not exhaustive. The lack of surety of action by-law enforcement agencies and the minimal penalties under the PCA Act prevent the effective implementation of the legal protections afforded to animals. Lacunae in the law are exploited without regard for the welfare of animals or the resultant impacts on human welfare.

The judiciary has used the constitutional provisions relating to animals to apply an eco-centric lens while adjudicating on matters relating to animal welfare. The judiciary has also in many instances recognised that animal welfare is often intrinsically linked to human welfare and has strived to ensure that the protections offered to animals under Indian law are enforced.

Improvement in animal welfare is intrinsically linked to the achievement of many of the Sustainable Development Goals (Keeling et al., 2019). The 'One Health' approach (World Health Organisation, 2017) is a global, multi-sectoral, coordinated, and transdisciplinary approach that recognises the interconnection between people, animals, plants, and their shared environment. This approach has been gaining traction across the world, with support from the WHO, OIE, and the FAO. Stakeholders in India including public health experts, veterinarians, health care providers and policy makers are placing greater weight on the "One Health" principle (Aggarwal and Ramachandran, 2020).

It is imperative that additional legislation and regulation are made to bridge the gaps in the law, improve mechanisms for enforcement and monitoring, and that penalties under the act are increased, to ensure the minimum welfare standards prescribed by-law are met, for the benefit of both humans and animals.

Note

1 Freedom from hunger, malnutrition, and thirst; freedom from fear and distress; freedom from heat stress or physical discomfort; freedom from pain, injury, and disease; and freedom to express normal patterns of behaviour.

References

Aggarwal D and Ramachandran A, 2020. One health approach to address zoonotic diseases. *Indian Journal of Community Medicine: Official Publication of Indian Association of Preventive & Social Medicine*, 45(Suppl 1), p. S6.

Animal Welfare Board of India, 2015. Guidelines with respect to pet & street dogs, their caregivers, and for Residents' welfare associations and apartment owners associations. 26 February, p. 6. Available at http://awbi.in/awbi-pdf/pet_dog_circular_26_2_2015.pdf [Accessed 9 January 2022].

Government of India, Census of India, 2014. Prevelence of vegetarianism and non vegetarianism amongst the population aged 15 years and above, India and bigger states 2014 (table 5.2) in *Sample Registration System baseline Survey 2014*. p. 22. Available at: https: censusindia.gov.in [Accessed 15.05.2022].

Kansal V, 2016. The curious case of Nagaraja in India: Are animals still regarded as "Property" with no claim rights? *Journal of International Wildlife Law & Policy*, 19(3), pp. 256–267.

Keeling L, Tunón H, Olmos Antillón G, Berg C, Jones M, Stuardo L, Swanson J, Wallenbeck A, Winckler C and Blokhuis H, 2019. Animal welfare and the United Nations sustainable development goals. *Frontiers in Veterinary Science*, 6, p. 336.

Law Commission of India, 2017. *Transportation and House-keeping of Egg-laying Hens (Layers) and Broiler Chickens*. Report No. 269. Available at https://lawcommissionofindia.nic.in/reports/Report269.pdf [Accessed 31 December 2021].

Ministry of Agriculture and Farmers' Welfare, 2017. National action plan for egg and poultry to 2022. Doubling farmer's income by 2022. *National Action Plan,* p. 21. Available at https://www.dahd.nic.in/sites/default/filess/Seeking%20Comments%20on%20National%20Action%20Plan-%20Poultry-%202022%20by%2012-12-2017.pdf [Accessed 30 December 2021].

Mittal, N, 2018. India has the 2nd least meat consumption per person in the world, what happens when 2000-year-old tradition of vegetarianism may be about to change? blog post, *Bergensia*, 2018. Available at https://bergensia.com/india-has-the-2nd-least-meat-consumption-per-person-in-the-world-what-happens-when-2000-year-old-tradition-of-vegetarianism-may-be-about-to-change/ [Accessed 12 December 2021].

Press Trust of India, 2021. How many stray cats, dogs loiter unsheltered in India? Here's the answer. *India Today*. Available at https://www.indiatoday.in/india/story/how-many-stray-cats-dogs-loiter-unsheltered-in-india-1881222-2021-11-26 [Accessed 12 December 2021].

Rich B, 2008. *To Uphold the World*. India: Penguin Books, pp. 191–192.
Shahbandeh M, 2021. Global chicken meat production 2020 & 2021, by selected country. *Statista*, 26 May. Available at https://www.statista.com/statistics/237597/leading-10-countries-worldwide-in-poultry-meat-production-in-2007/ [Accessed 12 December 2021].
Times News Network, 2017. India fastest growing egg producer in world. *The Times of India*, 20 November. Available at https://timesofindia.indiatimes.com/business/india-business/india-fastest-growing-egg-producer-in-world/articleshow/61730376.cms [Accessed 12 December 2021].
Vijairaghavan M, Swami S and Sinha D, 2021. Only carrots, no stick: Regulating the trade in exotic live animals in India. blog post, *Vidhi Centre for Legal Policy*, 5 June 2021. Available at https://vidhilegalpolicy.in/blog/only-carrots-no-stick-regulating-the-trade-in-exotic-live-animals-in-india/ [Accessed 12 December 2021].
World Health Organisation, 2017. One health. blog post, *News Room*. 21 September 2017. Available at https://www.who.int/news-room/questions-and-answers/item/one-health [Accessed 12 December 2021.

32
KEY ANIMAL LAW IN SOUTH AFRICA

David I Bilchitz and Amy P Wilson

A brief history of animals in South Africa

While there are some legal protections offered to non-human animals (henceforth, "animals") in South Africa, the country lacks an effective and holistic animal welfare regulatory regime. For the vast majority of animal uses and abuses, the government has failed to promulgate legally binding standards, paving the way for either self-regulation or no regulation at all. The law that exists has often been inadequate to address current abuses of animals and lacks adequate enforcement. Before delving into the country's specific animal regulatory regime, it is important to look back briefly at the history of human engagement and attitudes towards animals in South Africa.

South Africa is often known as the cradle of humankind and, since the emergence of the species *Homo sapiens*, the fate of animals has been intertwined with that of human beings. Traditional African communities developed a moral philosophy focused on the notion of "ubuntu", generally expressed by the aphorism that mandates a relational ethic: "a person is a person through other people" (for further information on this concept, see: Mnyaka and Motlhabi, 2005), which involves developing harmonious relationships between humans themselves as well as between humans and other facets of the environment. Whilst there was no doubt conflict between humans and animals and some exploitation, ultimately, the two coexisted relatively harmoniously and sustainably.

With the advent of colonialism, a different ethos took root: people arrived whose attitude was one of domination over other people, as well as other species. Hunting parties would go out and shoot every creature in sight, ultimately leading to a serious drop in the numbers of iconic animals such as elephants. The Roman-Dutch law that was applied (being the system of the colonisers) provided for two classifications: persons – who could have rights and duties – and things that could have neither (Njotini, 2017). Animals were classified as things, paving the way for their domination by humans.

Following similar laws passed in Britain, the first nation-wide animal welfare law (The Prevention of Cruelty to Animals Act, 8 of 1914) was passed when South Africa became a Union in 1910, highlighting the contradictory attitude to non-human animals. This approach of domination over other animals with some welfare protection (contained in The Animals Protection Act, 71 of 1962) continued to be applied throughout the tenure of the apartheid government – which imposed a harsh regime of racial discrimination and segregation – that

DOI: 10.4324/9781003182351-38

came to power in 1948. That government also established legislation allowing for the ownership and fencing-off of wild areas, thus promoting the notion of private rights over wildlife (Game Theft Act, 105 of 1991).

In 1994 however, South Africa fundamentally changed. Its Interim Constitution and later, its Final Constitution (Constitution of the Republic of South Africa, 1996), which came into force in 1997, sought to transform South Africa from a society based on division, exploitation, and oppression to one founded on human dignity, equality, and freedom. The focus of this transformation was on addressing the serious legacy of racial discrimination in the country; yet, in establishing the foundations of a new society, all forms of injustice needed to be considered. In a poetic speech when the Final Constitution was adopted, the Deputy President at the time (Thabo Mbeki) stated the following (emphasis added):

> [a]t times, and in fear, I have wondered whether I should concede *equal citizenship* of our country to the leopard and the lion, the elephant and the springbok, the hyena, the black mamba and the pestilential mosquito. A human presence amongst all these, a feature on the face of our native land thus defined, I know that none dare challenge me when I say – I am an African.
>
> *(SA People, 2016)*

In this quote, Mbeki suggests that a defining feature of African-ness is sharing the land with other creatures. He suggests an ethos not of domination but of mutual respect and cohabitation with other animals. In fact, there was a concerted campaign to include the protection of animals in the new Constitution; unfortunately, however, it was unsuccessful. There is very little direct mention of animals, other than in the schedules which designate which level of government has regulatory authority over certain animal-related issues. Nevertheless, the Constitution ushered in a new era where all facets of the legal system could be re-evaluated. The question thus arose as to what implications would this fundamental constitutional change have for non-human animals (Bilchitz, 2009). In this chapter, we seek to engage broadly with this question, briefly set out the framework of how non-human animals are governed in certain contexts in South Africa, and highlight opportunities for reform.

The Constitution is the foundation of the society and as such affects all other areas of the law. The second part of this chapter charts significant developments in the last few years that promise potentially ground-breaking changes for how animals are considered and treated. At the same time however, very few particular statutory laws have changed. The third part of this chapter will outline the main animal protection and related laws that exist in South Africa as well as some of the environmental laws that are relevant to wild animals. We will also briefly engage with other elements of the legal framework which provide the scaffolding for animal protection and use in South Africa. In this we do not aim to be comprehensive but to provide an outline of the main sources of law in this regard. There are many pressing practical challenges facing animals in the country, whether they are in the domestic agriculture setting or free roaming in the wild. Accordingly, we will also draw on a few key examples of how the regulatory regime interacts with animals and highlight some of the failures and opportunities in this regard.

The Constitution and animals: Changes and possibilities

The Constitution of South Africa is the supreme law, and all other law must be consistent with it. There are many facets of the Constitution which have implications for animals – we will elaborate on three of them.

The protection of animals and their intrinsic value

As was mentioned earlier, the Constitution itself does not include provisions directly protecting non-human animals. It is thus necessary to engage in constitutional interpretation in order to understand what protections animals are granted by the Constitution. Bilchitz has argued in a prior article that the constitutional rights in the Bill of Rights may be interpreted to apply to non-human animals (Bilchitz, 2010). This argument remains viable but, as yet, it has not been adjudicated on by courts. We thus turn to focus now on the actual concrete holdings of the courts.

The most important of these is a case decided in 2016 by the Constitutional Court (the highest court in South Africa) that had to consider whether the National Council of Societies for the Prevention of Cruelty to Animals ("NSPCA") a statutory body constituted in terms of the Societies for the Prevention of Cruelty to Animals, 169 of 1993 ("SPCA Act") was entitled to bring a private prosecution against individuals who had committed acts of cruelty against two camels when performing a religious sacrificial slaughter (*National Society for the Prevention of Cruelty to Animals v Minister of Justice and Constitutional Development and Another* [2016] ZACC 46 ("2016 NSPCA Case")). As will be described below, unnecessary cruelty to animals is criminalised in terms of the Animals Protection Act ("APA"). However, public prosecutors often decline or fail to prosecute cases of animal cruelty. This was a seminal case as, through a reading of the relevant legislation, the Constitutional Court reached the conclusion that the NSPCA was statutorily empowered to bring private prosecutions. In the course of arriving at its decision, the court examined the history of the protection of animals in South African law and how the Constitution had changed their position. It started off by recognising that early court decisions had seen cruelty laws not as protecting animals themselves but protecting the "finer feelings and sensibilities" of humans who were offended by animal cruelty (*R v Moato* 1947(1) SA 490 (0)). The court acknowledged the insufficiency of this rationale and referenced statements made by judges in other cases including one which stated that animals "are sentient beings that are capable of suffering and of experiencing pain" (*National Council of Societies for the Prevention of Cruelty to Animals v Openshaw* [2008] ZASCA 78). Another statement from the Supreme Court of Appeal when dealing with rhino poaching was also quoted, namely that "constitutional values dictate a more caring attitude towards fellow humans, animals and the environment more generally" (*S v Lemthongthai* [2014] ZASCA 131 para 20). Having examined this history, the court reached a ground-breaking conclusion: "[t]herefore, the rationale behind protecting animal welfare has shifted from merely safeguarding the moral status of humans to placing *intrinsic value* on animals as individuals" (2016 NSPCA Case, para 57).

This is one of the most far-reaching statements globally recognising a change in the way animals are conceived. Their status as "things" in the common law is challenged by the court's finding. If non-human animals have intrinsic value (which is similar to the notion of dignity that underlies *human* rights), it is possible then for them also to be protected directly by fundamental rights. The meaning and implications of this statement are far-reaching and still to be worked out in South African law – nevertheless, it opens the door for more extensive protection of animals as well as a deeper respect to be developed for their lives and welfare. It is hard to see how practices such as the live export of animals or hunting them simply for entertainment, both of which are currently occurring in South Africa, are consistent with the recognition of their intrinsic value.

The environmental right and animal welfare

The Constitutional Court statement about "intrinsic value" is a general broad claim that appears initially to lack grounding in the constitutional text itself. The Constitutional Court in this same case identified section 24 of the Constitution, the environmental right, as the relevant basis for such protection. The environmental right specifically mentions the importance of conservation – the court highlighted the importance of adopting an integrative approach which "links the suffering of individual animals to conservation" (2016 NSPCA Case, para 58) and held that "showing respect and concern for individual animals reinforces broader environmental protection efforts. Animal welfare and animal conservation together reflect two intertwined values".

These statements require some background to understand their full import. Conservation had been approached in South Africa through an "aggregative approach" which sees animals as resources to be used and that, as individuals, they do not matter. This approach seeks to ensure that animals as a "resource" are collectively not depleted for future uses – and so the focus is usually on the survival of a species. The Constitutional Court effectively rejects this approach and understands conservation to involve protection for animals as individuals which, in turn, entails treating them with respect for their welfare and intrinsic value. It recognises that protection of a species depends upon respect for individuals (for a more in-depth discussion of these two approaches, see Bilchitz, 2017).

The contrast between the two approaches can be illustrated by the example of trophy hunting. Some tourists travelling to South Africa wish to go on hunting expeditions and to bring back dead trophies of large animals such as lions, elephants, and rhinos. A large industry has developed surrounding hunting. Trophy hunting is defended often on the grounds that it can contribute to conservation through bringing significant funds into the coffers of local communities. That in turn, it is argued, gives people an incentive to conserve animals. Defenders of this view would accept restrictions on the numbers of animals that can be hunted for purposes of preserving the species for future generations to hunt. As long as current uses do not jeopardise future uses, the "aggregative approach" regards hunting as being justifiable. The integrative approach rejects the idea that killing an animal for entertainment – no matter how much money it brings in – can ever be a viable approach to conservation. It argues that instrumentalising the lives of animals in this way will contribute ultimately to the destruction of animal species – for instance, when hunting revenues are limited (as during COVID-19), individuals will have very little incentive to conserve animals.

At the time of writing, the South African government had released a Draft Policy Position on the Conservation and Sustainable Use of Elephant, Lion, Leopard, and Rhinoceros ("Draft Policy Position") in respect of wild animals. There are several promising sentiments and statements expressed in the Draft Policy Position including the phasing out of the captive lion industry. Nevertheless, as it stands, the Draft Policy Position enshrines a central contradiction: on the one hand, it has been forced by the Constitutional Court to recognise that the government policy relating to wild animals must be extended to include animal welfare and that it has a duty to encourage respect for individual animals; and, on the other hand, it seeks to build a large industry based on trophy hunting for economic purposes which is incompatible with these prior recognitions. The manner in which this debate is resolved will indicate the degree to which the protection of animals in South Africa will be enhanced in the new constitutional era.

Governance and administrative law

Given the wanton abuse of power and authority in the past, the Constitution provides that any exercise of government authority must take place according to set principles of administrative justice (in particular, section 33 of the Bill of Rights). These principles are outlined in a piece of legislation known as the Promotion of Administrative Justice Act ("PAJA"). Many activities relating to animals, including the hunting of certain species, involve the issuing of permits by various governmental authorities. In deciding whether to grant these permits, authorities are duty-bound to follow the rules of administrative justice. These rules at times have the potential to advance the interests of animals. An instance where this occurred is the case that dealt with the trade in lion bones, and the quota of lion skeletons which could legally be exported to satisfy this trade in terms of the international treaty, the Convention on International Trade of Endangered Species of Fauna and Flora ("CITES"). The relevant minister had in this case increased the export quota of lion skeletons to 800 skeletons in 2017 and to 1,500 skeletons in 2018 respectively. The decision of the minister was challenged by the NSPCA on the grounds that she had failed to consider the deleterious effects of her decision on the welfare of the animals. Drawing on the statements of the Constitutional Court which have already been referred to, Justice Kollapen found that the minister's decision was not justifiable given that animal welfare and conservation were deeply intertwined. He concluded: "it is inconceivable that State Respondents could have ignored welfare considerations of lions in captivity in setting the annual export quota" (*National Society for the Prevention of Cruelty to Animals v Minister of Environmental Affairs* [2019] ZAGPPHC 337 para 74) and deemed the quotas to be unlawful and constitutionally invalid. The case illustrates the way in which the duties of good governance included in administrative law can help advance the interests of animals. Accordingly, when a matter pertains to the environmental right and conservation, all officials will now have to take animal welfare into account in their decision-making. Administrative law is, however, a double-edged sword: it has also in the past, been successfully utilised by animal use industries in South Africa to reduce animal protections, in the context of the domestic rhino horn trade and canned lion hunting.

The question of governance surrounding animals raises important issues about which structures are created to protect them and their efficacy. In South Africa, the APA recognises that there exist societies for the prevention of cruelty to animals whose remit is to investigate and address potential cases of cruelty against animals. The SPCA Act creates the NSPCA, a national society that can develop policy and engages with all levels of government concerning animals. The Constitutional Court in the aforementioned 2016 NSPCA Case found that the NSPCA was ultimately the "special guardian" of animal welfare in South African law (2016 NSPCA Case, para 59). It is nevertheless a body that is not provided with government funding to fulfil its mandate. The manner in which it is constituted and functions also raises questions as to whether it is truly able to ensure animal interests are optimally protected. The live export of animals by sea – discussed later in this chapter – is an instance where the NSPCA has tried to stop that trade on grounds of severe cruelty caused to the animals. It has, however, been limited in what it can do by, for instance, a lack of funding and very limited legal powers which has meant it has thus far failed in several instances to be able to stop the shipments.

The statutory framework governing animals

Domesticated and captive animals

The Constitution and the cases discussed earlier, as we saw, have created a number of exciting possibilities for developing protections for animals. However, in general, there is no substitute for particular statutory protections for animals. We now elaborate upon selected contexts of animal use and the main regulatory regime that address the protection of animals in South African law (for a more detailed analysis of South African Animal Law, see: Wilson, 2019a).

As a general observation, the animal legal framework is fragmented at different levels – including international, national, provincial, and local, and contains a plethora of legislation, policy, regulations, norms and standards, and soft codes.

The Animals Protection Act

Despite various efforts by advocates, animal sentience is not explicitly recognised in legislation. The exception here is elephants, whose sentience has been acknowledged in the National Norms and Standards for the Management of Elephants in South Africa, 2008. The predominant animal legislation in South Africa is the Animals Protection Act of 1962. Unlike certain other jurisdictions, the APA includes all animals in all contexts provided they are subject to human captivity or control. The definition of "animal" in the APA is: "any equine, bovine, sheep, goat, pig, fowl, ostrich, dog, cat, or other domestic animal or bird, or any wild animal, wild animal, wild bird or reptile which is in captivity or under control of any person". The NSPCA has opted to interpret this definition utilising a wide approach indicating that fish are included in the "other … animal" catch all, and that wild animals interacting with persons fall under the final element of the definition. The APA is a criminal statute which sets out various detailed offences and prohibits generally any unnecessary cruelty to animals. It therefore could be applied to farmed animals, animals in all captive situations (from zoos to laboratories), working animals, and companion animals; however, the use of this law in these contexts is extremely rare. The statute also provides for powers of enforcement by the police, societies for the prevention of cruelty to animals and for criminal penalties for cruelty ranging from fines to imprisonment.

The APA has been criticised on several grounds, several of which are raised below. Firstly, the requirement that an animal must be subject to human control has led to uncertainty concerning the application of the APA to fish, other aquatic species, and free-roaming wild animals, and accordingly the extent to which the APA can be successfully used to protect them. Secondly, the APA's list of offences is vast, but the crimes are often qualified by terms such as "unnecessarily", or "cruelly". These vague terms cause uncertainty as to their meaning and application. Thirdly, the statute is a criminal statute requiring a high evidentiary burden of proof beyond a reasonable doubt. The statute is not well suited to proactively protecting animals from prospective cruelty. Fourthly, the penalties are relatively minor for committing animal cruelty acts, thus suggesting that the law does not consider such actions to be very serious. Fifthly, prosecutions rarely occur, and it is largely left up to the NSPCA to enforce the APA, with its limited resources and without government funding. This situation leads to countless abuses going uncharged, unprosecuted, and unenforced. Lastly, the APA is a general statute. Despite the APA empowering the minister to make regulations, to date no standards have been promulgated for animal uses in any context including particular farming practices. To date, no bans (other than for animal fighting) have been put in place even for egregious practices such as sow stalls and battery cages that other jurisdictions have taken steps to prohibit, restrict, or phase out. In fact, such practices are increasing, and generally animal use in all contexts is actively supported and promoted by government.

Despite these criticisms of the APA, the NSPCA maintains that, overall, it is a useful piece of legislation (CER & EWT Report; Fair Game, 2018). There have been proposals to further amend the Act, including two attempts in terms of private members' bills to have cosmetic testing on animals criminalised. The government is considering a complete overhaul of the APA and shortly seeks to release a "Draft Animal Welfare Bill". It is hoped that the new Act will address the shortcomings in the APA and current framework and not represent a back-tracking on animal welfare, though there is cause for concern given increasing animal exploitation being seen by some as a means to advance other economic and societal goals.

Other laws

Another piece of relevant legislation, particularly in the context of farmed animals, is the Meat Safety Act of 2000. While the Act predominantly relates to food safety for meat and animal products, certain regulations promulgated in respect thereof provide for the humane treatment of live animals including in relation to how they are transported to abattoirs and slaughtered. It thus provides untapped opportunities to tackle certain aspects of animals used for food production and problematic practices.

Apart from farmed animals, animals are also exploited for other purposes such as for entertainment – whether in circuses, zoos, aquaria, or in direct wildlife interactions such as lion cub-petting, walking with lions, or riding on elephants. Animals used in entertainment are generally regulated by the Performing Animals Protection Act. Despite its name, the Act in its initial form was predominantly a licensing statute requiring certain paperwork to use animals for these purposes, with no specific standards set for their welfare. It was only in 2016 that an amendment was introduced to require prospective licensees to include in their application an indication of how they would address certain animal welfare requirements.

Softer standards

Other than the laws mentioned (and others that cannot be dealt with within the constraints of this chapter), the failure by the relevant government departments to promulgate specific standards for different animal uses has meant that these issues have largely been addressed through non-binding standards set by exploitative industries themselves (self-regulation) or standards set by the South African Bureau of Standard ("SABS"). The SABS is a statutorily constituted body that sets national standards for products known as South African National Standards ("SANS"). A number of national standards have been produced with regard to animals – ranging from farmed animals (such as pigs, chickens, cows, crocodiles, ostriches and others) to captive wild animals, including those used in zoos, to those used in scientific research and laboratories.

These standards have been produced largely in consultation with the animal exploitation industries, and thus those with a vested interest in maintaining certain practices. Although the NSPCA is included as a stakeholder on the relevant drafting committees, and recently more animal welfare groups have been invited to participate, the SANS do little to protect animal interests or their welfare. These national standards are only binding if they are subsequently incorporated into legislation, and notably, very few, if any, SANS have been relating to animal welfare. Thus, on the whole, the standards relating to animals are voluntary, largely unenforceable, and also inaccessible to most people as they need to be purchased and have strict copyright provisions. It is deeply troubling that the SABS has had to step into the role of standard-setting for animal welfare and this fact demonstrates a failure of the government. In addition, the processes followed are different from those required by law-making or in governmental bodies, nor are they subject to the same scrutiny. At the time of drafting, SABS is currently drafting standards for poultry (allowing for battery cages until 2039) and aquaculture.

Apart from the SABS, industry bodies have also developed their own standards and codes, which are also non-binding and voluntary. These range from animals used in farming (with standards set by bodies such as the Livestock Welfare Coordinating Committee ("LWCC")) and wildlife including big cats (with standards set by the South African Predator Association ("SAPA")). As can be expected, they contain minimal animal "protections" and are not enforceable.

Wildlife and free-roaming animals

In contrast to legislation for domestic and captive animals, another distinct set of laws relate to wild animals who form part of the environment. As was suggested above, the main purpose of these laws relates to the conservation of species and biodiversity. Adding a layer of complexity, the environment is regulated at both a national and provincial level, with both having concurrent authority over this sphere in terms of the Constitution. South Africa has nine different provinces, and, problematically, each has different and inconsistent regulations in relation to wildlife and the environment (for a detailed summary of some of the issues with the current environmental framework in South Africa, particularly as it pertains to animal welfare, see CER & EWT Report, Fair Game, 2018).

At a national level, the National Environmental Management Act ("NEMA") is the umbrella piece of environmental legislation which is implemented through specific environmental management Acts or "SEMAs". Examples of SEMAs include those that relate to the management of biodiversity (National Environmental Management: Biodiversity Act ("NEMBA")) and terrestrial and aquatic protected areas ("National Environmental Management: Protected Areas Act"). Certain regulations can then be promulgated in terms of these Acts or SEMAs. For example, under NEMBA, regulations have been promulgated for invasive and alien species as well as threatened or protected terrestrial and aquatic species – referred to as the "TOPS Regulations". The latter regulations provide for certain restricted activities with regard to the relevant species in their ambit, set out offences and their penalties, as well as the required processes to obtain permits and licenses. The restricted activities range from hunting and breeding in the terrestrial context, and shark cage diving and dolphin watching in the aquatic context. There are additional levels of regulation in the form of Policy Documents, Norms and Standards, Biodiversity Management Plans, and others. As aforementioned, this national regulation operates in addition to the provincial legislation, making the regulatory landscape for wild animals extremely confusing, unclear, and problematic.

One relatively unique situation in South Africa is the intensive confinement and farming of traditionally wild animals. As a result, the country has some of the highest populations of species such as lions, ostriches, crocodiles, and rhinos in the world. While the international trade of rhino horn is essentially banned in terms of CITES, domestic trade is allowed within the country's borders. The intensive breeding of wildlife presents multiple unique problems and challenges (for further reading on some of the failures pertaining to captive wildlife, particularly lions, see: Wilson, 2019b). These include, but are not limited to, harmful ecological consequences and animal welfare violations. Due to increasing pressure from the public and civil society as well as increased awareness around the harms of these practices, there have recently been tangible efforts to regulate and enforce the welfare of wild animals kept in captivity – including in the recent Draft Policy Position highlighted above.

Lateral regulation impacting on animals

Several other laws indirectly impact on animal use and exploitation, such as in the realms of food safety (such as the Animal Diseases Act), consumer protection (such as the Consumer Protection Act), and property law (such as by-laws at a local level that deal with licensing and the number of animals one can have on a property), amongst others. These areas present opportunities to bring about reform to improve animal welfare and protection both in terms of legislation as well as in the courts, but cannot be dealt with in depth in this chapter.

International law

At an international level, South Africa is a party to a number of international treaties and bodies regulating animals and related matters. The country was one of the first to ratify CITES in 1975, although it has since found the treaty overly restrictive in its ambitions to profit from the wildlife trade. South Africa also ratified the Convention on Biological Diversity in 1996 and participates fully in the activities of the International Whaling Commission, being a non-whaling nation since 1975.

The country is also a member of the World Organisation for Animal Health ("OIE"), although it has not incorporated all the relevant animal welfare codes into its domestic law. This has become particularly relevant in the context of the live export of farm animals by sea. In recent years, an industry has grown to export tens of thousands of live animals over thousands of kilometres on ships across the oceans simply to be slaughtered in other countries. Among various other harms (including to the workers and the environment), this trade results in serious violations of animal welfare: such harms include tremendous heat build-up in the interiors of the ships leading some animals to overheat and die (particularly during the summer months); not having access to food given the high density of animals on board the ship (often surpassing 50,000 at one time); the disposal of waste leading to high ammonia concentration in the air and pollution of the seas; and many others (Animals Australia, 2018).

South Africa does not have appropriate domestic regulations to address this practice. Whilst the NSPCA has argued that the APA applies to this practice, a recent court judgment in August 2020 (*National Council of Societies for the Prevention of Cruelty to Animals v Al Mawashi (Pty) Ltd and Others* (995/2020) [2020] ZAECGHC 118), directed that the OIE standards be applied to live export, despite the OIE not being incorporated into domestic legislation. Unfortunately, this judgment was essentially ineffective due to weak enforcement by the relevant government departments. The government has since released "Draft Guidelines for the Transportation of Live Animals by Sea"; however, such guidelines are deficient in their content, have yet to be promulgated, and they are not binding in the same way regulations would be.

Conclusions

Recently, South Africa scored an "E" on the World Animal Protection Index, indicating that there is much to be desired in terms of its protection for animal welfare. On the whole, South Africa is a mixed bag in terms of animal law: as we sought to show, there are a series of laws that regulate animal welfare, but because of major flaws, vested interests, and governmental failure, the regulatory framework is largely inefficient at providing animals with any real protection. There exist a number of promising possibilities derived from the Constitution (for further information on the interaction and intersection between human rights and animal rights, see: Wilson, 2020); however, the constitutional developments thus far have taken place at a relatively abstract level. Animal protection work is also not a priority given the country's unique history and challenges in addressing the legacy of past discrimination and economic inequality.

Positively, there is a growing advocacy movement for animals in South Africa as well as increasing education on animal use and exploitation. There is also a rise in the availability of alternatives to the exploitation of animals. In the last few years, more than two companies were established in South Africa working towards cultured meats – including for beef and seafood (Mzansi Meat Co. and Sea-stematic respectively). These initiatives, together with the growth of plant- and fungi-based alternatives present opportunities to reform the food system, where so many animals are treated abysmally. In addition, the recently released Draft Policy Position presents a new direction in the management of wildlife that is promising. The document notes the interconnectedness of human beings and other animals and calls for recognition of animal welfare, and respect for individual animals and recognises the need to change the status quo.

In May 2022, animal law was taught for the first time in South African universities, meaning a new generation of lawyers will be empowered with knowledge to drive efforts for greater protection. Increasingly, organisations are collaborating on efforts, including across social justice and scientific realms, to improve the protection for animals, humans, and the environment. At an African level, in 2019 the African Union explicitly recognised the sentience of animals in its Animal Welfare Strategy for Africa (African Platform for Animal Welfare). With increased awareness, advocacy, and the growth of alternatives, there are undoubtedly positive actions happening in the animal law and policy space and South Africa will unquestionably be a country to watch in this regard in the future.

References

Africa Platform for Animal Welfare (APAW) with the secretariat at Africa Union InterAfrican Bureau for Animal Resources. *African Animal Welfare Strategy*. Available at: http://worldanimal.net/images/stories/documents/Africa/AWSA.pdf [Accessed on 15 October 2021].

Animals Australia, 2018. Not good is the short answer. [online] *Animalsaustralia.org*. Available at: https://www.animalsaustralia.org/features/whats-it-like-on-a-live-export-ship.php [Accessed on 15 October 2021].

Bilchitz D, 2009. Moving beyond arbitrariness: The legal personhood and dignity of non-human animals. *South African Journal on Human Rights*, 25(1), pp. 38–72.

Bilchitz D, 2010. Does transformative constitutionalism require the recognition of animal rights? *Southern African Public Law*, 25(2), pp. 267–300.

Bilchitz D, 2017. Exploring the relationship between the environmental right in the South African constitution and the protection of animals' interests. *South African Law Journal*, 134(4), pp. 740–777.

CER & EWT Report, 2018. *Centre for Environmental Rights & Endangered Wildlife Trust: Fair Game*. Available at: https://cer.org.za/wp-content/uploads/2018/06/CER-EWT-Regulation-of-Wildlife-Welfare-Report-25-June-2018.pdf [Accessed on 15 October 2021].

Constitution of the Republic of South Africa, 1996. Available at: http://www.justice.gov.za/legislation/constitution/SAConstitution-web-eng.pdf [Accessed on 15 October 2021].

South African Government, *Draft Policy Position on the Conservation and Sustainable Use of Elephant, Lion, Leopard and Rhinoceros ('Draft Policy Position')*, Notice R.566 of Government Gazette 44776, 28 June 2021 .

Mnyaka M and Motlhabi M, 2005. The African concept of ubuntu/botho and its socio-moral significance. *Black Theology*, 3(2), pp. 215–237. DOI: 10.1558/blth.3.2.215.65725.

Njotini M, 2017. Examining the "objects of property rights": Lessons from the Roman, Germanic and Dutch legal history. *De Jure Law Journal*, 50(1), pp. 136–155.

SAPeople, 2016. I am an African poem by Thabo Mbeki, former South African President VIDEO. [online] *SA People: Worldwide South African News*. Available at: https://www.sapeople.com/2016/05/25/i-am-an-african/. Emphasis added [Accessed on 15 October 2021].

Wilson AP, 2019a. Animal law in South Africa: Until the lions have their own lawyers, the law will continue to protect the hunter. *Derecho Animal. Forum of Animal Law Studies*, 10(1), p. 35.

Wilson AP, 2019b. South Africa's fallen pride: How law and government fail to protect lions. *The Revelator*. Available from: https://therevelator.org/lion-hunting-south-africa/ [Accessed on 15 October 2021].

Wilson AP, 2020. (Non)human(imal) rights: Dismantling the separateness in law and policy. *Society Register*, 3(3), pp. 39–65. https://doi.org/10.14746/sr.2019.3.3.03 [Accessed on 15 October 2021].

33

KEY ANIMAL LAW IN THE UNITED STATES

Matthew Liebman

Introduction

This chapter provides a general overview of animal law in the United States. It begins by offering a brief overview of the United States legal system. It then discusses animals' legal status as property and various efforts to challenge that status through law reform. Finally, it describes the principal state and federal laws that regulate human treatment of animals, including criminal anti-cruelty laws and laws regulating the use of animals in agriculture, scientific research, exhibition, the wild, and the pet trade.

Background

A discussion of animal law in the United States should begin with an acknowledgement of the influence of settler colonialism on the country's legal and social views about the non-human world. Pre-colonial Indigenous cosmologies entailed strong commitments to the ethical significance of animals and the non-human world, viewpoints that many contemporary Native American communities continue to hold (Deer and Murphy, 2017). While it is impossible to homogenise the diversity of tribal belief systems, and we should be sceptical about stereotypes of "the ecological Indian" (Smithers, 2015), Indigenous ontologies have historically conceived of the natural world non-dualistically and non-hierarchically.

The European colonisers of North America did not recognise Indigenous orders as valid legal or moral codes (Hermes, 2008). Instead, they rationalised exploitation of both Indigenous people and animals by appealing to the Great Chain of Being, or *scala naturae*, which valued life according to a scale of development that placed white European males above women, other races, animals, and the rest of the natural world (Deckha, 2008). This hierarchy continues to animate some of the fundamental issues and tensions in US law (animal law and otherwise), including the classification of animals as the property of humans.

Even early animal cruelty laws, like the Massachusetts Colony's 1641 Body of Liberties, which prohibited "any Tirranny or Crueltie towards any bruite Creature which are usuallie kept for man's use" (Massachusetts Colony 1641), were concerned less with animals' inherent moral value and more with protecting the property rights of animals' owners and preventing the coarsening of public morals that cruelty could engender (Unti, 2002, pp. 16–19).

 DOI: 10.4324/9781003182351-39

The United States legal system has long aligned itself with the anthropocentric perspectives of the country's colonial founders over more inclusive, less dualistic approaches. But with the growth of the animal protection movement in recent decades, courts and legislatures are increasingly forced to negotiate tensions between the long-standing instrumentalist, property-based view of animals and alternative visions of justice based on respect for animal flourishing (Nussbaum, 2004). Deckha (2020) argues that this reconfiguration of human–animal relations must acknowledge North America's legacy of colonialism, while looking to Indigenous legal orders for new pathways for conceiving of legal systems that respect the non-human world.

Overview of the United States legal system

Structurally speaking, the United States is a constitutional representative democracy that operates under a federalist system of government. As a federation of states, governmental power exists at both the national and state level. The United States Constitution provides for three coequal branches of the federal government: a bicameral legislature with a Senate and a House of Representatives, an executive overseen by the president, and a judiciary presided over by the Supreme Court. Under the federalist system, the federal government is one of limited powers – the US Congress may only legislate where the Constitution allows, such as regulating interstate commerce. Under the Tenth Amendment to the Constitution, all powers not delegated by the Constitution to the federal government are reserved to the states. State governments typically have plenary "police" power, which allows them to regulate all matters relating to public health, safety, and welfare.

This federalist system influences the development and nuances of animal law in the United States in important ways. The welfare of animals is within the traditional police powers of the states and so animal protection has historically been an issue of state regulation. However, the use of animals affects interstate commerce in numerous ways, especially in commercial areas such as agriculture, research, and exhibition. Because regulating interstate commerce is among Congress's enumerated powers, it has passed federal laws concerning these industries, as discussed below.

The primary value of federal animal protection laws, compared to state regulation, is their scope: they extend nationwide, including in states that are otherwise hostile to the interests of animals. As such, animal advocates have focused much energy on federal animal protections, both through legislation and administrative regulations.

The primary value of state animal protection laws, compared to federal regulation, is the possibility of passing protections for animals that are far more progressive than would be possible at the national level. Legislation often requires consensus-building, which may be difficult in a national legislative body. States can therefore serve as so-called "laboratories of democracy", where lawmakers and advocates can test innovative new means of protecting animals. In California, for example, the state legislature has taken steps that would be inconceivable in Congress: banning cosmetics testing, prohibiting the use of extreme confinement in animal agriculture, and outlawing the sale of dogs, cats, and rabbits at pet stores.

The patchwork of federal and state laws affecting animals is discussed in the following sections.

Animals' legal status

Animals as property

Animals are considered property in all 50 states and under federal law. Although some commentators have argued that such status inevitably dooms animals to exploitation (Francione,

1995), others have argued that there is much progress to be made in the US legal system within the property paradigm (Favre, 2010). Animals' status as property unquestionably supports their exploitation, as it subsumes animals' interests to those of their owners and facilitates their objectification. The culture of American individualism also tends to exalt private property rights as inviolable, which contributes to political and social hesitancy to regulate how people can use their property, especially in conservative states. But as discussed below, every state has an anti-cruelty law, which limits how owners can treat their animals, at least in some limited contexts. This protection sets animals apart as a unique form of property: no other form of property receives legal protections based on its own interests.

Animals' property status has vexed judges in cases where property-based rules would lead to unjust outcomes. For example, under general principles in tort law, compensation for destroyed property is the market value of such property. But in cases where the damaged or destroyed property is a companion animal, applying that rule leads to unjust outcomes: the market value of most dogs and cats is minimal. Recognising the injustice of failing to adequately compensate injured plaintiffs, some courts have relaxed the traditional property rules. In *Martinez v. Robledo*, in which the defendant had injured the plaintiffs' dog, the California Court of Appeals acknowledged that "animals are special, sentient beings, because unlike other forms of property, animals feel pain, suffer and die" (*Martinez v. Robledo*, 210 Cal.App. 4th 384, 392 (2012)). The court therefore held that "the usual standard of recovery for damages to personal property—market value—is inadequate when applied to injured pets". But other courts, including the Supreme Courts of Texas and Georgia, have been more conservative, treating companion animals like other forms of property in tort cases and limiting recovery when they are injured or killed (*Strickland v. Medlen*, 397 S.W.3d 184 (Tex. 2013); *Barking Hound Village LLC v. Monyak*, 787 S.E.2d 191 (Ga. 2016)). Courts have typically rejected emotional distress damages to human plaintiffs in cases where defendants negligently injured their companion animals (such as veterinary malpractice actions), but have allowed them in cases of intentional harm or abuse (*McMahon v. Craig*, 176 Cal. App. 4th 1502 (2009); *Womack v. Von Rardon*, 135 P.3d 542 (Wash. Ct. App. 2006)).

The tension between animals' legal status as property and their social status as family members also arises in custody disputes over companion animals. For example, when a married couple gets divorced, their property is typically divided based on the economic value of the property. But when the property is a beloved companion animal, courts must decide how to determine who gets custody and whether to consider the animals' interests or simply distribute them like other forms of marital property. In *Travis v. Murray*, a New York judge wrestled with how to resolve a custody dispute involving a dachshund named Joey, ultimately deciding to hold a hearing to determine what was "best for all concerned" (*Travis v. Murray*, 42 Misc. 3d 447 (N.Y. Sup. Ct. 2013)). Although the court declined to adopt a standard based on the "best interests of the animal", it also acknowledged the need to recognise Joey as more than mere property. Other courts have refused to consider animals as anything other than chattel, to be allocated according to traditional property rules (*Bennett v. Bennett*, 655 So. 2d 109 (Fla. Dist. Ct. App. 1995)).

Animals as persons

The animal protection movement has used several strategies to try to transform animals' legal status. One strategy for promoting a more just legal status for animals has been to legislatively recognise animals' sentience, that is, their capacity to feel pleasure and pain. Of course, every statute that prohibits cruelty to animals implicitly recognises that at least some animals are sentient: how could one cause unjustifiable suffering to animals if they were not sentient? Nevertheless, express legislative recognition of animal sentience and an acknowledgement of the relevance of sentience to legal protection may prove symbolically valuable and lay the groundwork for a more robust recognition of animals' legal rights (Blattner, 2019). In 2013, the Oregon legislature found and declared that "[a]nimals are sentient beings capable of experiencing pain, stress and fear" and that "[a]nimals should be cared for in ways that minimize pain, stress, fear and suffering" (Oregon Revised Statutes § 167.305). Although such findings are scientifically uncontroversial, they have proven to be legally significant, with the Oregon Supreme Court citing them as evidence that animals' legal status is changing and expanding (*State v. Newcomb*, 359 Or. 756, 758 (2016); *State v. Fessenden*, 355 Or. 759, 768 (2014)).

Animal advocates in the United States have also used litigation to try to establish animals' status as legal persons. The Nonhuman Rights Project has filed several petitions for writs of habeas corpus on behalf of chimpanzees and elephants. The writ of habeas corpus is an ancient common law writ that challenges unlawful detentions and deprivations of liberty. The Nonhuman Rights Project has filed cases in New York and Connecticut, arguing that the confinement of chimpanzees and elephants at research facilities, private homes, and zoos violates the animals' rights to bodily liberty and autonomy. The Nonhuman Rights Project contends that these animals' sophisticated and complex cognitive abilities entitle them to fundamental rights and that the American legal system's purported commitment to liberty and equality demands recognition of their personhood (Wise, 2019). So far, courts have rejected this theory, relying on social contract theory to exclude non-humans from the community of legal rights holders. In *Nonhuman Rights Project v. R. W. Commerford and Sons*, a case brought on behalf of elephants confined at a zoo, the Connecticut appellate court had

> little difficulty concluding that the elephants—who are incapable of bearing legal duties, submitting to societal responsibilities, or being held legally accountable for failing to uphold those duties and responsibilities—do not have standing to file a petition for a writ of habeas corpus because they have no legally protected interest that possibly can be adversely affected.
>
> (Nonhuman Rts. Project, Inc. v. R.W. Commerford & Sons, Inc., *192 Conn. App. 36, 48 (2019)*)

New York's intermediate appellate court reached the same conclusion in a case concerning chimpanzees, holding that their

> incapability to bear any legal responsibilities and societal duties … renders it inappropriate to confer upon chimpanzees the legal rights—such as the fundamental right to liberty protected by the writ of habeas corpus—that have been afforded to human beings.
>
> (People ex rel. Nonhuman Rts. Project, Inc. v. Lavery, *124 A.D.3d 148, 152, (N.Y. App. Div. 2014)*)

But one member of the New York Court of Appeals, the state's highest court, has expressed doubts about this reasoning (*Nonhuman Rts. Project, Inc., on Behalf of Tommy v. Lavery*, 31 N.Y.3d 1054 (2018) (Fahey, J, concurring), and at the time of this writing, the New York high court had granted review in *Nonhuman Rights Project v. Breheny*, a case concerning an elephant named Happy confined at the Bronx Zoo, to address the question of animals' personhood (*Nonhuman Rts. Project, Inc. v. Breheny*, 36 N.Y.3d 912 (2021)).

Another significant animal personhood case is *Justice v. Vercher*, which seeks to establish animals' right to sue their abusers for civil damages (*Justice v. Vercher*, No. 18CV17601 (Or. Cir. Ct., filed 1 May 2018)). The case, filed in Oregon by the Animal Legal Defense Fund on behalf of a horse named Justice who was a victim of animal cruelty, argues that the anti-cruelty law provides Justice with substantive rights to be free from cruelty. The lawsuit argues that Justice, like any other victim of a crime, should have the procedural right to initiate a civil suit to recover damages caused by the defendant. The trial court dismissed the case for lack of standing, holding that a horse cannot be a plaintiff. As of this writing, the case is awaiting a decision from the Oregon Court of Appeal. (The author discloses that he is the lead counsel on the case.)

Other animal personhood cases include two high-profile cases filed by PETA: *Tilikum v. Sea World*, which argued that SeaWorld's confinement of orcas violated the Thirteenth Amendment to the US Constitution's prohibition on slavery, and *Naruto v. Slater*, which argued that a macaque monkey held the copyright to a photograph he took of himself (*Tilikum ex rel. People for the Ethical Treatment of Animals, Inc. v. Sea World Parks & Ent., Inc.*, 842 F. Supp. 2d 1259 (S.D. Cal., 2012); *Naruto v. Slater*, 888 F.3d 418, 420 (9th Cir. 2018)). In each case, the court held that the operative law – the Thirteenth Amendment and the Copyright Act – applies only to humans and does not provide substantive rights to animals.

Thus far, efforts to establish animal personhood through legislation and litigation have not succeeded in changing animals' status, but with cases pending and more to come, the issue remains a lively one. As Judge Eugene Fahey of the New York Court of Appeals put it, "the question will have to be addressed eventually [:] Should [a non-human animal] be treated as a person or as property, in essence a thing?" (*Nonhuman Rts. Project, Inc., on Behalf of Tommy v. Lavery*, 31 N.Y.3d 1054, 1056 (2018) (Fahey, J, concurring)).

Criminal anti-cruelty law

Although courts have not yet recognised animals as legal persons, animals do receive legal protections against some forms of cruelty, a limitation on the absolute or unqualified property rights of animals' owners.

Protecting animals falls within individual states' "police power", that is, the plenary authority of state governments to regulate the health, safety, morals, and welfare of their citizens. As such, criminalising animal cruelty has traditionally been the province of the 50 states, each of which has passed an anti-cruelty law. In 2019, Congress passed the Preventing Animal Cruelty and Torture (PACT) Act, effectively creating a federal animal cruelty law, though one riddled with exceptions (18 U.S.C. § 48).

State criminal laws prohibiting cruelty to animals date back to the 1820s when Maine passed the country's first anti-cruelty law. Early anti-cruelty laws were primarily concerned with two human-centred interests: protecting the property rights of animals' owners and guarding against the coarsening of public morals that public cruelty could cause. Beginning in the 1860s and following the influence of developments in English anti-cruelty laws, however, a new breed of anti-cruelty law emerged (Favre and Tsang, 1993). Led by the activism of Henry Bergh and the newly formed American Society for the Prevention of Cruelty to Animals, New York became the first state to shift the focus towards the experiential welfare of individual animals and the wrongness of suffering itself, though concerns about property rights and public morals also continue to animate anti-cruelty laws, even to this day (Priest, 2019).

Animal cruelty laws vary significantly from state to state (Animal Legal Defense Fund, 2020). The important elements of anti-cruelty laws are: what animals are covered, what conduct is prohibited, what conduct is exempted, and what criminal sentences are possible.

The first key element of anti-cruelty laws is which animals they protect. Some states adopt broad definitions, such as California, which defines "animal" to include "every dumb creature" (Cal. Penal Code § 599b), and New York, which defines it to include "every living creature except a human being" (N.Y. Agric. & Mkts. Law § 350(1)). Other states are more limited, such as North Carolina, which restricts its protection to "every living vertebrate in the classes Amphibia, Reptilia, Aves, and Mammalia except human beings", excluding invertebrates and fish (N.C. Gen. Stat. Ann. § 14-360(c)). The federal PACT Act adopts a similarly narrow definition, applying to only "living non-human mammals, birds, reptiles, or amphibians" (18 U.S.C. § 48(f)(1)). Some states expressly exclude classes of animal from the definition of animal, such as Texas, which defines animal to leave out "an uncaptured wild living creature" (Tex. Penal Code Ann. § 42.092(2)). Where the term "animal" is not defined by statute, courts may have to determine the scope of the law, as in the 1981 Massachusetts case *Knox v. Massachusetts Society for the Prevention of Cruelty to Animals* (12 Mass. App. Ct. 407 (1981)), which interpreted "animal" to include goldfish, or the 1963 Oklahoma case *Lock v. Falkenstine* (380 P.2d 278 (Ok. Ct. Crim. App. 1963)), which interpreted "animal" to exclude roosters used in cockfighting.

The second key element of anti-cruelty laws is what conduct they proscribe. Most state anti-cruelty laws constellate around the norm of prohibiting activities that cause "unjustifiable" or "unnecessary" suffering. New York, for example, defines cruelty to include "every act, omission, or neglect, whereby unjustifiable physical pain, suffering or death is caused or permitted" (N.Y. Agric. & Mkts. Law § 350(2)). Anti-cruelty laws typically prohibit both acts of commission, such as beating or torturing an animal, as well as acts of omission, such as neglecting an animal by failing to provide them adequate food, water, shelter, or veterinary care. In California, for example, "every person who maliciously and intentionally maims, mutilates, tortures, or wounds a living animal, or maliciously and intentionally kills an animal, is guilty of a crime", illustrating the prohibition on affirmative acts of cruelty (Cal. Penal Code § 597(a)). California law also criminalises someone who "having the charge or custody of any animal, either as owner or otherwise, … fails to provide the animal with proper food, drink, or shelter or protection from the weather", illustrating the prohibition on omission-based acts of cruelty (Cal. Penal Code § 597(b)). The federal PACT Act applies only to affirmative acts of cruelty in which an animal is "purposely crushed, burned, drowned, suffocated, impaled, or otherwise subjected to serious bodily injury" (18 U.S.C. § 48(f)(1)). Whether a particular infliction of animal suffering is unnecessary or unjustifiably cruel is typically a question of fact to be decided by a jury, although some courts may conclude that certain forms of suffering are necessary or justifiable as a matter of law.

The third key element of anti-cruelty laws is what conduct they exclude from coverage. As many animal law scholars have noted, anti-cruelty laws have historically targeted aberrant, socially marginal, and irrational acts of abuse, not institutionalised and socially accepted forms of animal suffering, such as those caused by animal agriculture, biomedical research, hunting, or pest control (Francione, 1996). Animal cruelty laws typically include exemption sections, which delineate allowable forms of animal treatment, even if they would otherwise violate the statutes' prohibitions on causing unjustifiable or unnecessary cruelty. As Taimie Bryant observes, "current legal definitions of 'cruelty' allow institutional exploiters of animals to claim that their practices are not 'cruel' no matter how excruciatingly painful they may be for the animals" (Bryant, 2006, 72). In the context of animal agriculture, many states exempt common agricultural practices. In Texas, for example, although it is unlawful to "torture" "livestock", the anti-cruelty law creates an exception for conduct that is "a generally accepted and otherwise lawful … animal husbandry or agriculture practice involving livestock animals" (Tex. Penal Code Ann. § 42.09(f)(2)). Not all states, however, categorically exempt industrial abuse from their anti-cruelty laws.

California, for example, does not exclude animal agriculture, although prosecutions for farmed animal cruelty are still exceedingly rare.

The fourth key element of anti-cruelty laws is what the consequences are for violating them. In all 50 states, at least some forms of animal cruelty are punishable as a felony, meaning offenders could receive sentences that exceed a year in prison. States vary on which conduct is punishable as a felony: in some states, only repeat offenders may be charged with felonies; in other states, a first offense, if egregious enough, could be charged as a felony. Some states allow for felony charges in neglect cases, while others deem only intentional cruelty a felony. In addition to jail or prison sentences, people convicted of animal cruelty may be sentenced to community service, probation, restitution, and fines. Other sentencing options may focus less on punishing the offender and more on protecting future victims. For example, many states now permit or require judges to ban offenders from owning more animals in the future or to order offenders to undergo psychological evaluation and counselling.

Some scholars have criticised efforts to address animal abuse through tougher criminal anti-cruelty laws and stricter penalties. Lori Gruen and Justin Marceau (2022, Marceau, 2019) argue that this carceral approach to animal law has failed to deter crimes against animals, contributed to institutional racism in the criminal justice system, reduced accountability for institutional animal abusers such as factory farms, and put the animal protection movement out of step with other social justice movements, which are increasingly moving away from retributive and carceral approaches to crime.

Federal and state laws have not only criminalised certain forms of cruelty to animals, but – from the other direction – criminalised efforts by animal activists to contest animal exploitation. The federal Animal Enterprise Terrorism Act (AETA) prohibits anyone from damaging an animal enterprise, causing the loss of an animal enterprise's property (including liberating animals from labs or farms), or putting someone in reasonable fear of death or serious bodily injury with the intent to interfere with an enterprise's operations (18 U.S.C. § 43). Federal courts have upheld the AETA against constitutional challenge, holding that the statute does not prohibit protected free speech activity or campaigns, but merely prohibits unprotected conduct, such as property destruction and animal liberations (*Blum v. Holder*, 744 F.3d 790 (1st Cir. 2014)). States have also passed anti-activist laws, including so-called "Ag-Gag" laws, which criminalise undercover investigations at factory farms and slaughterhouses. Federal courts have struck down Ag-Gag laws – in whole or in part – as unconstitutional restraints on protected free speech rights established by the First Amendment, although the decisions have not been entirely consistent (*Animal Legal Def. Fund v. Wasden*, 878 F.3d 1184 (9th Cir. 2018); *Animal Legal Def. Fund v. Reynolds*, 8 F.4th 781 (8th Cir., 2021); *Animal Legal Def. Fund v. Kelly*, 9 F.4th 1219 (10th Cir., 2021)).

Laws regulating animal use

While the general animal cruelty statutes are the oldest and most common form of legal regulation of animal treatment, state and federal governments have also enacted statutes to regulate specific commercial and recreational uses of animals, including in agriculture, research, exhibition, the wild, and the pet trade.

Animals in agriculture

In the United States, there are no federal laws regulating the treatment of animals during their lives on farms. Federal law regulates only the transportation and slaughter of farmed animals.

The federal 28-Hour Law, passed in 1873, requires anyone transporting animals to unload them for food, water, and rest after 28 consecutive hours of confinement (49 U.S.C. § 80502). According to animal protection advocates, the law is virtually never enforced (Animal Welfare Institute, 2020).

The main federal law covering farmed animals is the Humane Methods of Slaughter Act (HMSA). The statute, initially passed in 1958, declares that it is "the policy of the United States that the slaughtering of livestock and the handling of livestock in connection with slaughter shall be carried out only by humane methods" (7 U.S.C. § 1901). In passing the law, Congress found that humane slaughter prevents needless animal suffering, is safer for workers, and improves meat products, thus "expedit[ing] an orderly flow of livestock and livestock products". (7 U.S.C. § 1901). Substantively, the HMSA declares slaughter humane if the animal is "rendered insensible to pain by a single blow or gunshot or an electrical, chemical or other means that is rapid and effective, before being shackled, hoisted, thrown, cast, or cut" (7 U.S.C. § 1902(a)). The HMSA also declares religious slaughter to be *per se* humane so long as "the animal suffers loss of consciousness by anemia of the brain caused by the simultaneous and instantaneous severance of the carotid arteries with a sharp instrument" (7 U.S.C. § 1902(b)). The HMSA initially did not ban inhumane slaughter outright, but merely prohibited the federal government from purchasing products from inhumanely slaughtered animals. But in 1978, Congress passed a new HMSA, which amended the Federal Meat Inspection Act to prohibit inhumane slaughter at all federally inspected slaughterhouses. The HMSA of 1978 tasked inspectors from the United States Department of Agriculture's (USDA) Food Safety Inspection Service (FSIS) with ensuring compliance with humane slaughter regulations.

Animal protection advocates have criticised the scope of the HMSA. The Act applies only to "livestock", which the USDA interprets to exclude birds. The United States slaughters more than 9 billion birds annually, more than 90% of the animals slaughtered for meat. This means that the federal law charged with ensuring the humane slaughter of farmed animals applies to less than 10% of the farmed animals who are slaughtered each year. Another major concern about the HMSA is its under-enforcement by the USDA. Undercover investigations by animal protection organisations have repeatedly exposed inhumane methods of slaughter, despite the presence of FSIS inspectors. Reports from the USDA's inspector general have also repeatedly identified significant gaps in the agency's enforcement of the HMSA (USDA OIG, 2013).

In the absence of federal on-farm regulations, some states have enacted laws regulating how animals are treated on farms. Most legislation concerns confinement methods that are common on factory farms – battery cages for egg-laying hens, gestation crates for pregnant pigs, and veal crates for calves. In 2008, California voters passed a ballot initiative (a form of direct democracy) to prohibit these forms of confinement by requiring that animals have sufficient space to be able to lie down, extend their limbs, and turn around. Voters amended the legislation in 2018, again by ballot initiative, to provide more specific numeric space requirements and to ban the sale of non-compliant products produced in other states (Cal. Health & Safety Code § 25990–25994). At the time of publication, however, the new law was pending review by the United States Supreme Court (*Nat'l Pork Producers Council v. Ross*, 142 S. Ct. 1413 (2022)). As of 2022, 14 states had passed laws limiting intensive confinement.

Animals in research

The federal law governing research on animals in the United States is the Animal Welfare Act (AWA), originally passed in 1966 as the Laboratory Animal Welfare Act in response to widespread public concern about the theft of companion animals and their sale to research facilities. In passing the AWA, Congress stated its policy "to insure that animals intended for use in research facilities … are provided humane care and treatment" (7 U.S.C. § 2131). The AWA creates a licensing and inspection scheme, whereby research facilities must register with the Department of Agriculture and submit to periodic inspections to ensure compliance with welfare standards promulgated by the Secretary of Agriculture. These standards concern "handling, housing, feeding, watering, sanitation, ventilation, shelter from extremes of weather and temperatures, adequate veterinary care, and separation by species" (7 U.S.C. § 2143(a)(2)(A)). The AWA also requires special consideration for dogs, who should receive exercise, and for primates, who should receive "a physical environment adequate to promote the[ir] psychological well-being" (7 U.S.C. § 2143(a)(2)(B)). The AWA requires researchers to endeavour to minimise animals' pain and distress when doing so is consistent with the experimental design of the research and to consider alternatives to procedures that cause pain and distress. (7 U.S.C. § 2143(a)(3)). The AWA requires research facilities to establish an internal review board, called an Institutional Animal Care and Use Committee (IACUC), to review and approve animal research protocols and ensure they comply with the standards of the AWA. (7 U.S.C. § 2143(b)).

Animal protection advocates have criticised the AWA on multiple grounds. First, the scope of the statute has been criticised as overly narrow. The AWA defines "animal" to exclude rats, mice, birds, and cold-blooded animals (including fish and invertebrates), who collectively make up the vast majority of animals used in research (7 U.S.C. § 2132(g)). (These excluded animals are covered by the Public Health Service Policy on Humane Care and Use of Laboratory Animals pursuant to the Health Research Extension Act of 1985, which applies to research facilities that receive federal funding. This Policy itself, however, lacks the force of law, and there are no penalties for its violation aside from the possibility of losing federal grants.) The AWA is also limited in the substantive protections it provides to animals. The AWA does not authorise the promulgation of regulations concerning the design or performance "of actual research or experimentation by a research facility as determined by such research facility". (7 U.S.C. § 2143(a)(6)(A)). In other words, aside from encouraging reductions in animal pain (unless required by the nature of the research) and the consideration of alternatives, the AWA does not affect the conduct, methods, or implementation of research projects involving animals. Advocates have also expressed concern about the oversight mechanisms of the AWA. Although the statute does require periodic inspections of facilities by the USDA, advocates complain that USDA's enforcement of the AWA against research facilities is lax and that penalties are insufficient to deter mistreatment of animals. Moreover, the AWA's reliance on internal review boards to authorise and oversee research protocols raises concerns about thoroughness and consistency, with some commentators expressing concern that many IACUCs are staffed primarily by animal researchers who too readily approve research protocols (Hansen et al., 2012).

At the state level, four states, California, New Jersey, New York, and Virginia, have banned cosmetics testing on animals. These statutes make it unlawful to test cosmetics on animals if there exists a scientifically validated alternative testing method approved by the Inter-Agency Coordinating Committee for the Validation of Alternative Methods (ICCVAM).

Animals in exhibitions

The Animal Welfare Act also governs the exhibition of animals at zoos and circuses (7 U.S.C. § 2133). The AWA requires exhibitors to register with the USDA and to submit to inspections to ensure compliance with the AWA's implementing regulations, which establish minimal standards of husbandry for exhibited animals. The USDA has promulgated some species-specific standards, but many animals are governed by vague and general standards, such as the requirement that enclosures have "sufficient space to allow each animal to make normal postural and social adjustments with adequate freedom of movement" or that animals receive food that is "wholesome, palatable, and free from contamination and of sufficient quantity and nutritive value to maintain all animals in good health" (9 C.F.R. §§ 3.128, 3.129). Critics complain that the vagueness of these standards leads to inconsistent enforcement and fails to adequately protect exhibited animals, especially at substandard roadside zoos.

State and local governments have also passed laws concerning animal exhibition. In 2016, for example, California passed the Orca Protection Act, which bans holding orcas in captivity, although it exempts those orcas already in captivity before 2017, limiting their use to "educational presentations" (Cal. Fish & Game Code § 4502.5). In 2018, New Jersey became the first state to ban the use of wild animals in circuses (N.J. Stat. Ann. § 23:2A-16). In 2019, California passed the Circus Cruelty Prevention Act, which bans the use of animals in circuses, except for dogs, cats, and horses (Cal. Fish & Game Code § 2207). Hawaii and Colorado have similar bans, as do more than 100 localities and municipalities across the United States.

Animals in the wild

The federal and state governments also regulate human use of wild animals.

Federally, the Lacey Act, passed in 1900, targets wildlife trafficking. The Act prohibits trade in any wildlife "taken, possessed, transported, or sold in violation of any law, treaty, or regulation", including those of the United States, individual states, Indian nations, foreign nations, or international agreements (16 U.S.C. § 3372). For instance, an individual transporting the body of an unlawfully hunted animal in interstate or foreign commerce would be subjected to federal prosecution or civil penalties under the law.

In 1956, Congress passed the Fish and Wildlife Act to federally manage the commercial and recreational use of fish and other wildlife, although the Act is concerned with the efficient exploitation of these animals as resources, rather than with their welfare or inherent value (16 U.S.C. § 742a).

In 1971, Congress unanimously enacted the Wild Free-Roaming Horses and Burros Act to protect wild horses and burros from capture, branding, harassment, and death, and to recognise them "as an integral part of the natural system of the public lands" (16 U.S.C. § 1331). Although the Act offers some legal protections to these wild animals, it also authorises the Department of the Interior and the Bureau of Land Management to capture, remove, sell, and, in some cases, kill "excess" wild horses and burros (16 U.S.C. § 1333).

In 1973, the United States passed the Endangered Species Act (ESA), which domestically implements the Convention on International Trade in Endangered Species of Wild Fauna and Flora (CITES). The ESA empowers the Secretary of the Interior through the United States Fish & Wildlife Service (and in the case of marine species, the Secretary of Commerce through the National Marine Fisheries Service) to list a species as endangered if it is "in danger of extinction throughout all or a significant portion of its range" or as threatened if it "is likely to become an endangered species within the foreseeable future" (16 U.S.C. § 1532(6),(20)). The Secretaries must make listing decision based solely on the best available scientific evidence, without regard to economic considerations (16 U.S.C. § 1533(b)(1)(A)). They may also designate "critical habitat", that is, habitat that must be conserved to avoid the extinction of the species, although critical habitat designations do include economic considerations (16 U.S.C. § 1533(b)(2)). Listed species receive special protection against governmental actions that jeopardise their continued existence or the habitats upon which they rely (16 U.S.C. § 1536(a)(2)). The Act further prohibits actions that may "take" members of listed species (16 U.S.C. § 1538(a)(1)(B)). "Take" means "to harass, harm, pursue, hunt, shoot, wound, kill, trap, capture, or collect, or to attempt to engage in any such conduct" (16 U.S.C. § 1532(19). The ESA provides robust protections to listed species and their habitats. Its protections, however, are not absolute: a federal committee may authorise governmental activities that jeopardise listed species (16 U.S.C. § 1536(h)). Individuals may also receive take permits that allow otherwise-prohibited activities for scientific purposes or to enhance the propagation of listed species, or where a take is only incidental (16 U.S.C. § 1539(a)). Although the ESA has traditionally been seen as a conservation-based statute, it also protects animals' individual welfare, by prohibiting activities that "harm" or "harass" animals (Waltz, 2020; Winders et al., 2021). The ESA applies to captive members of listed species, and animal protection advocates have used the ESA to target roadside zoos that inhumanely confine members of listed species (*Kuehl v. Sellner*, 887 F.3d 845 (8th Cir. 2018)).

The Marine Mammal Protection Act provides similar protections against "takes" of marine mammals (16 U.S.C. § 1361 *et seq.*).

The management of wild animals also falls within the police power of the states, who are obligated under the public trust doctrine to regulate the use of "natural resources" for the benefit of the public. Most states treat animal populations as exploitable resources, allowing for hunting and trapping (and in some cases, enshrining the right to hunt and trap in state constitutions). States place some limitations on uses of wild animals, such as by creating hunting and trapping seasons, restricting the quantity of animals that may be taken, and regulating methods of hunting and trapping. In 2019, California became the first state to ban commercial trapping (it had banned body-gripping traps by ballot initiative in 1998) (Cal. Fish & Game Code § 3003.1). In 2021, New Mexico banned trapping on public lands (N.M. Stat. Ann. § 17-11-3).

Animals in the pet trade

The AWA, in addition to governing research and exhibition, regulates the commercial pet trade (7 U.S.C. § 2133). The AWA applies to pet dealers and breeders with five or more breeding female animals, but it does not apply to retail pet stores (9 C.F.R. § 2.1(a)(3)(iii); 7 U.S.C. § 2132(f)). As with research facilities and exhibitors, pet breeders are subject to licensing and inspection to ensure compliance with standards promulgated by the USDA. But animal protection advocates have criticised these standards as insufficient, as they still allow wire flooring, small cages, and exposure to extreme temperatures, among other welfare concerns (HSUS, 2020). As a result of poor regulations and lax enforcement, puppy mills remain widespread through the United States.

At the state level, several states have banned the commercial sale of animals as pets. California became the first state to do so in 2017 when it banned retail sales of cats, dogs, and rabbits at pet stores (Cal. Health & Safety Code § 122354.5). (In case the pattern is not yet clear: California is often the first state to break new ground when it comes to animal protection legislation.) Maryland followed suit in 2018 (Md. Code Ann., Bus. Reg. § 19-703). More than 300 cities and counties have banned pet sales, including large cities such as Boston and Philadelphia.

Other state and local laws and regulations concerning companion animals include the following: cruelty-reporting laws, which permit (and in some cases, require) veterinarians and other animal professionals to report suspected animal cruelty to authorities; bans on certain veterinary procedures and mutilations, such as declawing cats or cropping the ears of dogs; laws requiring companion animals to be spayed or neutered; and breed-specific legislation, in which municipalities prohibit possession of certain breeds of dogs, typically pit-bulls. The proposal or enactment of such laws in a number of jurisdictions has occasioned important debates – and in some cases litigation – about the legal regulation of companion animals.

Conclusions

From ambitious efforts to transform animals' legal status to more moderate reforms of animal protection statutes, animal law is a rapidly growing field in the United States, deeply influenced by competing philosophical views of the proper relationship between humans and the non-human world. Despite these reform efforts by the animal protection movement, animal laws in the United States still assume that animals are exploitable resources that humans are allowed to use. Nevertheless, the states and the federal government have passed legislation to prohibit some forms of animal cruelty and to regulate human uses of animals across a broad range of areas, including the use of animals in agriculture, scientific research, exhibitions, the wild, and the pet trade. As societal concern for the well-being of animals increases, legal restraints on human use of animals are likely to become more pervasive and more consequential.

References

Animal Legal Defense Fund, 2020. *Animal Protection U.S. State Laws Rankings Report.* https://aldf.org/wp-content/uploads/2021/02/2020-Animal-Protection-US-State-Laws-Rankings-Report-Animal-Legal-Defense-Fund-1.pdf.

Animal Welfare Institute, 2020. A review: The twenty-eight hour law and its enforcement. https://awionline.org/sites/default/files/uploads/documents/20TwentyEightHourLawReport.pdf.1

Blattner CE, 2019. The recognition of animal sentience by the law. *Journal of Animal Ethics*, 9(2), pp. 121–136.

Bryant TL, 2006. Trauma, law, and advocacy for animals. *Journal of Animal Law & Ethics*, 1(1), pp. 63–138.

Deckha M, 2008. Intersectionality and posthumanist visions of equality. *Wisconsin Journal of Law, Gender and Society*, 23(2), pp. 249–267.

Deckha M, 2020. Unsettling anthropocentric legal systems: Reconciliation, indigenous laws, and animal personhood. *Journal of Intercultural Studies*, 41(1), pp. 77–97. DOI: 10.1080/07256868.2019.1704229.

Deer S and Murphy L, 2017. 'Animals May Take Pity on Us': Using traditional tribal beliefs to address animal abuse and family violence within tribal nations. *Mitchell Hamline Law Review*, 43(4), pp. 703–742.

Favre D, 2010. Living property: A new status for animals within the legal system. *Marquette Law Review*, 93(3), pp. 1021–1070.

Favre D and Tsang V, 1993. The development of anti-cruelty laws during the 1800's. *Detroit College of Law Review*, 1993(1), pp. 1–35.

Francione G, 1995. *Animals, Property, and the Law.* Philadelphia: Temple University Press.

Francione G, 1996. *Rain Without Thunder: The Ideology of the Animal Rights Movement.* Philadelphia: Temple University Press.

Gruen L and Marceau J (eds.), 2022. *Carceral Logics: Human Incarceration and Animal Captivity.* Cambridge: Cambridge University Press.

Hansen LA, Goodman JR and Chandna A, 2012. Analysis of animal research ethics committee membership at American Institutions. *Animals*, 2,(1), pp. 68–75. DOI: 10.3390/ani2010068.

Hermes K, 2008. The law of native Americans, to 1815. In M. Grossberg and C. Tomlins, eds. *The Cambridge History of Law in America.* Cambridge: Cambridge University Press. DOI: 10.1017/CHOL9780521803052.003

(HSUS) Humane Society of the United States, 2020. Puppy mills and the animal welfare act. https://www.humanesociety.org/sites/default/files/docs/puppy-mills-awa-booklet-lores_0.pdf.

Marceau J, 2019. *Beyond Cages: Animal Law and Criminal Punishment.* Cambridge: Cambridge University Press.

Massachusetts Colony, 1641. The body of liberties. https://www.mass.gov/doc/1641-massachusetts-body-of-liberties/download.

Nussbaum MC, 2004. Beyond 'Compassion and Humanity': Justice for nonhuman animals. In C. Sunstein and M. Nussbaum, eds. *Animal Rights: Current Debates and New Directions.* Oxford: Oxford University Press.

Priest C, 2019. Enforcing sympathy: Animal cruelty doctrine after the civil war. *Law & Social Inquiry*, 44(1), pp. 136–169.

Smithers GD, 2015. Beyond the 'Ecological Indian': Environmental politics and traditional ecological knowledge in Modern North America. *Environmental History*, 20(1), pp. 83–111.

(USDA OIG) United States Department of Agriculture, Office of Inspector General, 2013. Food safety and inspection service: Inspection and enforcement activities at swine slaughter plants. https://www.usda.gov/sites/default/files/24601-0001-41.pdf.

Unti B. 2002. *The quality of mercy: Organized animal protection in the United States 1866–1930*. Dissertation, American University, Washington DC. https://www.wellbeingintlstudiesrepository.org/acwp_awap/40.

Waltz D, 2020. The 'Embarrassing' endangered species act: Beyond collective rights for species. *Columbia Journal of Environmental Law*, 45(1), pp. 1–55.

Winders DJ, Goodman J and Rally H, 2021. Captive wildlife under the endangered species act. In DC Baur and Y-W Li, eds. *Endangered Species Act*. 3rd edn. IL: ALBA Book Publishing.

Wise SM, 2019. The struggle for the legal rights of nonhuman animals begins: The experience of the non-human rights project in New York and Connecticut. *Animal Law*, 25(3), pp. 367–393.

PART VII

Social change for animals

34

STAKEHOLDER GROUPS AND PERSPECTIVES

Joy M Verrinder and Clive JC Phillips

Who are the stakeholders in animal welfare?

A stakeholder is considered someone who can affect a decision or is affected by an action (Freeman, 1984). Stakeholders, from an animal welfare perspective, are identified by their interests in animal use or management, be it for food or fibre production, work, science, teaching, entertainment, protection of people, animals or environment. Stakeholders include not only those directly involved, such as farmers, scientists or teachers, but also consumers of animals, or products or services derived from animal use or management. Consumers' lives are affected, even though they may not be aware of this.

The animals themselves are also stakeholders, since their lives and well-being are affected by, and often depend on, human use or management. However, their lack of choice in participating in decisions that affect them in most instances means that their stake is of a fundamentally different nature to that of humans. At best, it constitutes a mutually beneficial arrangement with humans; at worst, it is one of servitude, cruel treatment, and a short life.

Animal advocates also form a unique group of stakeholders, since they attempt to represent the animals, rather than themselves. They have an important role as opinion leaders in the community, particularly if they have a good understanding of the animal use systems (Ross and Phillips, 2018). More knowledgeable advocates are better able to analyse different animal production systems, to approach those in government and the animal industries about welfare issues, and to discuss them with their friends, work colleagues, and product retailers.

Another important stakeholder group is the non-consumers of the output of animal industries. As our lives are interdependent, all animal use or management has some impact, directly or indirectly, on all living beings. Therefore, those who don't eat meat, wear wool, follow the sports involving animals, visit zoos, or use medication and cosmetics that have been tested on animals, are still impacted upon by others' choices and should logically be allowed to have a say in how animals are used and managed. As well, the animal kingdom is everyone's heritage, and therefore everyone's responsibility.

Although it is important to recognise all the stakeholders in animal welfare, it is also important to understand that not all have an equal stake. For some stakeholders, there may be marginal benefit, for others it is a matter of life or death.

DOI: 10.4324/9781003182351-41

Stakeholders' decision-making regarding animals

In relation to ethics, three levels of reasoning for decision-making have been identified:

Level 1

Personal Interest Reasoning involves making decisions based on direct consequences for yourself, i.e. obedience toward authority figures to avoid punishment, or acting in a certain way for personal benefit.

Level 2

Maintaining Norms Reasoning involves choosing those behaviours accepted by the society in which you live and following rules and laws.

Level 3

Principled Reasoning involves making decisions based on universal principles of justice, with compassion and respect for all (Kohlberg, 1984).

The various levels of reasoning may be used at any stage of life. Research has shown that Principled Reasoning is influenced by education which exposes people to different perspectives requiring critical reflection (Rest et al., 1999). It is therefore important to reflect on how the various stakeholder groups make decisions regarding animals from different perspectives, which are also influenced by personal characteristics of people within stakeholder groups.

A. Industry

The animal industries include those producing and/or using animals for food and other purposes, and their diverse and extensive ancillary industries. There have been two schools of thought regarding relationships between industries and other stakeholders. The first instrumentalises ethics, with the purpose of business being financial success and engagement in ethics only a subsidiary purpose, e.g. doing some good works with stakeholders. The second is that businesses are a network of relationships, and business' goals are logically to support all stakeholders' goals (Noland and Phillips, 2010). Most industries involved in animal use have been instrumentalising ethics by making some concessions to the interests of the animal stakeholders and their advocates for industry interests, i.e. Personal Interest Reasoning. For example, in Australia, following animal advocates' and the general public's concerns about training regimes and the extensive killing of greyhounds that are too slow, the greyhound racing industry introduced rehoming programmes for a limited number of greyhounds to enable the continuation of the industry. Similarly, supermarket chains monitor and respond to consumer choices, mirroring trends towards greater concern for animal welfare, to maintain and build their customer base. Such engagement with stakeholders, although apparently a good faith exercise, is affected by the relative power of the parties advancing their interests, along with their rhetorical skill (Noland and Phillips, 2010, p. 42).

Industries that use animals rarely consider them as stakeholders whose interests are just as important as human stakeholders. Animals are usually treated as property, which has been deemed culturally and legally acceptable, with some minimal adaptations required by law to modify some of the cruellest of treatments. Laws relating to animals often represent the lowest common denominator of social views, as they are subject to vested interests of stakeholders, inherent cultural biases, and are slow to create and adapt. As well, they are hard to apply to the great variety of animal uses extant today. Practices become entrenched, and hard to change because business investment is required to achieve returns over many years. The complex array of associated industries also has interests in maintaining the status quo of animal use to retain livelihoods. Industries continue doing what they know and have skills in, as the costs of retraining and re-equipping to move into more ethical industries is expensive.

Industries are therefore often not transparent about their products' impacts on animal survival and well-being. Animal industry staff may be required or encouraged to follow an industry directive on animal welfare issues, which may be contrary to their beliefs, but they support it, for fear of reprisals by the company if they do not. Honest reporting of various industries' impacts on animal survival and well-being may eventually be required, just as cigarette companies are now required to report their products' impacts on humans.

B. Consumers

In personal decision-making, individuals have the right to choose what they do in relation to their use of animals, providing it is within the law. For those with plenty, these decisions are based not just on survival needs as they are in low-income societies. People can choose, or not choose, to eat and wear animals, to use animals for companionship, hobbies or sport, with species and breed types based on popularity and status, and can kill animals, even cruelly, by a simple classification if we determine they are not aligned with our current view of what we need (e.g. governments' classifications of animals as feral or pests). While there is a growing number of people in the world who use Principled Reasoning and do not eat or wear animals and their products, who choose to help animals in need and strive for fair treatment of all animals, the majority, at the time of writing, still do use and contribute to animal abuse through their daily choices, demonstrating Personal Interest or Maintaining Norms Reasoning.

Consumers are often unaware or unwilling to admit that they are complicit in the choices made by industries and organisations who breed, rear, and/or kill animals. Principled Reasoning regarding animal products is hindered by consumers' limited understanding of production systems (Erian and Phillips, 2017). When animal use practices are made explicit through honest information and there are easily accessible alternative choices, consumers have an opportunity to change animal use and end abuse.

However, while honest and accurate labelling is crucial, Personal Interest Reasoning by consumers may still predominate. For example, product quality is often judged more important than humane treatment, even though the two are intrinsically linked. If the quality of meat produced in intensive systems is perceived to be adversely affected by the way animals have been treated, consumers are reluctant to buy it (Schröder and McEachern, 2004).

C. Professionals

Professionals involved in animal industries have the capacity to use Principled Reasoning and guide stakeholders to improve animal welfare. However, they often use Personal Interest and Maintaining Norms Reasoning, which limit or prevent improvements.

Veterinarians and veterinary associations, conscious of conflicts of interest in terms of their public role in helping individual animals whilst employed by industries whose use of animals is often harmful, have often avoided conflict rather than addressing the systemic harms caused by the industries. Although entering their training with a strong affiliation to animals, as their course progresses veterinarians adopt less benign attitudes to animals, particularly towards livestock (Paul and Podberscek, 2000). This is probably because of the reliance of many veterinarians on servicing the livestock industries, and the influence of lecturers reliant on those industries for their research funding. Students with extensive experience of the livestock industries are less likely to make choices supporting the rights of animals to life and bodily integrity, whereas those with experience of horses and companion animals are more likely (Verrinder and Phillips, 2018). Similarly, veterinary students are more likely than students of the humanities to support euthanasia of healthy animals (Verrinder and Phillips, 2018).

Animal scientists are often expected to use animals for research. Regrettably, many animal welfare scientists are dependent on the very industries that they are investigating for research funding, which influences their assessment of the welfare status of animals within their care (Van der Schott and Phillips, 2013). Some animal welfare scientists have been criticised for the conclusions that they draw about animal welfare issues due to such vested interests (Phillips and Petherick, 2014).

Growth in animal welfare science research has focused on improving animal welfare within the intensive systems, rather than looking at the system itself as an animal welfare issue to be addressed. Buller and Roe (2018) identify animal welfare science as a "peculiar hybrid" of applied ethology, animal production science and preventative veterinary medicine (p. 4). They argue that, due to social interest in food animal welfare, animal welfare scientists have had to constantly reflect on the interrogation and criticism of what may be considered as socially, politically, and ethically acceptable ways in which to treat livestock animals (Buller and Roe, 2018, p. 21).

D. Government and policy makers

Policy makers in democratic governments attempt to include all key stakeholders in decision-making through consultation processes, inviting public submissions and setting up advisory committees to share stakeholder perspectives. Parliamentary committees then weigh stakeholders' perspectives usually against the costs, in a cost–benefit assessment. Because no clear ethical criteria are used to compare stakeholder perspectives and design a course of action, the most powerful stakeholders are usually the most influential, i.e. governments support those industries which boost the economy in order to be re-elected.

Stakeholders whose interests are less powerful, such as non-human animals and their advocates, may be given token consideration. This leaves animals vulnerable to continued exploitation for economic gain. Unless there is a belief in, and a process to ensure, an ethical approach to animals' inclusion as stakeholders using Principled Reasoning, the economic and political benefits of maintaining the cultural status quo in the short-term often triumph. Rarely can sufficient weight be brought to the table by those representing the least powerful stakeholders.

E. Animal advocates

Decisions by animal advocates are based on the interests of animals. Because animals are rarely given a fair opportunity to make choices about if and how they are used, animal advocates usually take a principled stance on their behalf, sometimes at the expense of their own employment opportunities and incomes. Because the imbalance of power is so great between stakeholders who are animal users and animals as stakeholders, animal advocates sometimes show less regard for the interests of other stakeholders, and defy laws which enable animals' interests to remain hidden or ignored, just as other social justice advocates have done to overcome racial and gender discrimination.

F. People's personal characteristics

Within all the above human stakeholder groups, there are variations in perspectives based on gender, age, education, training, and cultural background.

Gender

In situations in which women have freedom of expression, they are more likely than men to say that they have positive attitudes towards animals. Women also report that they are more willing to buy welfare-friendly animal products (Erian and Phillips, 2017) and they have less support for contentious welfare practices, such as the live export of animals (Verrinder and Phillips, 2018), compared with men. This can perhaps be explained by women's traditional role in caring for children being generalised to animals, but it may also be that women are more willing than men to admit that they have these positive attitudes towards animals (Verrinder and Phillips, 2018). In situations in which women do not have freedom of expression, they are inclined to adopt a more masculine and less benign attitude towards animals (Phillips et al., 2010).

Age, education, and training

Consumers often do not have sufficient knowledge of animal production systems to determine whether they are acceptable or not (Erian and Phillips, 2017) and make decisions based on convenience to themselves. However, increased knowledge does not necessarily produce more ethical behaviour. Though understanding increases with age (Erian and Phillips, 2017), in the survey of Erian and Phillips (2017), more knowledgeable consumers reported that they ate more chicken compared with those with less knowledge, despite consumers generally wanting chickens to be treated humanely. This may be because people who are more knowledgeable about the production systems are also more aware of the reported health benefits of eating chicken, compared with red meats and less knowledgeable about non-animal alternatives. More knowledgeable people are also less inclined to provide veterinary care to their pets (Marinelli et al., 2007), perhaps because they have a weaker attachment to them, compared to less well-educated people (Johnson et al., 1992), but also probably because they have the knowledge to manage pets better themselves (Mariti et al., 2012). In support of the former explanation, better-educated people are less inclined to support animals' rights to life and bodily integrity (Verrinder and Phillips, 2018). Training in moral reasoning is possible, but often does not relate well to the choices made in relation to animal issues, which may be based more on intuition, and therefore resistant to change (Verrinder and Phillips, 2018).

Cultural differences

There are regional differences in people's attitudes towards animals that can largely be explained by their nationality (Phillips et al., 2012). Often these relate to socioeconomic factors. Thus, in the emerging economies of Asia, people are more likely to accept animal practices that would not be normally accepted in Europe (Phillips et al., 2012). Religion also plays an important role and needs to be considered when organising training for stakeholders.

To avoid the inequity and conflicts these differences in perspective within and between stakeholders bring to decision-making, stakeholders need knowledge of the philosophical and scientific foundations of ethics (i.e. what *should* be done) which justifies ethical sensitivity to all stakeholders' interests, including all conscious beings, and an ethical decision-making process for achieving ethical outcomes.

Philosophical and scientific arguments for valuing all stakeholders' perspectives

Philosophers for centuries have been theorising about the ethical basis for decision-making and action. Some of these theories were considered at odds with each other, rather than complementary. As described in Chapter 26, deontological ethics is based on each individual's duty to act according to what can be logically reasoned and universalised. Utilitarianism is based on weighing up harms and benefits to find the greatest good. Virtue ethics focuses on character development. Care ethics focuses on fostering relationships. The first three of these theoretical frameworks suggest ethical decisions can be made individually. The fourth suggests building caring, compassionate relationships is key. However, the other three frameworks provide a basis for choosing between competing interests – fundamental universal principles and virtues are needed, and where conflicts arise, the weighing up of harms and benefits of different actions to minimise harm.

Thanks to the development of neuroscience, the mechanisms of ethics are now being identified and analysed in the common structures of the brains of conscious beings. Our brains and hormones react to situations that impact on our survival and well-being. Through imaging studies that associate responses to moral situations with corresponding areas of the brain, we can now analyse the complexities of moral sensitivity and decision-making.

The same neural circuit governing experience of our own pain also governs anticipation, perception, and imagination of another individual in pain (Decety et al., 2008). This neural network constitutes a physiological mechanism that mobilises the organism to react — with heightened arousal and attention — to threatening situations, providing a strong signal that can promote empathic concern.

Three facets of empathy have been identified from brain images (Decety and Cowell, 2015, p. 4.):

1. Affective sharing – the natural capacity to become emotionally aroused by others' emotions;
2. Empathic concern – motivation to care for another's welfare;
3. Perspective taking (or cognitive empathy) – the ability to consciously put oneself into the mind of another individual and imagine what they think or feel.

Each of these emotional, motivational, and cognitive facets of empathy emerges from specific neurobiological processes and reflects evolved functions that allow humans to thrive by detecting and responding to significant social events necessary for surviving, reproducing, and maintaining well-being. Understanding all three components is important as each has a different relationship to morality and is swayed by both social context and interpersonal relationships (Decety and Cowell, 2014, p. 534).

Ethics, therefore, is no longer only based on the musings of philosophers' or individuals' opinions of what is right or wrong. Science shows it is grounded in the evolution of sentient beings. Ethics is central to our physical and neurological makeup.

Two common aspects of sentient life create the need for ethical understanding and action:

1. Our inbuilt desire to survive, experience well-being, and avoid suffering;
2. Our interdependence, which has enabled us to respond to and care for others.

Why we must give consideration to animals' interests

Historically, many philosophers have only included humans in their ethical deliberations, arguing this on the grounds that only humans had consciousness, emotions, feelings, and/or thought. However, in the last few decades, philosophical justification for recognising animals' sentience and overcoming speciesism has advanced (Gruen, 2021), along with scientific evidence justifying the inclusion of all sentient beings.

Before neuroimaging was possible, scientists observed the behaviour of other species. Darwin theorised that: "We are impelled to relieve the suffering of others in order to relieve our own painful feelings" and "those communities which included the greatest number of sympathetic members would flourish best; and rear the greatest number of offspring" (Darwin, 1871).

Frans De Waal, a primatologist, states that his research with non-human primates supports the view of Darwin and others that "morality is a direct outgrowth of the social instincts we share with other animals", where "morality is neither unique to us nor a conscious decision taken at any specific point in time: it is the product of social evolution" (De Waal, 2006, p. 6). Animals have been identified as not only being in the sphere of moral concern, but part of the moral community, relying on cooperation and demonstrating a range of retributive emotions such as resentment and anger, along with pro-social emotions such as empathy, sympathy, and altruism.

Jaques Panksepp, a neuroscientist and psychobiologist who coined the term "affective neuroscience" to mean the study of the neural mechanisms of emotion, stated that:

> - The human social brain, as well as all other mammalian brains, is fundamentally built upon ancient emotional and motivational value systems that generate affective states as indicators of potential fitness trajectories.
> - Basic affective states – and the neural mechanisms to support them – are homologous in all mammals.
>
> *(Panksepp, 1998)*

In 2012, a prominent international group of cognitive neuroscientists, neuropharmacologists, neurophysiologists, neuroanatomists, and computational neuroscientists reassessed the neurobiological substrates of conscious experience and related behaviours in human and non-human animals, and made a number of unequivocal observations. including:

> - Artificial arousal of the same brain regions generates corresponding behavior and feeling states in both humans and non-human animals
> - Neural circuits supporting behavioral/electrophysiological states of attentiveness, sleep and decision-making appear to have arisen in evolution as early as the invertebrate radiation, being evident in insects and cephalopod mollusks (e.g., octopus)

- Birds appear to offer, in their behavior, neurophysiology, and neuroanatomy a striking case of parallel evolution of consciousness
- Convergent evidence indicates that non-human animals have the neuroanatomical, neurochemical, and neurophysiological substrates of conscious states, along with the capacity to exhibit intentional behaviors
- Consequently, the weight of evidence indicates that humans are not unique in possessing the neurological substrates that generate consciousness. Nonhuman animals, including all mammals and birds, and many other creatures, including octopuses, also possess these neurological substrates.

(Lowe, 2012)

Our treatment of animals needs to change to reflect this weight of scientific evidence. Deliberately causing physical or emotional harm to animals is no longer justifiable for human benefit, i.e. in uses of animals for food or entertainment, or in research, or management of animals.

As well, animals' interest in living can no longer be ignored, particularly where animals are killed (after short, restricted lives) for food or medical research. Currently, animals' lives are only protected if beneficial to humans, e.g. as companions whose lives directly comfort their carers, and some native wild animals whose lives are protected by humans' concern for sustainability of the environment or aesthetic pleasure. Even then, most people's demand for their own space, resources, or entertainment is allowed to override these interests.

Albert Schweitzer, a German philosopher, musician, and medical doctor, argued that the basic principle of ethics is devotion to all life in the world resulting from "the reverence felt by my will-to-live for every other will-to-live" (Schweitzer, 1949, p. 325). Noel Preston, an Australian philosopher, argues that we have ethical obligations because our lives take place in a web of interdependent relationships understood in a biocentric (life-centred) rather than an anthropocentric (human-centred) way, i.e. "I am ultimately responsible to all living beings in the cosmos" (Preston, 2001). In his Ethic of Response, he synthesises the four main ethical frameworks, mentioned earlier, as complementary rather than competitive elements. To address the need for "appropriate values and principles" so that the ethic cannot be easily manipulated into relativism and subjectivism, Preston includes three values or principles which are widely endorsed by a range of ethical approaches:

a. The respect for life principle – this extends beyond human beings to other forms of life in our biosphere and, if relevant, the cosmos; such respect is especially considerate of the rights of sensate beings. This principle requires that conflicts involving choices about life (including the initiation and termination, or the environmental threat to earth's balance of life) are treated with the maximum possible care.
b. The justice principle, i.e. being fair by giving priority to the interests of the most disadvantaged and also future generations.
c. The covenantal integrity principle. This involves truthfulness and honesty in all our relationships, the importance of self-consistency as moral agents, with promises and loyalty serving the purposes of respecting life and seeking justice.

(Preston, 2001, p. 75)

Both philosophy and science provide a unifying justification for being ethically sensitive to all stakeholders' perspectives, including animals.

Developing ethical sensitivity to stakeholders' perspectives

We define ethical sensitivity as: "the ability to interpret, through thoughts and feelings, the moral aspect of situations, including the impact of situations and actions and their possible consequences on the lives and well-being of sentient creatures". Ethical sensitivity has been identified as one of the four main components of moral behaviour along with moral judgement, moral motivation. and moral character (Rest, 1994) and plays a role in the development of moral judgment (Jagger, 2011).

An important part of ethical sensitivity is recognition of emotions, and empathy, which includes emotional sharing. Until recently, empathy was not taken seriously, even regarding humans. Regarding animals, much resistance still exists (de Waal, 2009, p. 90) despite evolutionary evidence that sentient beings, at least all mammals, have the same basic set of emotions (Panksepp, 1998). Moral philosophy has traditionally distinguished cool reasoning, regarded as the source of practical rationality and moral knowledge, from emotions, regarded as an irrational, even distorting, influence on moral judgment (Demaree-Cotton and Kahane, 2019, p. 91). This ignores recent work in moral epistemology. "It has been argued, for example, that emotions are often needed to bring morally relevant features to our attention and may even be necessary for grasping their moral importance" (Demaree-Cotton and Kehane, 2019, p. 91). "Emotional" neural circuits seem to facilitate impartial, altruistic behaviour, with extraordinary altruists having enlarged right amygdala that are more active in response to other people's emotions, which they are better at identifying (Marsh et al., 2014).

Individuals' empathy is influenced by experience and learning in morality (Demaree-Cotton and Kahane, 2019, p. 95). Our innate mechanisms for empathy are moulded by the family, society. and culture in which we live and those social groups to which we aspire, leading to implicit bias. In general, males demonstrate empathy less than females (Chakrabarti and Baron-Cohen, 2006, pp 408–409), driven by hormone levels, particularly testosterone and oxytocin (Panksepp and Panksepp, 2013, p. 10).

Adopting the perspective of another, particularly someone from another social group, is cognitively demanding and hence requires additional attentional resources and working memory, and inhibitory control. (Decety and Cowell, 2015, p. 7). Because animals may suffer in a way that we have difficulty understanding, some stakeholders involved in cruel animal practices may have difficulty with empathy towards animals.

However, a structured approach to developing ethical sensitivity is possible. If social prejudice can be learned, then it should be possible to unlearn it, preferably by group activities that have a common goal, drawing upon the contributions of each person, and involving taking the perspective of others (Railton, 2017). Ten elements of ethical sensitivity have been identified:

1. Identification of physical responses of animals and people to the particular situation
2. Identification of emotional responses of animals and people
3. Recognition of own thoughts (perceptions, appraisal, interpretation) of the situation
4. Recognition of own feelings in relation to the observed responses of animals and people
5. Identification of why the issue is an ethical one
6. Recognition of all stakeholders' perspectives including animals
7. Expression of empathy for others' perspectives
8. Recognition of moral conflicts

9. Recognition of professional conflicts between legal, organisational and ethical responsibilities
10. Identification of alternative actions and their possible impacts on stakeholders.

(Verrinder et al., 2019, p. 304)

These elements can be developed with instruction and practice (Verrinder et al., 2019, p. 311).

Including stakeholders' perspectives in decision-making

Despite the scientific evidence that ethics is grounded in our mutual desire for survival and well-being and our interdependence, the belief that ethics is based on opinion and that one person's view is as equally justifiable as any others' (ethical relativism) is widespread. Some veterinary ethics textbooks take a pluralist approach offering a range of prominent ethical views regarding how animals can be considered because "professionals must now accept that there are different ethical views, and that his or her own view is not the only one that a person can reasonably hold" (Sandoe and Christiansen, 2008, p. xiii). However, a pluralist approach provides little guidance for professionals or policy makers involved in animal industries who want to address concerns or improve animal welfare. Both ethical relativism and pluralism are confusing and damaging to the ethical decision-making process, as each stakeholder feels entitled to cling to their existing attitudes and behaviours.

Protecting and enhancing survival and well-being are the basis on which to make judgements about what is right or wrong. An ethical decision-making framework therefore must be focussed on enhancing well-being and preventing harm and ensuring all stakeholders', including animals', interests are considered fairly. Fairness goes hand in hand with communal survival (de Waal, 2009, p. 187). Many humans and other animals show "inequity aversion", i.e. are sensitive to injustice and show scorn and anger if offered unfair proposals. The fairest offers occur in societies with the highest levels of cooperation; those in which every family takes care of itself are marked by unfair offers (de Waal, 2009, p. 186/187).

There are many decision-making models available, but few incorporate all the ethical frameworks and principles as complementary elements and provide a means for working towards the most ethical decision based on survival and well-being. Ethical decision-making differs from just using reasoning in decision-making. The most reasonable decision may not be ethical. It may be subject to bias and blind spots, especially when entirely severed from emotional input (Demaree-Cotton and Kahane, 2019, p. 91). Some decision-making models base selection of the best alternative on what most parties are satisfied with. This can lead to an unjust decision for the least powerful party, particularly non-human animals, e.g. animals' interests in staying alive are more important than consumers' interests in products not essential to life, and industries' interests in maximising profit. Mepham's Ethical Matrix used in bioethics issues allows ethical rating of a particular action choice but doesn't compare all alternatives (Mepham et al., 2006).

Based on Preston's Ethic of Response for ethical decision-making (Preston, 2001, pp. 69–88), a template has been developed (Verrinder, 2016) to provide a structure for stakeholders to work together to consider all possible alternative actions, from all stakeholder perspectives, using the main ethical frameworks and principles as sequential and complementary, rather than competitive, elements. See worked example in Appendix A and a blank template in Appendix B for practising this ethical decision-making process.

The use of this template in small groups facilitates sharing of thoughts, emotions, and intuitions, and a requirement to come to a the most fitting ethical decision and justify it to the whole group. University students provided with knowledge of moral development and moral principles and the opportunity to engage with others in small groups using this template showed improved moral judgment, which was not achieved by demonstrating use of the template without small group interaction (Verrinder and Phillips, 2015).

Enacting change

A detailed understanding of ethical sensitivity and ethical decision means nothing if not translated into action. Leadership is especially required from professionals who have both knowledge of animals' sentience and involvement in how animals are cared for in animal use industries, i.e. veterinarians, animal scientists, and their professional associations. By working with stakeholders using ethical sensitivity and ethical decision-making, veterinary and animal science professionals could have a significant impact on eliminating endemic suffering and significantly reducing the numbers of animals killed.

For industry, "good strategy properly understood must encompass what are typically recognized as moral concerns, because the very purpose of the firm and the capitalist system within which it operates is, when viewed rightly, the creation of value for all stakeholders" (Noland and Phillips, 2010, p. 39). The stakeholder interpretation of the firm highlights the need for stakeholder groups to recognise their interdependence, to embrace it, and to work together to meet the changing needs and expectations of each group (Wicks et al., 1994, p. 486). "The strategic direction of the firm should always be thought of and developed in terms of 'us' – the interests, desires, and needs of all stakeholder groups rather than a firm charting its path as a lone act" (Wicks et al., 1994, p. 490).

Communicating with stakeholders with ethical sensitivity and use of an ethical decision-making framework provides a process for this change of perspective. For consumers, this will mean more honesty and an opportunity to make more ethical choices. For animal advocates, this will create a greater opportunity for animals' interests to be met. For governments, legislation development should be easier as it enables governments to resist the demands of the most powerful and develop more ethical policies and actions.

Appendix A – Ethical decision-making using Preston's ETHIC OF RESPONSE TEMPLATE – JVerrinder©

Worked example: Request to Euthanise a Healthy Dog

A woman brings her lively, five-year-old kelpie/cattle cross dog in to see a veterinarian, Dr Benjamin, for euthanasia. She says she is moving into an apartment with her boyfriend who doesn't like the dog, and pets aren't allowed in the apartment building. Besides this, the dog is too active for her and is barking all the time. The veterinarian asks if she has tried to put the dog up for adoption, but she replies that the local pound already has too many working dogs and they would probably euthanise it anyway. She simply wants the dog humanely destroyed and, if the veterinarian doesn't euthanise it, her boyfriend will shoot it. Dr Benjamin wonders what to do.

STAKEHOLDERS	*Action: euthanise the dog*		*Action: veterinarian refuses to euthanise and owner goes elsewhere*		*Action: veterinarian persuades owner to surrender dog to be rehomed by shelter, rescue group, or vet clinic*	
	Respect life (desire to survive) ✓=Benefits, X=Harms	*Respect well-being (capacity to enjoy life, fulfil goals and capabilities) ✓=Benefits, X=Harms*	*Respect life ✓=Benefits, X=Harms*	*Respect well-being ✓=Benefits, X=Harms*	*Respect life ✓=Benefits, X=Harms*	*Respect well-being ✓=Benefits, X=Harms*
Dog	X loses life	X missed chance to fulfil capacities	X may still lose life	X possible reduced opportunities to experience well-being	✓ survives	✓ finds more suitable home
Owner	–	✓ lives with boyfriend X may regret later	–	✓ owner has more time to think/find alternatives	–	✓ lives with boyfriend ✓ less regret
Boyfriend	–	✓ no annoying dog	–	X may regret having the dog destroyed	–	✓ no annoying dog
Veterinarian	–	X upset, loss of integrity	–	✓ has not had to kill a healthy dog X concern for animal's fate	–	✓ maintains integrity X requires some effort/ expense to care for the dog and find best rescue group/ home
Veterinary practice	–	X not meeting Vet Code of Ethics: "Health, welfare & respectful treatment of animal"		✓ not directly affecting the welfare of the dog	–	✓ meets Vet Code ✓ builds compassionate community support
DEONTOLOGICAL ETHICS (1–5) *1 = Duty to respect life and well-being*	**5**		**4**		**1**	
UTILITARIAN ETHICS (1–5) *1 = Greatest benefit for all affected*	**4**		**3**		**2**	
JUSTICE AS FAIRNESS (1–5) *1 = Greatest benefit to the least advantaged*	**5**		**4**		**1**	
VIRTUE ETHICS/INTEGRITY (1–5) *1 = Most virtuous, consistent*	**5**		**4**		**1**	

Sample justification for decision

- The most fitting ethical decision for the request to euthanise a healthy dog is to persuade the owner to surrender the dog to the vet for rehoming.
- This decision shows respect for all stakeholders' desire for life and well-being (deontological ethics requires a duty to universal reciprocity). It satisfies utilitarian ethics by producing the greatest good and the least harm to all stakeholders. It satisfies justice as fairness, giving most support to the least advantaged – in this case the young healthy dog who has the capacity to live a happy life and therefore the most to lose. In comparison, the effort involved in working with others to find a suitable home for the dog is a small cost. It also satisfies virtue ethics as the vet can show courage and compassion and maintain her integrity, showing consistency with the other two fundamental universal ethical principles which reflect the biological structures of sentient beings.

For assistance with using this ethical decision-making template, contact jverrinder@awlqld.com.au

Appendix B – Preston's ETHIC OF RESPONSE – J Verrinder ©TEMPLATE V4 Ethical Issue:

STAKEHOLDERS	*Action:*		*Action:*		*Action:*	
	Respect life (desire to survive) ✓=Benefits, X=Harms	*Respect well-being (capacity to enjoy life, fulfil goals and capabilities) ✓=Benefits, X=Harms*	*Respect life ✓=Benefits, X=Harms*	*Respect well-being ✓=Benefits, X=Harms*	*Respect life ✓=Benefits, X=Harms*	*Respect well-being ✓=Benefits, X=Harms*
1						
2						
3						
4						
5						
6						
7						
8						
DEONTOLOGICAL ETHICS Rate Actions 1–5 (1 = Respect for the will to survive and thrive; 5 = Suffering and/or lack of thriving and/or death)						
UTILITARIAN ETHICS Rate Actions 1–5 (1 = Greatest benefit for all affected; 5 = Least benefit for all affected)						
JUSTICE AS FAIRNESS Rate Actions 1–5 (1 = Greatest benefit to the most disadvantaged; 5 = Least benefit to the most disadvantaged)						
VIRTUE ETHICS/INTEGRITY Rate Actions 1–5 (1 = Most virtuous, consistent with ethical principles; 5 = Least virtuous, least consistent)						

References

Buller H and Roe E. 2018. *Food and Animal Welfare*. Bloomsbury, London.

Chakrabarti B and Baron-Cohen S, 2006. Empathizing: Neurocognitive developmental mechanisms and individual differences. In Enders E, et al., eds. *Progress in Brain Research*. Elsevier, Netherlands vol. 156, p. 404.

Darwin C, 1871. *The Descent of Man and Selection in Relation to Sex*. John Murray, London.

de Waal F, 2006. Morally evolved. In Macedo S and Ober O, eds. *Primates and Philosophers: How Morality Evolved*. Princeton University Press, Princeton, NJ, pp. 3–58.

de Waal F, 2009. *The Age of Empathy*. Harmony Books, New York.

Decety J and Cowell JM, 2014. Friends or foes: Is empathy necessary for moral behaviour? *Perspectives on Psychological Science*, 9(5), pp. 525–537.

Decety J and Cowell JM, 2015. Empathy, justice and moral behaviour. *AJOB Neuroscience*, 6(3), pp. 3–14.

Decety J, Michalska KJ and Akitsuki Y, 2008. Who caused the pain? A functional MRI investigation of empathy and intentionality in children. *Neuroimage*, 24, pp. 771–779.

Demaree-Cotton J and Kahane G, 2019. The neuroscience of moral judgment. In Jones K, Timmons M and Zimmerman A, eds. *The Routledge Handbook of Moral Epistemology*. Routledge, New York.

Erian I and Phillips CJC, 2017. Public understanding and attitudes towards meat chicken production and relations to consumption. *Animals*, 7, p. 20. https://doi.org/10.3390/ani7030020.

Freeman RE, 1984. *Stakeholder Management: Framework and Philosophy*. Pitman, Mansfield.

Gruen L, 2021. The moral status of animals. In Zalta EN, ed. *The Stanford Encyclopedia of Philosophy*. Summer 2021 edn., viewed 25th September, 2021, https://plato.stanford.edu/archives/sum2021/entries/moral-animal/.

Jagger S, 2011. Ethical sensitivity: A foundation for moral judgment. *Journal of Business Ethics and Education*, 8(1), pp. 13–30.

Johnson T P, Garrity TF and Stallones L, 1992. Psychometric evaluation of the Lexington Attachment to Pets Scale (LAPS). *Anthrozoös*, 5(3), pp. 160–175. https://doi.org/10.2752/089279392787011395.

Kohlberg L, 1984. *The Psychology of Moral Development: The Nature and Validity of Moral Stages*. Harper & Row, San Francisco, CA.

Lowe, P, 2012. The cambridge declaration on consciousness. In Francis Crick Memorial Conference on Consciousness in Human and non-Human Animals, Churchill College, University of Cambridge, Cambridge, UK, viewed 24 June 2021. http://fcmconference.org/img/CambridgeDeclarationOnConsciousness.pdf.

Marinelli L, Adamelli S, Normando S and Bono G, 2007. Quality of life of the pet dog: Influence of owner and dog's characteristics. *Applied Animal Behaviour Science*, 108(1–2), pp. 143–156. https://doi.org/10.1016/j.applanim.2006.11.018.

Mariti C, Gazzano A, Moore JL, Baragli P, Chelli L and Sighieri C, 2012. Perception of dogs' stress by their owners. *Journal of Veterinary Behaviour*, 7, pp. 213–219. https://doi.org/10.1016/j.jveb.2011.09.004.

Marsh AA, Stoycos SA, Brethel-Haurwitz, KM, Robinson P, VanMeter, JW and Cardinale, EM, 2014. Neural and cognitive characteristics of extraordinary altruists. *Proceedings of the National Academy of Sciences*, 111, pp. 15036–15041.

Mepham B, Kaiser M, Thorestensen E, Tomkins S and Miller K, 2006. *Ethical Matrix Manual*. LEI, onderdeel van Wageningen UR. https://edepot.wur.nl/216589.

Noland J and Phillips R, 2010. Stakeholder engagement, discourse ethics and strategic management. *International Journal of Management Reviews*, 12, pp. 39–49. https://doi.org/10.1111/j.1468-2370.2009.00279.x.

Panksepp J, 1998. *Affective Neuroscience: The Foundations of Human and Animal Emotion*. Oxford University Press, New York.

Panksepp J and Panksepp JB, 2013. Toward a cross-species understanding of empathy. *Trends in Neuroscience*, 36(8), pp. 489–496.

Paul ES and Podberscek AL, 2000. Veterinary education and students' attitudes towards animal welfare. *Veterinary Record*, 146(10), pp. 269–272.

Phillips CJC and Petherick JC, 2014. The ethics of a co-regulatory model for farm animal welfare research. *Journal of Agricultural and Environmental Ethics*, 28, pp. 127–142. https://doi.org/10.1007/s10806-014-9524-9.

Phillips CJC, Izmirli S, Kennedy M, Lee GH, Lund V, Mejdell C, Pelagic VR, Rehn T, Aldavood J, Alonso M, Choe BI, Hanlon AJ, Handziska A, Illmann G and Keeling L, 2010. An international comparison of female and male students' attitudes to the use of animals. *Animals*, 1, pp. 7–26.

Phillips CJC, Izmirli, Aldavood SJ, Alonso M, Choe BI, Hanlon A, Handziska A, Illman G, Keeling L, Kennedy M, Lee GH, Lund V, Mejdell C, Pelagic VR and Rehn T, 2012. Students'attitudes to animal welfare and rights in Europe and Asia. *Animal Welfare*, 21, pp. 87–100.

Preston N, 2001. *Understanding Ethics*. 2nd edn. The Federation Press, Sydney.

Railton P, 2017. Moral Learning: Conceptual foundations and normative relevance. *Cognition*, 167, pp. 172–190. https://doi.org/10.1016/j.

Rest J, 1994. Background, theory and research. In Rest, J and Narvaez, D, eds. *Moral Development in the Professions: Psychology and Applied Ethics*. Lawrence Erlbaum Associates, Hove, UK, pp. 1–26.

Rest JR, Narvaez D, Bebeau MJ and Thoma SJ. 1999. *Postconventional Moral Thinking: A Neo-Kohlbergian Approach*. Lawrence Erlbaum Associates, Mahwah, NJ.

Ross T and Phillips CJC, 2018. Relationships between knowledge of chicken production systems and advocacy by animal protection workers. *Society and Animals*, 26, pp. 73–92.

Sandoe P and Christiansen SB. 2008. *The Ethics of Animal Use*. Blackwell Publishing, Oxford, UK.

Schröder MJA and McEachern MG, 2004. Consumer value conflicts surrounding ethical food purchase decisions: A focus on animal welfare. *International Journal of Consumer Studies*, 28, pp. 168–177.

Schweitzer A, 1949. *The Philosophy of Civilization*. Prometheus Books, Amhers, NY.

Van der Schott A and Phillips CJC, 2013. Publication bias in animal welfare scientific literature. *Journal of Agricultural and Environmental Ethics*, 26, pp. 945–948.

Verrinder J, 2016. *Identifying and developing capacity for veterinarians to address animal ethics issues*. PhD Thesis, School of Veterinary Science, The University of Queensland. https://doi.org/10.14264/uql.2016.792.

Verrinder JM and Phillips CJC, 2015. Assessing veterinary and animal science students' moral judgment development on animal ethics issues. *Journal of Veterinary Medical Education*, 42(3), pp. 206–216.

Verrinder JM and Phillips CJC, 2018. The relationship between intuitive action choices and moral reasoning on animal ethics issues in students of veterinary medicine and other relevant professions. *Journal of Veterinary Medical Education*, 45, pp. 269–292. https://doi.org/10.3138/jvme.0117-016.

Verrinder JM, Ostini R and Phillips CJC, 2019. Assessing veterinary students' ethical sensitivity to farm animal welfare issues. *Journal of Veterinary Medical Education*, 46(3), pp. 302–339. https://doi.org/10.3138/jvme.0617-083r.

Wicks AC, Gilbert DR and Freeman RE, 1994. A feminist reinterpretation of the stakeholder concept. *Business Ethics Quarterly*, 79, pp. 395–405.

35

ANIMAL ADVOCACY AND HUMAN BEHAVIOURAL CHANGE

Tamzin Furtado, Suzanne Rogers, and Jo White

Introduction

If you hear the words "animal welfare campaign", what do you think of? A common answer is hard-hitting animal rights posters, for example comparing the animals we eat to the animals we keep as pets. This type of imagery is designed to shock its recipients into rethinking their behaviour and making behavioural changes and is often remembered because of its shock factor. However, these tactics are just one type of campaign. No matter where you live in the world, and no matter your culture or your personal opinions, you've probably come across more attempts to influence your behaviour than you realise in relation to animal welfare. Campaigns are in our supermarkets, our advertising, our legislation, on our television streaming channels, and in our veterinary surgeries. They shape the way we think and how we behave in the world in relation to animals – and given how animals permeate our culture, that means how we behave in a broad sense. Animal welfare is embedded in the choices we make around what we eat, how we farm, medications we consume, our conservation programmes, our choice of pets; and yet, methods of bringing about change remain largely unknown, even though there is still a lot to improve in the field of animal welfare.

Campaigning is important in animal welfare because we cannot improve the lives of animals without first changing human behaviour, given that humans are often the cause of animal suffering. Which humans do we want to change? That's trickier. Every animal issue is unique; the subjects might be the consumers of an animal product, or consumers of animal entertainment or tourism; perhaps they are farmers, perhaps pet owners, or policy makers. Who they are will impact the type of campaigns that are likely to be effective in creating change, as will other important factors such as their cultures, beliefs, and the environment or context. Campaigning successfully involves a thorough understanding of those factors, to enable change.

This chapter will explore the different approaches to animal welfare campaigns; starting with why campaigning is necessary, before moving on to consider what we can learn from historical welfare campaigns and from the application of human behaviour change science. Finally, we will explore different approaches to advocacy campaigning, including legislation and policy change, awareness campaigns, motivation campaigns, and environmental change campaigns.

DOI: 10.4324/9781003182351-42

Background

Attitudes and behaviour towards animals vary dramatically across the globe according to country and culture, leading to wide disparities in levels of welfare. Animal welfare campaigns exist because of a perceived imbalance between one group and another in the way animals are treated or used. Those groups could be communities, cultures, religions, or countries, but the important thing in campaigning is that one side believes that the other side needs to change their practices.

The pervading research and frameworks around assessing animal welfare are a product of Western cultures, where views on welfare have been heavily influenced by a culture of reliance on scientific study. In Western cultures, views of animals as emotionless "machines" or "automata" promoted by leading philosophers such as Descartes, Aristotle, and Kant, as well as the pervading Christian beliefs at the time, began to change during the Enlightenment (Duncan 2019). At this point, the culmination of extensive wars in Europe led philosophers to consider ethics and morality in new ways, and the concept of "scientific study" was born. This reconceptualisation of life and the study of life led to changes in the way animals were considered. While animals were (and are still) considered primarily from a utilitarian perspective, increasingly, researchers and philosophers considered animals to be emotional, sentient beings, and texts such as Ruth Harrison's *Animal Machines* (1964), Singer's *Animal Liberation* (1975), and Griffin's *The Question of Animal Awareness* (1976) shaped later thinking.

The reliance on scientific study as a basis for policy has permeated Western cultures, and the sciences of veterinary studies, animal welfare, and anthrozoology (the combination of anthropology and zoology; the ways in which cultures think and behave towards animals) are legitimised fields of study. Western views around the role of animals are also driven by beliefs and our feelings about the way it is acceptable or ethical to treat other beings (Munro, 2012; Bryant and Sullivan, 2019). It is important to note that Western countries are no panacea for animal welfare, and many issues persist in relation to farming, animal entertainment, and companion animal care. For example, Western cultures are likely to keep pets as part of an "animal family", which can itself lead to serious welfare concerns, such as obesity (Bradshaw and Casey, 2007).

Other cultures approach animal well-being differently, influenced by history, religion, and practice. For example, India has an extensive set of animal laws covering animals at slaughter, in experiments, and in entertainment, and law suggests "*It shall be the duty of every citizen of India … to have compassion for living creatures*". This is likely a product of India's Jain, Hindu, and Buddhist religions, in which concern for all life is imperative, partly due to beliefs in animal reincarnation and Gods appearing in animal form. Nevertheless, issues remain, for example in slaughter practices, ritual slaughter, and methods of reducing animals considered to be pests, such as feral dogs and monkeys.

In contrast, the approaches shown across Africa towards animal welfare are intertwined with its complex historical and cultural context and result in a wide variation of practices across African countries, religions, and tribes. Many African countries make little reference to animal welfare in their legal frameworks, though some (e.g., Tanzania) legislate for the prevention of cruelty. However, in many African tribes and communities, animals are perceived as totems or spirit guides and cannot be eaten, although those cultures may still practise animal sacrifice. For example, the Shona culture of Zimbabwe recognises the relationship between humans and animals and as a result hunting is regulated by Shona custodians. However, in other African spaces and cultures, human–animal relationships are bound by racial and economic tensions. For example, apartheid, colonialism, and proselytism left South Africa with embedded hierarchical beliefs that led to the oppression of black South African people, and subsequently the abuse and oppression of animals (Pickover, 2005). In other African cultures, the welfare of animals is closely linked to the welfare and wealth of their owners; lack of food and medical care in humans leads to an inability to maintain animal welfare. Notably, South Africa, Tanzania, and Zimbabwe have made significant progress in animal welfare through increased awareness, education, and legislation.

An entirely different situation is seen in China; despite its Taoist and Buddhist roots and the fact it is the largest animal farming nation in the world, China has very few laws relating to animal welfare. China's animal welfare has been described as: "*[a]nimal suffering is unprecedented in China in magnitude in both numerical terms ... and in welfare conditions*" (Li, 2012). China's collectivist culture may be a factor, with the needs and goals of the group being given priority over those of the individual, whether that be human or other species, together with other factors such as economic and social productivity (Lu et al., 2013; You et al., 2014). The concept of "welfare" is not translatable in China, and the idea of animal welfare and animal rights was only introduced in the 1990s, following translations of texts such as *Animal Liberation*. Nevertheless, we should not forget China has a rich and long past which should be explored regarding human–animal relationships and is now gradually responding to changing global social license around animal use.

Central and South American culture presents a different picture again, with its ancient cultures holding animals in high regard, and efforts to institutionalise animal welfare present from the 19th century on the grounds of hygiene, as well as the effect of animal cruelty on humanity. Argentina has the longest standing history of animal welfare movements, with two of its well-known presidents playing key roles. However, as with other countries, issues with animal welfare are closely linked to the welfare of people, and hence disparities in income play a major role (Coleman and Hemsworth, 2014). Countries within Central and South America that have made moves towards improving welfare have primarily made changes within the areas of transport and slaughter, possibly due to the economic benefits of making those changes.

While this short narrative cannot even begin to do justice to the rich and varied cultural histories of those countries and continents described, even from this initial view it is evident that the colonialist view of animal welfare promoted by Western cultures may sometimes overlook the subtleties and complexities between human and animal welfare that can be present in other cultures, and particularly in low- and middle-income countries (Horta Duarte, 2013; Funes Monzote, 2013). An approach that simply imposes Westernised views of an issue onto other cultures will have limited impact because of the context in which animal welfare happens. As a result, it is vitally important that each animal welfare issue is explored with reference to its political, sociological, and cultural background.

Lessons from historical welfare campaigns

Traditionally, animal welfare campaigns have been a result of one group of people disliking the practices of another, and the campaigning group seeking to incite change – often by shocking the second group about their behaviour; the examples of blood-dripping imagery within animal rights advertisements are typical of such interventions.

These ideas follow a neoliberalist discourse, which has been also prevalent in public health narratives and campaigns. These discourses place behaviour in the hands of the individual consumer, who is judged to be making decisions considered by the campaigner to be inadvisable. The campaign therefore aims to correct this choice by presenting the subject with a shocking truth, with the assumption that change will follow. Examples of this in public health include the images present on many cigarette packets and anti-smoking campaigns, depicting cancerous lungs.

This approach *can* bring about change, or change attitudes towards a behaviour, but there are limitations to its application. It overlooks the complex interplay between individual human behaviours and the factors which impact them (our physical and social environment, our habits, emotions), as well as ignoring the drivers of behaviour at a cultural, political, socioeconomic, and societal level.

Consider a typical Western vegetarian campaign that juxtaposes a pet dog and a cow, suggesting that it is morally wrong to eat either (n.b., this approach could be aimed only at Western cultures, where dogs are considered "pets" but cows are commonly considered "food"; it would not work in a culture that eats dogs, or one that reveres cows). This campaign aims to make the subject feel uncomfortable about their eating behaviour, and to re-align their beliefs around which animals are food. Although this may sometimes be impactful, this approach overlooks the fact that eating habits are heavily embedded in our habits, society, and cultures. Many of us will have had the experience of learning something about a food type and thinking "I'll never eat that again". Even with the best of intentions, before long the strong drivers of habit, social practice, and availability override the uncomfortable feelings.

Moreover, shock approaches ignore the knotty issue that almost all consumerism is built upon cognitive dissonance at some level. Cognitive dissonance is a lack of alignment between beliefs and behaviour, meaning that we may believe one thing yet behave in a way which is incongruent with that belief (Kroesen et al., 2017). A devoted meat-eater might draw the line at eating a dog or an insect, but not be able to explain where or why the moral distinction exists between one type of animal or another, and a committed vegan may still own a pet dog or eat products containing palm-oil. This dissonance transcends animal welfare; public health discourses suffer from the same issue. In 2021, it's very unlikely that anybody smokes because they think it is a healthy thing to do or are unaware of the potential health impacts. Instead, people smoke because it's a social behaviour, because they think it is enjoyable, stress-relieving, and a habit as well as an addiction. The world is a complex place, and each of us has personal "lines in the sand" about what we consider to be morally acceptable or attractive behaviours. While the line might move over time, each of us is adept at holding multiple moral views at one time, and being confronted with our entangled moral compasses may not necessarily be enough to change that.

Another traditional approach to improving welfare focuses on legislation around an issue, with the assumption that laws will alter behaviour. Legislation can be extremely helpful in altering behaviour, but, just like the shock campaigns, legislative change needs careful consideration before application because of the messy reality of human lives. For example, many countries have banned the hunting and eating of wildlife for conservation reasons and to avoid potential disease outbreaks such as Coronavirus and Ebola, which are both thought to derive from the eating of wildlife (Ebola Leroy et al., 2009; Corona; Rothan and Byrareddy, 2020). However, eating wildlife is not a choice but a necessity for low-income communities who live near to those habitats, and have had their livelihoods altered by recession or climate change, causing them to hunt wildlife to eat (Brashares et al., 2011). Legislating against hunting wildlife is not, on its own, an appropriate response to targeting this issue for communities like these; alternative livelihoods or sustainable food sources would need to be incorporated into any behaviour change intervention.

A similar example of the complexity of legislative change is the Netherland's 2014 decision to ban the breeding of brachycephalic (short-nosed) dogs, by making it illegal to breed from a dog whose snout is less than one-third of the length of the dog's entire head. This is a risky legislative approach, given that it is very difficult to apply this law in practice – the application of dog-head measurements is not always straightforward, and is not one person or organisation's responsibility. Further, many European countries with thriving consumer demand for specific dog breeds have concerns regarding puppy smuggling from countries with less stringent animal welfare standards; the Netherlands may simply end up with puppies being brought in from other countries. However, if the legislation were supported with appropriate monitoring, awareness campaigns, and a broader application of the principles of behaviour change science, this legislation could be used to instigate a positive change to puppy breeding in the Netherlands.

The examples above highlight that no behaviour (or welfare issue) happens in isolation or in a vacuum; instead, they must be explored within their context, and within the systems which are shaped by cultural, political, and social practices. The science of behaviour change has developed over the past decades alongside our improved understanding of human psychology, public health, and consumer behaviour, among other fields. Collectively, these fields suggest that a holistic understanding of the issue itself, the reasons the behaviour happens in the first place, alternatives to the behaviour, and the systems in place around the stakeholders involved in the behaviour, must all be considered before planning interventions. Once the behaviour is fully understood in all its complexity, the manner for bringing about change can be explored.

Application of behaviour change science

"Effective altruism" and, in our field, "effective animal advocacy" describe the application of an evidence base to campaigning or charitable activities that are aimed at benefiting a specific group (most often animals, in the world of animal welfare, but sometimes the human animal too). Historically, campaigns have simply been run in response to issues perceived by the campaigning person or group. In recent years, it is more usual for those running campaigns to think more strategically; which campaigns are going to generate the most impact? What is the "most impact" – is it the highest number of animals impacted, or a reduction in issues with the most suffering? Quantitatively speaking, the amount of funding and awareness over the welfare of 2,300 captive cetaceans, or even of the 10,000 bears in bile-farming, makes little sense in comparison with the paltry volume of campaigns for improving the welfare of the 25.9 billion chickens living on any given day (Statista, 2021).

Effective animal advocacy as a concept also supports the idea that campaigners should think in terms of the systems within which animal welfare exists; as we have seen, human behaviour is impacted by numerous internal and external drivers. A good example of this is described by Garcés, whose work in the US broiler chicken industry uncovered the complexity of systems surrounding chicken farms (as described in her 2019 book, *Grilled*). Garcés broached her concerns about the well-being of broiler chickens in commercial units with the farmers in charge of those chickens, assuming that she needed to change the farmers' attitudes to the well-being of their birds; with this assumption, the target of any campaign aiming to improve chicken welfare would be the farmers. Garcés discovered that the farmers' behaviour was driven by an insidious cycle of investment and debt from commercial broiler companies who employed the farmers, and from whom farmers could not hope to escape. The issue of broiler welfare was as much an issue of farmer welfare as chicken welfare; the subject of Garcés' campaign for change needed to focus much more broadly than encouraging farmers to change their behaviour.

Before a campaign is initiated it is important to understand the interlinked aspects of its aim and how success will be measured. If the aim is to change behaviour, the intervention may be quite different to a campaign seeking to raise awareness, or a campaign aiming to change how people think about an issue. Similarly, "success" will look different; an awareness campaign might measure success by surveying a group to determine how much they know about a particular issue before and after a campaign; a behaviour change campaign would need to observe how behaviour has altered in real life, and whether that behavioural change has resulted in meaningful change for the animal.

A thorough understanding of the issue in question should therefore help to clarify whether the potential change should be in policy, including legislation; altering the environment in which the behaviour occurs; increasing awareness or education; or increasing motivation or social responsibility around an issue. No single approach will work for every issue or in every situation; each animal welfare campaign needs to be considered individually. The remainder of this chapter will consider each of those types of change in turn.

Campaigning strategies

Policy and legislation

Although we have already discussed legislation in this chapter; here we consider it in its broader context, including how legislative rules link to the wider animal welfare policy environment. For legislation to result in impactful change, it is important that it is created with careful thought and monitoring around how people behave in the real world. Legislation that alters a frequent, popular, or well-endorsed behaviour may simply drive that behaviour "underground", meaning that it is even less visible and has potentially even more serious welfare impacts. Legislative change needs to be carefully timed, and often benefits from being supported by other behaviour change strategies.

An example of a legislative change that needs to be supported in practice is China's removal of dogs from its list of permitted livestock animals. The timing of this change follows decades of campaigning from animal welfare groups both within and outside China, meaning that public awareness of the ethical and welfare issues of the dog meat trade are likely to be relatively high. This is good timing in that the legislative change is likely to receive more support now than it might have done a few years ago. However, given that dog meat is still available, the legislative change may need to be supported, for example through endorsement of alternatives to dog meat.

Changes to guidelines and codes of practice can be impactful when sensitively applied. These are less stringent than laws and allow for flexible interpretation. Here, a useful example is seen in the world of antimicrobial use. It is important for human and animal health alike that antimicrobial use is limited because of an increase in resistance to antimicrobials; as a result, a global policy on antimicrobial use was agreed in 2015, which has five key objectives aimed at reducing the need for and use of antimicrobials in both human and animal medicine (World Health Organization WHO, 2015). This policy is then adapted for use in each country and setting, which brings the benefit of flexibility, but the drawback that rigour and consistency might be lost (Rogers Van Katwyk et al., 2019), as well as the fact that individual users can simply ignore the guidance if they wish. Therefore, any formal guidance, policy, or legislative change needs to be appropriately timed and supported with additional interventions to support real-world change.

A further issue with both legislation and policy is the language used in those documents, and subsequent interpretation of them. For example, "animal cruelty" is legislated against in many countries. However, it is quite clear that what constitutes "cruelty" is constructed and quantified differently according to individual, social, and cultural practices; the matter is further muddied when translation is required across languages. Legislation, guidance, and policies need to be carefully and rigorously written in order to overcome this issue.

Awareness and education

Earlier in this chapter we discussed shock campaigns, which are designed to trigger an emotional reaction, usually a negative reaction. Other options for awareness campaigns include providing information and education around an issue. Importantly, awareness campaigns are just that: they increase awareness, which does not *necessarily* correlate to a change in behaviour (although it sometimes can). The gap between awareness, intention to change, and actual behaviour change has been well-studied (the effect is labelled the "intention-behaviour gap" (Sheeran and Webb, 2016); for example a campaign aiming to reduce the use of critically endangered Saiga antelope's horn in Singapore found that its messaging was effective in changing attitudes towards the product, but that this did not necessarily translate into altered buying behaviour (Doughy et al., 2021). The fact that this intention-action gap exists does not mean that awareness campaigns aren't important, because they absolutely are – it's just that their limitations need to be considered so that they can be adequately supported, where necessary, with additional interventions that help to bridge the gap between awareness and behaviour.

Awareness and educational campaigns are useful tools when something about the issue is little known, and the alternative behaviour can be easily achieved or supported through other interventions (e.g., by fostering a sense of social responsibility around the alternative behaviour). For instance, shark fin soup is a delicacy in many countries, but it is problematic in terms of welfare (sharks have their fins removed, and are then thrown back into the ocean and do not survive) and sustainability. According to WildAid, a charity working on this issue, when they began their work to reduce the demand for shark fin soup, awareness of the issues was very low and people believed that shark fins could grow back once removed. Following decades of awareness campaigns, including celebrity endorsement of the campaign, the demand for shark fin soup has dropped dramatically in China, with studies showing a high awareness for the campaigns themselves and a desire to protect sharks. Here, the awareness campaign was a useful tool because people were unaware of the issues, and extensive awareness and education (including the powerful social driver of role modelling from the celebrities), and a sense of fostering sustainability resulted from choosing not to eat the soup. Unfortunately, the demand for shark fin soup has increased in other countries, showing the need to consider changes at a holistic or global level when appropriate (Wildaid, 2018).

Documentaries are another example of awareness and educational campaigns. Insightful and inspiring documentaries around important animal welfare issues include *Pedigree Dogs Exposed*, *Blackfish*, *The Cove*, *Seaspiracy*, and *Blue Planet* to name but a few. Documentaries raise our awareness of an issue in a way that engages more deeply with our emotions than an advert could. They achieve this through the use of salient narratives and stories, which are powerful methods of sharing information (Boissat et al., 2021).

Educational campaigns can be more targeted to specific skills and/or specific communities, rather than the broad population-level campaigns described above. A good example of targeted in-depth educational campaigns is the work of charities to improve the lives of equids working in brick kilns. Equids may spend many hours each day pulling heavily laden carts of bricks in the heat, often with little health care, nutrition, or farriery. This is not due to intentional neglect; horse, mule, and donkey owners' income is directly related to their animal's ability, so it is in their interest to maintain their animal's health (Mitra and Valette, 2017). Interventions focused on increasing knowledge about the needs of those animals, and their ability to feel pain (Mitra and Valette, 2017) and providing training to the equid owners in skills such as harness care (to avoid sores), pain recognition, and nutrition. Welfare charities also train veterinary staff and farriers to provide support to the owners. In this setting, an educational and skills-based approach was most appropriate and has led to marked improvements in the welfare of brick kiln equids (Haddy et al., 2021; Mitra and Valette, 2017).

Motivation and social responsibility

Humans are a sociable species, and as such our desire to be seen to be acting in a morally respectable way is strongly embedded. As a result, social movements are a powerful driver of behaviour change because they encourage us to conform to a respected norm. The public health world has seen a move towards social movements for behaviour change; behaviours from stopping smoking to taking up exercise are often now encouraged in a visible, social form, including social media pledges and sharing of results.

Interventions that use our social nature to encourage behaviour change usually present us with opportunities to "badge" ourselves as socially responsible or moral agents and join with like-minded people. This approach is therefore useful in the animal welfare world, given that people can be extremely passionate about their animal welfare beliefs.

An example is Veganuary, a non-profit organisation that encourages people to try being vegan during the month of January each year, and to share their experiences on social media with the hashtag #Veganuary. Veganuary's main messaging is not about the benefits of being vegan, but instead simply reads: "*join the new year's revolution*", and shares figures about the number of people and places involved (Veganuary, 2021). Further, Veganuary has aligned with the manufacturers of vegan products and restaurants to launch new vegan foods and menu options in January, which not only fosters the feeling of an exciting "revolution" in eating habits but also ensures that its followers can more easily engage in vegan lifestyles. When signing up (which is encouraged through highlighting the existence of the social movement), potential joiners can choose whether the marketing and newsletters aimed at them are primarily in relation to veganism's role in improving animal welfare, sustainability, or health. This clever feature ensures that the triad of reasons people usually give for becoming vegan are front and centre of its campaign, but without actually *telling* people that there are three good reasons for becoming vegan; plus, of course, it allows potential joiners to engage in the information most relevant and salient to them.

Social movements are also important in the consumer industries including fashion and tourism, where we display ourselves through the choices we make (e.g., by choosing anti-fur, non-leather, and responsible tourism). This is also impactful on a larger scale; corporations use the examples of their own social responsibility to appeal to their customer base, which can in turn help to raise awareness of issues and shape products available to consumers. For example, the holiday company Virgin has stopped promoting or selling tickets to captive cetacean parks, is an "elephant-friendly" travel company (i.e., not encouraging elephant tourism, which has a host of concerning ethical practices) (World Animal Protection, 2019), and says it is committed to responsible whale-watching guidance (Virgin Holidays, 2019). Several large clothing and fashion manufacturers have signed up to a commitment to avoid fur, and wool which has been created from flocks where mulesing is practised (mulesing involves the removal of skin from the tail area of the sheep, with the aim of reducing fly strike) (Four Paws, 2021). The attention of these companies on each issue raises awareness with the public and also increases the perceived social responsibility of the company and those who consume its products.

Social responsibility is powerful in more targeted settings, and there are numerous examples of local communities inciting change through their passion for a certain local issue. The aforementioned Shona tribe's monitoring of hunting according to their religious beliefs about animals is a good example. Another example is shown by the organisation Paso Pacifico, who conducted extensive community engagement and research before working with communities in Nicaragua to develop a collaborative sea-turtle conservation approach, which was sensitive to community needs, as well as to the turtles (Smith and Otterstrom, 2009). Given that their research had shown that the local community was passionate about the local turtle communities and concerned about poaching and the impact of tourism, Paso Pacifico's approach was to find ways to facilitate the community's own sense of responsibility to the turtles.

Altering the environment

Although we don't always realise the extent of it, our environments are carefully structured to encourage us to behave in certain ways. For example, the physical structuring of supermarkets is a major field of study; we are influenced by product placement, colour, size, music, lighting, smell, and even the direction we turn when we enter a shop (Brinkworth, 2017).

Environmental changes are often referred to as "nudges" following a seminal text of the same name (Thaler and Sunstein, 2008) about the strong evidence base for environmental changes influencing behaviour. The potential of these sorts of interventions is exciting and can sometimes be useful in place of, or alongside, other types of campaigns. A simple way of thinking about this is considering how the environment could make the undesirable behaviour more difficult (add friction) or the desired behaviour easier to perform (add fuel) (Ariely, 2008). One example of this is the global move towards either charging people an additional fee for single-use plastics, or simply banning them outright (as has happened in Kenya (Bahuria, 2021); both approaches make it more difficult for people to use these plastics, and in this instance the "friction" approach is likely to be more impactful at a large scale than incentivising the use of re-usable products (e.g., by giving a reduced cost to people who bring their own straws and bags).

There are numerous examples of friction-based environmental change in the trade of animals, including exotic pets and puppies. Although legislative changes have had some impact, manipulations to the environment may contribute to an increased effect: for example, disallowing advertisements on sites where people frequently impulse-buy animals (e.g., Facebook) so that people have to actively look for new pets on specific sites, rather than simply seeing them on sites they already visit (Facebook, 2019). The need for a license before owning a certain species of animal is an additional level of friction.

Fuel-based interventions (e.g. the incentivisation of certain behaviours) can be just as impactful. For example, reducing the cost of neutering, or rewarding those with animals in optimum condition. Other examples are more complex: for example, in the illegal wildlife trade, encouraging the uptake by consumers of legal alternatives to popular illegal wildlife (e.g., legally harvested songbirds versus the catching of rare, illegal wild songbirds) can lead to a move away from catching illegal wildlife (Wallen and Daut, 2018). In this instance, an approach that simply hampers the catching or trade in illegal animals may be problematic for local communities relying on the current trade to survive and will therefore seek ways around any new "friction"-based interventions but using "fuel" to facilitate a more desired behaviour may yield better results.

Discussion

There are many ways to influence human behaviour around animal welfare, and no single method or approach is applicable to every welfare issue, nor can one approach be applied to every setting. Furthermore, no matter how experienced the campaigns team, no approach is likely to be successful unless it follows research that aims to understand the issue in-depth and from multiple viewpoints. Activities such as stakeholder mapping, theories of change, and systems identification are more likely to lead to a campaign that will yield success than any other factor, because those approaches will help to understand the drivers for the issue, and the means for change. The benefit of these approaches is that they can also facilitate the development of participatory or community-led initiatives, which are often impactful because they are responsive to local needs.

Most campaigns bring together aspects of different types of approaches, for example combining awareness and social responsibility, or legislative change with environmental restructuring. Given that we as individuals all respond differently to the materials and tactics employed in campaigning, there is certainly a place for all these diverse approaches, sensitively applied.

An example of a successful welfare campaign that brought together diverse behaviour change techniques is the campaign for improving conditions for equines travelling for slaughter in Europe. Historically, horses were regularly transported long distances into and across Europe for slaughter; suffering cramped travelling conditions and extreme hot and cold for many days with little regulation about the need for stops, water, or food provision (World Horse Welfare, 2011; Leadon, 2012). This was a politically difficult campaign given that no single organisation or governing body was responsible for these journeys, particularly given that they could begin or end anywhere across numerous countries.

World Horse Welfare (the organisation leading this campaign, which they had been active on since being founded in 1927) joined forces with charities and other interested parties (e.g., veterinarians, enforcement authorities) across Europe to act. This involved scoping out the issue; gaining a full understanding of the extent of the problem, the stakeholders involved (everyone from drivers to politicians), and possible practical solutions. The result of this process was a clear aim – a set of policy, including legislative, changes that would improve the welfare provisions of the equids during those journeys, and would contribute to the reduction in the number of live horses and donkeys being transported long distances to slaughter in Europe. With this aim in mind, a strategic campaigning plan was created to tackle the problem:

- Developing a dossier of evidence based upon field, scientific, and desk-based research that detailed the equine welfare issues and the means to address these;
- Liaising with European policy makers (e.g., European Commission), providing them with evidence of the issues and the potential solutions to improve legislation;
- Working with individual Member State country policy makers, local authorities, NGOs and the public to raise awareness of the issue, and work out how practical in-country changes could work;
- Engaging the support of the public and elected members of parliament in different Member State countries to apply pressure on policy makers for change;
- Awareness campaigns so that the public could put pressure on elected members of parliament (this included articles in the media, petitions, and talks).

A legislative change was agreed in 2004, which introduced criteria on fitness for transport of animals with specific requirements for equines, the introduction of individual partitions for animals travelling long distances (with the exception of mares travelling with foals), restriction on the movement of unbroken equines, and prevention of the use of multi-decked vehicles for the movement of horses (EC 2004). At this point the campaign did not stop; work continued pressing for finite journey limits and increased space allowance, but the focus also shifted to supporting and training the drivers, vets, and local authorities who would now need to change their practice, as well as monitoring to ensure that the legislative change was resulting in actual change (including the legislation being enforced in different countries). This example shows the need for tackling any issue with a strong evidence base around the issue, a clear aim, a strategy (which in this case necessitated the use of multiple campaigning angles), and following up with support for those carrying out the change, as well as ongoing monitoring and evaluation.

Conclusions

Animal advocacy has evolved since its origins and continues to evolve. In recent years, the increasing emphasis placed on strategy and evaluation of campaigns, together with an increased understanding of human behaviour change science, means that this is an especially interesting time in the development of the movement.

Human societies are becoming increasingly adept at changing the behaviour of individuals and groups of people – consider the tactics used by social media, website algorithms, supermarket design, and even the tactics used by political parties to gain votes, to see the breadth of scope for the behaviour change sciences. Although behaviour change science is often a tool used in globalisation, capitalism, and industrialisation (areas which drive the use and exploitation of animals), the world of animal welfare is catching up on harnessing these methods and using them to improve the lives of animals ranging from wildlife to pets.

The move to globalisation is also increasing our understanding and appreciation of the way that groups, cultures, religions, and countries interact with one another; we have a growing understanding of -isms (e.g., racism, sexism, speciesism). This is furthering our understanding of ethics and morality in relation to global issues and animal use. These apparently opposing forces have come together at a crucial time, as current threats to sustainability could have wide-reaching effects on our civilisation as a whole. Understanding drivers for the many groups involved in the systems surrounding animals, will help the animal welfare movement to hopefully tip the balance towards protecting animals, the environments they live in, and ultimately secure our future.

Acknowledgements

We are incredibly grateful to Anouk de Plaa who helped with the research and writing of this chapter.

References

Ariely D, 2008. *Predictably Irrational: The Hidden Forces that Shape Our Decisions*. London: HarperCollins.

Bahuria P, 2021. Ban the (plastic) bag? Explaining variation in the implementation of plastic bag bans in Rwanda, Kenya and Uganda. *Environment and Planning C: Politics and Space,* 39(8), 1791–1808. https://doi.org/10.1177/2399654421994836

Boissat L, Thomas-Walters L and Verissimo D, 2021 Nature documentaries as catalysts for change: Mapping out the 'Blackfish Effect'. *People and Nature*, 3(6), 1179–1192. https://doi.org/10.1002/pan3.10221

Bradshaw JWS and Casey RA, 2007. Anthropomorphism and anthropocentrism as influences in the quality of life of companion animals. *Animal Welfare*, 16(1), pp. 149–154.

Brashares JS, Golden CD, Weinbaum KZ, et al., 2011. Economic and geographic drivers of wildlife consumption in rural Africa. *Proceedings of the National Academy of Sciences of the United States of America*, 108(34), pp. 13931–13936. https://doi.org/10.1073/pnas.1011526108

Brinkworth C, 2017. Supermarket savvy: An analysis of psychological exploitation within grocery stores Independent Study Project. *Independent Study Project (ISP) Collection*, p. 2603. https://digitalcollections.sit.edu/isp_collection/2603

Bryant TL and Sullivan M, 2019. Why American animal-protective Legislation does not always "stick" and the path forward. In Hild S and Schweitzer L, eds. *Animal Welfare: Science to Law*. Paris: La Fondation Droit Animal Ethique & Sciences, pp. 77–87.

Coleman GJ and Hemsworth PH, 2014. Training to improve stockperson beliefs and behaviour towards livestock enhances welfare and productivity. *Revue Scientifique et Technique*, 33(1), pp. 131–137. https://doi.org/10.20506/rst.33.1.2257

Doughy H, Milner-Gulland EJ and Huay Lee JS, et al., 2021. Evaluating a large-scale online behaviour change intervention aimed at wildlife product consumers in Singapore. *PLoS ONE*, 16(3), p. e0248144. https://doi.org/10.1371/journal.pone.0248144

Duncan IJ, 2019. Animal welfare a brief history. In Hild S and Schweitzer L, eds. *Animal Welfare: Science to Law*. Paris: La Fondation Droit Animal Ethique & Sciences, pp. 13–21.

EC, 2004. *Council Regulation (EC) No 1/2005 of 22 December 2004 on the Protection of Animals during Transport and Related Operations and Amending Directives 64/432/EEC and 93/119/EC and Regulation (EC) No 1255/97*. Brussels: Europena Commission.

Four Paws, 2021. Brands against mulesing. Retrieved from https://woolwithabutt.four-paws.org/brands-against-mulesing (accessed 12th September 2021).

Garcés L, 2019. *Grilled: Turning Adversaries into Allies to Change the Chicken Industry*. London: Bloomsbury Publishing.

Griffin DR, 1976. *The Question of Animal Awareness: Evolutionary Continuity of Mental Experience*. New York: Rockefeller University Press.

Harrison R, 1964. *Animal Machines: The New Factory Farming Industry*. London: Vincent Stuart Publishers LTD.

Facebook, 2019. Prohibited content animals and animal products. Retrieved from https://www.facebook.com/policies_center/commerce/animals (accessed 03/11/2021).

Funes Monzote R, 2013. Animal labor and protection in Cuba: Changes in relationships with animals in the nineteenth century. In M Few and Z Tortorici, eds. *Centering Animals in Latin American History*. Durham: Duke University Press, pp. 209–242.

Haddy E, Burden F, Fernando-Martinez D, et al., 2021. Evaluation of long-term welfare initiatives on working equid welfare and social transmission of knowledge in Mexico. *PloS One*, 16(5), p. e0251002. https://doi.org/10.1371/journal.pone.0251002

Horta Duarte R, 2013. Birds and scientists in Brazil: In search of protection, 1894-1938. In M Few and Z Tortorici, eds. *Centering Animals in Latin American History*. Durham: Duke University Press, pp. 270–301.

Kroesen M, Handy S and Chorus C, 2017. Do attitudes cause behavior or vice versa? An alternative conceptualization of the attitude-behavior relationship in travel behavior modelling. *Transportation Research Part A: Policy and Practice*, 101, pp. 190–202.

Leadon DP, 2012. Unwanted and slaughter horses: A European and Irish perspective. *Animal Frontiers*, 2(3),pp. 72–75. https://doi.org/10.2527/af.2012-0053

Leroy EM, Epelboin A, Mondonge V, et al., 2009. Human ebola outbreak resulting from direct exposure to fruit bats in Luebo, Democratic Republic of Congo, 2007. *Vector-Borne and Zoonotic Diseases*, 9(6), pp. 723–728. https://doi.org/10.1089/vbz.2008.0167

Li PJ, November 2, 2012. Animal rights in China. Retrieved from http://www.forbes.com/sites/michael-tobias/2012/11/02/animal-rights-in-china/ (accessed 03/11/2021).

Lu J, Bayne K and Wang J, 2013. Current status of animal welfare and animal rights in China. *Alternatives to Laboratory Animals*, 41(5), pp. 351–357. https://doi.org/10.1177/026119291304100505

Mitra D and Valette D, 2017. Brick by brick: Unveiling the full picture of South Asia's brick kiln industry and building the blocks for change. International Labour Organization, Brooke and The Donkey Sanctuary Report. Retrieved from https://www.thebrooke.org/sites/default/files/Downloads/Brick%20by%20Brick%20report.pdf (access date 03/11/2021).

Munro L, 2012. The animal rights movement in theory and practice: A review of the sociological literature. *Sociology Compass*, 6(6), pp. 511–518. https://doi.org/10.1111/j.1751-9020.2012.00462.x

Pickover M, 2005. *Animal Rights in South Africa*. Cape Town: Double Storey Books, Juta & Co. Ltd.

Rogers Van Katwyk S, Grimshaw JM, Nkangu M, et al., 2019. Government policy interventions to reduce human antimicrobial use: A systematic review and evidence map. *PLoS Med*, 16 (6), p. e1002819. https://doi.org/10.1371/journal.pmed.1002819

Rothan HA and Byrareddy SN, 2020. The epidemiology and pathogenesis of coronavirus disease (COVID-19). *Outbreak Journal of Autoimmunity*, 109, p. 102433. https://doi.org/10.1016/j.jaut.2020.102433

Sheeran P and Webb TL, 2016. The intention-behavior gap. *Social and Personality Psychology Compass*, 10(9), pp. 503–518. https://doi.org/10.1111/spc3.12265

Singer P, 1975. *Animal Liberation: A New Ethics for Our Treatment of Animals*. New York: HarperCollins.

Smith R and Otterstrom S, 2009. Engaging local communities in sea turtle conservation: Strategies from Nicaragua. *George Wright Forum*, 26(2), pp. 39–50.

Statista, 2021. Number of chickens worldwide from 1990 to 2019. Retrieved from https://www.statista.com/statistics/263962/number-of-chickens-worldwide-since-1990/ (access date 03/11/2021).

Thaler RH and Sunstein CR, 2008. *NUDGE: Improving Decisions about Health, Wealth, and Happiness*. London: Yale University Press.

Veganuary, 2021. Veganuary 2021 campaign in review. Retrieved from https://veganuary.com/wp-content/uploads/2021/03/Veganuary-2021-Campaign-in-Review.pdf (access date 03/11/2021).

Virgin Holidays, 2019. Cetaceans. Retrieved from https://www.virginholidays.co.uk/responsible-tourism/supply-chain/cetaceans (access date 03/11/2021).

Wallen KE and Daut EF, 2018. The challenge and opportunity of behaviour change methods and frameworks to reduce demand for illegal wildlife. *Nature Conservation*, 26, pp. 55–75. https://doi.org/10.3897/natureconservation.26.22725

WildAid 2018. Appetite for shark fin drives threatened shark population decline. Retrieved from https://wildaid.org/appetite-for-shark-fin-drives-threatened/ (accessed 16.04.2022].

World Animal Protection, 2019. Elephant-friendly travel companies. Retrieved from https://www.worldanimalprotection.org/sites/default/files/media/int_files/elephant_friendly_travel_companies_10.12.18.pdf?_ga=2.208089751.421810659.1631103723-1019923557.1631103723 (access date 03/11/2021).

World Health Organization (WHO), 2015. Global action plan on antimicrobial resistance. retrieved from https://ahpsr.who.int/publications/i/item/global-action-plan-on-antimicrobial-resistance (access date 03/11/2021).

World Horse Welfare, 2011. *Dossier of Evidence: Recommendations to EU Council Regulation.* (EC) No1/2005. 2nd edn. Norfolk, UK: World Horse Welfare.

You X, Li Y, Zhang M, Yan H and Zhao R, 2014. A survey of Chinese citizens' perceptions on farm animal welfare. *PloS One*, 9(10), p. e109177. https://doi.org/10.1371/journal.pone.0109177

36

ANIMAL WELFARE EDUCATION AND COMMUNICATION

Ruth De Vere

Introduction

We are gradually moving away from animal welfare occupying a niche space in the communal psyche. The facts of animal sentience are indisputable – animals have the capacity to suffer and a desire to feel joy and live good lives. Successful campaigning from a variety of voices has influenced legislative bodies and decision makers to amend some of the most egregious policies and procedures that directly affect animals in a positive way. However, the change is not systemic yet and neither is it understood or agreed by all as a necessary requirement. Social norms that cause and enable animal suffering still pervade, and knowledge of animal sentience is still the reserve of academia. A corporation will struggle to find the motivation to change their policies unless there is consumer demand for it, and a government won't introduce legislation unless it feels certain that the majority of citizens agree and can abide by it. Therefore, the concept of animal welfare must continue to move beyond the realms of the elite and into the hands of the many.

The phrase "animal welfare" doesn't translate all that well into many languages, giving rise to awkward conversations, and underpinning the need for carefully considered and effective communication to audiences. When it does break through the surface into dinner table conversations, it is often seen as a luxury and something sometimes laughable in areas of the world where human welfare is thin on the ground. It is often seen as a trade-off, an issue mutually exclusive to others, rather than a fundamental strand in the interconnectedness of issues as wide-ranging as climate change, food (in)security, disaster preparedness, pandemic prevention, right through to bullying in children and signs of domestic abuse. Recognition of the potential for this antipathy is vital if communication on the issue is to be effective. Those passionately working in the field of animal welfare can be viewed with disdain and easily brushed off, whilst those enabling or implementing systems and actions that perpetuate suffering are seen as villains, giving rise to an unhelpful "them" and "us" scenario.

Poor animal welfare often arises because of human behaviour. That behaviour may be intentional, but in the vast majority of cases there is an unintentional status quo that enables the behaviour to continue. It's not always the case that outright cruelty, deliberately and intentionally inflicted, is the cause of the suffering. It's also equally wrong to simply cite ignorance as the principal factor – "*If only we could educate more people*". We live in a complex system whereby our behaviour arises from a number of influences. Our knowledge is a fundamental pillar. Education

 DOI: 10.4324/9781003182351-43

– as discussed in detail in this chapter – has a vital role to play. However, what we see others do – especially those we look up to – and how our society behaves as a whole affects our actions and can override what we have been taught and what we know. Societal norms enable us to take the mental shortcuts in decision making, and not exhaust ourselves with ethical debates as to the rights and wrongs of how we should act, which pet we should buy, which, if any, animal we should consume, or which venue we should visit on our holidays. We can change behaviour faster than we can change attitudes with the right approach. However, partnering effective behaviour-change approaches to tackle the most pressing animal welfare issues with longer-term generational change as an insurance policy to ensure sustainability, is hugely powerful. Provision of animal welfare education in a formal sense, therefore, is critical for the long term.

Some of the greatest causes of animal suffering on an industrial scale have occurred because of globalisation in recent decades. The intensive farming industry gives rise to some of the largest numbers of animals affected, and that has only grown and expanded in the last century. Animals are seen as commodities and products or assets, and therefore the industrialisation of the industry has grown because there have been no checks and balances in place. Let's not make the mistake of talking about the industry as an amorphous mass. It is driven by people. No farmers, policy makers, or senior veterinarians were willing or able to recommend or require the incorporation of animal-based outcomes (animal-centred measures based on meeting their very real needs) rather than human-centred measures, which are by far the most widely used measures in our "dominion-over-all" approach to the world. In developed countries, where industrialisation is at its most advanced, even insurance policies for farms affected by disasters leave farmers in a dilemma as to just how far to go to protect their animals when disaster strikes, and provide very real disincentives to work to reduce losses (Linnerooth-Bayer and Mechler, 2009). In these and other scenarios, the needs of the animals are very far down the list of priorities.

This chapter will explore formal education programmes aimed at children and young people as well as vocational training for professionals working with animals that have the aim of raising the needs of animals higher up that list of priorities. It will investigate the skills needed for effective communication and the roles that those who benefit from formal education and qualifications in the field of animal welfare have in changing the status quo. It will also draw links with human behavioural science and the field of behavioural economics to highlight the fact that all educational and knowledge-sharing approaches benefit from application of these methodologies to result in sustainable change for the benefit of animals, people, and the planet.

Humane Education

Since the 1800s there has been a conscious effort to teach children kindness – often through exploring their relationships with animals, and in many instances by bringing a companion animal into the home with the express desire to give adolescents responsibility for them (Grier, 1999). Humane Education (HE) is a more recent and more formalised manifestation of this effort. It teaches social justice, citizenship, environmental issues, and the welfare of animals and it recognises the interdependence of all living things. It is based on values that develop sensitivity to all life, appreciation of diversity, and tolerance of difference. It encourages children to become more compassionate and learn to live with greater respect for everyone, as well as provides opportunities for children to develop a sense of responsibility and a duty of care for their surroundings and the natural world (World Animal Protection, 2012).

In addition to introducing animal welfare through the five domains (Mellor DJ et al., 2020), and the ethical considerations around keeping, using, and farming animals for our benefit, animal welfare education (AWE) as a sub-group within HE provides important opportunities to study issues that are often locally relevant and multi-faceted. For example, children living in communities with roaming dog populations and a fear of rabies need to be taught the essentials of good dog population management, disease control through vaccination, and the skills for effective dog bite prevention – a potentially life-saving skill when we consider rabies is responsible for 59,000 deaths worldwide each year, with 99% of these caused by dog bites, and 40% of deaths occurring in children under the age of 15 ((FAO), et al., 2018). Children whose communities are vulnerable in times of disaster learn about disaster preparedness and ways families and communities as a whole can work to improve their own resilience, and improve the outcomes for their animals. Many of the world's poorest communities rely on working equines whose welfare can be extraordinarily poor. Children can apply the five domains and meet the needs of the animals instantly. Children can be taught to apply the five domains to captive animals and their own choices of pets in the home, leading to an understanding of the global nature of the wildlife trade, and the harrowing journeys undertaken by extraordinary numbers of "exotic" (non-native and non-domesticated) species to provide a reliable supply to the pet trade. With a stark number of children across the world unable to accurately state where their food comes from and how it grows (Hamilton and Surman, 2018), understanding agriculture, and the intensification of farming practices is an important subject area for a variety of reasons. Being able to explore this subject through the lens of animal welfare as well as climate change and environmental protection provides a valuable opportunity to explore personal agency and the role of conflicting consumer choices.

The list of subjects is long, and limited only by the skill and knowledge of the individual providing the instruction. Very young children benefit from topics that encompass animals they see and encounter in their everyday lives. As children grow and experience more of the world around them, so the subject matter can broaden to include animals and issues they don't find on their own shores. The basics of the five domains remain a fundamental platform from which to analyse and debate a given scenario, and the subject as a whole encourages a message of interconnectivity, safeguarding our planet, living with compassion and respect for living things, and the acceptance that we all have a role to play and we can all make a difference. The importance of instilling these five domains or animal-based outcomes as the significant measure rather than human-based outcomes cannot be overstated and this remains a fundamental goal of AWE. Once we have a generation who instinctively apply animal outcomes to any given life choice or professional decision concerning animals, we are closer to sustainably transforming long-term social norms around what animal use is acceptable.

In the 1990s the link between animal abuse and societal violence, whether that be domestic abuse or violent crime, began to be well documented, with significant studies showing links between children who perpetrated cruelty to animals that then go on to perform aggressive, antisocial behaviour towards humans (Ascione, 1993). The same review showed that witnesses and victims of violence in the home may themselves be cruel towards animals and therefore an additional benefit of AWE is that widespread attention to the treatment of animals, and taking the abuse of animals seriously by professionals in positions of power (including teachers), can lead to interventions at a critical stage. Teaching subjects that foster compassion can build empathy, promote prosocial behaviour, diffuse violence and potentially reduce bullying.

Providing HE and AWE specifically to children and young people is powerful because you get three times the return on your investment – if you're willing to play the long game. Although it is an alien concept to social change organisations, many parallels can (and should) be drawn with the marketing and retail sector when it comes to understanding audiences, and getting them to act in the way you desire. As corporations and marketers know all too well, and as McNeal outlined in his book *Kids as Customers* ((McNeal, 1992), young people occupy three categories of interest. First, they are a primary audience, with agency intrinsic to themselves. They are willing and able to take action, and in the case of brands wanting to sell products – they have disposable income of their own and are discerning and brand-conscious when they make decisions about where to spend it. Second, they occupy an incredibly powerful "influencer" role – both in terms of "upwards education" on welfare issues, where they share their newfound knowledge with their families and wider communities (Vaughan et al., 1999), and in terms of influencing upwards of $130 billion of adults' spending each year in the USA alone (McNeal, 1992). This is an important factor to remember when designing AWE programmes and campaigns, as family decisions influenced by children can range from which pet to get, which holiday venue or day out to choose, to what items make it into the family shopping basket. Thirdly, and most importantly, young people occupy that space favoured by marketers – the future market. Brand loyalty begins with some of the clever marketing approaches you might find on branded back-to-school stationery for example, or in big brands that offer great kids parties at their restaurants, or in banks that offer a child their first current account (McNeal, 1992). These tactics often result in limited short-term return on investment, but what they do provide is comfort, familiarity, and affection for brands which is likely to result in loyalty, and longer-term gains that outweigh the initial costs. This is fundamental to remember, both in terms of non-governmental organisations (NGOs) wanting to secure a potential future supporter base loyal to the cause (and the brand), but it's also significant in terms of creating a knowledge base and set of foundational values that children will find it harder to move away from as they progress into adulthood and occupy that future space. Educating the youth of today is a win for the decision makers of the future.

Approaches to implementation

If the end goal is to ensure that young people are exposed to, and effectively taught AWE specifically, and HE in general, then there are a multitude of ways this can be achieved. To date, animal welfare organisations take on the lion's share of implementation, and as a result diverse approaches are applied.

Commonly, local animal protection organisations will run school visits where they bring one of their shelter animals (usually a well-behaved dog) and give talks to raise awareness and discuss welfare, with classes or whole year groups. These are often a delight to the students, and they can be transformative for individuals in the audience. However, school assemblies are a highly sought-after resource, with all manner of issue-based NGOs, drama groups, and other services vying for a spot. If an animal NGO is lucky enough to secure one, return visits to follow up or build on messaging, or to carry out effective impact assessments are extraordinarily difficult to obtain.

Measuring the impact of education programmes like these has been notoriously difficult to do – not only because of the limitations of return visits to schools. There are also some fundamental flaws in the design of awareness-raising programmes that make measurement impractical. With a gradual increase in emphasis being placed on measuring impact within the not-for-profit sector, the importance of monitoring and evaluating education programmes is gaining traction. In addition, greater prominence is beginning to be placed on education programmes being designed with specific human behaviour-change goals included from the outset, meaning that measurement of the presence or absence of these behaviours is more attainable. Many animal NGOs are working to review their education programmes and their monitoring and evaluation approaches in order to measurably improve the outcomes for animals.

For the precise reason that metrics are hard to come by, alternative approaches have been undertaken directly by academic institutions where researchers have delivered prolonged AWE programmes in schools to document and measure their efficacy. These are of a set duration, cover a specific syllabus, and often involve traditional evaluation methodologies and control groups to determine impact. These studies enable effective monitoring and often return very positive results that are maintained after the study has finished (Samuels et al., 2016). They demonstrate the positive impact of teaching the subject (something other subjects aren't expected to demonstrate other than through formal exams), but without definitive inclusion in the taught curriculum, the teaching of the subject remains the exception rather than the norm in classrooms worldwide.

There are benefits in working at a more systemic level, seeking to ensure that animal welfare is included within the national curriculum. Teachers also need to be trained to teach the subject, either whilst they are in-service or through Initial Teacher Training Institutes. World Animal Protection (known as the World Society for the Protection of Animals, or WSPA, during the heyday of the education programmes) ran an education programme that began life as "Respect for all forms of Life" in Costa Rica in 1989, and expanded globally to become the International Animal Welfare Education (IN AWE) programme, and latterly "First Concepts in Animal Welfare" (FCAW) until the programme closed in 2016. The programme gradually evolved into a year-long programme of Continuous Professional Development (CPD), with assignments and criteria for passing the course. According to the internal Project Close Report (overseen by the author of this chapter), this approach resulted in 1,500 in-service teachers qualifying. In those countries where the curriculum did not specify that animal welfare was included, then these teachers created after-school clubs and groups in order to successfully complete their training. These clubs were popular with students and remained in place after the training course had ended. In addition to training teachers directly, the programme sought to ensure that Initial Teacher Training Institutes included the topic to guarantee that newly qualified teachers were capable of teaching the subject. According to the aforementioned report, approximately 30,000 teachers will have now graduated having been trained in AWE principles during their professional training.

By the time World Animal Protection closed its education programme Uganda had incorporated Animal Welfare into its Lower Primary, Primary, and Secondary curricula, making it the first country to mainstream the subject. Kenya soon followed suit with a comprehensive inclusion throughout the curriculum. In Vietnam, the Hanoi district also included the subject in its Primary and Secondary curricula. As a result of these official changes, the aforementioned report states that approximately 20 million children are taught animal welfare each year.

The more opportunities young people have to come into contact with the subject of AWE and apply it to real-life scenarios that affect them the better. All of these approaches have merit, and as a holistic approach they all serve to contribute to building a knowledge base around consideration of animals and our behaviour towards them. However, focus must be maintained on reaching the greatest numbers of young people if NGOs are to put what are usually donors' contributions to the best use. Achieving curriculum change, and ensuring that the education system of a given country takes responsibility for teaching the subject is an important strategy in shifting social norms.

Animal welfare education and animal industry professionals

Despite the efforts of animal protection NGOs across the world, the inclusion of AWE and HE within taught curricula is sporadic, which means that the plethora of animal-based professions have a mountain to climb to redress that balance given that most of the student admissions (globally) will arrive with little or no animal welfare knowledge or understanding. There are a wide variety of professions that require direct interaction with animals and the range of training and qualifications required to fulfil those roles varies just as much.

Veterinarians

The most obvious place to begin a discussion on the selection, training, and quality assurance of animal-related professions must surely be the veterinary profession. Vets are seen as role models in many parts of the world, and a veterinary career is in many cases an aspirational one. In many countries, veterinary undergraduate courses are hugely popular and oversubscribed, leading to admission being based on acquisition of exemplary grades and in some cases, substantial work experience. Much like human medicine admissions, the very high grades required narrows down the field of entry, even though the academic rigour of the subject matter itself may not require such high grades and high-performing students in order for them to succeed.

This isn't the case in every locality. Across Asia the veterinary profession doesn't always enjoy the same status as in other parts of the globe, with applicants seeking out human medicine, law, and other sciences as preferences over and above veterinary training. This brings with it a different set of obstacles to integration of compassion and welfare teaching.

Once admitted into a veterinary faculty, students will be exposed to a wide variety (and quality) in animal welfare teaching. In 2012, the World Organisation for Animal Health (OIE) included knowledge and practice of animal welfare in its recommendations for what it calls "Day 1 Competencies" for graduating vets across the world (World Organisation for Animal Health (OIE) 2012). Veterinary associations have also followed suit and taken significant steps to support the inclusion of animal welfare education within veterinary training approaches. Whilst many veterinary faculties and indeed Chief Veterinary Officers look to the OIE and their respective associations for precisely this kind of guidance, and whilst the recognition and inclusion of this topic within the OIE framework was of great significance and much celebrated, there is no single overarching body responsible for veterinary training to ensure it is embedded and taught effectively.

For many years World Animal Protection worked to fill the knowledge gap and encourage the inclusion of animal welfare into veterinary curricula by providing an education resource for veterinary faculties worldwide. In partnership with the UK's University of Bristol School of Veterinary Sciences the first edition of Concepts in Animal Welfare was launched in 2003 (De Boo and Knight, 2005). The resource itself was hugely popular and went through a further two revisions. The latest edition was launched in 2013 (World Animal Protection, 2013). This resource supported the wider education programme known as "Advanced Concepts in Animal welfare" or ACAW (De Vere, 2014). This programme saw partnerships with veterinary faculties and staff worldwide to build a community of vet schools delivering high quality animal welfare teaching to undergraduate vets.

Due to widely recognised constraints on an overcrowded curriculum, as well as cultural differences and willingness across the faculties worldwide, the inclusion of the subject and its implementation varied from faculty to faculty. Vet schools were able to include the subject at their own discretion and therefore there were broad differences as to whether the subject was embedded within the existing curriculum for veterinary students, or as a separate subject. Either way, the learnings may well be confined to the lecture hall and not transferred when the students physically move to their next subject for the day – to practise clinical and surgical procedures on healthy animals, for example.

There is a broad range of evidence to suggest that students may well arrive with high levels of empathy and compassion; however, during the course of their studies and precisely because of the cultures they are learning within, these decline over the course of the training (Self D et al., 1991). This decline can be measured using a number of important indicators such as lower levels of analgesia being administered in later years of the course (Hellyer P et al., 1999) – a finding mirrored in human medicine students (Neumann et al., 2011). This desensitisation towards animals reflects a commodification of the animals themselves, possibly a response to the clients being the human "owners" rather than the animals, and potentially a mechanism of self-preservation on behalf of the students themselves in order to deal with the emotional burden of the profession – something referred to as compassion fatigue (AVMA, 2021).

The culture and learning environment within the vet school – and when in placements with community vets – have a fundamental effect on this desensitisation. There are significant gender differences – with trends towards 80% of today's trainees across the globe being female (RCVS, 2018), whilst most of the generation before them (and therefore those doing the teaching and role modelling) are predominantly male. Given that female students inherently ascribe significantly higher sentience to certain animals than their male counterparts (Clarke and Paul, 2019), there are very real experiences of trainees feeling as though they need to "man up", resulting in the need for students to hide their empathetic responses to scenarios in order to pass the course and fit in. It is also widely understood that as humans we learn predominantly through imitation, and so regardless of what we are taught, we will behave the way others behave, and that approach will be learned as the "true north". If leaders and role models behave with little regard for the welfare of the animals in their care, then students will follow suit – regardless of whether they have just completed a course in animal welfare down the hall. If the faculty offers a course on animal welfare, and yet the faculty's own animal shelter houses animals in squalid conditions, then an unwritten lesson is learned by the students. Regardless of whether they can recite the five domains, if the faculty requires students to practise invasive procedures on healthy animals when they have little skill and experience in doing so, then there is indeed a "hidden curriculum". In his seminal paper exploring human medicine faculties, Hafferty defines the hidden curriculum as the structural and cultural influences within a department that often stand in contrast to the taught curriculum, suggesting that there is a fundamental distinction between what students are taught, and what they learn (Hafferty, 1998). He concludes by making a number of recommendations, including to restructure learning environments to be more consistent, rather than focusing on curriculum change.

Recognising the parallels within veterinary training, and in order to address this shortfall, World Animal Protection embarked on a more holistic approach. In 2016 they developed the "Standards of Excellence in Animal Welfare: A Guide for Veterinary Schools" (World Animal Protection, 2016), a set of clear guidelines to enable the inclusion of animal welfare practice and principles across all areas of vet schools. They show a clear line of progression for schools just starting out on the journey towards improvement, as well as offering a target for those schools already demonstrating leadership in the field. It recommends recognition for schools that achieve *centre of excellence* status and encourages veterinary associations and the veterinary community to come together and create such an award.

Veterinary nurses and paraprofessionals

There are wide-ranging jobs globally that fall under this umbrella. Similar titles can confer wide-ranging job descriptions and entry requirements. Veterinary nurses in one country may be classed as veterinary technicians elsewhere, although veterinary technician roles can vary widely with those in Australia forming more theoretical and research specialties, for example.

Veterinary nurses or technicians provide indispensable support to veterinary surgeons and an important focus on the welfare of the animals in their care. With duties ranging from general husbandry and care and comfort of patients, to assisting with clinical procedures, it is a skilled profession. As a career path, veterinary nursing is gaining traction and maturing in terms of professionalisation worldwide, although countries vary widely with regard to what qualifications (if any) are required to fulfil the role. The UK and Ireland have standardised the qualification and have made a qualification compulsory to obtain the title of "Registered Veterinary Nurse", whilst other countries are working towards establishing standardised credentials (Yagi, 2019). In the UK, many colleges offer a diploma accredited by the Royal College of Veterinary Surgeons (RCVS), and there are also a number of universities that offer degrees in veterinary nursing too. There are minimum entry requirements based on high school grades or previous work experience.

As an alternative, a veterinary paraprofessional has training in certain elements of animal care and husbandry – and in some instances has training in carrying out simple acts of veterinary surgery. They operate under the direction of a veterinarian and their role is to protect animal health and welfare. As with veterinary nursing, there are wide variations in regulation and accreditation for paraprofessionals, and it is a maturing profession with the UK's RCVS Council approving regulation and accreditation relatively recently for example (RCVS, 2019). The world Organisation for Animal Health (OIE) recognise the important value of paraprofessionals worldwide and have published Competency Guidelines for Veterinary Paraprofessionals in line with their Day 1 Competencies for Graduating Vets, in order to assist in the standardisation and quality assurance of practising paraprofessionals (World Organisation for Animal Health (OIE), 2018). Animal welfare is included within these guidelines.

Paraprofessionals hold a particularly important role in many countries in Africa, where qualified vets are few in number, and most occupy roles in government or in urban centres. In these cases, paraprofessionals outnumber vets in rural areas and they deliver most of the much-needed interaction between animal owners and any form of veterinary expertise for treatment or advice. As with other countries, they are required by law to work under the authority of a registered vet, but they perform almost all of the veterinary care. Students train to become paraprofessionals in Agricultural and Livestock Training Institutes (ALTIs) and Animal Science faculties, and thanks to World Animal Protection's partnership and train-the-trainer programme, effective incorporation of animal welfare principles and practice has been achieved in diploma and certificate courses in Kenya, Uganda, Tanzania, Namibia, and Sierra Leone.

Other animal-facing roles

Beyond veterinary and animal care services providing treatment for and prevention of sick animals, there are many roles that require or oversee interactions with and care of animals. Animal care staff can have job titles as varied as their duties and they can be found in all manner of places, from farms, zoos, sanctuaries, kennels and laboratories, to police and enforcement departments. The one unifying factor in the majority of these functions is the provision of on-the-job training, rather than entry requirements, and a few exemplars are mentioned here.

Farmers are frequently trained through hands-on, practical experience. Degrees in agriculture are available, as are courses in farming and land management or crop and livestock production, although these are not compulsory.

Animal care workers generally are not required to have any formal qualifications, with many positions offering the opportunity to enrol in work-based courses to learn the skills required. In many countries animal care courses are available to enhance an individual's prospects of securing the position, but these are not mandatory and there is little or no regulation, accreditation, or quality assurance from licencing bodies. "Animal handling" (specialising in humane handling and restraint to reduce stress and risk of injury in the animal) can be determined as the primary role, or a duty within a wider animal care role, and similarly, courses are available for on-the-job training, or apprenticeships may be available.

Care staff, focusing specifically on research and laboratory animals, are not necessarily expected to have gained certification or specific training prior to obtaining the position, although if they do arrive without formal training, they will often be expected to complete training before they can handle animals, and undergo certified or accredited CPD throughout their tenure. The specific requirements are determined by the institution's own ethics committee. Ethical concerns over the treatment of animals used in research have led to a variety of legal frameworks across the globe and the oversight and approval of research is the remit of the ethics committees to ensure these legal frameworks – and the welfare of the animals concerned – is adhered to. Most of these frameworks are built around the 3Rs (Replace, Reduce, and Refine methodologies to avoid or minimise the use of and impact on animal "subjects", although there is a growing chorus of support for collaborations such as "BioMed21" involving animal protection organisations, research funding bodies, academic, corporate, and regulatory bodies, who are providing the research and evidence to support the replacement of animal models entirely (BiomMed21, 2021)). Ethics committees vary in size and composition from one institution to another, and will usually require representation from a qualified veterinary surgeon and someone with qualifications and experience in laboratory science and research. An independent member is also expected to sit on the panel, although panel membership may often be unbalanced, with insufficient representation of animal interests (Hansen, et al., 2012).

Animal transporters

Those responsible for the transport of animals have varying levels of regulations to jump through depending on their geographic location. Some countries in the European Union, and the UK, have included mandatory training for livestock transporters, such as the requirement for a certificate of competence and training dependent on species (UK Government, 2021). A number of other countries, including Canada and the USA, are seeing voluntary commitments to transporter certification which has led to a variety of training programmes.

The role of animal welfare specialists

Whether through formal education in the classroom, through veterinary practitioner training in all its forms, or through the myriad CPD courses, diplomas, and certificate courses, there is a wider societal function required of specialists qualifying in animal welfare or related fields. This chapter began with the recognition that animal welfare concerns arise as a result of human behaviour, and behaviour arises out of social norms. The animal welfare movement requires each and every person graduating from these courses to play a vital role in challenging the environment that enables the status quo.

Animal care professionals will find themselves advising colleagues, animal owners, and others on appropriate care for the animals and the methods by which good, positive welfare can and should be achieved. The more that animal-based outcomes can become the norm in the course of these conversations, the more that mindsets begin to shift. Holding a specialist qualification in animal welfare provides an ideal opportunity to contribute to policy setting in all manner of forums – from organisational and institutional policies, to government policies. All practitioners can and should participate in the associations and professional bodies that represent their specialties, and the more senior among them may well find opportunities to define and update legislation and regulation as it pertains to animal welfare in their respective countries.

One overarching responsibility of these professionals (second only to upholding the welfare of animals in their care), is to communicate about animal welfare to a wide variety of audiences. Communicating about positive animal welfare can take a variety of forms – and the wider the variety, the better. It includes traditional scientific channels such as peer-reviewed journals and reports for academic and professional audiences. But it must necessarily include communication with different stakeholder groups and the public at large through diverse channels and using a variety of means.

Framing is key, and the use of language can sow seeds as much as it can reveal an unconscious bias, or worse, result in the opposite outcome to what was intended. Frames are the words we use that are associated with broader themes. The people we are communicating with use them as a shortcut – whether we mean them to be or not – and therefore we need to be careful not to accidentally reinforce undesirable actions by using damaging phrases. One really simple example is to look out for when we or our friends in the media refer to animals as "he", "she", and "they". When we hear or read about animals described as "it", the idea that they are commodities or assets is reinforced. Even if we're desperately trying to communicate about respecting "its" needs within the five domains, we undermine our own efforts. A superbly useful tool is the Framing Nature Toolkit (Public Interest Research Centre, 2018) and whilst it focuses on conservation and wildlife, the learnings are transformative and applicable to communication about animal welfare.

The skills required for effective communication go far beyond knowledge of the subject and sensitivity to framing. The best tools to communicate a subject that can be contentious and confrontational such as animal welfare are humility and the ability to listen. It is our responsibility to acknowledge when we're communicating something based on assumptions, and replace it with evidence, based on what we have heard from the person, group, or community we need to communicate with. Employ market research or focus groups if necessary (McKenzie-Mohr, 2011), or simple conversations free of judgement or reproach in order to listen. Offering pragmatism, and a focus on problem-solving, rather than demonising is essential – remembering that being part of the problem can sometimes help one to become part of the solution. It is recognising that providing information (often misconstrued as imparting knowledge or educating others) is not the same and not as effective as understanding why a person or industry behaves the way they do, and identifying the need that they are trying to meet. In the midst of all this understanding, the animal care professional must adhere to the principal that animal-based outcomes should become the norm in decision making that involves animals, but pragmatism and understanding the human need that underpins the behaviour (and the decision making) can lead to a very positive outcome for all concerned.

In a field of animal welfarists often brimming with knowledge and expertise in animal behaviour, understanding human behaviour is just as fundamental to achieve change. There is a growing field of "human behaviour change", and a growing recognition of its importance in achieving positive social change for animals. Understanding that we're sometimes predictably irrational and will follow the crowd, even when it is against our own better judgement (and against what we've been taught), is an essential leveller in the effort to mainstream animal welfare – and a fundamental starting point when we're attempting to communicate on the subject. Recognising that behaviour can be changed rapidly, and can subsequently facilitate attitudinal change is an essential ingredient for change makers, and they can use hugely effective behavioural tools (collectively known as behavioural economics) such as "nudges" and "prompts" to bring decisions back into the conscious mind rather than letting habitual decisions continue. This enables people to refer to new information they might have received through education or other means, and determine whether a different course of action is in order. Equally, "choice architecture" (or the means by which environments are laid out to enable people to make different consumer choices) is an essential mechanism for altering human behaviour in the very short term (Sunstein and Thaler, 2008). When partnered with effective education that begins in the early years and is both values-based and founded on the indisputable academic principles of animal sentience, we have a winning formula.

Conclusions

When behaviour-change approaches are overlaid onto a backdrop of effective AWE that delivers on the requirement that any decision that impacts the quality of life for an animal is measured with animal-based outcomes, we stand a very real chance of sustainably transforming the outlook for animals.

Behavioural economics are a hugely powerful and underutilised tool and can deliver instantaneous results when applied appropriately. Many of them can result in sustainable shifts in behavioural patterns. However, we must remember that we find ourselves in a world that needs to change because humans used innovation and creativity to solve big problems (such as feeding the world and making more money) and didn't have the foundational values of animal-based outcomes to check their decision making. We need effective AWE in all of its glorious forms to foster an understanding of sentience and a duty of care towards animals in order to prevent us from continually falling into this trap.

References

Ascione F, 1993. Children who are cruel to animals: A review of research and implications for developmental psycopathology. *Anthrozoos*, 6, pp. 226–247.

AVMA, 2021. Work and compassion fatigue. https://www.avma.org/resources-tools/wellbeing/work-and-compassion-fatigue

BiomMed21, 2021. https://biomed21.org/publications/

Clarke N and Paul ES, 2019. Veterinary students' beliefs about animal sentience: What role does gender play?" *Anthrozoös*, 32(5), pp. 581–595.

De Boo J and Knight A, 2005. Concepts in animal welfare: A syllabus in animal welfare science and ethics for veterinary schools. *Journal of Veterinary Medical Education*, 32(4), pp. 451–453.

De Vere R, 2014. World society for protection of animals (WSPA): A strategic partner in animal welfare for the veterinary profession. *Journal of the Commonwealth Veterinary Association*, 30(1), pp. 26–29.

FAO, OIE, WHO, et al., 2018. Zero by 30: The global strategic plan to end human deaths from dog-mediated rabies by 2030. https://www.oie.int/fileadmin/Home/eng/Media_Center/docs/Zero_by_30_FINAL_online_version.pdf

Grier KC, 1999. Childhood socialization and companion animals: United States, 1820–1870. *Society and Animals*, 7, pp. 95–120.

Hafferty FW, 1998. Beyond curriculum reform. *Academic Medicine*, 73(4), pp. 403–407.

Hamilton L and Surman E, 2018. How to teach kids where food comes from: Get them gardening. https://theconversation.com/how-to-teach-kids-where-food-comes-from-get-them-gardening-103277

Hansen LA, Goodman JR and Chandna A, 2012. Analysis of animal research ethics committee membership at american institutions. *Animals*, 2, pp. 68–75.

Hellyer P, Frederick C, Lacy M, et al., 1999. Attitudes of veterinary medical students, house officers, clinical faculty, and staff toward pain management in animals. *Journal of the American Veterinary Medical Association*, 214, pp. 238–244.

Linnerooth-Bayer J and Mechler R. 2009. *Insurance against Losses from Natural Disasters in Developing Countries*. United Nations, Department of Economic and Social Affairs Working Paper No. 85, October.

McKenzie-Mohr D 2011. *Fostering Sustainable Behaviour Change: An Introduction to Community Based Social Marketing*. 3rd edn. New Society Publishers.

McNeal JU 1992. *Kids as Customers: A Handbook of Marketing to Children*. Lexington Books.

Mellor DJ, Beausoleil NJ, Littlewood KE, et al., 2020. The 2020 five domains model: Including human–animal interactions in assessments of animal welfare. *Animals*, 10(10), pp. 1870.

Neumann M, Edelhäuser F, Tauschel D, et al. 2011. Empathy decline and its reasons: A systematic review of studies with medical students and residents. *Academic Medicine*, 86(8), pp. 996–1009.

Public Interest Research Centre, 2018. Framing nature toolkit: A guide to how words can help wildlife. https://publicinterest.org.uk/FramingNatureToolkit.pdf

RCVS, 2018. RCVS facts. https://www.rcvs.org.uk/news-and-views/publications/rcvs-facts-2018/

RCVS, 2019. RCVS Council opens the path for paraprofessional regulation. https://www.rcvs.org.uk/news-and-views/news/rcvs-council-opens-the-path-for-paraprofessionals-to-become/

Samuels WE, Meers LL and Normando S, 2016. Improving upper elementary students' humane attitudes and prosocial behaviors through an in-class humane education program. *Anthrozoös*, 29(4), pp. 597–610.

Self DJ, Schrader DE, Baldwin SK, et al., 1991. Study of the influence of veterinary medical education on the moral development of veterinary students. *Journal of the American Veterinary Medical Association*, 198, pp. 782–787.

Sunstein CR and Thaler RH, 2008. *Nudge*. Yale University Press.

UK Government, 2021. Guidance on animal welfare during transport. https://www.gov.uk/guidance/animal-welfare#animal-welfare-during-transport

Vaughan C, Gack J, Solorazano H, et al., 1999. The effect of environmental education on schoolchildren, their parents, and community members: A study of intergenerational and intercommunity learning. *Journal of Environmental Education*, 32(2), pp. 5–8.

World Animal Protection, 2012. First concepts in animal welfare: A guide for teachers. https://dkt6rvnu67rqj.cloudfront.net/sites/default/files/media/FCAW_Modules_1-7.pdf

World Animal Protection, 2013. Concepts in animal welfare. https://www.worldanimalprotection.org/animal-welfare-training-resources

World Animal Protection, 2016. Standards of excellence in animal welfare: A guide for veterinary schools. https://www.worldanimalprotection.org/animal-welfare-training-resources

World Organisation for Animal Health (OIE), 2012. OIE recommendations on the competencies of graduating veterinarians ('Day 1 graduates') to assure national veterinary services of quality. https://www.oie.int/app/uploads/2021/03/dayone-b-ang-vc.pdf

World Organisation for Animal Health (OIE), 2018. OIE competency guidelines for veterinary paraprofessionals. https://www.oie.int/app/uploads/2021/03/a-competence-1.pdf

Yagi K, 2019. A global perspective on veterinary nursing. https://todaysveterinarynurse.com/articles/a-global-perspective-on-veterinary-nursing/

INDEX

Page numbers in *italics* represent figures, while page numbers in **bold** represent tables.

The law referred to in this book is for discussion purposes and not all legal sources will be separately indexed.